SALTERS

Tansy Standen.

Higher *Chemistry*

Project Team

Gwen Pilling

Douglas Buchanan

Margaret Ferguson

Wilson Flood

David Hawley

John Lazonby

Brian Ratcliff

David Waddington

Bob Wilson

Heinemann

CONTENTS

'Chemical Ideas' chapters

CONTRIBUTORS

Project Team

Gwen Pilling (Director)	University of York
Douglas Buchanan (Adviser)	Higher Still Project
Margaret Ferguson	Lews Castle School, Stornoway
Wilson Flood	formerly, Science Adviser, Dumfries and Galloway
David Hawley	formerly, Dumfries Academy
John Lazonby	University of York
Brian Ratcliff	formerly, Long Road Sixth Form College, Cambridge
David Waddington	University of York
Bob Wilson	Forrester High School, Edinburgh

Central Team for Salters Advanced Chemistry

George Burton	Cranleigh School and University of York
John Holman (Project Director 1988–94)	Watford Grammar School and University of York
Margaret Ferguson	Lews Castle School, Stornoway
Gwen Pilling (Project Director 1994–)	University of York
David Waddington (Chairman of Steering Committee)	University of York

Advisory Committee

Dr Peter Doyle	Zeneca Group
Dr Tony Kirby, FRS	University of Cambridge
Professor The Lord Lewis, FRS (Chairman)	University of Cambridge
Sir Richard Norman, FRS	University of Oxford
Mr John Raffan	University of Cambridge

Sponsors

Many industrial companies have contributed time and expertise to the development of the Salters Advanced Chemistry course. The work has been made possible by generous donations from the following:

The Salters Institute for Industrial Chemistry
The Association of the British Pharmaceutical Industry
BP Chemicals
British Steel
Esso UK
Zeneca Agrochemicals
The Royal Society of Chemistry
Shell UK

Many other organisations have been very generous with their time and expertise in the development of the Salters Higher Chemistry course.

We thank the following for their considerable help with various sections:
3.3, Nutrasweet Kelco; **8.6**, Scot Power; **8.7**, BP Chemicals and the Science in the Environemnt Project, University of York; **9.2**, Great Lakes Chemical (Europe).

INTRODUCTION

The Salters Higher Chemistry Course is made up of three self-contained units. These correspond to the three units in the content statements for Higher Chemistry.

Unit 1 Energy Matters
Unit 2 The World of Carbon
Unit 3 Chemical Reactions

Each unit contains three **Chemical Storyline** chapters, each of which is linked to **Chemical Ideas** chapters in the same unit. You work through a Storyline chapter referring to the Ideas at points which are indicated in the book.

Chemical Storylines

Each of the Storyline chapters (chapters 1–9) focuses on an interesting application of chemistry. You will find that there are numbered *Assignments* at intervals. These are designed to help you understand what you have read and sometimes to revise earlier work. It is best to do the assignments as you go along.

Chemical Ideas

As you work through the Storylines, you will find that there are references to sections in related Ideas chapters (chapters 10–25). These sections cover the chemical principles you need to understand that particular part of the Storyline, and you will probably need to study them before you can go much further. There are also references to the Chemical Storylines from the Chemical Ideas so that you can look back to any previous work that is relevant.

As you study the Chemical Ideas you will find sections of *Problems* to do. These are designed to check and consolidate your understanding of the chemical principles involved.

What do you need to remember?

The content statements in the Arrangements for Higher Chemistry define what you have to remember and be able to do. The Chemical Storylines and Chemical Ideas together cover all the content statements for each unit. It is important to remember that the Chemical Ideas chapters are not the only places where key concepts in chemistry are covered. Concepts related to specific applications are often explained in the Chemical Storylines.

There is an *Overview* at the start of each unit that summarises what is covered in the chapters of the unit. A table of content statements tells you whether the topic is covered in the Chemical Storylines or in the Chemical Ideas. (Entries in brackets refer to further use of these ideas in other units.) It is important to remember that the order in which the content statements are covered is governed by the Storylines and so is different from the way they are listed in the Arrangements.

Assessment

Internal assessment: At the end of each of the three units of the Higher course you will sit a test which covers the basic knowledge and understanding and problem-solving associated with the content of the unit. The pass mark for each test is 18 out of 30. To achieve a unit, you must also write a satisfactory report on one of the three practical activities which are prescribed for the unit.

External assessment: To gain an award in Higher Chemistry, you must pass the end-of-course assessment as well as all the component units of the course. External assessment also provides the basis for grading at three different bands. The examination consists of one paper of 2 hours 30 minutes with a total allocation of 100 marks. You should answer all the questions. The paper has two sections: Section **A** 40 marks and Section **B** 60 marks. There will be three types of question: multiple-choice questions, grid questions and structured questions.

At the end of the book there is a set of practice examination questions. The questions are similar to those you will meet in your examination for Higher Chemistry. They cover the basic content of each of the units. Some of the questions illustrate how you will be required to:

- integrate knowledge and understanding across the three componenet units of the course,
- display problem-solving skills in less familiar contexts,
- show experience of the prescribed practical activities.

ACKNOWLEDGEMENTS

The authors and publishers are grateful to the following for permission to reproduce text extracts:

2.10 and **2.11:** GM and MJ Pilling, 'What is photochemical smog?', *Chemistry Review 5:5*, 1997, Philip Allan Publishers. **2.14:** AET Technology, Maps of Ozone Antarctic Monitoring Sites. **6.9:** J C Farman et al, *Nature Magazine volume 315, No. 6016, page 207*, 16 May 1985, Macmillan Journals Ltd. **6.12:** NASA Ozone Processing Team at Goddard Space Flight Center. **6.16a:** *The Guardian*, 30 June 1990. **6.16b:** *The Independent*, 30 September 1998. **6.17:** *Chemistry Review 1:1*, D Powell, 'The Rise and Fall of CFCs', September 1996, Philip Allan Publishers. **9.5:** Zeneca Agrochemicals, 'Fertilizer use in modern agriculture'. **8.17:** Institute of Petroleum, *Fossils into Fuels*. **12.2/3/4/5:** Scottish Qualifications Authority. **12.8/9/10/11 and 13.2:** J G Stark and H G Wallace, *Chemistry Data Book*, John Murray (Publishers) Ltd. **25.2:** P A Cox, *The Elements: Their origin, abundance and distribution*, 1989, Oxford University Press.

The authors and publishers are grateful to the Scottish Qualifications Authority and OCR for permission to reproduce examination questions in the book.

The authors and publishers would like to thank the following for permission to use photographs:
Cover photo by Science Photo Library/NASA

1.2 SPL/Royal Observatory Edinburgh/AATB **1.4** SPL/NOAO **1.6** SPL/Leonard Lessin/Peter Arnold **1.9** SPL/Jerry Mason **1.10** Bubbles **1.11** Ann Marie McDonald **1.15a** Science Museum Library **1.15b** GT Woods **1.16** SPL/Roger Resmeyer/Starlight **2.1** SPL/Susan McCartney **2.2** Collections/Liz Stares **2.4** Philip Parkhouse **2.6** Philip Parkhouse **2.7** Quadrant Picture Library **2.9** SPL/Conor Caffrey **2.12** Roger Scruton **2.15** Dr Harald Geiger, Physikalische Chemie, Bergische Universitat-Gesamthoch schule, Wuppertal, Germany **2.20** Johnson Matthay **3.1** SPL **3.5** David Williams Picture Library **3.8a** SPL/Robert de Gugliemo **3.8b** Ace Photo Agency **3.12** Natural History Museum **3.14** GSF **3.15** Natural History Museum **3.17** SPL/Tom van Sant/Geosphere Project Santa Monica **3.18** J Allan Cash **3.19** Robert Harding Picture Library **3.20** Margaret Ferguson **3.21** National Museum of Scotland **3.22a** Nutrasweet Kelco **3.22b** Jim Mackintosh /Nutrasweet Kelco **3.25** Salt Union **3.26** National Museum of Scotland **3.27** J Allan Cash **3.28** Institute of Oceanographic Sciences **3.34** J Allan Cash **4.1** Daimler Benz **4.3** Eta processing plant **4.4** Shell Canada **4.6** Brian & Cherry Alexander **4.8** Philip Parkhouse **4.10** Shell International **4.17** South American Pictures **4.18** ICI **4.19** Glasgow Herald/Edward Jones **4.20** Mercedes Benz **4.21a** BP Solar **4.21 b** Roger Scruton **4.21 c** J Allan Cash **4.21 d** SPL/Hank Morgan **4.23** Daimler Chrysler Aerospace Airbus **5.1** Bridgeman Art Library **5.2** Colin J Williamson **5.3** Popperfoto **5.4** Archiv fur Kunst und Geschichte, Berlin **5.7a** BP Chemicals **5.7b** BP Chemicals **5.8** Fischer Price **5.10** SPL/BlairSeitz **5.12** Zipperling Kessler & Co **5.14** Topham Picture Library **5.15** Fell/Hurworth Photography/ICI Chemicals & Polymers **5.16** Sylvia Katz/ICI Fabrics **5.20** Dupont **5.22** Monsanto **5.24** Enak Ltd **5.26** Mary Aitken **5.28** J Allan Cash **5.32** British Diabetic Association **5.34** Dr RE Hubbard/Dept of Chemistry/York **5.35** Dr RE Hubbard/Dept of Chemistry/York **5.36** Dept of Chemistry/University of York **5.37** Dr RE Hubbard/Dept of Chemistry/York **5.39** Dr RE Hubbard/Dept of Chemistry/York **5.40** Stuart Priest/University of York **5.41** SPL/Chris Priest and Mark Clarke **5.43** Barnabys Picture Library **6.1** Popperfoto **6.4** Philip Parkhouse **6.6** Sherry Rowland **6.9** SPL/Simon Fraser **6.10** SPL/Doug Allan **6.11** SPL/Philippe Plaily **6.14** Hutchison Library **6.15** SPL/Jeremy Burgess **7.3** Rex Features/Sipa **7.4** SPL/NASA/Goddard Institute for Space Studies **7.5** SPL/Graham Ewans **7.9** J. Allan Cash **7.10** J Allan Cash **7.11** Scottish Natural Heritage **7.13** Hutchison Picture Library **8.2** BP **8.4** AgrEvo UK Ltd **8.5** AgrEvo UK Ltd **8.6** ICI **8.7** BP Chemicals **8.8** SPL/Will & Deni McIntyre **8.9** BP **8.10** ICI **8.11** Westpoint Studios **8.12** Bayer AG/Leverkusen, Germany **8.13** The Glasgow Picture Library **8.14** BP Chemicals **8.15** BP **8.16** BP **9.1** Holt Studios International **9.7** SPL/Nelson Medina **9.8** Rothamsted Experimental Station **9.9** J Allan Cash **9.11/12** BASF **9.17** Zefa **9.18** Brian Simpson **9.19** Brian Simpson **9.20** Brian Simpson **9.21a** Brian Simpson **9.22** Brian Simpson **9.23** Brian Simpson **9.26** Brian Simpson **9.27** Michelin **9.30** Eileen Barratt **9.33** Copper Development Association **9.34** Mary Evans Picture Library **9.36** Nuclear Medicine, Conquest Hospital, Hastings **9.37** SPL/Martin Dohrn **9.39** AEA Technology **9.40** Chris Butlin/British Airways NDT unit **9.42** JET/AEA **15.17** Dr RE Hubbard/Dept of Chemistry/University of York **15.19** Novo Nordisk **15.20** Barnaby's Picture Library **16.11b** BP Solar **16.11b** BP Solar

The publishers have made every effort to trace the copyright holders, but if they have inadvertently overlooked any, they will be pleased to make the necessary arrangements at the first opportunity.

Picture research by Thelma Gilbert

Heinemann Educational Publishers
Halley Court, Jordan Hill, Oxford, OX2 8EJ
a division of Reed Educational & Professional Publishing Ltd
Heinemann is a registered trademark of Reed Educational & Professional Publishing Ltd

OXFORD MELBOURNE AUCKLAND
JOHANNESBURG BLANTYRE GABARONE
IBADAN PORTSMOUTH NH (USA) CHICAGO

First published 1999

ISBN 0 435 63098 9

03 02 01 00 99
10 9 8 7 6 5 4 3 2 1

Edited by Tim Jackson

Designed, typeset and illustrated by Gecko Limited, Bicester, Oxon

Cover design by Gecko Limited, Bicester, Oxon

Printed and bound in Spain by Edelvives

UNIT I ENERGY MATTERS

OVERVIEW

Chemical Storyline chapters

Chapter 1 The Elements of Life

This chapter tells the story of the elements of life: what they are and how they originated. It shows how studying the composition of stars can throw light on the formation of elements that make up our own bodies. It then looks in more depth at two elements – iron and calcium – and this leads into learning about patterns in the properties of elements and the Periodic Table. The chapter concludes with a brief look at how elements combine to form compounds such as the 'molecules of life' which form the body.

Chapter 2 Making Use of Energy

This chapter begins by summarising the importance to us of fossil fuels and considers where the energy comes from when a fuel burns. It goes on to look at the problems of emissions from car engines and the formation of photochemical smog. It concludes by looking at what is being done to tackle the problem of emissions, by changing engine technology, using catalytic converters or by changing the fuel.

Chapter 3 Our Planet

This chapter examines the wide variety of substances found on the Earth: the gases in the atmosphere, the soil and rocks that make up the land and the substances dissolved in the oceans. It describes how the Earth comes to have its present atmosphere and how rocks are weathered to produce the soil. It looks at the composition of sea water and the role of the oceans as a resource for humans. It concludes by describing how the oceans transport energy around the surface of the Earth and the effect this has on our climate.

Chemical Ideas chapters

Each Chemical Storyline chapter is linked to relevant sections in the Chemical Ideas:

Chapter 10 Measuring Amounts of Substance
Chapter 11 Atomic Structure
Chapter 12 The Periodic Table
Chapter 13 Chemical Bonding
Chapter 14 Energy Changes
Chapter 15 Rates of Reactions
Chapter 16 Structure and Properties

Overview of chemical principles

Unit 1 links with the chemistry you have covered in Standard Grade. The early chapters allow you to revisit and consolidate some of this work, such as basic atomic structure, writing balanced equations and simple calculations involving moles.

From previous work you should:

- understand simple ideas about atomic structure
- be able to carry out simple calculations involving moles
- be able to work out the concentration of a solution in mol l^{-1}
- understand and be able to write balanced chemical equations
- know about the Periodic Table as a means of classifying chemical elements
- know that chemical elements combine together to form compounds.

At the end of this unit you will have covered all the content statements for Unit 1 in the Arrangements for Higher Chemistry. The table below shows where each topic is covered. Remember that the *order* in which the content statements are done is governed by the Chemical Storyline and so is different from the way they are listed in the Arrangements.

UNIT 1 ENERGY MATTERS			
Areas in the content statements for Higher Chemistry		**Chemical Storylines**	**Chemical Ideas**
(a) Reaction rates			
	Following the course of a reaction		15.2
	Factors affecting rate	2.3	15.1–15.5
	The idea of excess		10.4, 10.5
	Catalysts	1.5, 2.4 (4.3, 5.10)	15.5 (23.3)
(b) Enthalpy			
	Potential energy diagrams	(6.3)	14.1, 15.1, 15.4, 15.5
	Enthalpy changes	2.2	14.1, 14.2
(c) Patterns in the Periodic Table			
	Trends in properties	1.2–1.4	12.1, 12.2
	Electronegativity		13.2, 16.1–16.3, 16.5
(d) Bonding, structure and properties			
	Types of bonding	1.5	13.1-13.3, 16.2, 16.5
	Intermolecular forces of attraction		16.1, 16.4
	Structure	3.2	13.3, 16.2–16.5
	Properties	3.1–3.4	16.1-16.5
(e) The mole			
	The Avogadro constant	1.2	10.1
	Molar volume	2.3, 2.4	10.5
	Reacting volumes	2.3, 2.4	10.5

(The entries in brackets refer to further use of these ideas in Units 2 and 3.)

1 THE ELEMENTS OF LIFE

1.1 *Where do the chemical elements come from?*

A star is born

In the beginning, there was hydrogen – and a lot of it! There still is: hydrogen is the most common element in the universe. Humans also contain quite a lot of hydrogen, but we also contain other, heavier, elements as well. The theory of the evolution of the stars shows how heavy elements can be formed from lighter ones, and helps to explain the way elements are distributed throughout the universe.

The theory of how stars form is one of the major scientific achievements of this century. It was developed through observation of stars at different stages in their development, studying them as they changed over time.

The two elements whose atoms are most abundant in space are hydrogen and helium. But the atoms are few and far between. There is about one atom per centimetre cube in the space between the stars, compared with over 10^{19} atoms per centimetre cube in the air you are breathing now. With a density of hydrogen atoms in space as low as this, there is almost no chance that hydrogen atoms will come together to form hydrogen molecules.

However, in some regions between the stars called **interstellar gas clouds**, there may be up to 100 atoms per cm^3. The gas clouds also contain **dust** – debris from earlier stars which have been spread out in space.

Interstellar gas clouds are held together by gravitational forces. Within the gas clouds are regions where the gravitational field makes the cloud contract in on itself, and the gases become compressed. 'Clumps' of denser gas are formed.

The densest part of the 'clump' is its centre. Here the gases are most compressed and become very hot, up to 10 million °C. Such temperatures are high enough to trigger nuclear reactions.

A nuclear reaction is different from a chemical reaction. A chemical reaction involves a rearrangement of an atom's outer electrons, while a nuclear reaction involves a change in its nucleus. In a nuclear reaction one element can turn into another element – something which would be impossible in a chemical reaction. You will find out more about nuclear reactions in **Unit 3, Chapter 25**.

Sections 11.1 and **11.2** provide you with information about the structure of atoms. This will help you with your study of this and later sections.

One nuclear reaction that takes place in the centre of 'clumps' is **fusion**, when lighter nuclei are fused together to form heavier nuclei. Massive quantities of energy are needed to overcome the repulsions between the positive charges on the two nuclei, but in the centre of stars, where temperatures can reach hundreds of millions of degrees, fusion is common.

The nuclei of hydrogen atoms present in the gas cloud join together by nuclear fusion and the hydrogen turns into helium. Here is just one example of such a reaction:

$${}^{1}_{1}H + {}^{2}_{1}H \rightarrow {}^{3}_{2}He$$

Figure 1 shows more clearly what is happening in this nuclear reaction.

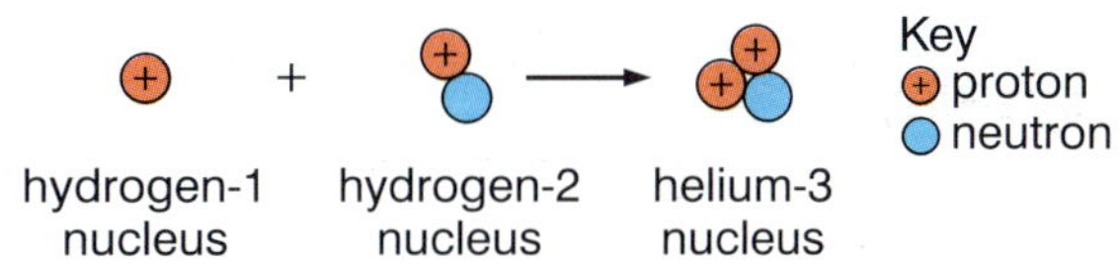

Figure 1 An example of a fusion reaction

The nuclear reactions release vast quantities of energy, which causes the dense gas cloud to glow (Figure 2). The nuclear reactions also generate a hot wind that drives away some of the dust and gas, leaving behind the new star. This is often surrounded by planets which have condensed out of the remaining dust cloud.

Heavyweight stars

What happens next to a star depends on its mass. All stars turn hydrogen into helium by nuclear fusion. This process occurs fastest in the heaviest stars because their centres are the hottest and the most compressed. These heavyweight stars have very dramatic lives. The temperatures and pressures at the centre of the star are so great that further fusion reactions take place to produce elements heavier than helium.

Layers of elements form within the star, with the heaviest elements near the centre where it is hottest and where the most advanced fusion can take place.

Figure 3 shows an example of the composition of the core of a typical heavyweight star after a few million years – long enough for extensive fusion to have taken place.

Figure 2 The glow of star formation in the Orion nebula

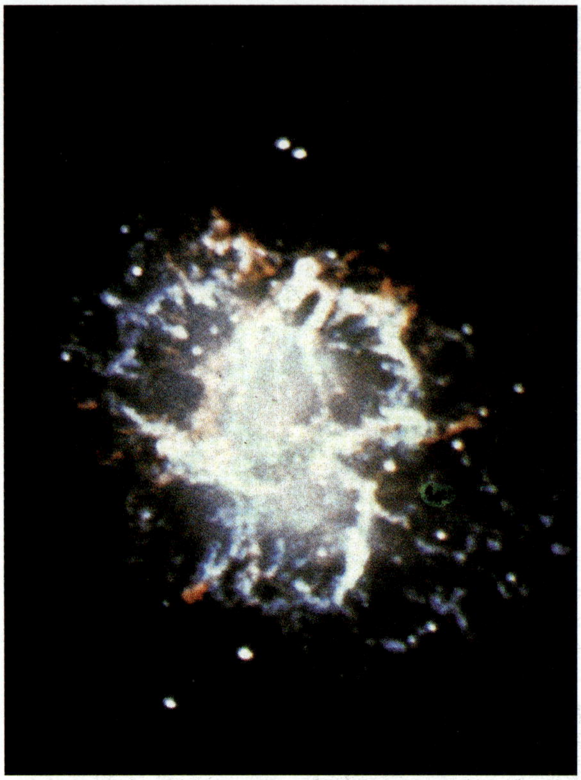

Figure 4 Violence in the Crab nebula – scientists believe that such pictures are evidence of a supernova

The element at the centre of the core is iron. When iron nuclei fuse they do not release energy but they *absorb* it. When the core of a heavyweight star reaches the stage where it contains mainly iron, it becomes unstable and explodes. These explosions are called **supernovae** – the most violent events in the universe.

Figure 3 A model of the core of a typical heavyweight star

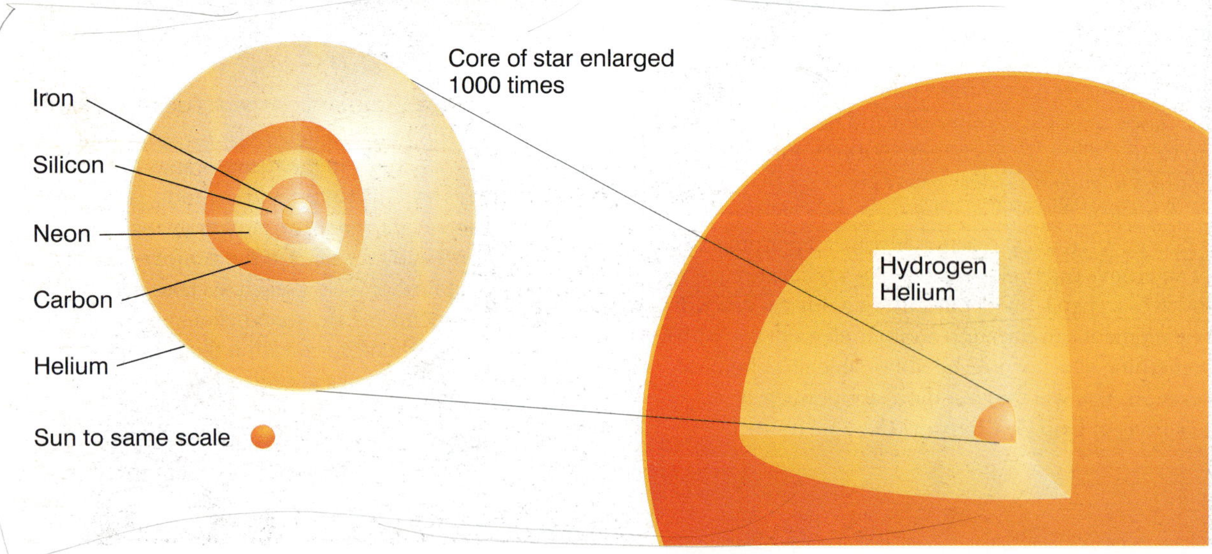

As a result of the supernova, the elements in the star are dispersed into the universe as clouds of dust and gas, and so the life cycle begins again.

The Sun – a lightweight among stars

The Sun is a **lightweight** star: it is not as hot as most other stars and will last longer than heavyweight stars. It will keep on shining until all the hydrogen has been used up and the core stops producing energy: there will be no supernova. Once the hydrogen is used up, the Sun will expand into a **red giant**, swallowing up the planets Mercury and Venus. The oceans on Earth will start to boil and eventually it too will be engulfed by the Sun. The good news for Earth is that the Sun still has an estimated 5000 million years' supply of hydrogen left!

As red giants get bigger, they also become unstable, and the outer gases drift off into space, leaving behind a small core called a **white dwarf**, about one-hundredth of the size of the original star.

Our solar system

Our solar system probably condensed from a huge gas cloud which gradually contracted under the force of gravity. As rings of gas and dust condensed around the Sun, the planets were formed. This material originated from a supernova and therefore contained a range of elements. The non-volatile elements condensed near to the Sun, where temperatures were greatest, while the more volatile elements condensed further away from the Sun at lower temperatures.

So, our solar system is made up of small, dense, rocky planets near to the Sun, and giant fluid planets further away from the Sun. Conditions on all the planets are very different and some of their chemistry seems very unusual when compared with our experiences on Earth. Figure 5 illustrates how unusual some of the chemistry is.

1.2 *What are we made of?*

Elements and the body

If you asked a number of people the question, 'What are you made of?' you would get a variety of different answers. Some people would use biological terms and talk about organs, bones and so on. Others might answer in more detail and mention proteins, fats and DNA. A chemist would be most likely to talk about atoms and molecules, or elements and compounds.

Figure 5 Did you know ...?

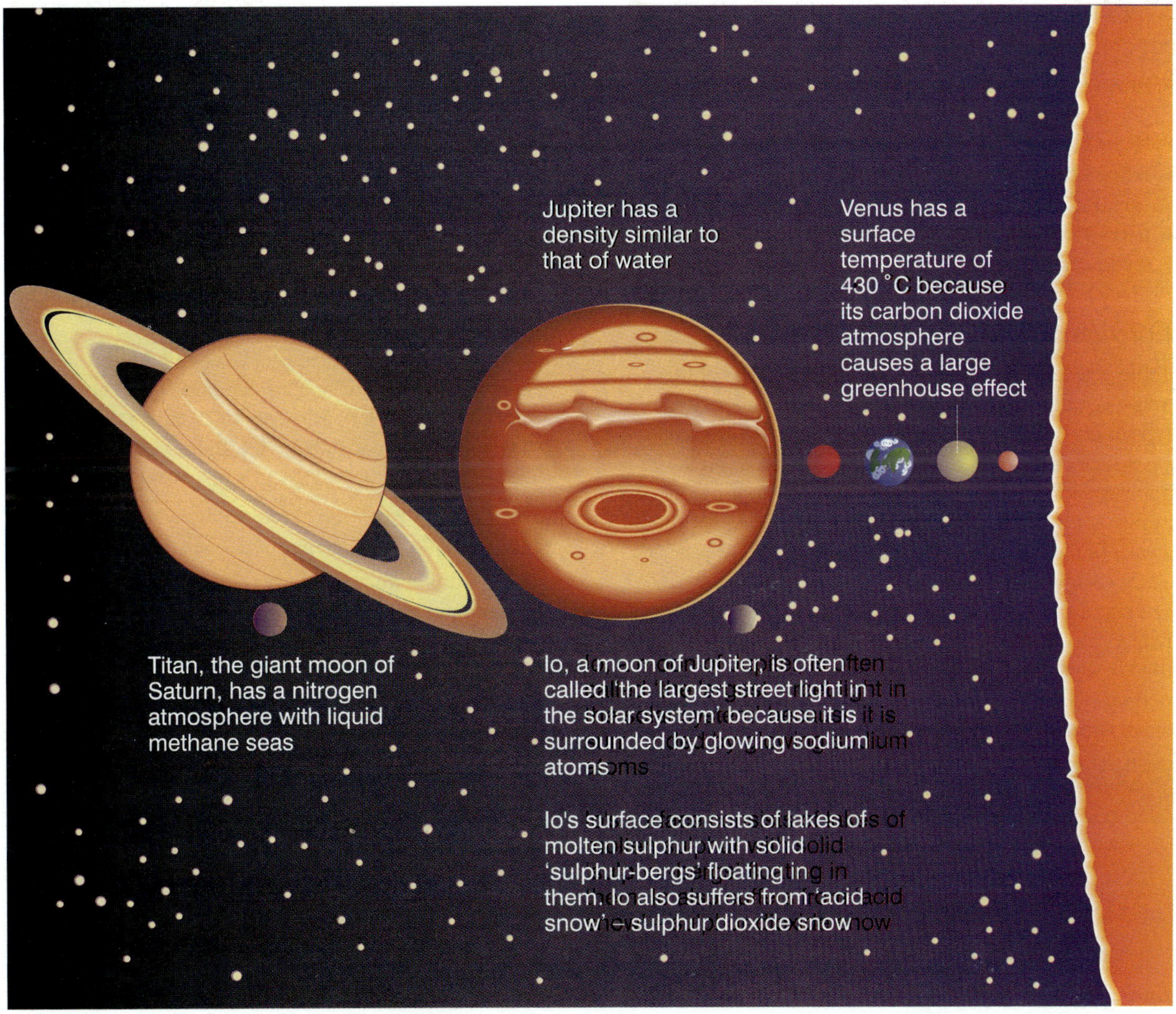

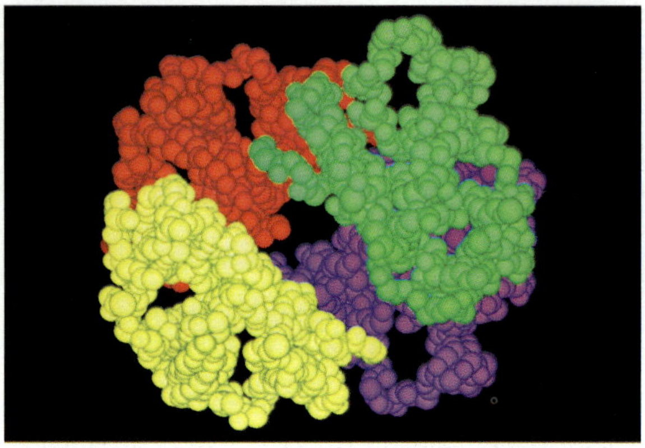

Figure 6 A model of haemoglobin – a biological molecule essential to human life

As a chemist, you know that you are not really made up of a mixture of elements but rather a mixture of compounds, many of which appear quite complicated. You will be finding out more about some of these compounds later in the chapter. To begin with, however, you will look at the elements which are most likely to be in the compounds in your body.

Elements in the body are classified as one of three types:

- **major constituent elements**, which make up 2%–60% of all the atoms present, eg carbon and hydrogen
- **trace elements**, which make up 0.01%–1%, eg calcium and phosphorus
- **ultra-trace elements**, which make up less than 0.01%, eg iron and iodine.

Counting atoms of elements

Table 1 gives you the masses and proportions of the major constituent elements in a person of average mass (about 60 kg).

Notice that in the two columns of numbers in Table 1 different elements appear to be more important in the body. For example, there are more atoms of hydrogen in your body than atoms of any other element, but hydrogen contributes far less than carbon or oxygen to the mass of your body. So a list of masses alone does not allow us to decide which of the three categories of 'elements of life' an element belongs to. To do this, we need to determine how many **atoms** of each element there are.

Element	Mass in 60 kg person/g	Percentage of atoms
oxygen	36 800	25.9
carbon	11 688	11.0
hydrogen	5326	59.4
nitrogen	2962	2.39

Table 1 The major constituent elements in the human body

Chemists can convert masses of elements into a measure of the number of atoms they contain by making use of **moles**. When we are dealing with elements, 1 mole is the amount of an element which contains the same number of atoms as 12 g of carbon.

Because atoms have exceedingly small masses, the number of atoms in 1 mole of an element is large. In fact it is very, very large – approximately 6×10^{23} atoms per mole. There are almost 1000 moles of carbon in a 60 kg person, so there are an awful lot of carbon atoms in the human body.

Once the number of moles of atoms of each element in the body has been calculated from its mass, these can be added to give the total number of atoms in the body. The percentage of atoms in the body of each element can then be worked out using the total.

Section 10.1 tells you more about moles and how to use them in calculations.

Table 1 shows you that the ratio of hydrogen atoms to oxygen atoms in your body is almost 2:1. That's because 65% of the mass of your body is water and the chemical formula of water is H_2O. You have probably known that formula for a long time. But how did you know it? Did you ever work it out? It's not difficult – provided that you know about moles.

ASSIGNMENT 1

Putting information in a table like Table 1 is often not the most striking way to present it. Pie charts or bar charts, for example, can be better.

Draw a pie chart to represent the proportions by mass of the four major constituent elements in the body. Label the fifth 'slice' of the chart to represent the contribution of all the trace and ultra-trace elements.

Now draw another pie chart, this time to represent the percentages of atoms.

Do you think that pie charts provide a better way of seeing and comparing the information in this case?

1.3 *Take two elements ...*

A trace is all you need

Section 1.2 shows that about 99% of your body is made up of only four elements: hydrogen, oxygen, carbon and nitrogen. These elements are obviously vital for life. But the trace and ultra-trace elements, although they make up only the remaining 1% or so of the body, are also essential for good health. You could not live without them!

Table 2 lists the proportions of the trace elements in a 60 kg person. The ultra-trace elements – which include cobalt, copper, iodine, iron, manganese, molybdenum, silicon, vanadium and zinc – are not given because their quantities are so small.

Element	Mass in 60 kg person/g	Percentage of atoms
calcium	780	0.22
sulphur	366	0.13
phosphorus	354	0.13
potassium	126	0.04
chlorine	103	0.03
sodium	61	0.03
magnesium	24	0.01

Table 2 The trace elements in the human body

Table 3 shows what some of the trace and ultra-trace elements do. You can see just how essential these elements are for proper body function.

The human body is often deficient in iron and calcium. The next section looks in more detail at the role and chemistry of these elements.

Figure 7 These foods are all rich sources of calcium compounds — bread and eggs also contain iron compounds, and fish is a good source of iodine compounds

Element	Function
calcium	major component of bone; required in some enzymes
phosphorus	essential for the synthesis of chemicals in the body and for energy transfer
sulphur	required in proteins and other compounds
copper	essential in enzymes involved in oxidation processes
iodine	an essential component of thyroid hormones
iron	contained in haemoglobin and many enzymes
zinc	required for the activity of many enzymes

Table 3 The functions of some trace and ultra-trace elements

An iron story

The body can easily become deficient in iron, particularly in young teenagers who are growing fairly rapidly. Iron deficiency, or anaemia, can usually be remedied with a course of iron tablets and an alteration in diet.

Iron carries out a vital role in the body: as part of the substance **haemoglobin**, present in blood, it is responsible for the transport of oxygen.

You saw a model of the haemoglobin molecule in Figure 6. Figure 8 shows part of the molecule in more detail. It consists of an iron atom at the centre of a ring structure. The ring is the *haem* part of the molecule. Also associated with the central iron atom is the protein, *globin*.

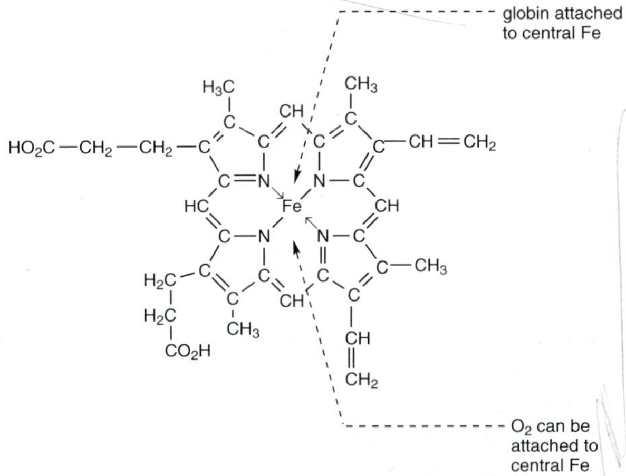

Figure 8 The haem group of a haemoglobin molecule

Haemoglobin is able to do its job because oxygen molecules become attached to the iron (Fe) atom. Oxygen is not the only substance to behave in this way: carbon monoxide can become similarly attached. Unfortunately, it is attracted more strongly than oxygen. This is why it acts as a poison if it gets into the body: it prevents the haemoglobin from carrying oxygen. One of the reasons why people who smoke are particularly prone to heart disease is that the carbon monoxide they inhale from cigarettes reduces the amount of oxygen their blood can carry. Their hearts therefore have to work much harder to maintain an adequate supply of oxygen to the body.

Once you are 18 years old you can become a blood donor. Because it is inadvisable for people with iron deficiency to give blood, a small sample of blood is always tested first to find the iron concentration. A drop of blood is taken from the donor's thumb and placed in a solution of copper sulphate. The copper sulphate reacts with the haemoglobin to form an insoluble compound, which appears as a white-coloured 'blob' in the solution.

The density of the copper sulphate solution is carefully arranged so that if there is sufficient

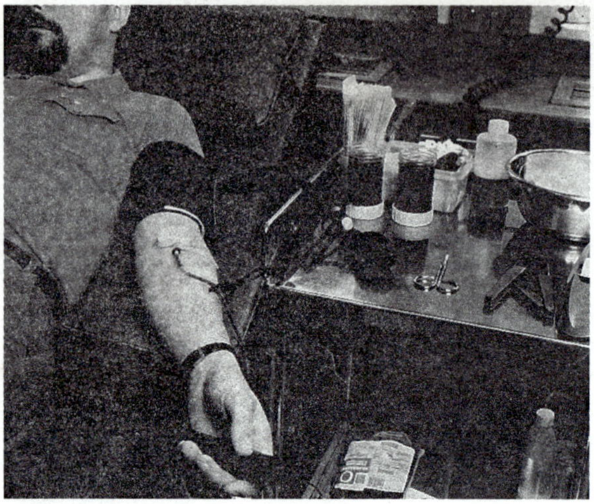

Figure 9 Anyone over 18 years of age can volunteer to become a blood donor

haemoglobin (and therefore iron) in the blood, the density of the blob is enough to make it sink. If this happens, the person can go on to make a blood donation.

However, if the blob floats there is insufficient haemoglobin in the blood, and the person is not allowed to make a blood donation. A sample of about 10 cm^3 of blood is taken for more detailed analysis from people found to have low haemoglobin levels.

A calcium story

Calcium is essential for the healthy development of bones. It is particularly important for a pregnant woman to eat a calcium-rich diet to ensure that her bones remain healthy and her baby's bones are properly formed.

Bones containing insufficient calcium are referred to as **demineralised** bones. One of the consequences of people living longer is that a condition called **osteoporosis** is becoming far more common. People

Figure 10 Diet is important during pregnancy, for both the mother and her baby. Among other things, a regular intake of calcium is essential

Susan's story

Susan works in a medical laboratory where samples of blood are sent for analysis. She has been working at the laboratory for 3 years since doing Highers, followed by a degree in Biochemistry. Susan describes here exactly what happens to blood sent to her for analysis.

Figure 11 Susan relies on chemistry every day in her work at a medical laboratory

"A 10 cm^3 blood sample is sent to our laboratories in a tube which is impregnated with heparin – a substance which stops the blood coagulating. I split the sample into two halves, adding anticoagulant to the first half of the sample to prevent it from clotting, but leaving the second half to clot.

The tests I do on the uncoagulated half are called full blood count (FBC) tests and involve counting the number of red blood cells, the number of white blood cells and the number of platelets. I don't actually count them one by one! An electronic cell counter does the job for me once I have loaded the sample.

The machine also measures the size of the red blood cells. If the cells are smaller than normal, this indicates that the person has anaemia. If, on the other hand, they are bigger than normal, this suggests that the person might have a vitamin deficiency.

I have to leave the second half of the sample to clot. Then the concentration of iron can be measured. To do this I first centrifuge the sample. This separates out the cellular material in the blood from the serum. The serum is a pale straw-coloured liquid, and this is the part I use for the second set of tests. The iron I am looking for is dissolved in this serum. I find the concentration of iron using a colorimeter, a machine which measures the light absorbed by iron in solution. It compares the amount of light absorbed by a solution made up from the serum with the absorption of solutions containing known concentrations of iron.

The results of my tests and those done by my colleagues are then sent back to the patient's doctor. I also have to send a reminder to the patient to go and see the doctor if they require any treatment as a result of having the tests carried out."

with osteoporosis have demineralised bones and tend to suffer from frequent bone fractures and curvature of the spine. The condition is usually observed in women who have passed through the menopause.

An increased intake of calcium can help the condition, and scientists have been trying to develop a technique which would allow them to find out how much of the calcium which is eaten is taken up by the bloodstream and becomes part of the bones. One way is to use a **tracer technique**. Patients are given a *meal* containing a radioactive calcium isotope, and the amount of radioactive calcium which is then absorbed into the bloodstream from the gut is measured.

However, radioactive isotopes can be dangerous and scientists have been working with strontium, an element very similar to calcium, to develop an alternative technique. The approach developed from observations made in the 1960s that some children had strontium in their bones. The quantities were small, but were still much larger than normal.

The strontium was present as the radioactive isotope strontium-90, a product of nuclear fission. Its uptake into children's bones was thought to result from the increased atmospheric concentration of strontium-90 which accompanied the testing of nuclear weapons in the 1950s.

It revealed a very important fact for scientists, which is that strontium, just like calcium, is taken up into bones during their formation. Normal, non-radioactive strontium can be analysed far more easily than calcium because strontium compounds give out a characteristic red light when they are placed in a flame. The intensity of this red light is a measure of how much strontium is present.

Scientists can therefore use *non-radioactive* strontium to detect the rate at which strontium is absorbed into the blood from the gut. They found that the rate of uptake of strontium was the same as the rate at which calcium was absorbed in tracer experiments. Using strontium is far safer than using a radioactive calcium isotope.

The ability to monitor patterns of absorption of calcium through the use of non-radioactive strontium should now lead to a greater understanding of calcium deficiencies and improve the treatment of bone disorders such as osteoporosis.

1.4 *Looking for patterns in elements*

When the elements were being discovered, and more was being learned about their properties, chemists looked for patterns in the information they had assembled.

You have seen that there are close similarities between calcium and strontium. Magnesium and barium can be added to them to make a 'family' of four elements. Your earlier studies probably introduced you to two other 'families' – lithium, sodium and potassium, and fluorine, chlorine, bromine and iodine.

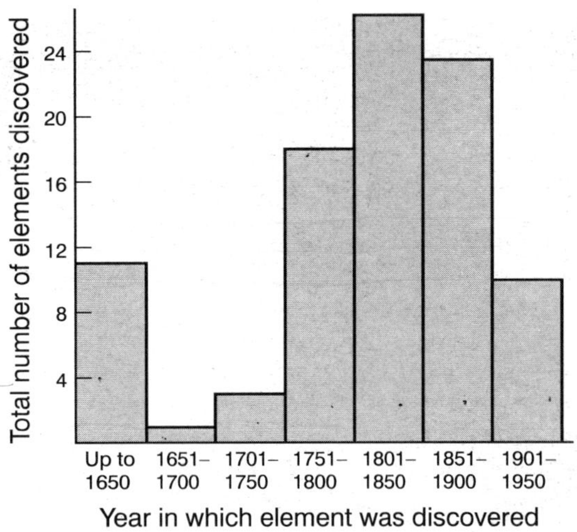

Figure 12 Historical pattern of the discovery of elements – many were discovered in the 19th century

Figure 12 shows you the historical pattern of the discovery of elements. Fifty-nine of the 92 naturally occurring elements had been discovered by 1850, so the search for patterns among the elements was particularly fruitful in the mid-19th century.

Much of the work was done by Johann Döbereiner and Lothar Meyer in Germany, John Newlands in England, and Dmitri Mendeleev in Russia. These chemists looked at similarities in the chemical reactions of the elements they knew about, and also patterns in physical properties, such as melting points, boiling points and densities.

Reactions to their suggestions were not always favourable. The March 1866 edition of *Chemical News*, a journal of the Chemical Society, shows you what Professor G. F. Foster had to say about Newlands' 'Law of Octaves' (Figure 13). Some of Newlands' symbols will look unfamiliar. They are listed below with their modern equivalents:

- G is now Be
- Bo is now B
- Di is now Ti
- Ro is now Rh.

Just 3 years later, however, Mendeleev's groupings (Figure 14) were seen as much more credible. The known elements were arranged in order of increasing atomic masses, so that those with similar properties came in the same vertical group.

Unlike Newlands, Mendeleev left gaps in his table of elements. These gaps were very important: they allowed for the discovery of new elements. (Look at Dr Gladstone's comments in Figure 13.)

PROCEEDINGS OF SOCIETIES

CHEMICAL SOCIETY

Thursday, March 1

Professor A. W. WILLIAMSON, *Ph.D.*, F.R.S., *Vice-President, in the Chair*

Mr. JOHN A. R. NEWLANDS read a paper entitled '*The Law of Octaves, and the Causes of Numerical Relations among the Atomic Weights*.' The author claims the discovery of a law according to which the elements analogous in their properties exhibit peculiar relationships, similar to those subsisting in music between a note and its octave. Starting from the atomic weights on Cannizzaro's system, the author arranges the known elements in order of succession, beginning with the lowest atomic weight (hydrogen) and ending with thorium (=231.5); placing, however, nickel and cobalt, platinum and iridium, cerium and lanthanum, &c., in positions of absolute equality or in the same line. The fifty-six elements so arranged are said to form the compass of eight octaves, and the author finds that chlorine, bromine, iodine, and fluorine are thus brought into the same line, or occupy corresponding places in his scale. Nitrogen and phosphorus, oxygen and sulphur, &c., are also considered as forming true octaves. The author's supposition will be exemplified in Table II., shown to the meeting, and here subjoined:–

Dr. GLADSTONE made objection on the score of its having been assumed that no elements remain to be discovered. The last few years had brought forth thallium, indium, caesium, and rubidium, and now the finding of one more would throw out the whole system. The speaker believed there was as close an analogy subsisting between the metals named in the last vertical column as in any of the elements standing on the same horizontal line.

Professor G. F. FOSTER humorously inquired of Mr. Newlands whether he had ever examined the elements according to the order of their initial letters? For he believed that any arrangement would present occasional coincidences, but he condemned one which placed so far apart manganese and chromium, or iron from nickel and cobalt.

Mr. NEWLANDS said that he had tried several other schemes before arriving at that now proposed. One founded upon the specific gravity of the elements had altogether failed, and no relation could be worked out of the atomic weights under any other system than that of Cannizzaro.

Table II – Elements arranged in Octaves

H	1	F	8	Cl	15	Co & Ni	22	Br	29	Pd	36	I	42	Pt & Ir	50
Li	2	Na	9	K	16	Cu	23	Rb	30	Ag	37	Ca	44	Os	51
G	3	Mg	10	Ca	17	Zn	24	Sr	31	Cd	38	Ba & V	45	Hg	52
Bo	4	Al	11	Cr	19	Y	25	Ce & La	33	U	40	Ta	46	Tl	53
C	5	Si	12	Ti	18	In	26	Zr	32	Sn	39	W	47	Pb	54
N	6	P	13	Mn	20	As	27	Di & Mo	34	Sb	41	Nb	48	Bi	55
O	7	S	14	Fe	21	Se	28	Ro & Ru	35	Te	43	Au	49	Th	56

Figure 13 An extract from Chemical News, *March 1866*

Mendeleev was so confident of the basis upon which he had drawn up his table that he made predictions about elements which had yet to be discovered. In 1871, he predicted the properties of an element he called **eka-silicon**, which he was confident would eventually be discovered to fill the gap between silicon and tin in his Periodic Table. Mendeleev's predictions are shown in Table 4. The element was discovered in 1886 and called **germanium**: its properties are in excellent agreement with Mendeleev's predictions.

	Group I	Group II	Group III	Group IV	Group V	Group VI	Group VII	Group VIII
Period 1	H							
Period 2	Li	Be	B	C	N	O	F	
Period 3	Na	Mg	Al	Si	P	S	Cl	
Period 4	K Cu	Ca Zn	* *	Ti *	V As	Cr Se	Mn Br	Fe, Co Ni
Period 5	Rb Ag	Sr Cd	Y In	Zr Sn	Nb Sb	Mo Te	* I	Ru, Rh Pd

Figure 14 Mendeleev's Periodic Table (the asterisks denote elements which he thought were yet to be discovered)

Figure 15 Dmitri Mendeleev's (1834–1907) ideas form the basis of our modern classification of the elements. In 1934, the centenary of his birth was marked with this 15-m-high version of Mendeleev's Periodic Table, unveiled at the building in St Petersburg where he had worked

Property	Prediction
appearance	dark-grey solid
relative atomic mass	72
density	5.5 g cm^{-3}
reaction with water	none
reaction with acid	very little
reaction with alkali	more than with acid
oxide	basic, reacts with acid
chloride	liquid, boiling point $<100\,^{\circ}C$

Table 4 Mendeleev's predictions for the properties of eka-silicon

ASSIGNMENT 2

Table 4 shows some of the predictions Mendeleev made about the properties of eka-silicon. Use reference books to find out about the properties of silicon and tin, and then make your own predictions about the element which would fill the gap between them.

Find out about some of the properties of germanium and see how well these compare with your predictions and Mendeleev's predictions about eka-silicon.

The modern Periodic Table is based on the one originally drawn up by Mendeleev. It is one of the most amazingly compact stores of information ever produced: with a copy of the Periodic Table in front of you, and some knowledge of how it was put together, you have thousands of facts at your fingertips!

Sections 12.1 and **12.2** tell you more about the modern Periodic Table and the patterns you can find in it.

ASSIGNMENT 3

In **Section 1.3**, you read that strontium in the diet is taken up into bones in a similar way to calcium.

a Use your knowledge of the Periodic Table and atomic structure to explain why calcium and strontium have similar properties.

b Both elements are present in bones as the metal carbonate. Write down the formula of strontium carbonate.

1.5 The molecules of life

Chemicals between the stars

You have heard a lot about atoms so far, but humans are made up of molecules and some ions rather than single atoms. So what are the molecules of life, and how did they come into existence?

In the **interstellar medium (ISM)** – the space between stars which contains clouds of gas and dust – there are many chemical species. They form when individual atoms in the ISM happen to meet and bond to one another. Table 5 shows some of these; some will look familiar, but many will look strange.

Sections 13.1, **13.2** and **13.3** tell you about the ways in which elements can combine with each other to form different types of compounds and how the type of bonding formed is related to the position of the constituent elements in the Periodic Table.

Many of the substances in Table 5 can be described as **organic** species. This means that they contain carbon atoms bonded to elements other than just oxygen.

You should notice something familiar about the elements in these species – they are the elements which are the major constituents of the human body.

Where did the molecules of life come from?

Some scientists have suggested that the molecules in the ISM were the building blocks which reacted together to make the molecules which form the basis of life on Earth. They believe that the energy needed to make these reactions take place came from ultra-violet radiation, X-rays and cosmic rays, and closer to Earth from lightning flashes.

In 1950 an American scientist, Stanley Miller, put methane (CH_4), ammonia (NH_3), carbon dioxide (CO_2) and water – simple molecules like those present in the ISM – into a flask and heated them. He also subjected the mixture to an electrical discharge to simulate the effect of lightning. On analysing the products, Miller found that a small amount of the reaction mixture had been converted into amino acids. Amino acids are the building blocks of proteins, which are formed from long chains of amino acids linked together.

In a separate experiment Leslie Orgel, another scientist in the US, made a very dilute solution of ammonia and hydrogen cyanide (HCN) and froze it for several days. When he analysed the 'ice' he identified amino acids and the compound adenine. Adenine is one of four compounds which, together with phosphate units and a sugar called deoxyribose, make up DNA (deoxyribonucleic acid) – the substance which contains the genetic code for reproduction.

Monatomics	Diatomics	Triatomics	Tetra-atomics	Penta-atomics
C^+	H_2	H_2O	NH_3	HCOOH
Ca^{2+}	OH	H_2S	H_2CO	CH_2CO
H^+	CO	HCN	HNCO	NH_2CN
	CN	HNC	HNCS	HC_3N
	CS	SO_2	C_3N	C_4H
	NS	OCS		CH_2NH
	SO	N_2H^+		CH_4
	SiO	HCS^+		
	SiS	HCO^+		
	C_2	NaOH		
	CH^+			
	NO			
Hexa-atomics	**Hepta-atomics**	**Octa-atomics**	**Nona-atomics**	**Others**
CH_3OH	CH_3CHO	$HCOOCH_3$	CH_3CH_2OH	HC_9N
NH_2CHO	CH_3NH_2		CH_3OCH_3	$HC_{11}N$
CH_3CN	H_2CCHCN		CH_3CH_2CN	
CH_3SH	CH_3C_2H		HC_7N	
CH_2CCH				

Table 5 Some chemical species in the interstellar medium

Figure 16 Stanley Miller used this apparatus to make amino acids from simple molecules

Both Miller and Orgel showed that molecules like those in the ISM could react together, under conditions similar to the ones which existed during the early history of the Earth, to form some of the molecules of life. Their experiments give added weight to the suggestion that life on Earth has its origin in molecules from outer space.

Of all the molecules of life, none are more remarkable or more essential than **enzymes**.

Enzymes are proteins which *catalyse* chemical reactions in the living cells of plants and animals. As catalysts, enzymes are highly efficient, increasing reaction rates more than a million fold. They are also highly specific. It has been estimated that a living cell contains about 3000 different enzymes and each one catalyses a specific reaction (or group of reactions).

You will take a more detailed look at the structure of proteins, enzymes and other 'molecules of life' in **Unit 2**, **Chapter 5**. Later in this unit, in **Chapter 15**, you will learn more about the action of catalysts, including enzymes.

1.6 *Summary*

This Chemical Storyline chapter is linked to Chemical Ideas Section 10.1 and Chapters 11–13.

You began by reading about the origin of the chemical elements in stars, which allowed you to revise your ideas on atomic structure.

Next, you looked at the elements present in human beings, which led to a more detailed study of two of the elements found in the body. To work out the proportions of the different kinds of atoms in a human body, you needed to know about moles of atoms and this led to a revision of your previous work on amounts of substance.

Having come across a number of different elements, you then found out about the way in which chemists developed a system for classifying the elements and looked in detail at the modern Periodic Table.

Finally, you studied the ways chemical elements combine together to form compounds and saw how the chemical elements present in outer space could have combined to form 'the molecules of life'.

2 MAKING USE OF ENERGY

2.1 *Fuels are important*

How did you get to school or college today? Unless you are lucky and live near enough to walk or cycle, you will have come by some form of motorised transport – by bus or car, or perhaps by train. Each of these forms of transport relies on burning **fuels** to supply the energy needed. The fuel might be petrol or diesel, both of which are obtained from crude oil, a **fossil fuel**. If the train was electric, the chances are that the electricity was generated by burning a fossil fuel such as coal or gas, though it might have been generated in an alternative way, such as using the energy from nuclear reactions or from hydroelectric power.

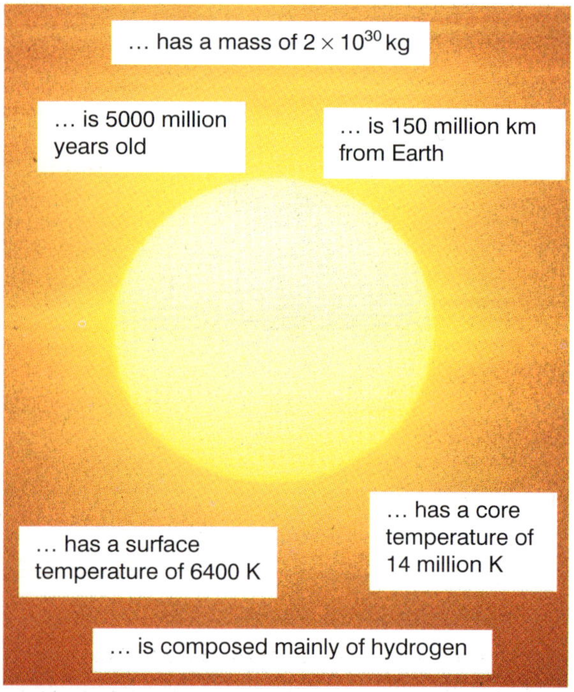

Figure 1 Without the Sun, life on Earth could not exist. We need its regular input of energy to maintain the Earth's surface temperature and to drive the chemical reactions essential for life. Most of our energy needs are met by the products of past photosynthesis – fossil fuels such as coal, oil and gas

In almost every aspect of modern life we rely on fuels to supply our energy needs. But there are problems. All carbon-containing fuels produce carbon dioxide when they burn and that's a major contributor to the greenhouse effect (see **Unit 3, Chapter 7**). They also produce other kinds of polluting emissions, so, even if you walked or cycled this morning, you would still have been breathing in air containing waste materials generated by other forms of transport.

In the first part of this chapter you will look at the chemistry behind combustion, where the energy comes from when a fuel burns, and how chemists can measure such energy changes. You will then go on to investigate some of the problems that burning fuels can cause and how chemists and chemistry are helping to solve these problems.

Figure 2 Motor vehicles release carbon dioxide and other kinds of polluting emissions into the atmosphere

---

To understand the sections that follow, you will need to find out how to deal with energy changes in chemistry. **Sections 14.1** and **14.2** will help you with this.

You will need to use balanced equations and chemical calculations with confidence throughout your course. If you feel in need of revision, **Sections 10.2** and **10.3** will help.

2.2 *Getting energy from fuels*

The role of oxygen

If you have had the opportunity to measure enthalpy changes of combustion in the laboratory, you will have found that the values you obtain are different for different fuels. Let's compare five important fuels – look at Figure 3.

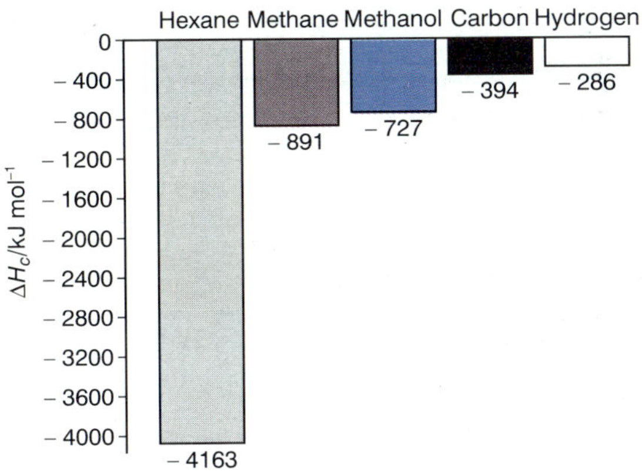

Figure 3 Enthalpy changes of combustion for some important fuels

Note that the enthalpy changes are *negative* – this is because during combustion the reactants *lose* energy to their surroundings.

You can see that the enthalpy changes of combustion vary widely. Why should this be? What decides how much energy you get when you burn a mole of a fuel?

We think of fuels as energy sources, but they can't release any energy until they have combined with oxygen. So really we should think of *fuel–oxygen systems* as the energy sources. As you can see in **Section 14.1**, when substances burn the energy released comes as bonds with oxygen are formed.

ASSIGNMENT 1

a Write down balanced equations for the complete combustion of methane (CH_4), hexane (C_6H_{14}) and methanol (CH_3OH).

b Use Figure 3 to find out how much energy is released when 1 mole of each substance in part **a** burns completely in oxygen.

c Use the ideas of bond making and bond breaking to explain the following. (Think in terms of the bonds that have to be broken and the new bonds that have to be made when the fuel burns in oxygen.)

- **i** Why is less energy released when 1 mole of methane burns than when 1 mole of hexane burns?
- **ii** Why is less energy released when 1 mole of methanol burns than when 1 mole of methane burns? After all, they both have the same number of C and H atoms.

The enthalpy change of combustion of a fuel depends on two things. First, there is the *number* of bonds to be broken and made – and that depends on the size of the molecule involved. That's why larger molecules such as hexane have a more negative ΔH_c than smaller ones such as methane.

But ΔH_c also depends on the *type* of bonds involved. Let's take your answer to part c(ii) in **Assignment 1** a bit further. The equations below show the bonds involved in the burning of methane and methanol. The products are the same, but the key difference is that *methanol already has an O–H bond*. In other words, one of the bonds to oxygen is already made – unlike with methane, where all the new bonds have to be made, starting from scratch. Another way of looking at this is to say that methanol is already partly oxidised.

The energy released during combustion comes from the making of bonds to oxygen. If methanol already has one bond made, it will give out less energy when it burns.

(a) Combustion of methane

$$CH_4 + 2O_2 \longrightarrow CO_2 + 2H_2O$$

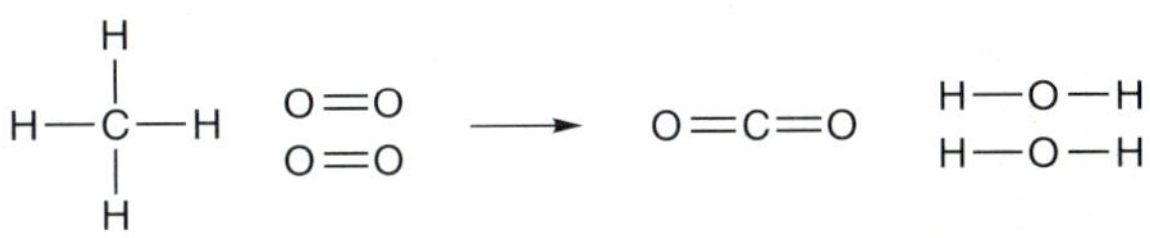

(b) Combustion of methanol

$$CH_3OH + 1\tfrac{1}{2}O_2 \longrightarrow CO_2 + 2H_2O$$

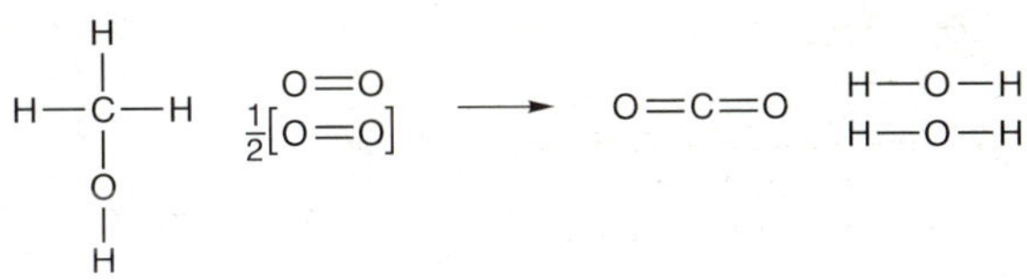

As a general rule, the more oxygen a fuel has in its molecule, the less energy it will give out when 1 mole of it burns. Oxygenated fuels such as alcohols and ethers are less energy-rich than hydrocarbon fuels. However, that's not to say that they are poor fuels. In some ways oxygenated fuels are better, because they are less polluting than hydrocarbons, and they often have a high octane number. There is more about octane numbers in **Unit 2, Section 4.3**.

Important news for slimmers

When you eat too much of an energy food, the excess energy gets stored in your body as fat. The more energy-rich the food, the more fattening it is.

Compare a carbohydrate such as glucose with a fat such as olive oil. Here are the formulae:

glucose	$C_6H_{12}O_6$
glyceryl trioleate	$C_{57}H_{104}O_6$
(the main constituent of olive oil)	

Per carbon atom, glucose is much more oxygenated than olive oil, so it is much *less* energy-rich. From 1 gram of a carbohydrate such as glucose you can get about 17 kJ. From 1 gram of a fat such as olive oil you can get about 39 kJ. Gram for gram, fats are more than twice as fattening as carbohydrates.

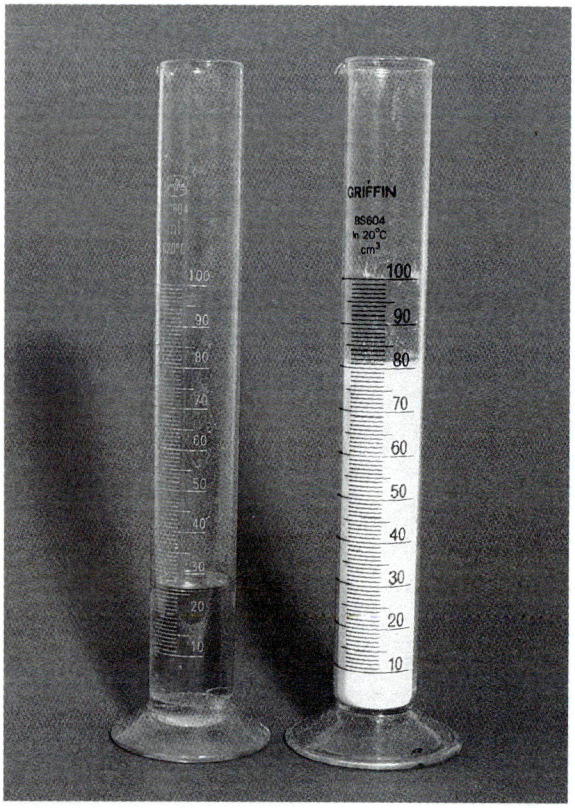

Figure 4 Fats and oils are more energy-rich than carbohydrates – the oil on the left will provide the same quantity of energy as the glucose on the right

Alcohol is neither a fat nor a carbohydrate – but it too is bad news for slimmers. In fact there is a whole series of related compounds called **alcohols**, but the particular alcohol present in drinks is ethanol, C_2H_5OH. The same substance is used as an alternative to petrol for cars in some countries. It burns in the car engine releasing energy – and it also releases energy when metabolised in the body (Figure 5).

But be thankful that you are not a potato. Humans store energy in fat, but potatoes store it in starch. To store a given amount of energy, you need to have more than twice the mass of starch compared with fat. Think how oversize *that* would make you!

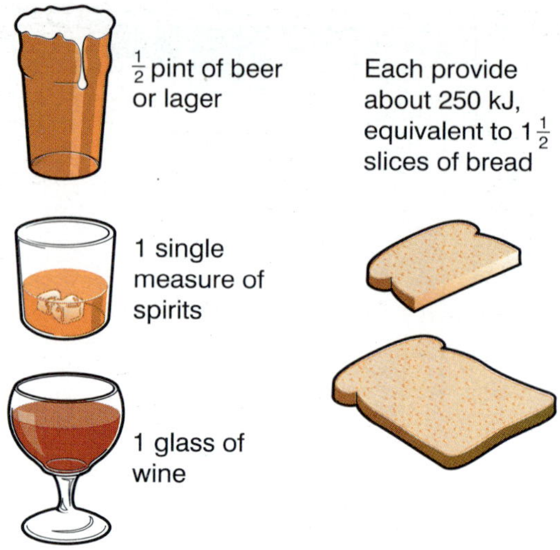

Figure 5 Alcohol can be fattening

Figure 6 How are these foods made to be less fattening?

Carrying fuels around

For a practical fuel, the enthalpy change of combustion may not be the most important thing to consider. What really matters is the **energy density** – how much energy you get per kilogram of fuel. After all, you have to carry the stuff around with you. We can work this

Figure 7 The energy density of coal is relatively low – but it is a lot higher than for wood

Fuel	Formula	Enthalpy change of combustion, ΔH_c/kJ mol^{-1}	Relative formula mass	Energy density (energy transferred on burning 1 kg of fuel)/kJ kg^{-1}
hexane	$C_6H_{14}(l)$	−4163	86	−48 400
methane	$CH_4(g)$	−891	16	−55 700
methanol	$CH_3OH(l)$	−727	32	−22 700
carbon	$C(s)$	−394	12	−32 800
hydrogen	$H_2(g)$	−286	2	−143 000

Table 1 Energy densities of some important fuels

out from the enthalpy change of combustion, using the relative formula mass. We've done this for five fuels and the results are in Table 1.

ASSIGNMENT 2

Look at the values for energy density of the different fuels in Table 1.

a On the basis of energy density, which is the best fuel in the table? What are the practical difficulties involved in using this particular fuel?

b Compare hydrogen and hexane. Explain why hydrogen has the higher energy density, even though it has the lower (less negative) enthalpy change of combustion.

c Here are some data for octane, C_8H_{18}, and decane, $C_{10}H_{22}$, both of which are components of petrol:

	ΔH_c /kJ mol^{-1}	Relative formula mass
octane, C_8H_{18}	−5470	114
decane, $C_{10}H_{22}$	−6778	142

Use the data to calculate the energy density for each of these compounds.

Compare your two answers. How do they compare with the values of energy density given for the fuels in Table 1?

2.3 *Trouble with emissions*

A serious problem

We'll look next at a problem of motor fuels which is causing worldwide concern: exhaust emissions. Figure 8 shows what goes into – and what comes out of – a car engine. The nitrogen oxides, NO and NO_2, are grouped together as NO_x, although the NO_x emitted by vehicle exhausts contains mostly NO (about 95%) with only small amounts of NO_2. Similarly, SO_x represents the oxides of sulphur, SO_2 and SO_3, and C_xH_y represents the various hydrocarbons present in the exhaust fumes.

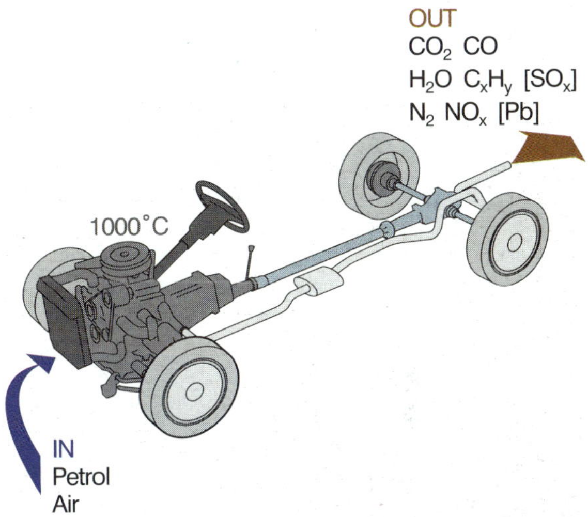

Figure 8 What goes into – and comes out of – a car engine

ASSIGNMENT 3

a For each of the substances emitted in the exhaust, as shown in Figure 8, use your knowledge of chemistry to explain where the substance comes from. Write chemical equations where appropriate.

Some will be easier to explain than others! There are two things to bear in mind: the temperature inside a car engine's combustion chamber is about 1000 °C, and air is a *mixture* of gases.

b For each of the emissions, say which of the following environmental problems you think it may contribute to:

A acid rain

B the greenhouse effect

C damage to the ozone layer

D toxic effects on humans

E photochemical smog.

You will need to use your knowledge of air pollution from your earlier studies, or find the information in a textbook.

Another problem arises because petrol is so volatile. A parked car on a warm day gives off hydrocarbon fumes, mostly butane, from the petrol tank and the carburettor. This is called *evaporative emission* and accounts for about 10% of emissions of volatile organic compounds from vehicles.

Photochemical smog

Unfortunately, these are not the only pollutants caused by motor vehicles. **Ozone** is a *secondary pollutant*, because it is not released directly into the atmosphere. It is formed as a result of chemical reactions that take place when sunlight shines on a mixture of two *primary pollutants*, nitrogen oxides (NO_x) and hydrocarbons (C_xH_y), together with oxygen and water vapour. Also produced are irritating and eye-watering compounds formed by the breakdown and further reaction of the hydrocarbons. These reactions all occur in **photochemical smogs**, which are a great cause for concern. (A photochemical reaction occurs when a molecule absorbs light energy and then undergoes a chemical reaction.)

Ozone – a molecule of many parts

Ozone is a highly reactive substance whose molecules contain three oxygen atoms. Its formula is O_3. Ozone occurs in the atmosphere, both high up in the *stratosphere* and close to the ground in the *troposphere*. Its presence in these two regions affects us in very different ways.

In the stratosphere, ozone acts as a sunscreen, filtering out dangerous ultra-violet light from the incoming solar radiation. Its presence there is vital to our survival, though there is now great concern because chlorofluorocarbons (CFCs) are reaching the stratosphere and are causing depletion of the protective ozone layer. You will find out more about this in **Unit 2, Chapter 6**.

Meanwhile, in the troposphere, ozone has an important role in the production of hydroxyl radicals (OH). These are short-lived species, formed from water, which bring about the breakdown of many substances that would otherwise build up to hazardous levels in the atmosphere. However, ozone is an irritant toxic gas and high concentrations near ground level are detrimental to human health. It weakens the body's immune system and attacks lung tissue. Also, ozone in the troposphere acts as a greenhouse gas and so contributes to global warming.

Figure 9 Mexico City, like many large cities, suffers from poor air quality and reduced visibility. Photochemical smogs are a global problem

Photochemical smog contains a mixture of primary and secondary pollutants (Figure 10). Its exact composition varies enormously and depends, for example, on the nature of the primary pollutants, the local geography, weather conditions, the time of day and the length of the smog episode.

Figure 10 The main constituents of photochemical smog

Photochemical smogs normally occur in the summer during high pressure (anticyclonic) conditions. The still air means that there is much less mixing with high altitude air and the pollutants are trapped near ground level. Figure 11 summarises the

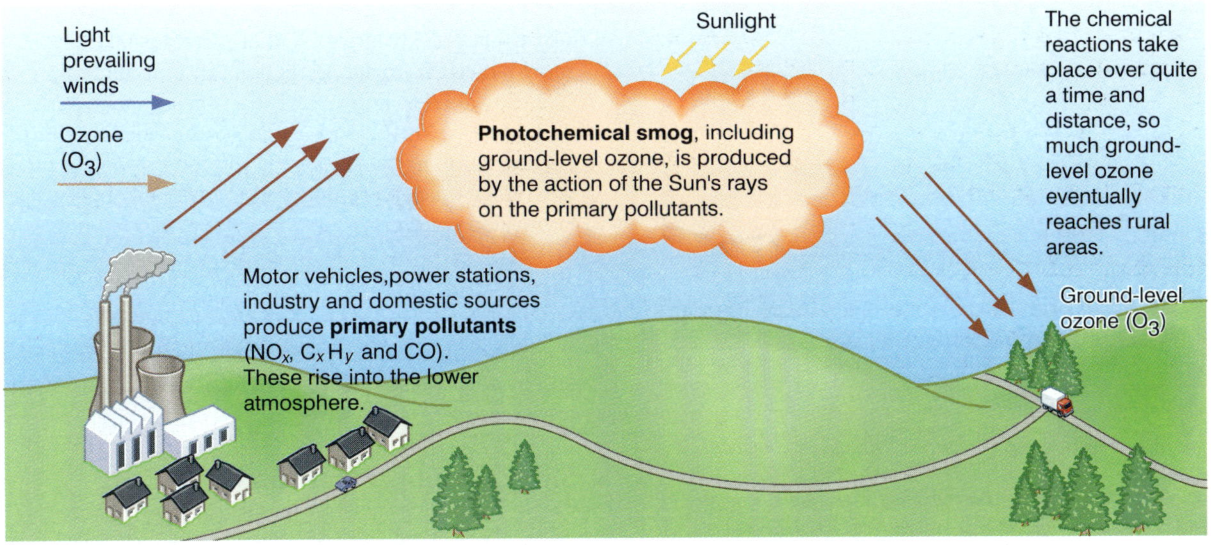

Figure 11 Formation of photochemical smog

formation of photochemical smog. Note that, even in clean relatively 'unpolluted' air, there is a small background concentration of ozone, and this ozone is involved in the series of reactions that produces the photochemical smog; its concentration is *increased* during smog formation.

If there is a light prevailing wind, the polluted air mass will be transported from the built-up urban area where it was generated and will move across rural areas. In fact the highest ozone readings are often recorded over rural areas because the chemical reactions producing the secondary pollutants take place relatively slowly.

What are the effects of photochemical smog?

Photochemical smogs cause haziness and reduced visibility in the air close to the ground. For many people, they can cause eye and nose irritation and some difficulty in breathing, but for vulnerable groups – people such as asthmatics, who already have respiratory problems, very young children and many old people – the effects can be more serious.

Ozone is not the only substance in smog to cause health problems. The polluted air contains a whole cocktail of harmful chemicals but the exact links between photochemical smog and health problems such as asthma are very difficult to establish and are still the subject of much research. There is no simple relationship between the two because so many other factors are involved.

People are not the only ones to suffer. High ozone concentrations affect animals and plants too. Ozone is a highly reactive molecule which attacks most organic matter. Compounds with carbon–carbon double bonds are particularly vulnerable, so many materials, such as plastics, rubbers, textiles and paints, can be damaged.

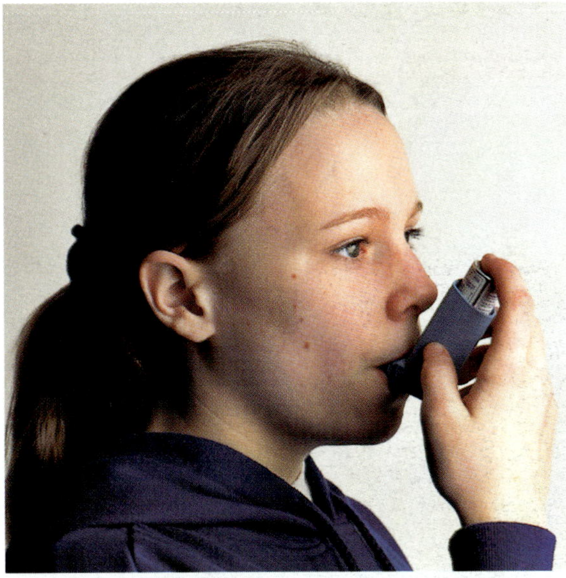

Figure 12 Air pollution can trigger an asthma attack, but there is no conclusive evidence to date that air pollution causes new cases of asthma

How can chemists help?

Research work is taking place on several fronts to unravel and explain the complex chemistry involved.

Monitoring pollutants

An important first step is to find out exactly which pollutants are present in the troposphere and how their concentrations vary. For example, ozone concentrations in urban areas show regular fluctuations which reflect changes in sunlight and vehicle emissions during the day (Figure 13). There are now several networks of monitoring stations across the UK recording pollutant concentrations. Figure 14 shows the location of the ozone monitoring sites in the UK.

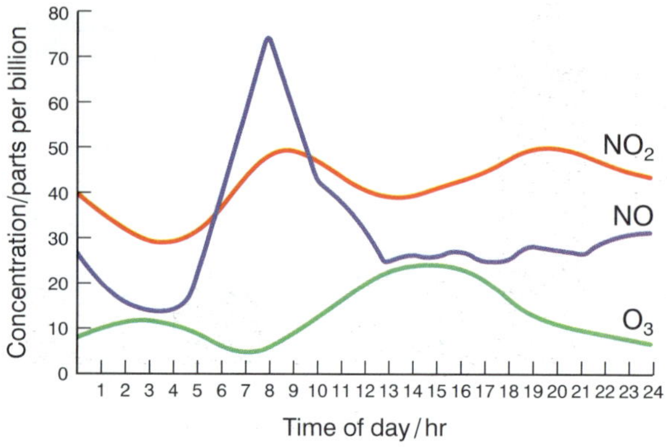

Figure 13 Variation of the concentrations of ozone and oxides of nitrogen on a summer weekday in London. The high NO peak coincides with the early morning traffic rush hour, and is quickly followed by a build-up of NO_2, as reactions in the air convert NO into NO_2. Photochemical reaction of NO_2 leads to formation of O_3, which reaches its peak in the early afternoon. The evening rush hour generates more NO and NO_2, but decreasing sunlight slows down photochemical reactions and there is no corresponding build-up of ozone

Figure 14 A map showing ozone automatic monitoring sites in the UK

Studying individual reactions in the laboratory

The air around us may look very still and unchanging but this is deceptive. The troposphere acts as a huge reaction mixture with a vast array of interrelated chemical reactions taking place all the time. Many of these reactions involve 'broken down' fragments of molecules, called **radicals**, and the reactions can take place very quickly indeed. Other reactions take place more slowly. To make predictions about pollution, chemists need to know what reactions can take place and how quickly they occur. This means investigating each reaction in the laboratory and measuring the rate at which it takes place under a variety of carefully controlled conditions.

Modelling studies

Much of the information on rates of reactions is used in computer simulation studies which aim to reproduce and predict the behaviour of pollutants during a smog episode.

Smog chamber simulations

These are laboratory experiments on a grand scale. Primary pollutants are mixed in a huge plastic bag called a smog chamber and exposed to sunlight under carefully controlled conditions (Figure 15). Analytical probes monitor the concentrations of various species as the photochemical smog builds up.

Figure 15 The EU smog chamber in Valencia, Spain. The reaction chamber has a volume of approximately 200 m^3 (200 000 litres). It has to be this big to reduce to a minimum any 'surface effects' where reactions take place on the walls of the container instead of in the gas phase

---

Studying the rates of chemical reactions is vital for chemists working to understand the many reactions taking place in the atmosphere. It is important in many other areas of chemistry too.

You can find out about the factors that affect the rate of a chemical reaction and how chemists measure rates of reaction in **Sections 15.1**, **15.2**, **15.3** and **15.4**.

The reactions you will study in the laboratory will take place in solution rather than in the gas phase. You may find it helpful to remind yourself about how chemists measure concentrations of substances in solution by reading **Section 10.4** before you study **Chapter 15**.

2.4 *Tackling the emissions problem*

Concern about air pollution from motor vehicles is mounting worldwide and many countries are bringing in legislation to limit emissions. The US has led the way, and cars are put through a rigorous emissions test cycle before they are allowed on the road. Figure 16 shows how emission limits have become increasingly severe.

There are two ways of tackling the emissions problem directly. One involves changing the technology of the car engine, including the exhaust system. The other involves changing the fuel used by the car.

Of course, there are important indirect methods for tackling the emissions problem, such as limiting the traffic entering towns and encouraging car-sharing schemes for people travelling to and from work. However, here we shall concentrate on methods of reducing the emissions from a single car.

Method 1: Changing the engine technology

How much air does a petrol engine need? You can work this out by doing **Assignment 4**.

ASSIGNMENT 4

For the purposes of this assignment, assume that petrol is pure heptane, C_7H_{16}.

a Write an equation for the complete combustion of heptane vapour.

b How many moles of oxygen are needed for the combustion of 1 mole of heptane?

c How many moles of C_7H_{16} are there in 1 g of heptane?

d What mass of oxygen is needed to burn 1 g of heptane (O = 16)?

e What mass of air is needed to burn 1 g of heptane? (Assume that air is 22% oxygen and 78% nitrogen by *mass*.)

f What *volume* of air (measured at 25 °C and 1 atmosphere pressure) is needed to burn 1 g of heptane? (Assume that 1 mole of oxygen has a volume of 24 litres under these conditions and that air is 21% oxygen and 79% nitrogen by *volume*.)

Section 10.5 will help you to carry out calculations involving gases.

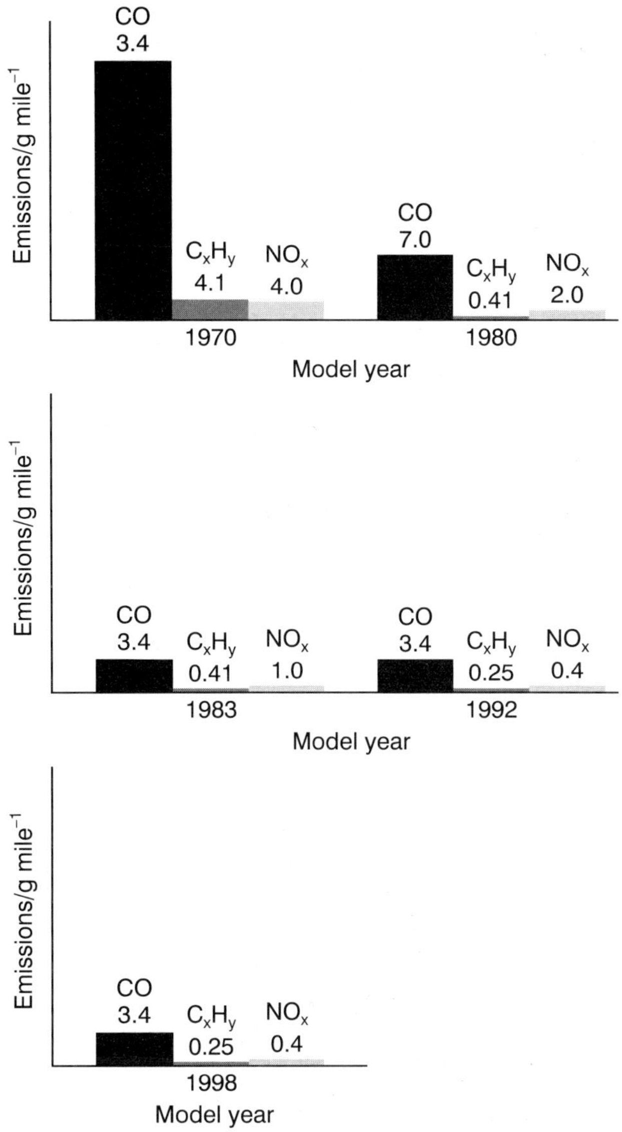

Figure 16 US Federal emission limits for new vehicles have become increasingly severe

The ratio of air to fuel which you worked out in Assignment 4 is the ratio needed to make the petrol burn completely. It is called the **stoichiometric ratio**. (*Stoichiometric* means *involving the exact amounts shown in the chemical equation*.) For an average car engine, the stoichiometric ratio is usually taken to be about 15:1 (by mass).

If you have *less* air than the stoichiometric ratio, it is called a 'rich' mixture, because it's rich in petrol. You burn a rich mixture when you use the choke to start a car on a cold morning. A 'lean' mixture has *more* air than the stoichiometric ratio: it produces less NO_x and CO but the levels of C_xH_y can actually increase (Figure 17). The trouble is, if the mixture is *too* lean, the engine misfires and emissions increase.

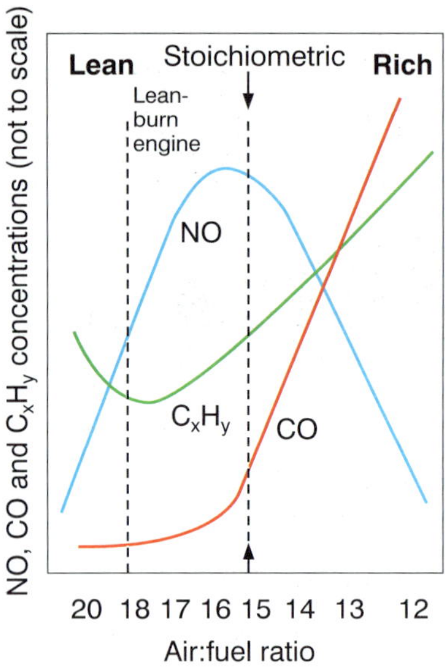

Figure 17 How gases in exhaust emissions vary with the composition of the air–fuel mixture

'Lean-burn' engines use an air:fuel ratio of around 18:1 and have specially designed combustion chambers to get over the problem of misfiring.

Another advantage of a lean-burn engine is that it gives better fuel economy, because you are using less fuel for each firing of the cylinder.

Catalysts can help

Catalysts speed up chemical reactions without getting used up themselves. We can use them to speed up reactions which involve pollutants in car exhausts. Look at these reactions:

$2CO(g) + O_2(g) \rightarrow 2CO_2(g)$ (reaction 1)
$C_7H_{16}(g) + 11O_2(g) \rightarrow 7CO_2(g) + 8H_2O(g)$ (reaction 2)
$2NO(g) + 2CO(g) \rightarrow N_2(g) + 2CO_2(g)$ (reaction 3)

(Here we've used C_7H_{16} as an example of an unburnt hydrocarbon and NO as the main component of NO_x in car emissions.)

In these reactions, pollutants are being converted to CO_2, H_2O and N_2, which are all naturally present in the air. These reactions go of their own accord, but under the conditions inside an exhaust system they go too slowly to get rid of the pollutants.

A catalyst made of a precious metal like platinum or rhodium speeds up these reactions in the exhaust system. Such catalysts are used in **catalytic converters**.

You can find out more about catalysts in **Section 15.5**.

ASSIGNMENT 5

Look at reactions 1–3.

a For each of the pollutants CO, C_7H_{16} and NO, say whether it is being oxidised (gaining oxygen and/or losing hydrogen) or reduced (the reverse).

b Which of these reactions would be important in controlling pollutants from
 i an ordinary engine?
 ii a lean-burn engine? (Look at Figure 17.)

A lean-burn engine uses an *oxidation catalyst system* which removes CO and C_xH_y. The exhaust gases are rich in oxygen, so CO and C_xH_y are oxidised to CO_2 and H_2O on the surface of the catalyst. This kind of catalyst system does little to lessen NO_x, because NO_x needs reducing, not oxidising, to turn it to harmless N_2. But that doesn't matter too much in a lean-burn engine, because this type of engine produces less NO_x anyway.

The three-way catalyst system

Catalytic converters can be fitted to ordinary engines too. But for an ordinary engine, a simple oxidation catalyst won't do – it wouldn't remove the NO_x. Here, a *three-way catalyst system* is needed, which both oxidises CO and C_xH_y, *and* reduces NO_x. This kind of catalyst system *only* works if the air–petrol mixture is carefully controlled so that it's exactly the stoichiometric mixture for the fuel. If the mixture is too rich, there is not enough oxygen in the exhaust fumes to remove CO and C_xH_y.

This means that cars fitted with three-way catalyst systems need to have oxygen sensors in the exhaust gases, linked back to electronically controlled fuel injection systems. You can see the arrangement in Figure 18.

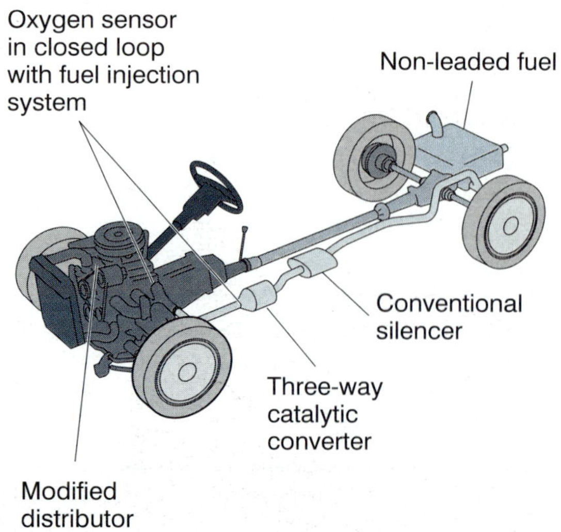

Figure 18 The three-way catalyst system

Figure 19 shows the effect of catalysts on exhaust emissions from ordinary and lean-burn engines. A three-way catalyst system is the most efficient way of simultaneously lowering all three emissions.

All catalytic converters work only when they are hot. A platinum catalyst starts working around 240 °C, but by alloying the platinum with rhodium you can get the catalyst to start working at about 150 °C. These catalysts are poisoned by lead, so the converters can only be used with lead-free fuel.

The catalyst is used in the form of a fine powder spread over a ceramic support whose surface has a network of tiny holes (Figure 20). The surface area of the catalyst exposed to the exhaust gases is about the same as two or three football fields.

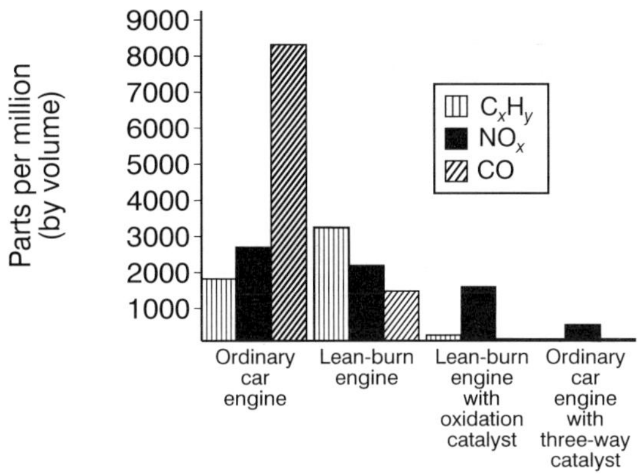

Figure 19 The effect of catalysts on the exhaust emissions from ordinary and lean-burn engines

ASSIGNMENT 6

a Why is it important that catalytic converters start working at as low a temperature as possible?

b What is meant by a *catalyst poison*?

c Why is the catalyst used in the form of a fine powder?

d Suggest a reason why the catalytic converter has eventually to be replaced.

e Catalytic converters convert the pollutants CO, C_xH_y and NO_x into harmless gases. This is still only a partial solution to the emissions problem. Why?

f Why is it possible to fit oxidation catalytic converters to existing cars, but not three-way converters?

Method 2: Changing the fuel

The other approach to tackling the emissions problem is to change the fuel used by cars.

Aromatic hydrocarbons make up as much as 40% of lead-free petrol. Aromatic hydrocarbons may cause higher CO, C_xH_y and NO_x emissions, and some of them may cause cancer. Benzene is the worst, and is strictly controlled, but others may also be controlled in the future.

Butane content too will probably be lowered in the future. Butane is volatile and is responsible for evaporative emissions leading to ozone formation and photochemical smogs.

However, both butane and aromatic hydrocarbons are high octane components of petrol. If they are removed, their octane quality must be replaced by something else. This is why the petrol companies are looking to the oxygenates as a possible solution. You will read more about this in **Unit 2, Chapter 4**.

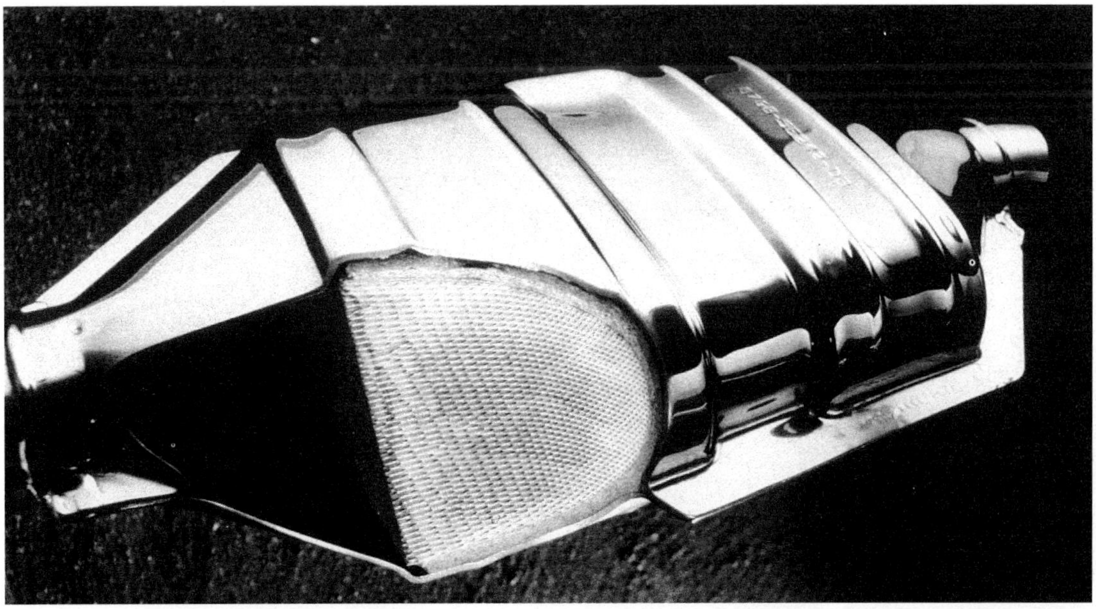

Figure 20 A catalytic converter for a car exhaust. Note the honeycomb structure, giving a high surface area

2.5 Summary

This Chemical Storylines chapter is linked to Chemical Ideas Sections 10.2–10.5 and Chapters 14 and 15.

You began by examining the nature of fuels and their central importance in our lives. This led you to compare the energy released when different fuels are burned in oxygen and to a general study of energy changes in chemical reactions.

Secondly, you considered the problem of emissions from the exhausts of motor vehicles, the difference between primary and secondary pollutants and the formation of photochemical smog. Photochemical smogs are a relatively recent problem caused by release into the atmosphere of two primary pollutants, oxides of nitrogen and hydrocarbons. Each year, emissions of the primary pollutants increase worldwide and future legislation to impose limits seems inevitable if our health is not to suffer. But photochemical smog does not recognise national boundaries and international cooperation will be essential. It is vital that any decisions are made on the basis of firm scientific understanding of the processes involved. Central to such studies is a study of the reactions involved and the rates at which these reactions occur. This led to a general study of rates of reactions, how these are measured in the laboratory and the factors that affect the rate.

Finally, you examined ways in which car manufacturers and oil companies are tackling the problem of emissions – by changing engine technology to control the way the fuel burns, by changing the composition of the fuel, and by using catalytic converters to speed up reactions in the exhaust that convert the pollutants to less harmful gases. You then looked at how catalysts work and considered other examples of catalysts in action, in industry and in biological systems.

3
OUR PLANET

Life on our planet is possible because of its surface temperature and the chemical substances present. In part, the surface temperature is determined by the distance of the Earth from the Sun, but it also depends on processes within the Earth, and in the atmosphere and the oceans. Whether substances are present as gases in the air, solids we can walk on or water in the clouds, rivers and oceans depends on their chemical structure, and the nature and strength of the bonding present. In this chapter you will learn more about the chemical nature of the world we live in.

3.1 *The air above us*

The atmosphere is a relatively thin layer of gas extending about 100 km above the Earth's surface. If the world were a blown-up balloon, the rubber would be thick enough to contain nearly all of the atmosphere (see Figure 1). Thin though it is, this layer of gas has an enormous influence on the Earth.

The atmosphere becomes less dense the higher you go. In fact, 90% of all the molecules in the atmosphere are within the first 15 km from the Earth's surface, a region called the **troposphere**. In this region, mixing is easy because hot gases can rise and cold gases can fall.

Figure 1 The Earth seen from space

Gas		Concentration (by volume)
		Per cent
nitrogen		78
oxygen		21
argon		1
		Parts per million
carbon dioxide		367*
neon		18.0
helium		5.0
methane		1.8*
krypton		1.1
hydrogen		0.5
dinitrogen oxide, N_2O		0.3*
carbon monoxide		0.1*
xenon		0.09
nitrogen monoxide, NO nitrogen dioxide, NO_2	NO_x	0.003*

*Table 1 Composition by volume of dry tropospheric air from an unpolluted environment (*variable)*

Table 1 shows the average composition by volume of dry air from an unpolluted environment and is typical of the troposphere.

The concentrations of some of these substances are measured in **parts per million (ppm)** by volume. This measure is often used where small concentrations are involved; 367 ppm corresponds to a percentage concentration of 0.0367%.

The atmosphere hasn't always had this composition. The first atmosphere was lost altogether during the upheavals in the early life of the solar system. The next atmosphere consisted of compounds such as carbon dioxide, methane and ammonia which bubbled out of the Earth itself.

You can find out why some covalent molecules are gases at the temperatures of our atmosphere, but others are liquids or solids, by studying **Section 16.1**.

Three thousand million years ago there was very little oxygen in the atmosphere. But when the first simple plants appeared, they began to produce oxygen through photosynthesis. For more than 1000 million years very little of this oxygen reached the atmosphere. It was used up as quickly as it was formed in oxidising sulphur and iron compounds and other chemicals in the Earth's crust. It wasn't until this process was largely complete that oxygen began to collect in the atmosphere.

When the oxygen concentration reached about 10%, there was enough for the first animals to evolve, using oxygen for respiration. Eventually there was enough respiration and enough other processes going on to remove the oxygen as fast as it formed. Since then, the oxygen concentration has remained at about 21%.

As the oxygen concentration built up, so did the amount of ozone, O_3, produced by the reaction of oxygen molecules with oxygen atoms:

$$O_2 + O \rightarrow O_3$$

The oxygen atoms came from the breakdown of oxygen molecules by ultra-violet radiation:

$$O_2 \rightarrow O + O$$

(You will meet these reactions in more detail in **Unit 2, Chapter 6**.)

Eventually there was enough ozone in the upper atmosphere, the **stratosphere**, to absorb most of the ultra-violet radiation arriving from the Sun. Then life on the surface of the land could develop, protected from the dangers of high-energy ultra-violet radiation. The planet was nearly 4 billion years old, and life had existed for 3 billion years before life moved on to the land (Table 2).

Look again at Table 1. All the gases listed are produced as a result of natural processes. Human activities add more gases to the atmosphere.

The first plants

Most early life forms were resistant to ultra-violet radiation. The first plants appeared about 2800 million years ago and were probably single-celled organisms called *cyanobacteria*, or *blue-green bacteria*.

They lived in the surface waters of the oceans under anaerobic conditions, and used sunlight as their energy source to drive the chemical reactions needed to maintain growth.

Close relatives of cyanobacteria are alive today (there are plenty of places to hide away from the oxygen – in the mud of salt marshes or in a clump of dead seaweed, for example) and they are still very tolerant of ultra-violet radiation.

Gases always mix together completely (given time) and this natural diffusion process is greatly speeded up by air currents. So in time, pollutant gases spread throughout the atmosphere. Atmospheric pollution is a global problem: it affects us all.

Million years ago		
4600	origin of Earth	first atmosphere lost from hot planet
3800	oldest sedimentary rocks organisms probably present	atmosphere consists of CO_2, NH_3, CH_4 and H_2S
3500	beginning of fossil record bacteria leave fossil remains	life probably uses energy released during fermentation, or from oxidation of organic molecules by SO_4^{2-} and NO_3^- ions
2800	first photosynthetic bacteria (cyanobacteria or blue-green bacteria)	oxygen produced; for 1 million years it is used up in the oxidation of S^{2-} and Fe^{2+} ions; no ozone is able to form in the atmosphere so life on the surface is impossible due to ultra-violet radiation; bacteria live in the top 200 m of the ocean, screened from ultra-violet radiation by water
2000	first significant $CaCO_3$ deposits	removal of two greenhouse gases (CO_2 to form $CaCO_3$, and CH_4 by reaction with O_2) leads to major ice age
1800	oxygen begins to build up in the atmosphere	
1500	first green algae – plants with chlorophyll in their cells	
800	first sea animals	1% oxygen in atmosphere; formation of ozone begins
400	first land plants	10% oxygen in atmosphere; ozone layer protects plant cells from ultra-violet radiation
300	first land animals	21% oxygen in atmosphere

Table 2 The origins of life on Earth

ASSIGNMENT I

Use Table 1 to answer the following questions.

a How many parts per million (by volume) of argon are there in a typical sample of air?

b In 1 litre of air, what is

i the volume of methane present?

ii the percentage of methane *molecules* in the sample?

c Human activities add more gases to the atmosphere. For each of the gases marked with an asterisk, suggest *one* way in which human activities increase its concentration.

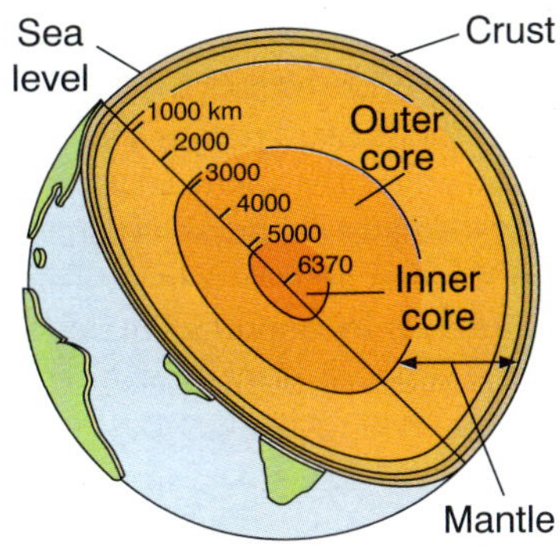

Figure 2 The layers of the Earth

3.2 *The Earth below us*

The Earth's crust is just 1.5% of the volume of the planet (Figure 2). Table 3 gives the composition of the Earth's crust by mass. A thin layer, typically 1 m–2 m thick, on the top of the crust forms the soil, vital for the survival of life on Earth.

Road cuttings often reveal to us the underlying layers of a soil (Figure 3). Plants take nutrients and water from the soil which also supports them as they grow.

Element	% by mass
oxygen	47
silicon	28
aluminium	8
iron	5
calcium, sodium, potassium, magnesium	2–4 each

Table 3 Composition of the Earth's crust by mass

Figure 3 Layers of soil exposed in a road cutting

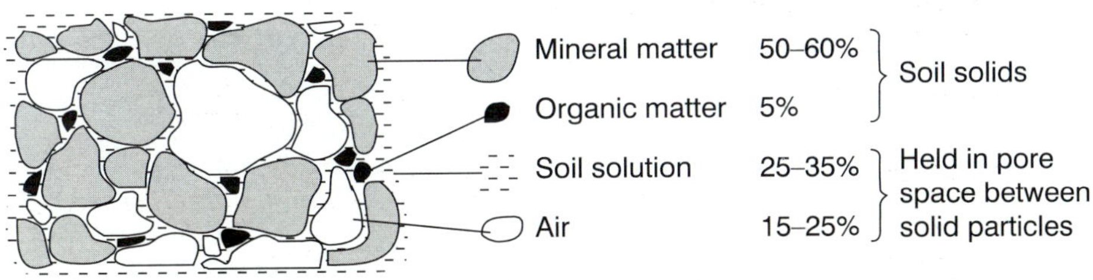

Figure 4 The four major components of soils, showing the approximate proportions by volume for a typical moist soil

What is soil?

Soil is a mixture of weathered rock fragments and organic matter. Together they make a porous fabric which can hold both air and water. The approximate proportions by volume for a typical moist soil are shown in Figure 4.

The quantities of air and water in a soil are variable and their proportions are important in determining how well the soil supports plant growth. The minerals play a vital part, both in retaining water and in making nutrient ions available to plants.

Soil minerals are produced by weathering of the rocks which make up the Earth's crust. The fragments of rock and minerals vary enormously in size, from coarse stones and gravel to fine clay particles.

The coarse fragments are relatively inactive chemically. It is the very fine clay particles that are the most active because clay minerals can bind positive ions to their surfaces. In this way, essential ions are held in the soil.

Figure 5 A view of the Scottish coastline at Sheigra, Sutherland, showing sand and fertile soil produced by the weathering of rocks

Weathering

Weathering is the action of wind, rain, frost and sunlight – and nothing on the surface of the Earth's crust escapes it. Weathering breaks up the rocks. It changes their physical structure and sometimes their chemical composition. It carries away soluble materials and some of the solid fragments as well. But weathering is creative as well as destructive: it makes a **soil** from the uppermost layers of weathered rock.

Figure 6 summarises the main pathways of weathering in the moderately acid conditions of humid temperate climates, like ours. In different climates the rates of the reactions alter, and this alters the compositions of the soils.

ASSIGNMENT 2

In different climates soil compositions can be very different. For example, many tropical soils are high in iron and aluminium oxides, which come from the chemical decomposition of rocks, as shown in Figure 6.

Use your knowledge of the factors affecting reaction rates to explain why the iron oxide and aluminium oxide concentrations are higher in tropical soils than in temperate soils.

Quartz and silicates

There are many different types of rock, of different ages, such as limestone, sandstone, granite and gneiss. Limestone is mainly one substance, calcium carbonate $CaCO_3$, but most rocks contain a variety of minerals. Many, such as sandstone and granite, contain quartz (SiO_2) or silicates (containing mainly silicon, oxygen and aluminium). If you examine granite, for example, you will see crystals of different colours, including a lot of clear quartz, which gives the rock its glistening appearance.

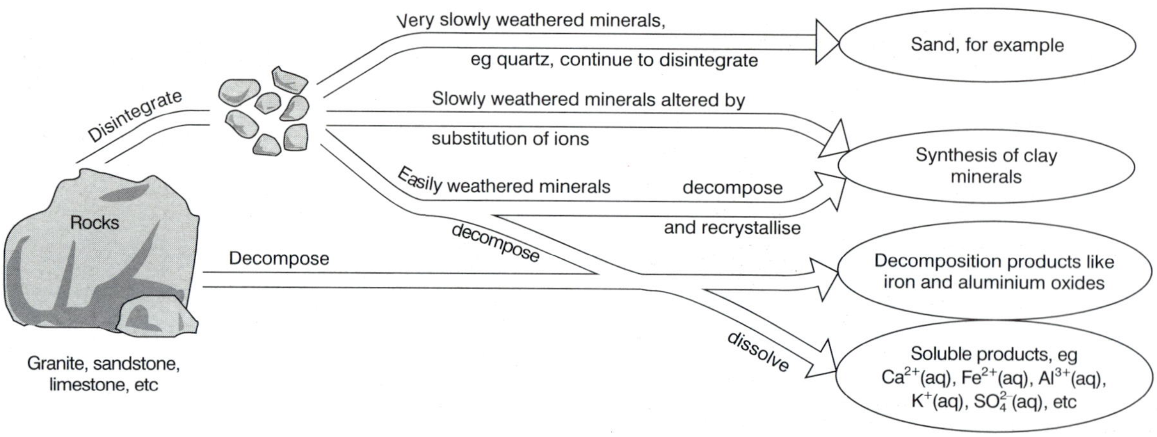

Figure 6 Processes of weathering in moderately acid conditions common in humid temperate climates (granite and sandstone are mixtures of silicate minerals such as quartz, feldspars and micas)

Pure quartz is a continuous three-dimensional network of silicon and oxygen atoms. Each silicon atom is covalently bonded to four oxygen atoms at the corners of a tetrahedron. Each oxygen atom is bonded to two silicon atoms and is shared between two adjacent tetrahedra, as shown in Figure 7.

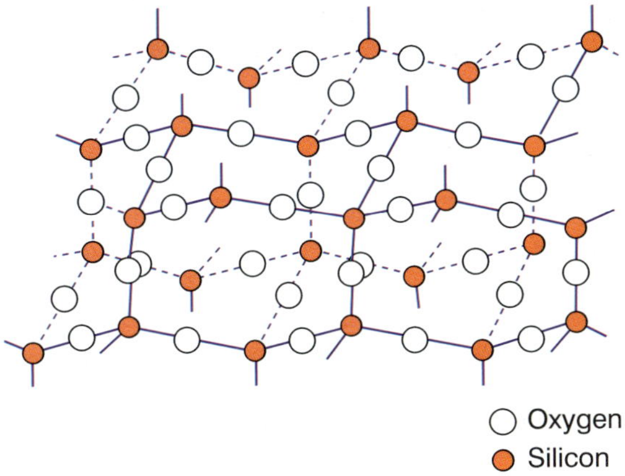

Figure 7 The structure of quartz

In quartz, the number of oxygen atoms per silicon atom is

$\frac{1}{2}+\frac{1}{2}+\frac{1}{2}+\frac{1}{2}=2$

and quartz has the overall composition SiO_2. It is very slowly weathered to sand. This type of weathering is a purely physical process. The resulting smaller crystals of sand still consist of a three-dimensional covalent network structure, giving a hard substance which is insoluble in water (Figure 8).

Carbon and silicon are both in Group 4 of the Periodic Table. To understand why CO_2 is a gas in the atmosphere, whereas SiO_2 is a solid found in the rocks in the Earth's crust, you will need to read about the structure of covalent substances in **Section 16.2**, and find out how their structure is related to their properties.

Many minerals in the Earth's crust are metal **silicates**. Silicon is covalently bonded to four oxygens at the corners of a tetrahedron, as in quartz, but only some of the oxygens are shared between tetrahedra. Each *unshared* oxygen carries a negative charge of −1. Figure 9 shows a silicate tetrahedron.

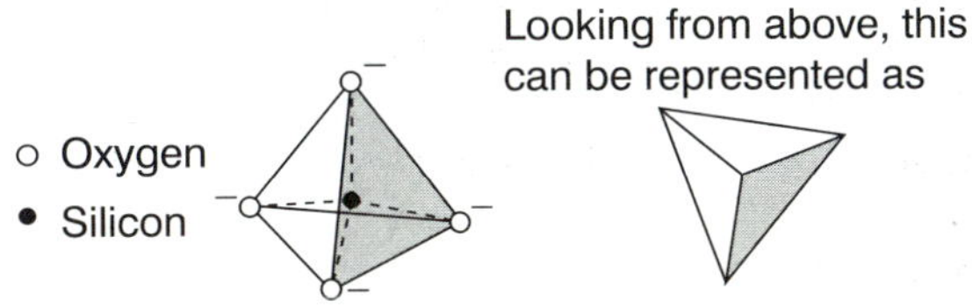

Figure 9 The silicate ion SiO_4^{4-}

The silicate tetrahedra can join up in a number of ways. Remember that the corners represent oxygen atoms. The tetrahedra can join up to form **chains** by sharing two corner oxygens (Figure 10).

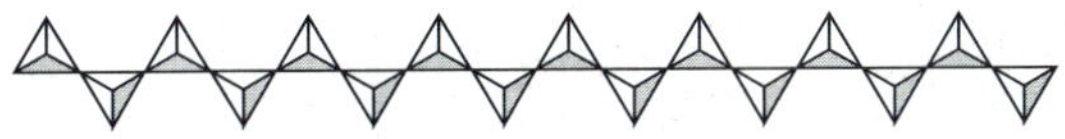

Figure 10 A single chain

Figure 8 (a) Crystals of quartz, SiO_2; (b) sand is made up of small grains of SiO_2 (the pale brown colour is due to iron impurities)

They can also form double-stranded chains in which two or three oxygens are shared (Figure 11). Asbestos is an example of this kind of arrangement (Figure 12).

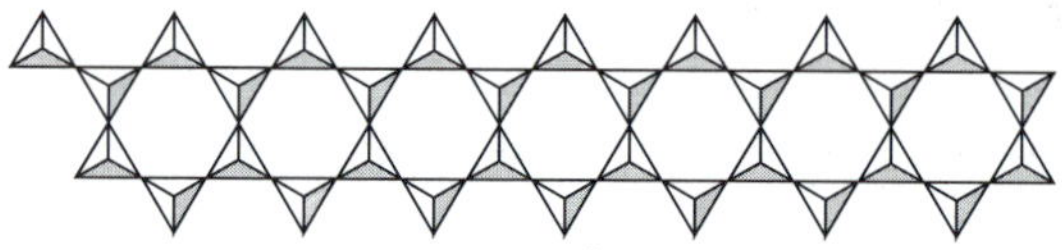

Figure 11 A double-stranded chain

Figure 12 Fibres of asbestos are made up of double-stranded silicate chains

Sheets of silicate tetrahedra can be built up by joining more strands together (Figure 13). These occur in micas and clays (Figure 14).

In sheets like this all the unshared, negatively charged oxygens point in one direction (upwards, out of the plane of the paper in Figure 13).

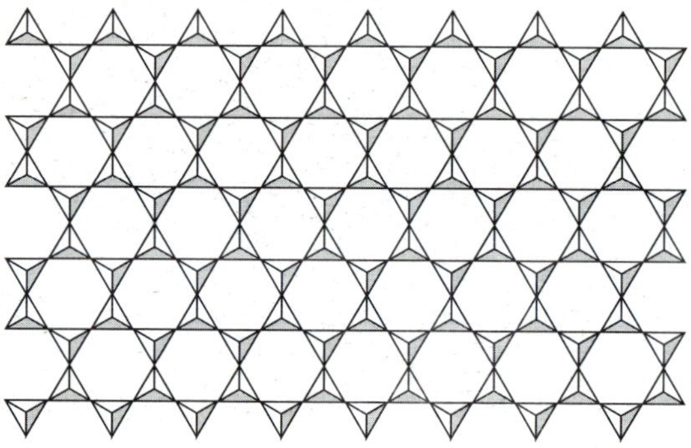

Figure 13 A silicate sheet

Figure 14 Mica is made up of sheets of silicate tetrahedra

What balances the negative charges on the unshared oxygen atoms? Many positive ions such as Na^+, Ca^{2+}, Mg^{2+} and Al^{3+} are held to the silicate sheets to balance the negative charge.

During weathering, aluminium can replace some of the silicon atoms at the centre of the tetrahedra to give a variety of different minerals, which are chemically different and have different physical properties. The layered structure of a clay, produced from rocks by weathering, is shown clearly in Figure 15.

Figure 15 Scanning electron microscope picture of kaolinite, a clay mineral, showing the crystalline layer structure

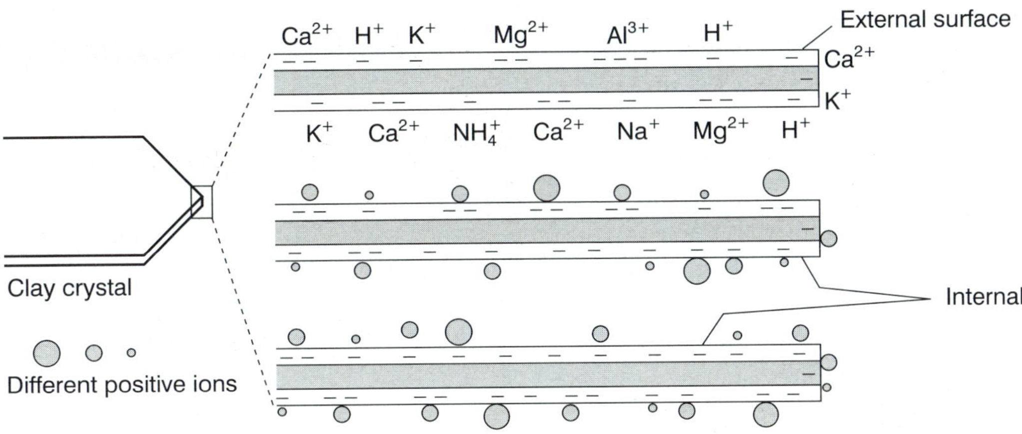

Figure 16 Positive ions held by ionic bonding to the internal surfaces of a typical clay mineral

Clays become slippery when wet because water can enter between the crystals, and sometimes between the layers within a crystal. Forcing the layers apart like this also exposes a large internal surface, negatively charged. A variety of positive ions are held by ionic bonds to these surfaces, and so are not washed away out of the soil by rainfall (Figure 16). Clay minerals provide a source of nutrient ions for use by plant roots, and are very important for soil fertility and agriculture.

ASSIGNMENT 3

The diagram of a clay crystal in Figure 16 shows the negative charges on the surfaces of the silicate layers.

a Explain why a variety of positive ions can bond to the surfaces and, from what you know about ionic bonds, explain why these positive ions are not easily washed away.

b When ammonium nitrate is added to soil as a fertiliser, a problem is caused by the leaching of nitrate ions into rivers. The nitrate ions act as a nutrient and cause excessive growth of algae. Why is there a leaching problem with the nitrate ions present in the fertilizer, but not the ammonium ions?

3.3 The surrounding sea

Approximately 70% of our planet's surface is covered by sea. The oceans have a surface area of about 360 million km^2 and transport enormous quantities of energy and materials around the surface of the planet. They play an important role in controlling our climate through their absorption of solar energy and the consequent production of water vapour and flows of warm water, which help drive currents in the air and the seas. The oceans help to make the Earth hospitable to life and have kept our planet habitable for over 3.5 billion years.

They that go down to the sea in ships ...

In a French restaurant, a plate of shellfish is known as *fruits de mer*, or 'fruits of the sea'. For thousands of years mankind has farmed the sea for food, and the easiest creatures to catch must have been the shellfish. But the archaeological records of every seaside culture from Neolithic times onwards contain the remains of marine fish, and the hooks and harpoons used to catch them. Classical literature tells us that the Ancient Greeks devised complicated traps for catching fish and that the Phoenicians and Carthaginians founded many cities around the Mediterranean as fishing settlements

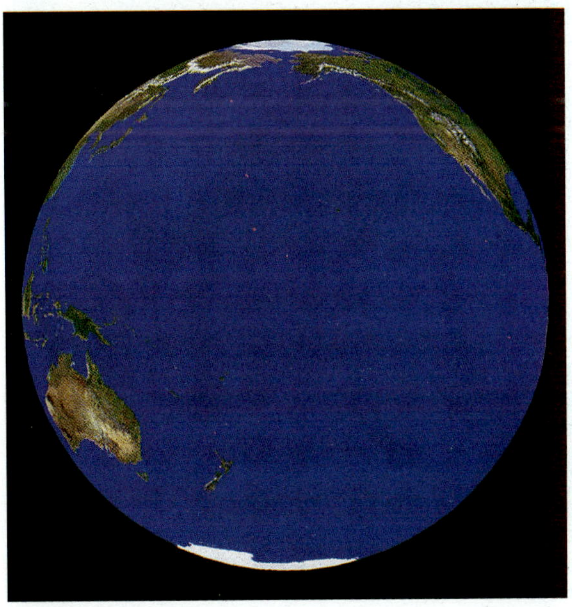

Figure 17 The watery planet: a view of the Earth showing the vast expanse of the Pacific Ocean

Figure 18 Indonesian fishermen; fishing is one of the earliest skills and people have farmed the sea for food for thousands of years

– Sidon, for instance, a Phoenician city now in Lebanon, means 'the fishing place'.

Sea creatures provided these ancient cultures with more than just food. The name Phoenicia comes from the Greek word *phoenix* meaning 'purple'. One of the foundations of this country's vast commercial empire, from around 1500 BC to 800 BC, was trade in a brilliant purple dye, extracted from the shells of a species of marine snail. Tyrian Purple, as the dye was called, was used to colour the attire of kings and emperors, and was a symbol of great wealth and power. Tyrian Purple is chemically similar to Indigo, the dye that the Ancient Britons extracted from the woad plant. What a coincidence that the two cultures should have used these two related dyes at about the same time – perhaps blues were in fashion!

Less extensive use has been made of marine plants, but in Japan several species of marine algae (seaweed) are cultivated for food. There are other uses for seaweed: it can be spread on the ground as fertiliser and used as a source of chemicals.

Figure 19 The ancient dye, Tyrian Purple, was extracted from the shells of a species of sea snail. The Roman Empire decreed that only members of the ruling family (such as Julius Caesar, above) could wear robes dyed with Tyrian Purple

Figure 20 Seaweed on the shore at Tarskavaig, Isle of Skye

From the early 1700s seaweed was used as a source of **alkali**. It took 20 tons of seaweed, collected, transported and burned to produce 1 ton of **kelp**, a brittle, multi-coloured material high in soda ash and potash (sodium carbonate and potassium carbonate) for use in the soap and glass-making industries of Britain. Kelp production from seaweed provided a substantial income to landowners, mainly in the Hebrides and Orkney (although not to the labourers who were paid £1 to £3 per ton of kelp). In a season, the kelper and his family would make at the most 4 tons of kelp.

In the most profitable years kelp sold for an average of £20 per ton and at the peak of the industry Scotland produced 20 000 tons in a year, processing 400 000 tons of wet weed. The white acrid smoke of burning kelp rose on a calm day in columns visible for miles, giving the kelp islands the appearance of a series of active volcanoes.

However, by the 1820s the industry was in trouble. The price fell for a number of reasons: the end of war with France allowed the import of higher quality material, import duty was cut, and a chemical process for making alkali from salt was developed. Mackenzie of Seaforth, the then owner of Lewis, was perceptive when he stated in 1827 that 'Chemistry is the true enemy of kelp, and we can only last as long as its advance to perfection is going on.'

Production continued on a smaller scale into the 19th century, with the seaweed used as a source of **iodine**.

Present day uses of seaweed include the extraction of **alginate**, a thickening or gelling agent used in a wide variety of food products, including pet foods. Other uses range from maintaining the head on a pint of beer, to medical dressings, paper, textiles and pharmaceuticals.

Figure 21 Burning seaweed for alkali production around 1889

Figure 22 The modern kelp industry. (a) Drying seaweed (tangle) before transport to the processing plant. (b) The seaweed processing plant at Girvan in Ayrshire, capable of processing 6 tonnes of dried seaweed per hour to extract the alginate

Seaweed is rich in sodium, potassium and iodine, and the marine snail used to make Tyrian Purple is rich in bromine. Both the seaweed and the snail concentrate certain elements found in the sea water they inhabit.

You will find it useful at this point to refer to **Section 16.3** on the chemistry of ions in solids and in solution.

Everyone knows that the sea is salty. Salts are ionic compounds and over 99% of all the dissolved substances in sea water are ionic. Whilst they remain dissolved, of course, the ions are free in solution so – strictly speaking – the sea contains not salts, but a mixture of ions. Figure 23 shows the abundance of these ions. The proportion of one to another is remarkably constant whichever part of whichever sea you choose to sample.

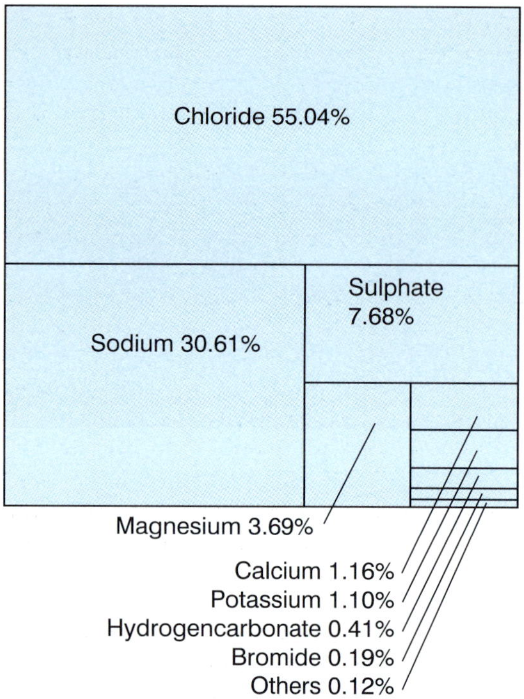

Figure 23 Percentages by mass of different ions in the total dissolved salts in sea water

ASSIGNMENT 4

Use the information in Figure 23 to work out

a the amount in moles of positive charge present in 100 g of the dissolved salts in sea water

b the amount in moles of negative charge present in 100 g of the dissolved salts in sea water.

(Remember that 1 mole of doubly charged ions contains 2 moles of charge.)

Comment on the relative magnitudes of the answers you obtain in parts **a** and **b**.

The composition of sea water has been known for over 100 years, but it is only recently that we have begun to understand in any detail how the sea became salty.

Salt of the Earth

There is an old Norse myth which tells of a magic salt mill grinding away at the bottom of the ocean, making the sea salt. Strangely enough, this notion is not so far from the truth.

It has been known for a long time that some of the salt comes from the land. Rainwater leaches salts from the soil and rivers wash them into the sea. But the sea contains in abundance some elements that are not found to any great extent in river water – chlorine, bromine and sulphur, for example. The source of these elements remained a mystery until we learned more about the structure of the ocean floors.

Underneath the sediments on the ocean floor are lavas, generated by long, thin, underwater volcanoes called *mid-ocean ridges*. The gases given off from these volcanoes are rich in chlorine, bromine and sulphur. Also, as molten lavas meet cold sea water they solidify and shatter. Water streams down through the cracks, scouring out soluble minerals. Figure 24 shows the sources of the dissolved ions in sea water.

Scientists think that the water itself escaped from these deep-seated rocks and that sea water was salty right from the beginning. The balance of ions is kept constant by a complicated geochemical cycle that scientists are only just beginning to understand.

Figure 24 The sources of dissolved ions in sea water

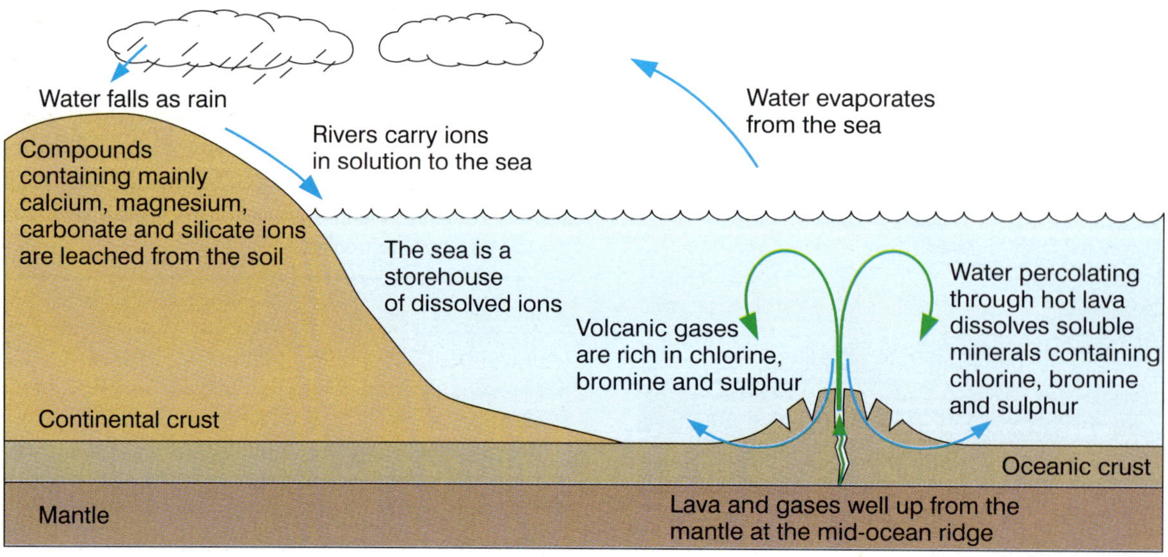

ASSIGNMENT 5

Bottled mineral water contains dissolved ions from the soil and rock through which the water has percolated. Most brands of bottled water give an analysis of the contents on their labels. Look at the labels of as many types of mineral water as you can find. Choose the still varieties as the fizzy ones have had carbon dioxide added to them. Compare the contents with the analysis of sea water shown in the table below.

Ion	Contents/mg l^{-1}
Cl^-	19 300
Na^+	10 700
SO_4^{2-}	2700
Mg^{2+}	1300
Ca^{2+}	400
K^+	380
HCO_3^-	140
Br^-	65
Others	40

a What is the total mass of dissolved solids in 1 litre of
  i sea water
  ii the different sorts of mineral water?

b Compare the analyses for the individual ions in the mineral waters with the figures given for sea water. Comment on any differences. Which of the ions in sea water cannot be accounted for by run-off of water from the land?

Salt sellers

Sea salt has almost the same composition wherever you get it from. However, the concentration of the salt in the water does vary from place to place.

The two most abundant ions in sea water are sodium (Na^+) and chloride (Cl^-). These ions are also present in the body fluids of all land creatures. An adult human contains about 300 g of dissolved sodium chloride (common salt). Some of this is excreted each day in urine and sweat, and must be replaced if the body is to function normally.

Our need for salt has been recognised for thousands of years and the substance has entered the language, superstition and history of all societies. The word *salary* means, literally, a payment in salt – as was the custom in the Roman legions. In medieval times, important guests sat 'above the salt' at a banquet – and the saying lives on. Spilling salt brings bad luck – a measure of its value hundreds of years ago. Not only was it important to eat enough salt; salt was also used to preserve food in the days before refrigerators and freezers had been invented.

The obvious place to get salt from was the sea. Along the east coast of Britain around 600 BC was a string of small salt-works. When the Romans occupied much of Britain, they imported salt from France (Gaul), where it was possible to evaporate the water by the heat of the Sun. After the Romans left, the British salt industry thrived again. In Scotland, grants of salt pans or salt-making rights were given to monasteries, and as early as 1286 the office of 'Master of the Royal Saltworks' existed. The simple process of boiling salt water in pans could be practised wherever fuel was available, and so coal-mining and salt production often occurred together. In the 1600s there were salt pans in large numbers along both shores of the Firth of Forth and little towns such as Culross and Prestonpans became Scotland's first industrial communities. The coal and salt masters built several new harbours such as Wemyss in Fife and Saltcoats in Ayrshire for the coal and salt export trade from Scotland in the late 17th century.

Great quarrels broke out during the last weeks of the Scottish Parliament in December 1706 about the level at which English salt taxes were to be imposed after the Union of Scotland and England in 1707. A duty was imposed on Scottish-made salt in 1713 and it struggled to compete with cheaper and purer salt from England. Salt produced by evaporating water over coal fires contained impurities from the coal, and the purer salt from France, or rock salt from Cheshire, was always preferred for commercial preservation of fish such as herring, packed in barrels of salt. Salt production in Scotland continued for domestic use, but declined in the 19th century. Today we get most of our salt in the UK from underground deposits.

Figure 25 Extracting rock salt from a mine in Cheshire

... and occupy their business in great waters

The sea is a storehouse for fish and for minerals, and communities which have lived by the sea have always used it as such. Hannah Glasse wrote in 1747, 'Newcastle is a famous place for salted haddocks. They come in barrels and keep a great while'. In the winter months, when it was too rough to put to sea, the salted fish provided the community with an assured supply of protein. These people used the coastal waters very much as others used the land. They exploited its resources, but, at the time, their enterprises were small and inefficient enough to avoid over-fishing. Arbroath and Aberdeen are still renowned for smoked fish, another method of preservation, although the Scottish herring industry collapsed from over-fishing.

Figure 26 Salting and smoking fish at a kippering house on the east coast of Scotland around 1900

Nowadays, fishing is big business and many traditional fishing stocks have been depleted. Factory ships from Russia and other Eastern European countries travel as far as the waters outside Argentina, accompanied by fuel tankers, salvage tugs, repair boats and refrigerator ships that will hold 10 000 tonnes of fish. The pursuit of the whale, for food and for chemicals, has also led to the near extinction of some species.

Figure 27 Fishing on a large scale; catches must now be regulated to ensure adequate supplies for the future

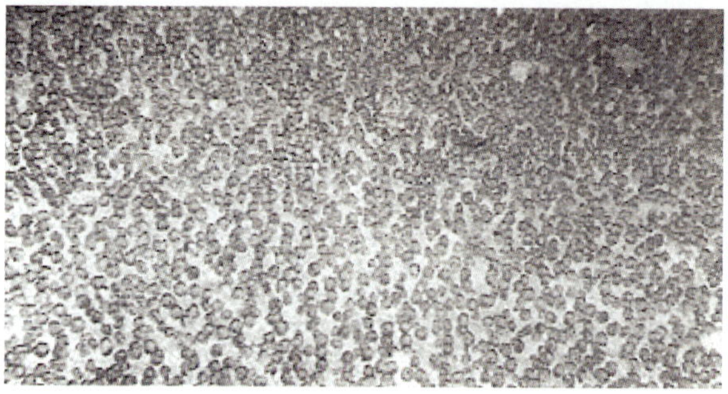

Figure 28 Manganese nodules on a deep ocean floor. The nodules (diameter about 5 cm) are mainly manganese and iron, but also contain traces of nickel, cobalt and copper. They could provide a valuable source of these metals in the future

Sea water provides us with bromine, chlorine and magnesium. More recently, manganese-rich nodules have been collected from the ocean floors (Figure 28).

In future, scientists hope to use the sea as a source of more complex, organic compounds. Many sea creatures, especially those which have no shell to protect them, have evolved poisons which could perhaps be used as medicines.

The oceans, however, are not infinitely rich in food and chemicals, nor are they an ever-open bin into which we can throw our waste.

The North Sea, for example, is polluted by industrial effluent, sewage and excess of fertiliser that the major rivers of Northern Europe carry into it. Heavy metal ions poison fish stocks, and the increase in nitrate concentration leads to rapid growth of algae.

3.4 *Oceans of energy*

The global central heating system

With the exception of *tidal power*, all our energy comes ultimately from nuclear sources. On Earth, we can mine radioactive compounds and produce from them fuel for *nuclear power stations*. Energy from the nuclear processes which go on in the core of the Earth can be tapped at suitable locations as *geothermal energy*.

But most of our energy comes from outside our planet – from a great nuclear furnace in the Sun. The Sun's rays heat the Earth directly, and through this drive the *winds* and *waves*. *Solar cells* convert sunlight into electricity. *Photosynthesis* uses sunlight to build up fuels: some are used almost immediately, and others have been changed over millions of years into *coal*, *oil* and *gas*.

When the Sun's rays reach the Earth they can be

- reflected
- absorbed by the atmosphere
- absorbed by the Earth's surface.

Roughly half the energy we receive from the Sun is absorbed by the land and the oceans, and this causes

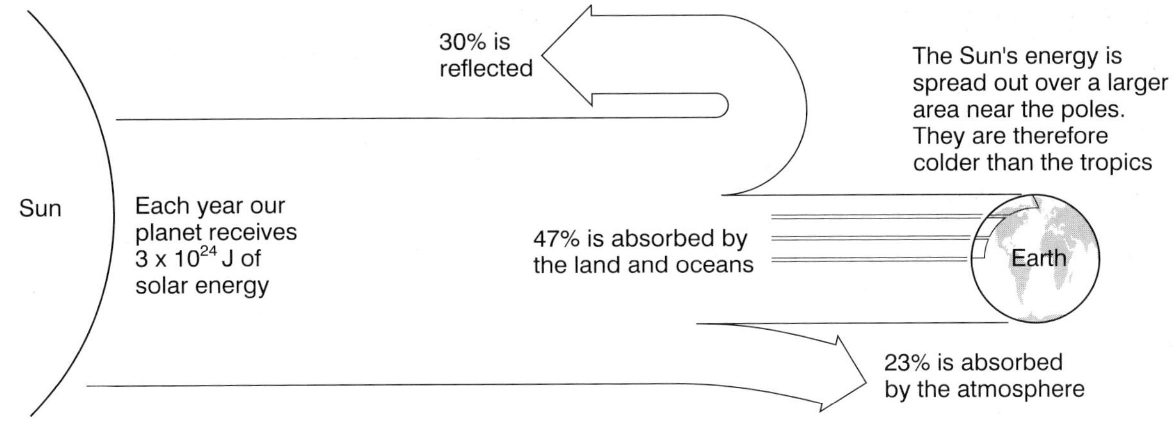

Figure 29 The fate of the Sun's energy reaching the Earth

the Earth's surface to warm up. The Earth in turn radiates energy back into space (Figure 29).

If the Earth was a dry lump of rock with no atmosphere, each part of its surface would soon settle down to a situation in which the energy being received from the Sun would, on average, be balanced by energy lost through radiation (Figure 30). The tropics would be much warmer and the poles even colder than they are – and the Earth would be far less hospitable to life.

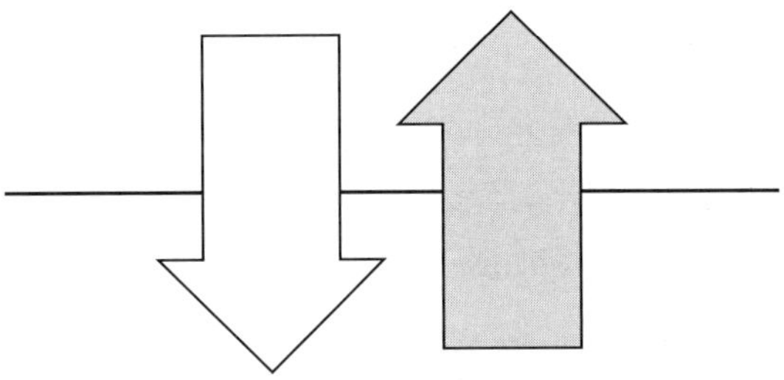

Figure 30 A planet with neither atmosphere nor oceans

But the Earth is surrounded by water and gas – both fluids. Temperature differences set up currents in the oceans and atmosphere which spread out the heating effect of the Sun more evenly. Just like warm air from a radiator spreads around a room, currents in the sea and air take thermal energy from the tropics to the colder regions of the Earth.

In fact, the ocean/atmosphere system is even more effective at spreading out energy. Warm water can do more than circulate – it can *evaporate.* Energy is taken in when water evaporates, so the situation must be reversed, and energy must be released, when water condenses. The tropics are cooled by evaporation, and currents in the atmosphere carry the water vapour to colder, high latitude regions where condensation releases energy (Figure 31).

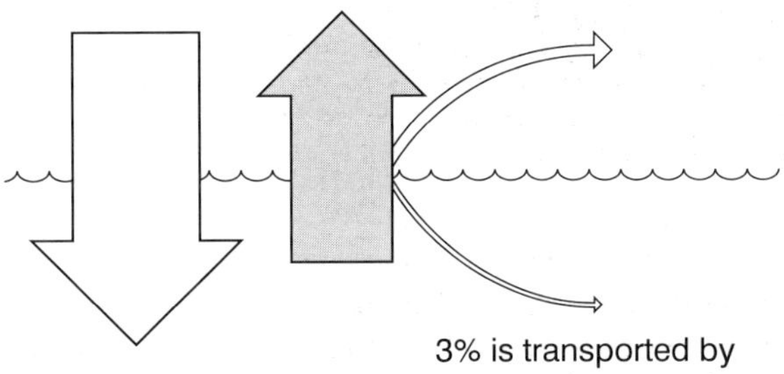

Figure 31 At the tropics of a planet with an atmosphere and oceans, about 20% of the incoming solar energy is transported to colder regions

High latitude regions receive more energy than is provided by the Sun alone: they are wetter – but warmer. Figure 32 shows the balance of condensation and evaporation around the world.

In the North Atlantic region, the winds and warm water currents flow from SW to NE. Northern Europe, including the UK, is warmed by energy which has been transported from the tropics and the Caribbean. In winter, as much as 25% of our thermal energy may come this way. Eastern North America does not receive this energy. So winters are much less cold in Glasgow than in Halifax, Nova Scotia, although Glasgow is further north (Figure 33).

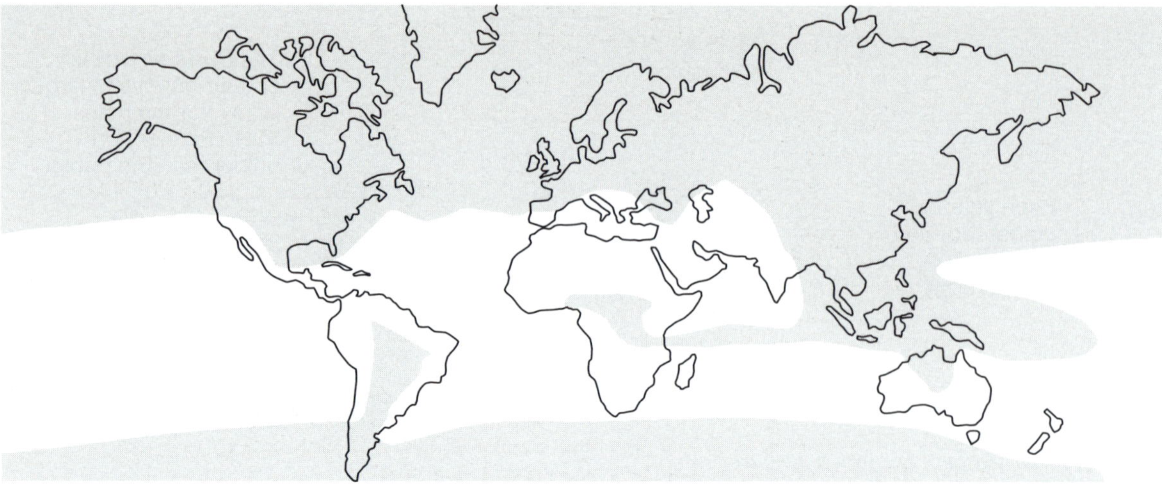

Figure 32 The condensation–evaporation balance of the Earth (shaded areas denote regions where condensation exceeds evaporation)

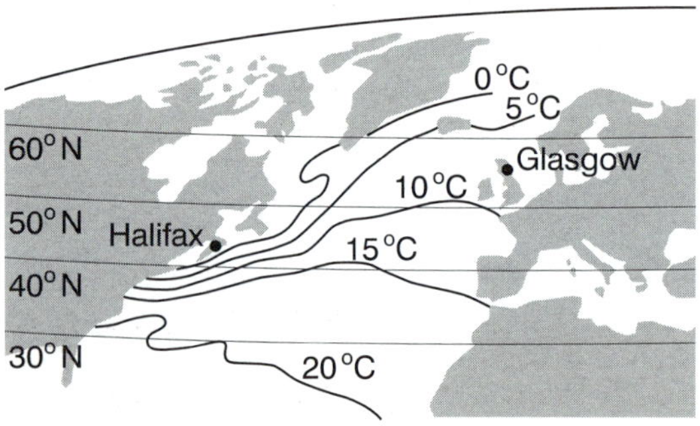

Figure 33 Average surface temperatures in the North Atlantic in February

ASSIGNMENT 6

The positions of three Canadian cities are marked on the map below. Explain the patterns in their winter and summer temperatures.

	Victoria	Winnipeg	St. John's
July maximum/°C	17	23	17
January maximum/°C	2	–19	–6

Energy in the clouds

The molecules in liquid water and water vapour differ in one important aspect. In the liquid, there are attractive forces between the molecules – **intermolecular forces** – which hold the molecules onto one another and keep them quite close together. These forces are much less important in water vapour where molecules are free to separate and take up far more volume.

When we evaporate water, changing it from liquid to vapour, the intermolecular attractions must be overcome – a process which takes in energy. The enthalpy change of vaporisation, ΔH_{vap}, is our measure of this energy. In the reverse process, condensation, molecules come together again, intermolecular attractions re-form and an equal quantity of energy is released. In other words, condensation is **exothermic** and the enthalpy change is $-\Delta H_{vap}$.

Figure 34 Energy is taken in from the surroundings when sea water evaporates, and is released when the water vapour condenses in clouds

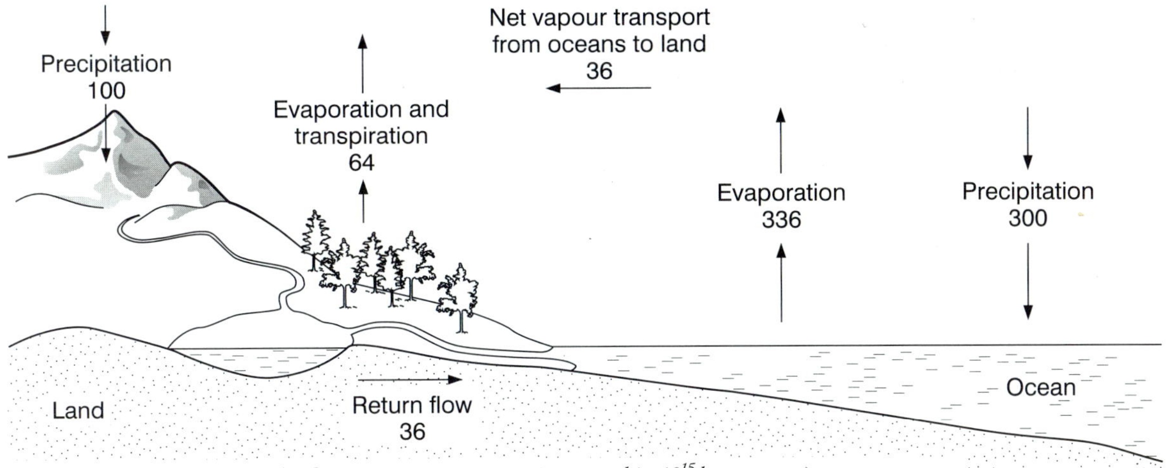

Figure 35 The global water cycle (figures represent quantities moved in 10^{15} kg per year)

You can find out about the attractions between water molecules in **Section 16.4**.

Figure 35 illustrates the global water cycle. It summarises the major processes by which water circulates around the world.

Notice that more water evaporates from the oceans than is directly returned to them as precipitation (ie rain and snow). Each year, 36×10^{15} kg of water vapour produced from the sea falls as precipitation over the land. The process makes the land wetter and keeps the rivers flowing. It also makes the land warmer.

So evaporation and condensation of water affect the temperature of different parts of the Earth in two ways:

- by transferring energy from low latitudes to high latitudes
- by warming the land through condensation of water which comes from the oceans.

The UK may seem cold enough and wet enough most of the time, but imagine a world with hexane seas. The enthalpy changes of vaporisation of hexane, water and some other liquids are given in Table 4 (notice that kJ kg^{-1} units are used).

Substance	Formula	ΔH_{vap}/kJ kg^{-1}
water	H_2O	+2260
ethanol	C_2H_5OH	+840
hexane	C_6H_{14}	+330
mercury	Hg	+300

Table 4 Some enthalpy changes of vaporisation

If our present rain were replaced by the same mass of 'hexane-rain' it would release only about one-seventh of the energy. For enough 'hexane-rain' to fall to keep our temperature the same, there would have to be about seven times more rain. Then there would be other problems – an increased fire risk, for example!

Perhaps it is a lucky chance, perhaps it is much more than that, but the properties of water make it an ideal liquid for spreading the Sun's energy around the world. Without it, the pattern of evolution and human development would probably have been very different.

ASSIGNMENT 7

a Use the figures in Figure 35 and Table 4 to calculate the thermal energy released each year when water vapour carried from the oceans condenses out over the land.

b The output of a typical power station is about 2000 MW, in other words about 6×10^{16} J per year. Approximately how many power stations would be needed to produce the same energy as that transferred from oceans to land by evaporation and condensation?

The oceans are a crucial factor in controlling the Earth's climate. We are learning more about the ocean circulatory system (Figure 36) and the oceans' role in controlling the Earth's heat energy balance. But we need to know much more before they can be included with confidence in global climate models – models which explain our present climate and predict how it may change in the future. We know that ocean currents are important but we cannot predict how they may change, and therefore how our climate may be affected, if the Earth warms up.

We are however sure that the consequences of global warming would be uneven. So parts of Northern Europe could become colder while the rest of the world was getting hotter.

Figure 36 Ocean currents in the Atlantic

3.5 *Summary*

This Chemical Storylines chapter is linked to Chemical Ideas Chapter 16.

The theme of these chapters has been the chemical nature of the world around us; the composition of the atmosphere, the minerals present in the rocks that make up the Earth's crust and the ions present in sea water.

This led you to think about why some substances are gases, some are liquids and others solids at the surface temperature of the Earth. To do this, you needed to learn about the structure and properties of ionic and covalent compounds. For covalent compounds, the structure may be a giant network (as in SiO_2) so that the compound is a solid, or the substance could exist as discrete molecules (as in CO_2). If it exists as discrete molecules, we need to think about the attractive forces *between the molecules*, intermolecular forces, to understand why the substance is a gas, liquid or a solid.

These chapters ended by considering the role of the oceans in transporting materials and energy around the surface of the planet and keeping the planet hospitable to life. Water is an ideal substance to carry out this role because of its special properties. This led you to a study of hydrogen bonding and its effect on the properties of water.

Section 16.5 draws together all these ideas in a summary of the relationship between the bonding and structure of a substance and its properties.

UNIT 2 THE WORLD OF CARBON

OVERVIEW

Chemical Storyline chapters

Chapter 4 Developing Fuels

This chapter links back to Chapter 2 and tells the story of petrol: what it is and how it is made. It describes and explains the processes used to convert naphtha from crude oil into petrol (isomerisation, cracking and reforming) and the use of oxygenates to increase the octane rating of the blended petrol. It also examines alternative fuels for motor vehicles, and looks to the future with a possible hydrogen economy.

Chapter 5 Polymers in Action

This chapter begins with the discovery of poly(ethene) and traces the story of polymers, through the development of nylons and polyesters, to the 'designer' polymers of the present day. These include a conducting polymer, a heat-resistant polymer that is as strong as steel but five-times lighter, and a dissolving polymer. It goes on to consider the issues involved in the disposal or recycling of plastics and concludes by looking at proteins as examples of natural polymers, in particular, at insulin (a protein hormone) and enzymes (protein catalysts).

Chapter 6 Ozone Depletion in the Stratosphere

This chapter describes some of the simple chemical reactions that take place in the upper atmosphere (stratosphere). It looks at the reactions by which ozone is formed and destroyed and how these lead to a 'steady state' in which the ozone concentration remains fairly constant. It tells the story of the discovery of the 'ozone hole' over Antarctica and how this was linked to our use chlorofluorocarbons (CFCs). Finally, it describes the international collaboration to phase out the use of CFCs and replace them with less environmentally-damaging alternatives.

Chemical Ideas chapters

Each Chemical Storyline chapter is linked to relevant sections in the Chemical Ideas:

Chapter 17 *Organic Chemistry: Hydrocarbons*
Chapter 18 *Organic Chemistry: Oxygen and Nitrogen Compounds*
Chapter 19 *The Structure and Properties of Polymers*

Overview of chemical principles

From previous work you should:

- know that crude oil is distilled to give fractions with different boiling points
- know that organic chemistry is the chemistry of carbon compounds
- understand and be able to draw structural formulae for simple organic compounds
- be able to apply ideas about intermolecular forces to polymer chains.

At the end of this unit you will have covered all the content statements for Unit 2 in the Arrangements for Higher Chemistry. The table below shows where each topic is covered. Remember that the *order* in which the content statements are done is governed by the Chemical Storyline and so is different from the way they are listed in the Arrangements.

UNIT 2 THE WORLD OF CARBON		
Areas in the content statements for Higher Chemistry	**Chemical Storylines**	**Chemical Ideas**
(a) Fuels		
Petrol	4.2, 4.3	17.1, 17.2
Alternative fuels	4.4, 4.5	
(b) Nomenclature and structural formulae		
Hydrocarbons		17.1, 17.3, 17.4
Substituted alkanes		18.1, 18.2, 18.4
Esters		18.5
Aromatic hydrocarbons		17.5
(c) Reactions of carbon compounds		
Addition	5.2, 5.4	17.3–17.5, 18.1
Oxidation	4.4 (2.2)	18.1–18.3
Making and breaking esters	5.6	18.4, 18.5
Percentage yields		18.8
(d) Uses of carbon compounds		
Competing demands on use crude oil	4.1	
Carboxylic acids and esters		18.4, 18.5
Halogenoalkanes	6.3	
Depletion of the ozone layer	6.1–6.4	
Benzene and related compounds		17.5
(e) Polymers		
Early plastics and fibres	5.2, 5.3, 5.5, 5.6	17.1, 17.3, 18.1, 18.5, 18.6, 19.1–19.4 (20.2, 23.2)
Recent developments	5.1, 5.4, 5.7, 5.8	
(f) Natural products		
Fats and oils	(2.2)	17.3, 18.5
Proteins	5.9, 5.10	18.6, 18.7, 19.3 (1.5, 15.5)

(The entries in brackets refer to further use of these ideas in Units 1 and 3.)

4 DEVELOPING FUELS

4.1 *Petrol is popular...*

Why don't we all drive electric vehicles? After all, they are clean, quiet and cause no exhaust pollution. Despite these advantages, out of the UK's 30 000 electric vehicles, 27 000 are milk-floats!

One problem with electric vehicles is their poor performance. This is mainly because of the heavy batteries they have to carry around. When were you last overtaken by a milk-float?

Another problem is that it takes so long to recharge their batteries. An electric vehicle has to be recharged overnight for several hours. Electrical energy trickles into the batteries at a rate of about 55 joules per second (55 W).

Figure 1 This is what car parks for electric cars may look like in the future – while the driver is out shopping or at work, the car can be 'filled up'

Now compare this with the rate at which you can fill up a petrol-engined car with its source of energy. A petrol pump delivers petrol at about 1 litre per second, and a litre of petrol transfers 34 000 000 J of energy when you burn it. So the petrol pump replaces the car's energy supply over 600 000 times faster than the battery charger. These numbers are illustrated in a different way in Figure 2.

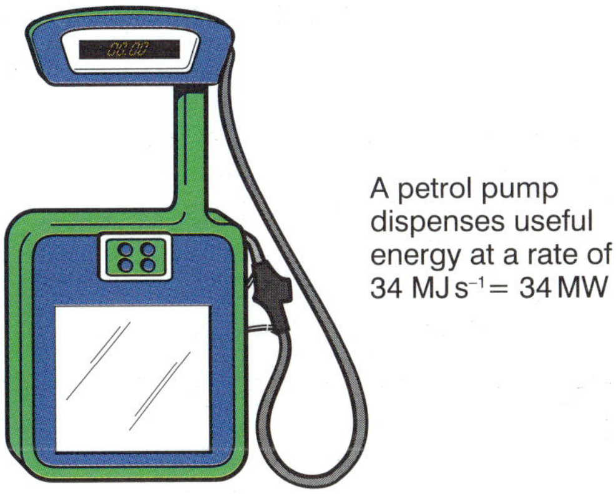

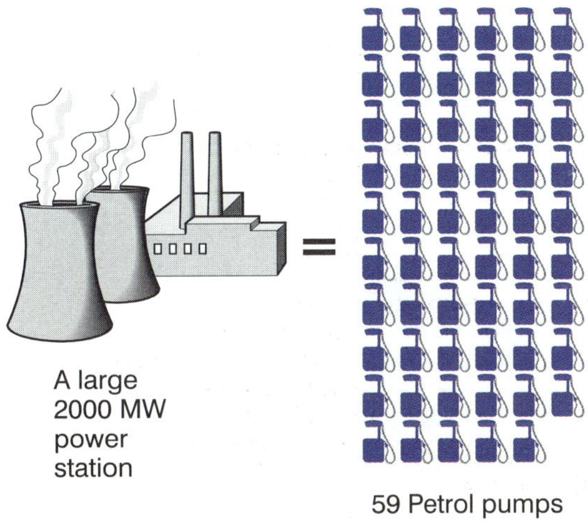

Figure 2 Just 59 petrol pumps can match the power output of a large power station

Petrol is a highly concentrated energy source. Couple that with the ease and cheapness of building petrol engines, and it's not surprising that petrol is popular.

..... but there are problems

One problem with petrol is that it's a finite resource, because crude oil supplies are limited. They probably won't last more than about another 100 years.

We need crude oil for more than petrol alone. Crude oil provides the starting materials or **feedstocks** for the petrochemicals industry. For example, it is needed for making synthetic fibres, detergents and pharmaceuticals. As we use up supplies, oil products may become too valuable to burn in car engines.

Another problem is pollution. Petrol produces carbon dioxide when it burns, and that's a major contributor to the greenhouse effect. As you saw in **Unit 1**, **Chapter 2**, petrol also produces other kinds of polluting emissions.

So there's a real need to improve the performance of petrol in car engines so that it burns as cleanly and efficiently as possible – and to find suitable fuels to replace petrol for road transport in the future. How chemists and chemistry can help is one of the major themes of this chapter.

4.2 *Focus on petrol*

"Sorry I'm late, the car wouldn't start ..."

What do you blame if the car won't start on a cold morning? Almost certainly the car itself – probably not the driver and certainly not the petrol. Yet the right petrol is actually very important.

What is petrol?

Petrol is a complex mixture of at least 100 different compounds, mostly hydrocarbons. The main source of petrol is the *gasoline fraction* obtained from the primary distillation of crude oil in a fractionating column. Usually this fraction accounts for 15%–30% of the crude oil. It is a mixture of liquids, mostly **alkanes** with between five and 10 carbon atoms, and boils in the range 20 °C–180 °C.

You can find out more about alkanes in **Section 17.1**.

The gasoline fraction is in great demand. Some of it goes to refineries to be used for producing motor fuels. The rest is used in the manufacture of organic chemicals – this part of the gasoline fraction is called **naphtha**.

You can find information about the other fractions produced in the distillation of crude oil in Figure 3 and Table 1.

The '*straight-run*' gasoline from the primary distillation makes poor petrol. The job of a refinery is to convert it into useful petrol components with the right composition and properties. To do this, the structure of the alkanes present must be modified in various ways. In addition, the heavier fractions, kerosene and diesel oil, are 'cracked' to produce smaller molecules for use as extra petrol components.

The residue can be used to make petrol too, but first it is distilled again, this time under reduced pressure, in *vacuum distillation*. This avoids the high temperatures that would be needed for distillation at

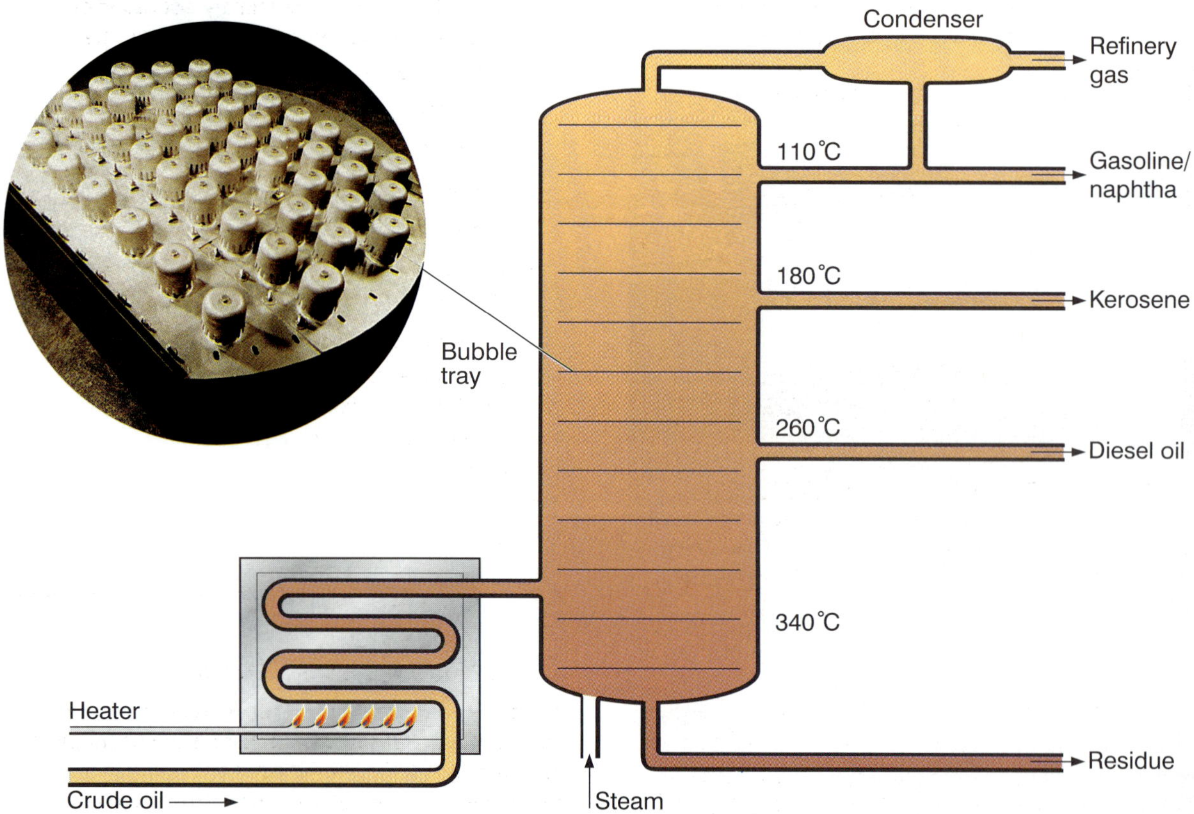

Figure 3 Primary fractional distillation of crude oil

Table 1 Fractions obtained from the fractional distillation of crude oil

Name of fraction	Boiling range/°C	Composition	% Crude oil	Use
refinery gas	<20	C_1-C_4	1–2	gaseous fuels and blending with petrol feedstock for organic chemicals
gasoline/naphtha	20–180	C_5-C_{10}	15–30	petrol for cars (gasoline) production of organic chemicals (naphtha)
kerosene (paraffin)	160–250	$C_{10}-C_{16}$	10–15	jet fuel heating fuel
diesel oil	220–370	$C_{13}-C_{25}$	15–20	diesel fuel heating fuel
residue	>370	$>C_{25}$	40–50	fuel oil (eg power stations) lubricating oils and waxes bitumen or asphalt for roads and roofing

atmospheric pressure, since such temperatures would tend to crack the hydrocarbons. The more volatile oils distil over, leaving behind a tarry residue. The fuel oils obtained can be used in power stations or ships' boilers – or cracked to produce more petrol.

As a result of these processes in the refinery, the final petrol contains not only alkanes but other types of hydrocarbons as well – such as **cycloalkanes** which contain rings, **alkenes** which contain a C=C double bond, and **aromatic hydrocarbons** which contain a benzene ring.

You can find out more about these other types of hydrocarbons in **Sections 17.1** (cycloalkanes), **17.3** (alkenes) and **17.5** (aromatic hydrocarbons).

Figure 4 The Athabasca oil sands in Canada: a large reserve of crude oil, soaked into sands

Vacuum distillation

This is a technique for distilling a liquid which decomposes when heated to its boiling point. Distillation cannot be carried out at atmospheric pressure, but if the external pressure is reduced the boiling point can be brought below the decomposition temperature.

ASSIGNMENT 1

a Which fraction from the primary distillation of crude oil would be most likely to contain the following hydrocarbons?

i $C_{35}H_{72}$

ii C_4H_{10}

iii $C_{20}H_{42}$

iv C_8H_{18}

b What type of hydrocarbon is each of the above? Explain your answer.

c The following hydrocarbons are all found in petrol. In each case, draw out one possible structural formula and state the type of hydrocarbon.

i C_7H_8

ii C_4H_8

iii C_5H_{12}

Winter and summer petrol

About 30%–40% of each barrel of crude oil goes to make petrol. But you can see that it's not as simple as just distilling off the right bit at the refinery and sending it to the petrol stations. Petrol has to be blended to get the right properties. One important property is the **volatility**.

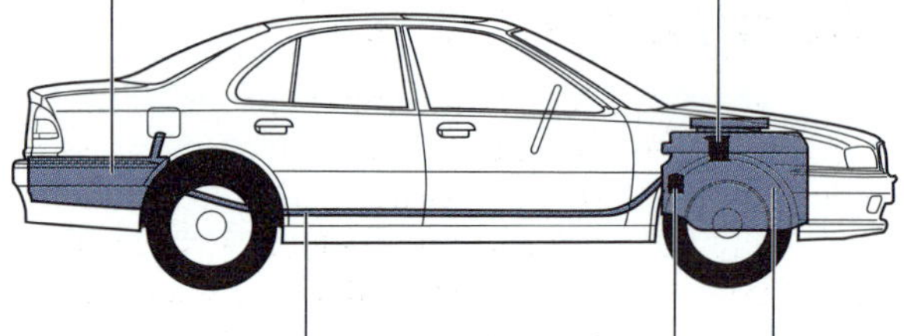

Figure 5 The fuel supply system. When you start a car engine, the fuel pump sucks petrol out of the petrol tank and pushes it to the carburettor. The carburettor partially vaporises the petrol, mixes it with air and sends it to the cylinders in the car engine (nowadays the carburettor is often replaced by an electronic fuel injection system)

In a car engine, a mixture of petrol vapour and air is ignited in a cylinder. The vapour–air mixture is made in the carburettor (Figure 5). When the weather is very cold, the petrol is difficult to vaporise, so the car is difficult to start.

To get over this problem, petrol companies make different blends for different times of a year. In winter, they put more volatile components in the petrol so it vaporises more readily. This means putting in more of the hydrocarbons with small molecules, such as butane and pentane.

On the other hand, in hot weather you don't want too many of these more volatile components, or the petrol would vaporise too easily. For one thing, you'd lose petrol from your tank by evaporation – a process which is costly and polluting. Also, if the fuel vaporises too readily, pockets of vapour form in the fuel supply system. The fuel pump then delivers a mixture of liquid and vapour to the carburettor, instead of mainly liquid. This means that not enough fuel gets through to keep the engine running. It's called *vapour lock*.

Any petrol is a blend of hydrocarbons of high, medium and low volatility. The colder the climate, the more volatile are the components added to the blend. Petrol companies change their blend four times a year – and you don't even notice. But you'd notice if they didn't!

Figure 6 In what way will the petrol blend in Alaska be different from that in Australia?

ASSIGNMENT 2

Petrol blenders talk about the 'front-end' (high volatility), 'mid-range' and 'tail-end' (low volatility) components of a blend.

a What differences will there be between
- **i** a winter and a spring blend for the UK?
- **ii** a summer blend for Spain and a summer blend for The Netherlands?

b The densities of some alkenes are as follows: butane, 0.57 g cm^{-3}; pentane, 0.63 g cm^{-3}; hexane, 0.65 g cm^{-3}; heptane, 0.68 g cm^{-3}; octane, 0.70 g cm^{-3}.
- **i** How does the density of alkanes change as their molecular mass increases?
- **ii** Which will have the greater mass at a given temperature: 1 litre of petrol bought in The Netherlands or 1 litre bought in Spain? Explain your answer.

c 'Spanish people get a better bargain than Dutch people when they buy petrol.' Discuss.

The problem of knocking

Another important property which blenders must take into account is the **octane** rating of the petrol (see box). This is a measure of the tendency of the petrol to cause a problem known as '**knock**'.

In a petrol engine, the petrol–air mixture has to ignite at the right time, usually just before the piston reaches the top of the cylinder.

Look at Figure 7. As the fuel–air mixture is compressed it heats up and the more it is compressed the hotter it gets. Modern cars achieve greater efficiency than in the past by using *higher compression ratios*, often compressing the gases in the cylinder by about a factor of 10.

Many hydrocarbons **auto-ignite** under these conditions. The fuel–air mixture catches fire as it is compressed. When this happens, *two* explosions occur: one due to the compression and another when the

Octane numbers

The tendency of a fuel to auto-ignite is measured by its **octane number**. 2,2,4-Trimethylpentane (which used to be called '*iso*-octane' – hence the name of the scale) is a branched alkane with a low tendency to auto-ignite. It is given an octane number of 100. Heptane, a straight-chain alkane, auto-ignites easily and is given an octane number of 0.

The octane number of any fuel is the percentage of 2,2,4-trimethylpentane in a mixture of 2,2,4-trimethylpentane and heptane which knocks at the same compression ratio as the given fuel. For example, four-star petrol has an octane number of 97 and knocks at the same compression ratio as a mixture of 97% 2,2,4-trimethylpentane and 3% heptane.

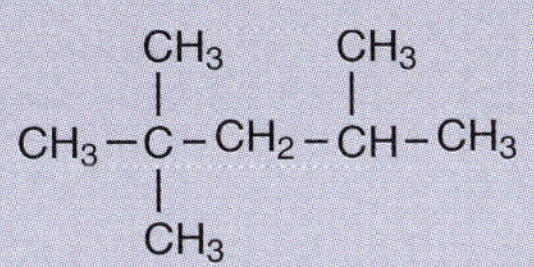

2,2,4-trimethylpentane low tendency to auto-ignition scores 100

$CH_3-CH_2-CH_2-CH_2-CH_2-CH_2-CH_3$

heptane high tendency to auto-ignition scores 0

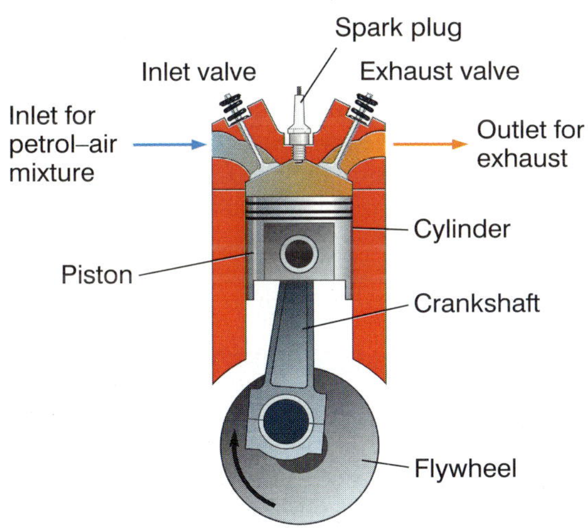

Figure 7 How a four-stroke petrol engine works. The compression stroke is shown here. The piston compresses the petrol–air mixture, then a spark makes the mixture explode, pushing the piston down and turning the crankshaft

spark occurs. This produces a 'knocking' or 'pinking' sound in the engine. The thrust from the expanding gases is no longer occurring at the proper time so engine performance is lowered, and the inside of the combustion cylinder can be damaged.

4.3 *Making petrol – getting the right octane rating*

Different cars have different compression ratios. High performance engines usually have a high compression ratio, and they need a high octane fuel – otherwise there would be knocking and the engine would suffer. There are two ways of dealing with the problem of knock. One is to put special additives in the petrol which discourage auto-ignition. The other is to blend high-octane compounds in with the ordinary petrol.

Figure 8 Auto-ignition is damaging in petrol engines, but in diesel engines it's an advantage. Diesels have no spark plugs – they use high compression to make the mixture of diesel fuel and air auto-ignite

Look – no lead

Anti-knock additives are substances which reduce the tendency of alkanes to auto-ignite. They increase the octane rating of the petrol.

Since the 1920s, small amounts of lead compounds have been used as economical and effective anti-knock additives. Exactly how they work is still not fully understood, but they help to prevent the reactions which cause knocking.

However, concern over environmental effects has led to a gradual phasing-out of leaded petrol. For one thing, the lead compounds present in the exhaust fumes are toxic. They also poison the metal catalysts in the catalytic converters installed to reduce the levels of other pollutants in exhaust fumes.

Rather than developing alternative lead-free additives, the petrol companies have turned their attention to using refining and blending to get high octane ratings without using lead.

Refining and blending

The hydrocarbons which give the best performance in a petrol engine are not the ones which are most plentiful in crude oil. So it is the job of the refinery to 'doctor' the hydrocarbons to suit our needs.

Mostly, it is a case of getting the right kind of alkane – although other types of hydrocarbons and some oxygenated compounds are also important, as you'll see later.

Which alkanes?

The structure of an alkane has an important influence on its tendency to auto-ignite – in other words, on its octane number. In general, the shorter the alkane chain, the higher the octane number. Short-chain alkanes are also more volatile, of course, so they can be used both to increase the octane number and to improve cold-starting. Even gaseous alkanes like butane can be used – they just dissolve in the petrol.

But the petrol blenders are limited in the proportion of short-chain alkanes they can include in the blend. Too much and the petrol blend would be *too* volatile.

The other factor that affects octane number is the *degree of branching* in the alkane chain. Quite simply, the more branched the chain, the higher the octane number.

Crude oil contains both straight-chain and branched alkanes. Unfortunately, it does not contain enough of the branched isomers to give it a naturally high octane number. The octane number of 'straight-run' gasoline is about 70.

To get around this, the petrol companies have a number of clever ways to increase the octane number of petrol. They include **isomerisation**, **reforming** and **cracking**.

The idea of isomerism is important in this section. You can find out about it in **Section 17.2**.

Isomerisation

Isomerisation basically involves taking straight-chain alkanes, heating them in the presence of a suitable catalyst so the chains break and letting them join together again (Figure 9). When the fragments join again, they are more likely to do so as *branched* rather than as *straight* chains.

Oil refineries do this on a large scale with pentane (C_5H_{12}) and hexane (C_6H_{14}), both products of the distillation of crude oil. It is one important way of increasing the octane quality of petrol.

One isomerisation reaction which can happen with pentane is shown below:

$$CH_3-CH_2-CH_2-CH_2-CH_3 \rightleftharpoons CH_3-\underset{}{\overset{\overset{\displaystyle CH_3}{|}}{CH}}-CH_2-CH_3$$

pentane octane number 62 | *2-methylbutane* octane number 93

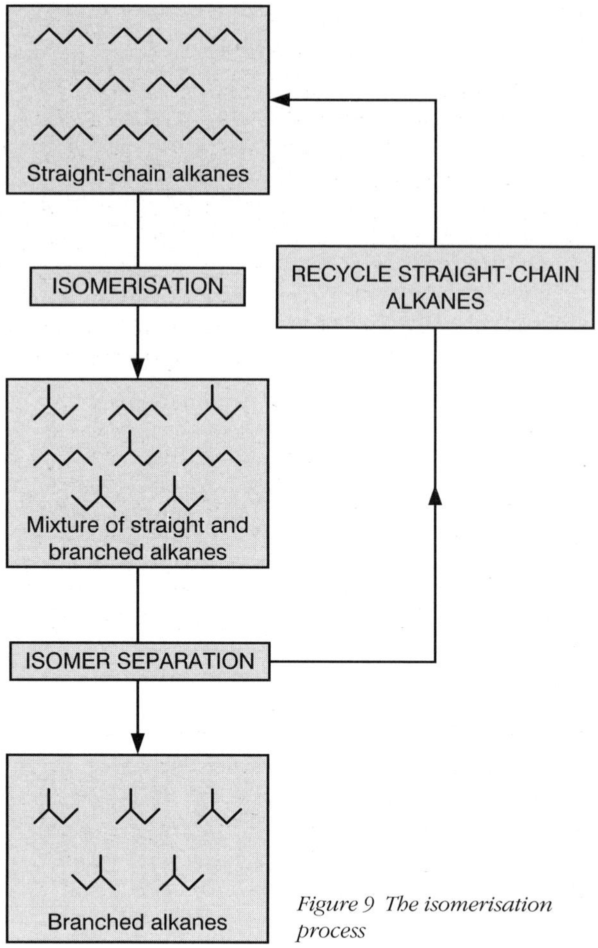

Figure 9 The isomerisation process

Isomerisation reactions like this do not go to completion and a mixture is formed containing all the possible isomers. We say the reaction has come to a **state of equilibrium** when no further change is possible under the reaction conditions. The arrows ($\rightleftharpoons$) indicate that the reaction is reversible and forms an equilibrium mixture.

ASSIGNMENT 3

a Explain why 2-methylbutane is an isomer of pentane.

b A second isomer of pentane is also formed in the isomerisation process. Draw out its structural formula.

In a modern plant, the isomerisation takes place in the presence of a platinum catalyst. The products then pass over a form of *zeolite* which acts as a *molecular sieve* and separates the straight-chain from the branched isomers. The straight-chain alkanes are then recycled over the platinum catalyst. Figure 9 gives a flow diagram for the process.

Zeolites

Zeolites belong to a large family of naturally occuring minerals containing mainly aluminium, silicon and oxygen. They can also be made synthetically.

There are many different zeolites because of the different ways the atoms can be arranged in the crystal structure. They all contain an extensive network of interlocking pores and channels, which provide a large surface area. These pores and channels are of different sizes in different zeolites.

Zeolites are used widely in industry as **catalysts** and also as **molecular sieves** to sort out molecules by size and shape (Figure 10).

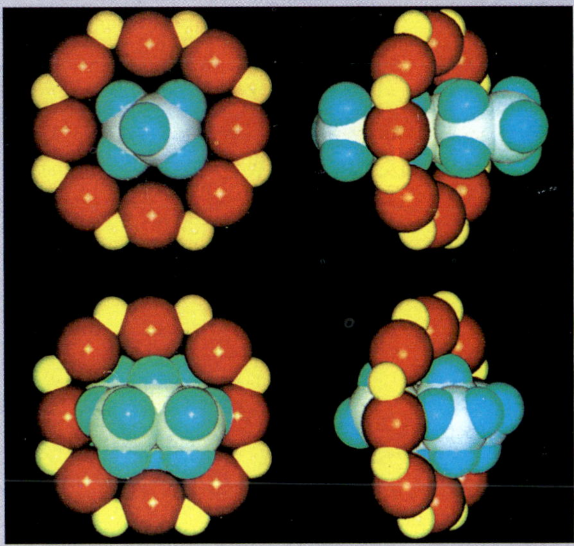

Figure 10 This computer-generated picture shows a molecular sieve in action. In the top pictures, a straight-chain molecule of hexane passes through a molecular sieve. In the bottom pictures, a branched-chain molecule of dimethylbutane cannot pass through. Zeolite molecular sieves are used to separate straight-chain and branched-chain isomers in the isomerisation process.

Cracking: using the whole barrel

Cracking is one of the most important reactions in the petroleum industry. It starts with alkanes that have large molecules and are too big to use in petrol – for example alkanes from the fuel oil fraction. These large molecules are broken down to give alkanes with shorter chains that can be used in petrol. What's more, these shorter-chain alkanes tend to be highly branched, so petrol made by cracking has a higher octane number.

There's another benefit from cracking: it helps solve the supply and demand problem. Figure 11 illustrates the supply and demand for different fractions of crude oil.

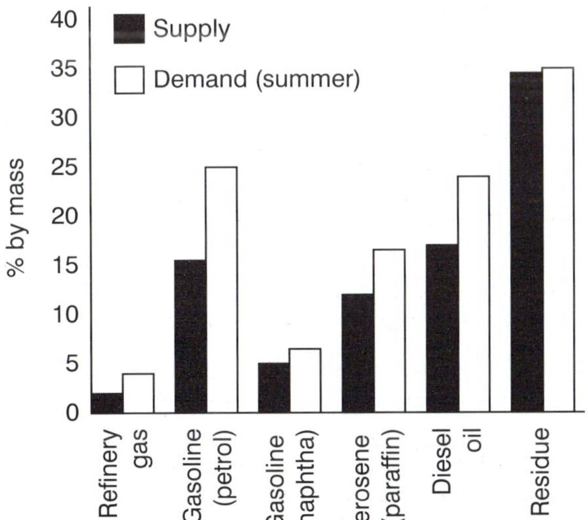

Figure 11 Supply and demand for different crude oil fractions from the North Sea (the supply figures are given as per cent by mass of crude oil; the demand figures are given as per cent by mass of the total demand)

ASSIGNMENT 4

a Demand for petrol exceeds supply. Which of the other fractions would be most suitable for converting to petrol by cracking, in order to meet the shortfall?

b How would the demand figures for *winter* compare with those given for summer in Figure 11?

Cracking: how is it done?

Most of the cracking carried out to produce petrol is done by heating heavy oils in the presence of a catalyst. It is called **catalytic cracking**, or cat cracking for short. The feedstock can be kerosene, diesel oil or the heavier fuel oils from the residue. The molecules in the feedstock can have 25–100 carbon atoms, although most will usually have 30–40 carbon atoms.

Cracking reactions are quite varied. Some of the types of reactions are

- alkanes → branched alkanes + branched alkenes

An example of this is:

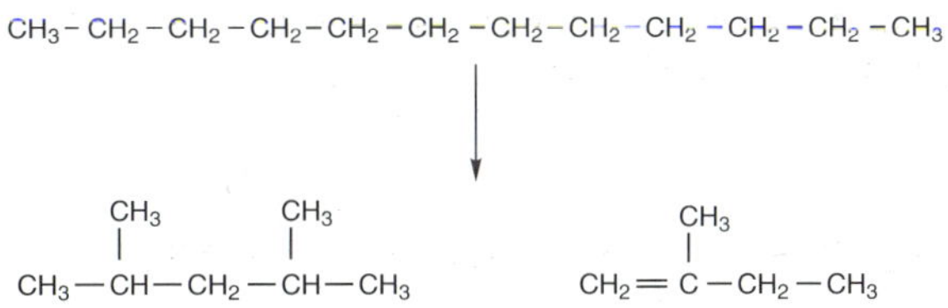

- alkanes → smaller alkanes + cycloalkanes
- cycloalkanes → alkenes + branched alkenes
- alkenes → smaller alkenes.

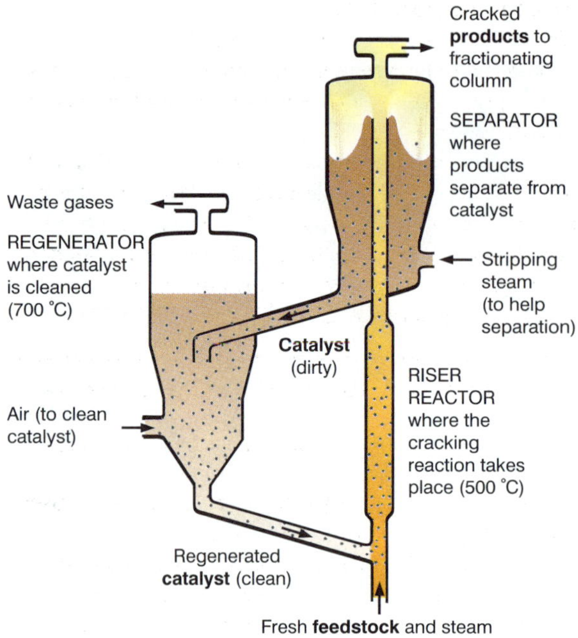

Figure 12 How a cat cracker works

The alkenes which are produced are important feedstocks for other parts of the petrochemicals industry.

Cracking always produces many different products, which need to be separated in a fractionating column.

Once again, zeolites play an important role. Type Y zeolite is particularly effective in producing good yields of high octane number products.

You can read about the importance of industrial catalysts, such as zeolites, in **Unit 1**, **Section 15.5**.

In a modern cat cracker the cracking takes place in a 60 m-high vertical tube about 2 m in diameter (Figure 12). It is called a *riser reactor* because the hot vaporised hydrocarbons and zeolite catalyst are fed into the bottom of the tube and forced upwards by steam. The mixture is a seething *fluidised bed* in which the solid particles flow like a liquid.

It takes the mixture about 2 seconds to flow from the bottom to the top of the tube – so the hydrocarbons are in contact with the catalyst for a very short period of time.

One of the problems with cat cracking is that, in addition to all the reactions you have already met, coke (carbon from the decomposition of hydrocarbon molecules) forms on the catalyst surface so that the catalyst becomes inactive. To overcome this, the powdery catalyst needs to be **regenerated**.

After the riser reactor, the mixture passes into a separator where jets of steam carry away the cracked products leaving behind the solid catalyst. This then goes into the regenerator, where it takes about 10 minutes for the hot coke to burn off in the stream of air which is blown through the regenerator.

The catalyst is then reintroduced into the base of the reactor ready to repeat the cycle.

The energy released from the burning coke heats up the catalyst. The catalyst transfers the energy to the feedstock so that cracking can occur without additional heating.

Cat crackers have been in operation since the late 1940s – and have become very flexible and adaptable. They can handle a wide range of different feedstocks. The conditions and catalyst can be varied to give the maximum amount of the desired product – in this case branched alkanes for blending in petrol. Other cat crackers are geared to produce large amounts of ethene ($CH_2{=}CH_2$), an important raw material for the petrochemicals industry.

ASSIGNMENT 5

The riser reactor uses a fluidised bed of solid catalyst and reactants. Fluidised beds have the following useful properties:

- the solid can flow along pipes
- as the solid can flow, it is easier to make the reaction into a continuous process
- there is very good contact between the gas and the large surface area of the solid catalyst
- energy transfer between the solid and the gas is very efficient.

Suggest why each of these properties is useful or important in the cat cracker.

Reforming

Reforming of naphtha is another trick used by oil refiners to increase the octane quality of petrol components. Reforming involves ring compounds: alkanes are converted to cycloalkanes, and cycloalkanes to aromatic hydrocarbons. Some typical reforming reactions are shown in Figure 13. Note that reforming alters the arrangement of atoms in molecules without necessarily changing the number of carbon atoms per molecule. The products of reforming are branched-chain alkanes, cycloalkanes and aromatic hydrocarbons, as well as some straight-chain alkanes. Aromatic hydrocarbons, such as benzene and methylbenzene (Figure 13), have high octane numbers and are important components of unleaded petrol.

Reforming reactions produce hydrogen as well as high octane gasoline components. The hydrogen is a valuable product and is piped away to be used in other refinery processes.

A platinum/rhenium catalyst is often used for reforming – the process is then called *platforming*. There can be £5 million worth of platinum inside one platformer, so it is very important to regenerate the catalyst and keep it in prime working order.

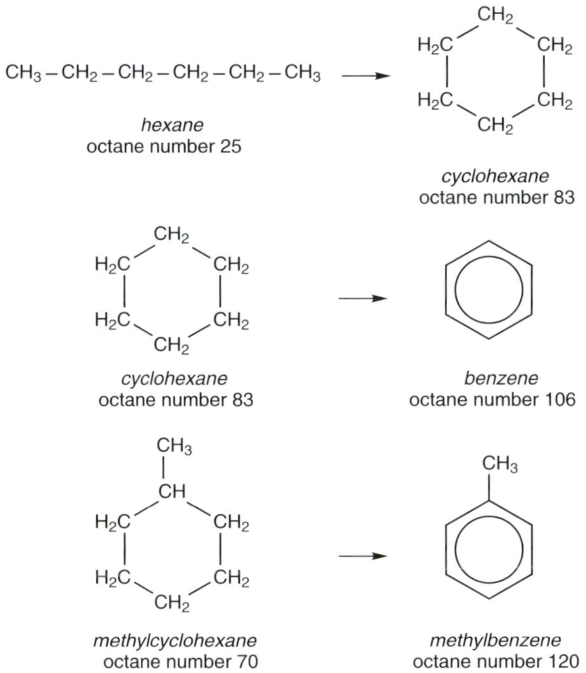

Figure 13 Some typical reforming reactions, showing octane numbers of reactants and products

ASSIGNMENT 6

a Using molecular formulae, write out a balanced equation for one of the conversions shown in Figure 13.

b Explain why these reactions are not isomerisations.

c Below is a list of alkanes. For each one, say whether you think

A It could be used unchanged in petrol

B It should go through a reforming process before blending to form petrol

C It should be cracked before blending to form petrol.

i $CH_3CH_2CH_2CH_2CH_2CH_3$

ii $CH_3(CH_2)_{15}CH_3$

iii $(CH_3)_3CCH(CH_3)CH_3$

iv $CH_3CH_2CH_2CH_3$

Adding oxygenates

'Oxygenates' is the name the petrol blenders use for fuels containing oxygen in their molecules. Two types of compounds are commonly used: alcohols and ethers (see box).

Members of the homologous series of **alcohols** all have an OH group in their molecule. You have already met one example – ethanol, C_2H_5OH (commonly just called 'alcohol'). **Ethers** have an oxygen atom bonded to two carbon atoms, for example ethoxyethane (sometimes just called 'ether'), $CH_3CH_2OCH_2CH_3$.

You can find out more about alcohols and ethers in **Section 18.1**.

Table 2 shows the oxygenates most commonly used for blending with petrol. The most commonly used oxygenate is MTBE (MTBE stands for 'methyl tertiary butyl ether', the name that was once used for this compound; its modern systematic name is 2-methoxy-2-methylpropane).

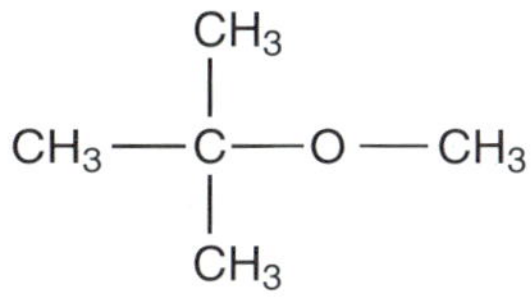

MTBE

Figure 14 shows the percentages of MTBE added to different grades of petrol.

There are two advantages to adding oxygenates to petrol. First, they increase the octane number, as you can see from Table 2. Secondly, they tend to cause less pollution when they burn. In particular, levels of carbon monoxide in the exhaust are reduced.

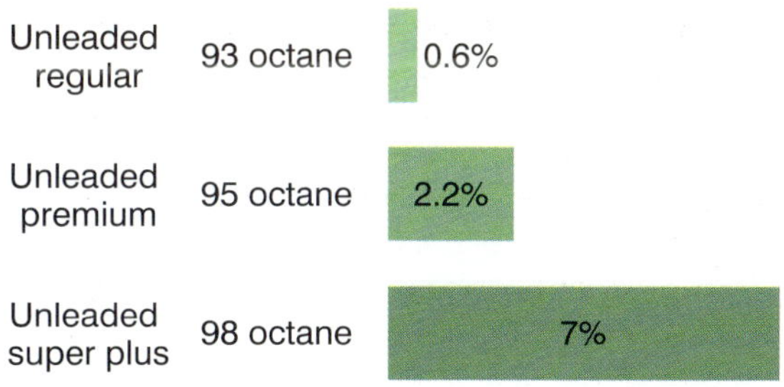

Figure 14 Percentages by volume of MTBE typically used in different blends of petrol

Name	Formula	Homologous series	Octane number	Boiling point/°C	Relative cost per litre (approx)
methanol	CH_3OH	alcohols	114	65	1.0
ethanol	CH_3CH_2OH	alcohols	111	79	5.8
MTBE	$CH_3OC(CH_3)_3$	ethers	118	55	1.6
'straight-run' gasoline	—	—	70	—	1.1

Table 2 Oxygenates commonly used for blending with petrol; the relative costs fluctuate with demand ('straight-run' gasoline is given for comparison)

Alcohols and ethers

Alcohols all have an OH group in their molecule. The best known alcohol is ethanol, commonly called just 'alcohol':

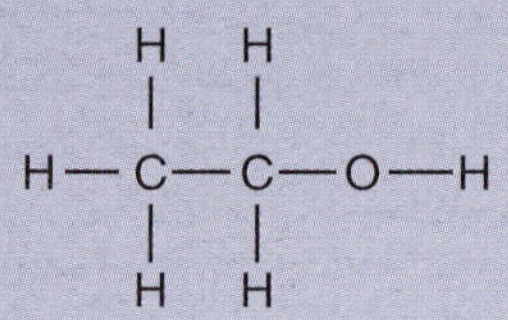

Ethers all have an oxygen atom bonded to two carbons. An example is ethoxyethane (sometimes just called 'ether'):

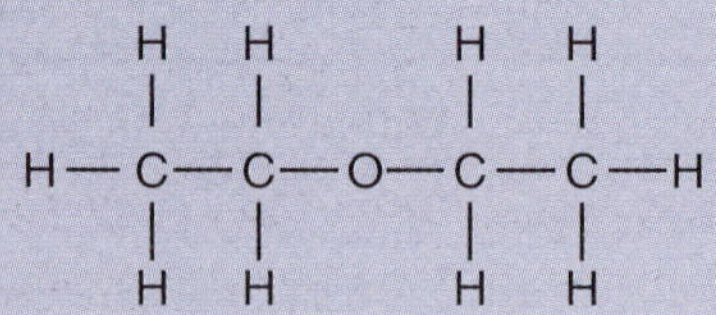

The perfect blend

Figure 15 shows the octane numbers of some of the key ingredients at the petrol blender's disposal.

The blender's job is to produce petrol from these ingredients at minimum cost. Of course, the petrol must meet specifications about volatility, octane number, density, etc. Prices and availability fluctuate and refiners are helped in their decisions by computer models of their refinery. They must take into account not only the prices of crude oil and the final petrol, but prices of their other products from the refinery and energy costs too.

Blending is done in batches of around 20 000 000 litres at a time, and can take as long as 20 hours to complete. Thorough mixing is important to give a homogenous liquid.

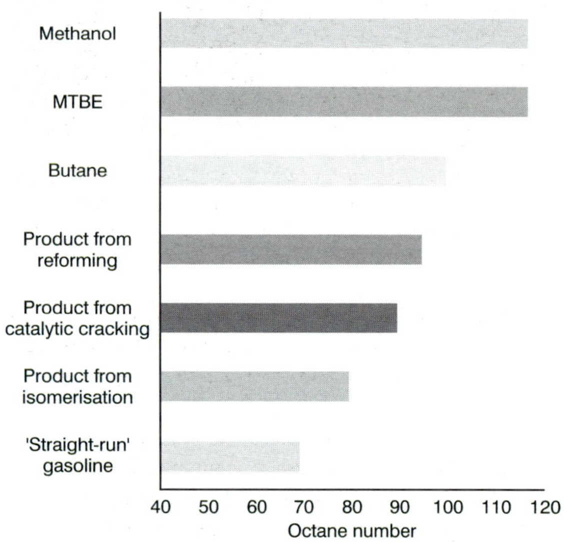

Figure 15 Octane numbers of some of the key ingredients used in petrol blending

4.4 *Changing the fuel*

Petrol has been developed to work well as a transport fuel. But problems remain. You studied the problem of emissions in **Unit 1**, **Chapter 2**. Changing the fuel used by cars could help to solve the emissions problem. Figure 16 shows past and projected rates of world consumption of oil and coal. The use of petrol will decline in the future as oil supplies run out, so a search for alternative fuels could be vital to our transport system.

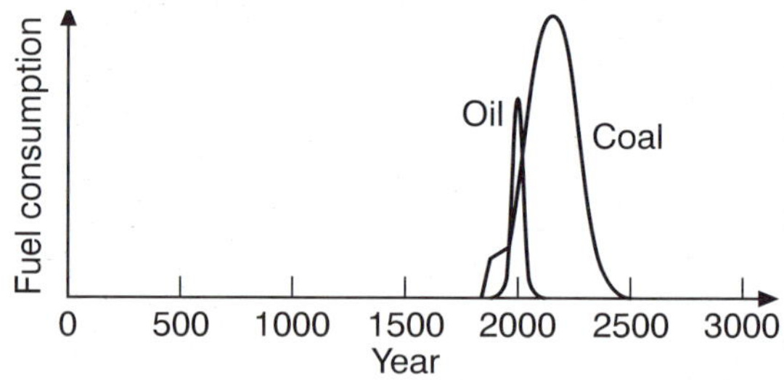

Figure 16 Past and projected rates of world consumption of oil and coal: the use of fossil fuels will decline in the future as supplies run out

Adding ethanol

One liquid fuel from a renewable source is **ethanol** from the fermentation of cane sugar juice. Brazil has to import oil, but the climate is good for growing cane sugar. A programme was set up in the 1970s to produce ethanol as an additiive to petrol. By the 1990s Brazil made enough ethanol to power about one-third of the country's 12 million cars – either using gasohol (Figure 17) or ethanol alone. In 1985 80% of the new cars made in Brazil were designed to run on ethanol alone.

Burning ethanol produces less carbon monoxide, sulphur dioxide and nitrogen oxides than petrol, and unburned ethanol does not contribute to photochemical smogs. The carbon dioxide produced is balanced by absorption of carbon dioxide in new cane, and so using ethanol as a fuel adds less to the greenhouse effect.

Problems arise when oil prices drop because ethanol from fermentation becomes less competitive. Large amounts of energy are needed for intensive cultivation of the crop, and there are queries about the overall energy efficiency of the process. The production cost of **biofuels** can be three times higher than the production costs of fuels from crude oil.

Biomass and biofuels

Biomass is organic matter, mostly from plants such as wood or sugar cane but also from animal waste, that can be used as a source of energy.

Biomass fuels (**biofuels**), such as ethanol produced from sugar cane, provide a renewable source of energy.

Figure 17 A gasohol station in Santos, Brazil. The fuel is a mixture of petrol and ethanol (78% petrol, 22% ethanol)

Methanol – the key to future fuels?

The pros

Methanol (CH_3OH) burns cleanly in a car engine. Its combustion is more complete than that of petrol so emissions, particularly of carbon monoxide, are reduced. Nor does it release carcinogenic benzene vapour or other aromatic hydrocarbons into the air. It has a high octane number (114), and its use requires only small changes to engine and petrol pump design. What's more, it's quite cheap.

So, why don't we use methanol as a fuel? After all, Formula 1 racing drivers have been using it for years. Methanol is less volatile than petrol and less likely to explode in a collision.

In fact, some countries, such as Germany and parts of the US, have used methanol, but as part of a mixture with petrol.

The cons

The main problem when using a mixture of methanol and petrol is getting the two liquids to mix. They tend to separate into layers, like oil and vinegar, unless a co-solvent is used. In Germany, a higher alcohol than methanol, called TBA, is mixed with the methanol for this purpose. TBA dissolves in both methanol and petrol.

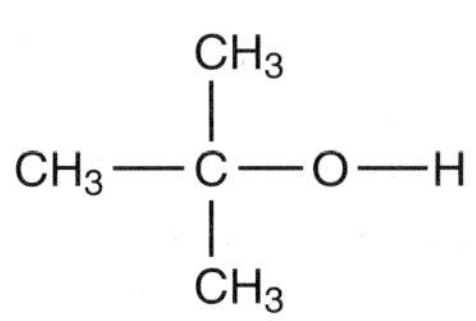

TBA

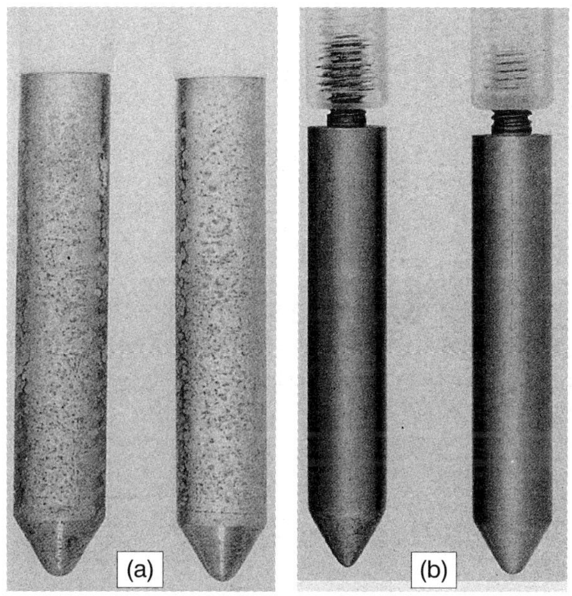

Figure 18 Methanol absorbs water, making it corrosive. The steel rods in (a) were immersed in petrol containing methanol. Compare them with the rods in (b), which were immersed in normal unleaded petrol

Another problem is that the petrol–alcohol mixture is *hygroscopic* and absorbs water vapour from the air. The wet methanol corrodes parts of the engine (Figure 18).

When the water content reaches a critical level (about 0.5%) two layers separate, giving a layer of petrol sitting on top of a water–alcohol mixture.

Although polluting emissions are reduced, methanol itself is toxic. Long-term exposure can cause blindness and brain damage – that's why 'meths' drinkers go blind. So there is concern over its possible effect on mechanics and petrol-pump attendants at filling stations. Also, disturbingly high levels of methanal have been found in exhaust fumes. Methanal is the first member of a homologous series of **aldehydes**. It is formed by the partial oxidation of methanol and is known to be carcinogenic in humans.

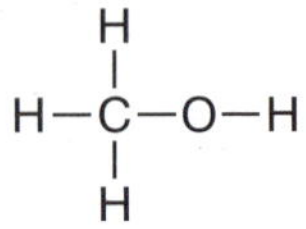

methanol
(an alcohol)

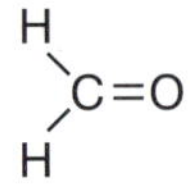

methanal
(an aldehyde)

You can learn more about the chemistry of aldehydes and ketones in **Section 18.2**.

One litre of methanol produces about 40% less energy than 1 litre of petrol. This means larger and heavier fuel tanks and more refuelling trips.

Biogas

Under anaerobic conditions many biological materials ferment to produce **methane**, another example of a biofuel. This gas forms an explosive mixture with air and has in the past been a hazard as it can seep from waste dumps. Strict safety regulations now govern landfill sites and many are being recognised as useful sources of fuel (Figure 19).

Figure 19 Methane from landfill sites is now used as a fuel

Gas extraction plants connected to landfill sites can burn the methane to produce electricity to feed into the National Grid. One site is already operating at Greengairs in Lanarkshire, and another at Summerston in Glasgow holding three million tonnes of waste is scheduled to begin generating electricity in early 1999.

Methane has a greenhouse factor 30 times greater than the same amount of carbon dioxide, so burning it rather than letting it seep will help to limit global warming. Any substitution of alternative fuels also extends the time over which fossil fuels remain available for our use.

Gases like methane are not obvious candidates for transport fuels because of storage problems, but one gas in particular has attracted a lot of attention – hydrogen.

4.5 *The hydrogen economy*

Why hydrogen?

People who favour a **hydrogen economy** see water as a plentiful source of hydrogen. If hydrogen could be extracted from water without consuming fossil fuels it would reduce our dependence on these fuels, and help to reduce the amount of carbon dioxide released into the atmosphere.

Hydrogen could be distributed as we now distribute natural gas, and burned as a heating fuel, or used in internal combustion engines, or converted into electricity in a **fuel cell**.

Figure 20 This car is powered by hydrogen

But how can we make the hydrogen? The most likely large-scale method of producing hydrogen seems to be by electrolysis of water, obtaining the energy needed for this from some renewable source, such as solar cells. But why produce electricity and then use it to make hydrogen? There are two major advantages:

- hydrogen can be stored
- it can be used in the internal combustion engine.

A possible scheme is shown in Figure 21. All the technologies needed already exist. We have large-scale electrolysis plants for other purposes, large storage tanks, and a network of gas pipelines.

The hydrogen economy would use hydrogen as a way of storing and distributing energy. If systems are costed over whole lifetime use in terms of money and energy, then distributing hydrogen by pipeline may be cheaper than transmitting electricity.

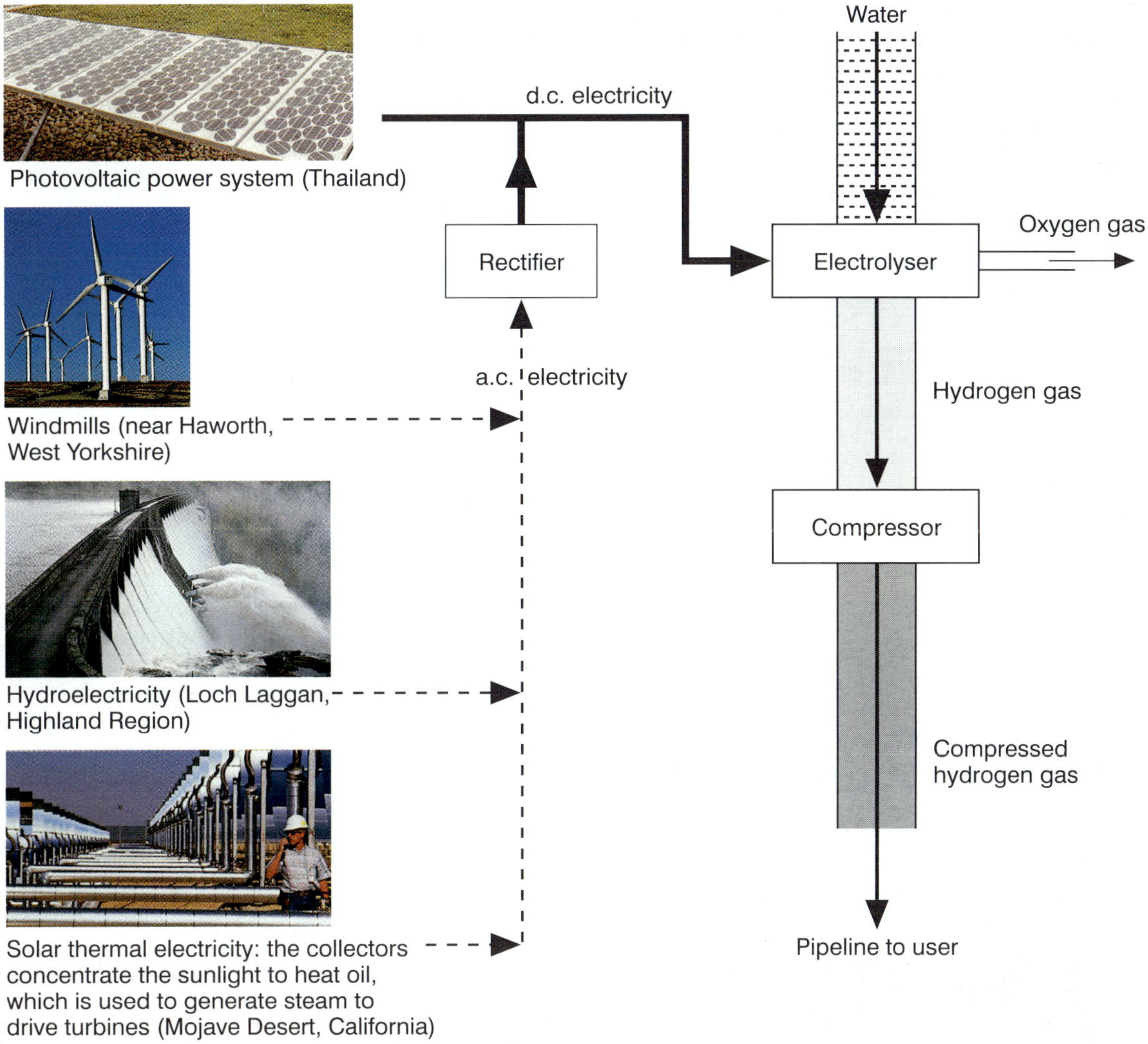

Figure 21 Hydrogen production in the future

Locally, fuel cells could be used to generate electricity on a small scale where it is needed, and where photovoltaic systems are not practical.

The main problem in the design of a hydrogen car is the storage of hydrogen. A large volume of gaseous hydrogen is required to get equivalent mileage to a fuel tank of petrol. We need to find some way to store it more compactly.

A high-pressure fuel tank could be used, as shown in the BMW experimental model (Figure 22).

Hydrogen turns liquid if cooled to –253 °C, and liquid hydrogen is being explored as a fuel for aircraft (Figure 23). Part of the attraction is that hydrogen gives three times as much energy as the same mass of kerosene. But there are design problems in storing the large volume of cooled fuel required. In a Daimler Benz prototype plane the insulated fuel tank is carried above the main passenger cabin.

Schemes for developing alternative fuel use on a large scale depend on long-term and large-scale investments in new infrastructures, and so success will depend on political as well as economic factors. The reward could be cleaner and renewable fuels.

An experimental BMW uses liquid hydrogen as a fuel. Mixed with air, it is injected into an internal combustion engine. A hydrogen car is not expected to be practical until early in the 21st century.

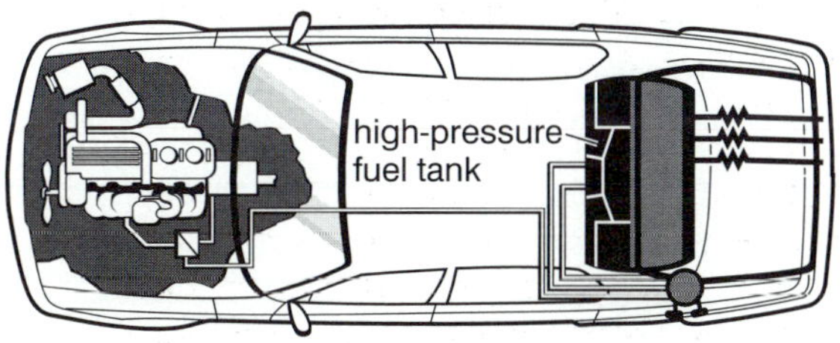

Figure 22 An experimental model of a hydrogen-powered car

Figure 23 An experimental plane powered by liquid hydrogen. The fuel tanks are situated above the fuselage

ASSIGNMENT 7

The petrol tank of a typical car holds about 45 litres of petrol (approximately 10 gallons).

a Calculate the amount of energy that is released by burning 45 litres of petrol. Use the following information to help you.
- Assume that petrol is octane (C_8H_{18}).
- The enthalpy of combustion of C_8H_{18} is $-5500\,kJ\,mol^{-1}$.
- The density of octane is $0.70\,g\,cm^{-3}$.
- 1 litre is $1000\,cm^3$.

b Calculate the mass and volume (at 20°C and 1 atmosphere pressure) of hydrogen needed to provide the same amount of energy as 45 litres of octane.
- The enthalpy of combustion of H_2 is $-286\,kJ\,mol^{-1}$.
- 1 mole of a gas at 20°C and atmospheric pressure has a volume of about 24 litres.

(You can remind yourself about calculations involving gases by reading **Unit 1**, **Section 10.5**)

Hydrogen engines are efficient. In motorway driving conditions a hydrogen engine can be over 20% more efficient than a petrol engine. In city 'stop-go' driving conditions the hydrogen engine is about 50% more efficient than a petrol engine.

c Taking efficiencies into account, what mass and volume of hydrogen are needed to give the same mileage as 45 litres of petrol in
- **i** motorway driving conditions?
- **ii** city driving conditions?

4.6 *Summary*

This Chemical Storylines chapter is linked to Chemical Ideas Sections 17.1, 17.2, 17.3, 17.5, 18.1 and 18.2.

You began by considering some of the desirable properties of a good fuel, and found out how petrol is made and how chemists and chemical engineers are helping to develop better fuels for motor vehicles. This means producing fuels which give improved performance and greater fuel economy, so that we use our precious reserves of crude oil as economically as possible. But it is also vital that we drastically reduce the polluting emissions from burning petrol. Quite a challenge!

Petrol is a mixture of hydrocarbons, many of which are alkanes. This led you to find out more about the properties and reactions of alkanes. But the hydrocarbons which give the best performance in the petrol engine are not the ones which are most plentiful in crude oil. So, after the primary distillation of crude oil into fractions, there is a whole range of chemical processes carried out in the refinery to 'doctor' the hydrocarbons to suit our needs. Petrol must have the correct octane number to avoid auto-ignition. Branched-chain alkanes have higher octane numbers than straight-chain alkanes, which led you to study structural isomerism in more detail.

For some molecules it is enough to rearrange the position of the atoms, but large molecules must be broken down into more useful smaller ones, and some smaller ones are joined together. To understand these processes, you needed to read about the structure and properties of alkenes and aromatic hydrocarbons, such as benzene. The final petrol is a blend of many hydrocarbons and possibly some oxygenates too.

In the immediate future, it seems that the use of oxygenates will play a large part in fuel development. Here you needed to find out more about the chemistry of alcohols and ethers and of aldehydes and ketones.

However, the story did not end with petrol. Our supplies of crude oil are finite and are needed for more than petrol alone, so you were invited to think about alternative fuels for motor vehicles for the more distant future. You looked at some biofuels (ethanol from sugar cane and methane from landfill sites) and at methanol and hydrogen as possible transport fuels.

5

POLYMERS IN ACTION

5.1 *Designer polymers*

Producing a perfect copy

If you had lived in medieval times and visited a monastery, you might have found that one of the most important rooms was the one in which monks made copies of religious documents, painstakingly copying out the original by hand.

Fortunately, today we don't have to copy out by hand everything we want duplicates of: we can use a photocopier. Unlike some of the medieval bibles, the results are not works of art, but they are very good copies and it is almost impossible to tell the difference between the copy and the original. You probably accept this as a normal, everyday fact. But do you know how the photocopier works?

At the heart of the photocopying process is a photoreceptor surface which is sensitive to light. In many machines, the photoreceptor is in the form of a metal drum coated with a very thin layer (about 10^{-5} m thick) of a special polymer called **polyvinyl carbazole**.

The polymer is made from a compound, vinyl carbazole, with the chemical structure

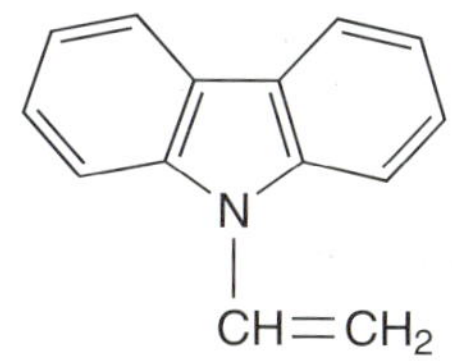

vinyl carbazole

A remarkable property of polyvinyl carbazole is that it exhibits **photoconductivity** – it conducts electricity much better when light shines on it than when it is in the dark.

Figure 1 Part of a decorative initial letter in the Lindisfarne Gospels, written by Northumbrian monks in the late 7th or 8th century; religious documents like this were laboriously copied by hand

In a photocopier, the surface of the drum is first given an electrostatic charge so that the polymer becomes positively charged. The interesting behaviour starts when light shines on the drum. If it is in the light, the polymer becomes conducting and so its charge disappears. But if it is in the dark, the polymer retains most of its charge.

A piece of paper with a picture on it is laid on a glass plate and illuminated, so that an *image* of the picture is projected onto the drum.

Can you imagine what happens? The polymer stays charged where there were dark areas in the picture, but loses its charge where the picture was white. So an electrically charged image of the picture is held on the drum. Negatively charged ink powder or toner is dusted over the drum and sticks to the charged image. The toner is then transferred from the drum to a sheet of ordinary paper. Finally, a hot roller fuses the image to the paper to make the photocopy permanent.

Something new

There are many other 'designer polymers' with useful properties. Take *poly(1,1-difluoroethene)*: it's **piezoelectric**, that is it generates electricity when it is bent or twisted. If a sheet of poly(1,1-difluoroethene) is made to shake and wobble by a sound wave, the electrical signals can be processed and the sound wave detected. In reverse, electrical signals sent to a piece of this polymer can make it wobble and act as a type of loudspeaker.

In this chapter you will learn about *Kevlar*, a heat-resistant polymer that is as strong as steel but five times lighter, *poly(ethyne)*, a plastic that conducts electricity, and *poly(ethenol)*, a polymer designed to dissolve in water.

All these polymers have properties which are not possessed by any single natural material, and this gives them some fascinating applications. Making them poses many chemical and technological problems. Chemists have to develop ways of making the required quantities of starting compounds at a reasonable cost, and have to combine them together in the right way to produce a usable material; engineers have to understand how to use and process the polymers.

Some early polymers were discovered almost by accident. But as our knowledge and understanding of the polymerisation process and the properties of polymers have grown, the creation of new polymers has become more systematic. This chapter shows you how our knowledge of polymers and the processes which produce them has developed.

You can learn about the nature of polymer molecules by studying **Chapter 19**.

5.2 *The polythene story*

An accidental discovery

Imperial Chemical Industries (ICI) was formed in 1926 by the amalgamation of a number of smaller chemical companies. The prime aim of the merger was to form a strong competitor to the huge German chemical company IG Farben.

To understand this section you will need to know about the chemistry of **alkenes** such as ethene. You can remind yourself of this by looking at **Section 17.3**.

In 1930 Eric Fawcett, who was working for ICI, got the go-ahead to carry out research at high pressures and temperatures aimed at producing new dyestuffs. His results were disappointing and his project was eventually abandoned.

The team then moved into the field of high-pressure gas reactions and was joined by Reginald Gibson. On Friday 24 March 1933, Gibson and Fawcett carried out a reaction between ethene and benzaldehyde using a pressure of about 2000 atm. They were hoping to make the two chemicals add together to produce a ketone:

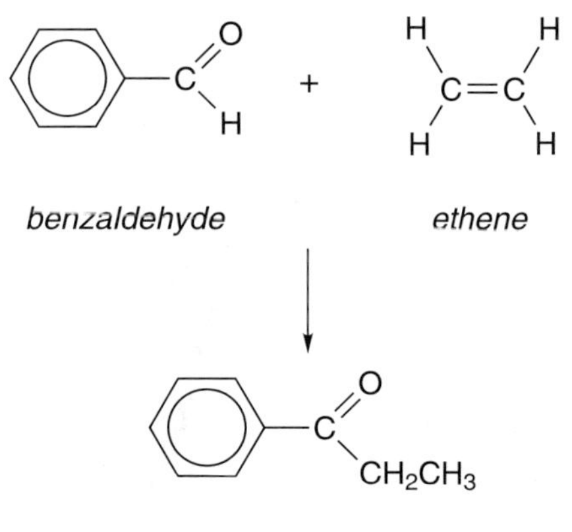

They left the mixture to react over the weekend, but their apparatus leaked and at one point they had to add extra ethene.

They opened the vessel on the following Monday and found a white waxy solid. When they analysed it they found that it had an empirical formula CH_2. They were not always able to obtain the same results from their experiment: sometimes they got the white solid, on other occasions they had less success, and sometimes their mixture exploded leaving them with just soot.

The work was halted in July 1933 because of the irreproducible and dangerous nature of the reaction.

Learning to control the process

In December 1935 the work was restarted. Fawcett and Gibson found that they could control the heat given out during the reaction if they added cold ethene at the correct rate. This kept the mixture cooler and prevented an explosion. They also found that they could control the reaction rate and relative formula mass of the solid formed by varying the pressure.

A month later they had enough of the material to show that it could be melted, moulded and used as an insulator.

Most crucial of all was the identification of the role of oxygen in the process, by Michael Perrin who took charge of the programme in 1935. If no oxygen is present, the polymerisation cannot occur. Too much oxygen causes the reaction to run out of control. The trick is to add just enough oxygen. The leak in Fawcett and Gibson's original apparatus had accidentally let in a small amount of oxygen. Without this, the discovery of poly(ethene) might not have been made. It was also Perrin who showed that even if the benzaldehyde is left out of the reaction mixture, the polymer still forms.

Poly(ethene) is an example of an **addition polymer**. You can find out more about the formation of polymers like poly(ethene) in the section on addition polymerisation in **Section 19.2**.

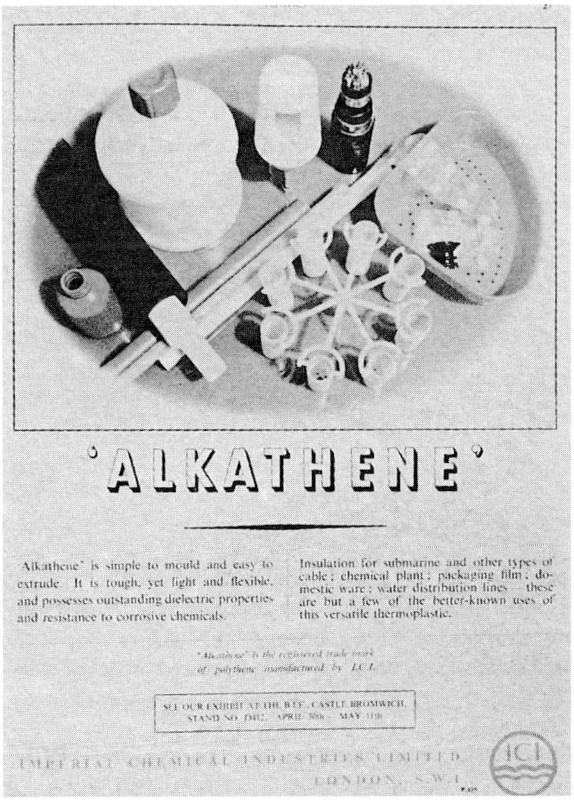

Figure 2 An early advert for polythene

Poly(ethene) – or polythene as it is commonly called – is tough and durable, and has excellent electrical insulating properties. Unlike rubber which had previously been used for insulating cables, poly(ethene) is not adversely affected by weather or water (Figure 2).

It also has almost no tendency to absorb electrical signals. Its first important use was for insulating a telephone cable laid between the UK mainland and the Isle of Wight in 1939. Its unique electrical properties were again essential during the Second World War in the development of radar (Figure 3).

Figure 3 Poly(ethene) was essential in the development of radar equipment in the Second World War

The first poly(ethene) washing-up bowls appeared in the shops in 1948, and were soon followed by carrier bags, squeezy bottles and sandwich bags. Sadly, poly(ethene) and some other early polymer materials were overexploited. They were used for all manner of novelties and as cheap but poor substitutes for many natural materials. This gave *plastic* a bad name – the word is often used to describe something which looks cheap and does not last. The reputation still sticks, as an undeserved slur on many of today's excellent materials.

ASSIGNMENT 1

a Write an equation for the formation of poly(ethene) from ethene.

b Draw the structural formula of part of a poly(ethene) chain. Explain why poly(ethene) can be thought of as a very large alkane.

c Explain why polymerisation of ethene does not occur when there is no oxygen present, but with too much it gets out of control.

Section 19.4 explains the factors, such as chain length, which affect the physical properties of polymers.

5.3 *Towards high density polymers*

Karl Ziegler

The poly(ethene) produced by Fawcett and Gibson was what we would today call **low density poly(ethene) (ldpe)**. Although they had some control over the product of their polymerisation process, the low density poly(ethene) they made was still quite messy at a molecular level. The polymer chains were extensively branched. This makes it impossible for the chains to fit together in an organised way: they coil around randomly taking up a lot of room, and hence lower the density of the material.

This relatively disorganised and open structure also lowers the strength of the poly(ethene). The next major advance in poly(ethene) production came with the development of **high density poly(ethene) (hdpe)** which resulted from discoveries made by Karl Ziegler.

The German scientist Karl Ziegler was born in Helsa, near Kassel, in 1898. Encouraged to work hard by his father, the young Ziegler set up a chemical laboratory at home, where he became so advanced in his chemistry studies that he was allowed to omit the first year of his degree course.

At the age of 23 he became a professor and started on a research path which was to have a great influence on the future development of polymerisation processes.

Ziegler catalysts

Ziegler was studying the catalytic effects of *organometallic* compounds. These are compounds which contain covalent metal–carbon bonds. Strange things happened in some of his experiments when he was using an aluminium organometallic compound.

Figure 4 Karl Ziegler

He tracked down the unexpected behaviour to tiny traces of nickel compounds left over as impurities in his apparatus after cleaning. His research group then tried putting as many different transition metal salts as possible with the alkylaluminium compound to see what would happen.

In 1953 Ziegler, with his colleagues Holzkamp and Breil, found that adding titanium compounds easily led to the production of very long chain polymers. Simply passing ethene at atmospheric pressure into a solution of a tiny amount of $TiCl_4$ and $(C_2H_5)_3Al$ in a liquid alkane caused the immediate production of poly(ethene).

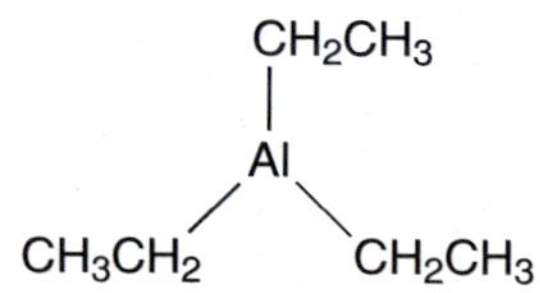

Figure 5 The structure of triethylaluminium, $(C_2H_5)_3Al$

So, like Fawcett and Gibson's discovery of poly(ethene), Ziegler's achievements began accidentally with an impurity.

The poly(ethene) which Ziegler produced had an average relative formula mass of 3 000 000 with very little branching along the polymer chain. The chains could therefore pack more closely than those made by the original high-pressure process. This gives the polymer its higher density and greater strength. Figure 6 shows an illustration of how poly(ethene) chains might be packed in hdpe.

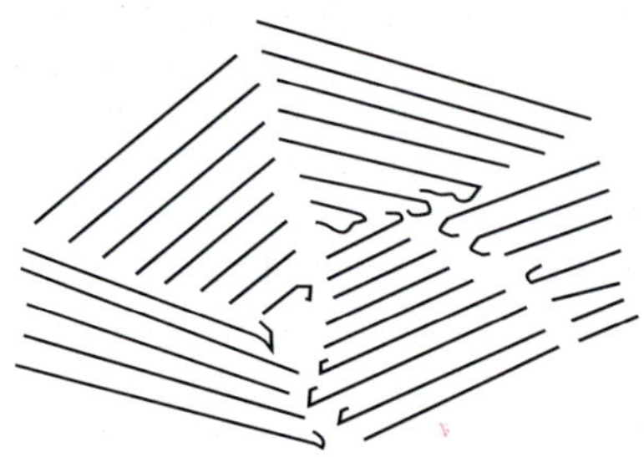

Figure 6 Chain packing in hdpe. When the polymer chains are aligned and packed closely together, the material is stronger

High density poly(ethene) is often used to make washing-up bowls, water tanks and piping. It is strong and can easily be moulded into complicated shapes: car petrol tanks, for example, can be made to fit neatly into the spaces under the car, something which was impossible with metal tanks. Hdpe is not as easily deformed by heat as ldpe and an early use was as Tupperware food storage containers. The ability to retain shape during heating means that hdpe articles can be heat-sterilised, making hdpe an important material for hospital equipment such as buckets and bed-pans.

Figure 7 Piping and many containers are made from hdpe

Ziegler patented hdpe. As a result he became a multimillionaire, and on his 70th birthday he gave $10 million to support further research at the Max Planck Institute where he worked.

Natta and stereoregular polymerisation

Giulio Natta was born in Imperia, Italy, in 1903. He studied chemical engineering at the Milan Polytechnic Institute, and after working at various universities he returned to Milan in 1938 to become Professor of Industrial Chemistry.

Natta was convinced that alkylaluminium catalysts were the key to making polymers with a regular structure. In March 1954 he used Ziegler's catalyst to polymerise propene. His reaction mixture contained two forms of poly(propene). He was able to separate them.

Figure 8 Isotactic poly(propene) is one of the plastics used to make this toy. Its strength and rigidity, combined with its lightness and ease of moulding, make it ideal for this application

In one form, the methyl groups all have the same orientation along the polymer chain. Natta called this the **isotactic** form. (Iso means *the same* – as in isotope or isobar.) In the other polymer, the methyl groups are randomly orientated and this was called the **atactic** form. Figure 9 tells you more about these two forms of poly(propene).

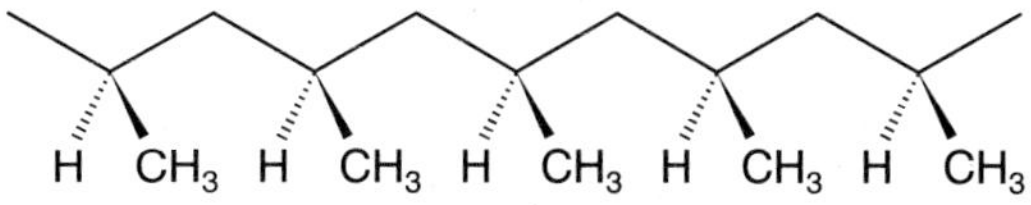

isotactic poly(propene)

Regular structure, chains pack together closely so it is tough - like hdpe.

Used in sheet and film form for packaging and containers; used to make fibres for carpets.

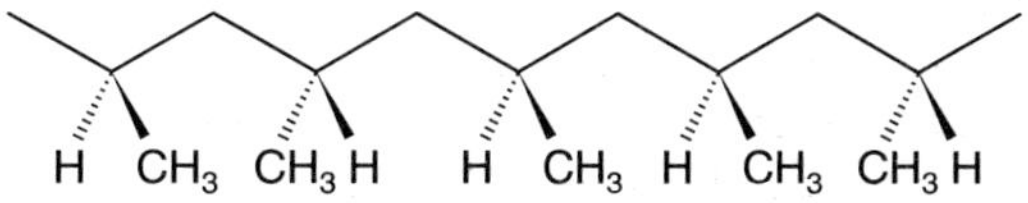

atactic poly(propene)

Irregular structure: chains loosely held, so it is soft and flexible.

Used to make roofing materials, sealants and other weatherproof coatings.

Figure 9 Isotactic and atactic poly(propene)

The enormous contribution of both Ziegler and Natta to chemistry was rewarded when they jointly received the Nobel Prize for Chemistry in 1963.

5.4 *Conducting polymers – breaking the rules*

Conducting polymers bring us almost up to the present day in the story of addition polymers. Several kinds of conducting polymer have been made, but so far their properties have not been exploited. When they are we shall probably see the next significant step forward in the applications of plastics in our lives.

Figure 10 Polymers protect you from lethal electric shocks every time you use electricity. Most polymers are excellent insulators because they have covalent molecular structures. So a conducting polymer is a surprise

ASSIGNMENT 2

On 8 January 1982 *The Wall Street Journal* stated:

> "If you had a stable conducting plastic, the number of uses you could ferret out would be astonishing."

Conducting polymers are light, so they are likely to be used in situations where saving weight is important. Suggest some of the situations in which a conducting plastic could be used to good effect.

Geometric isomerism

Molecules such as but-2-ene, $CH_3CH{=}CHCH_3$, exist in two different forms. The two forms look like this:

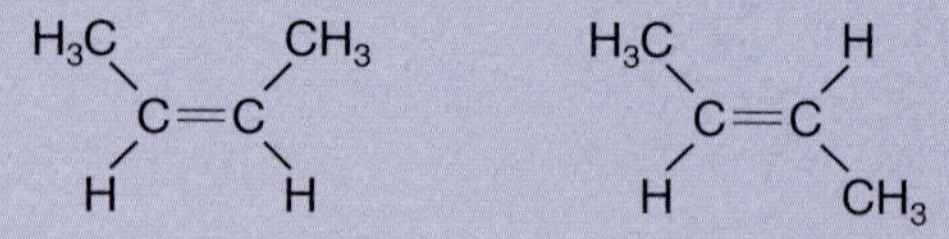

In order to turn the second form into the first form you would have to spin one end of the molecule round in relation to the other end. This cannot be done without breaking one of the bonds in the double bond and so the two forms are two different chemicals. They have different shapes and different properties.

We need names for the two different forms. The one with both methyl groups on the same side of the double bond is called *cis*-but-2-ene, and the one with the methyl groups on opposite sides of the double bond is called *trans*-but-2-ene. *Cis* and *trans* isomers will occur whenever there are C=C bonds that have one group substituted on each carbon atom.

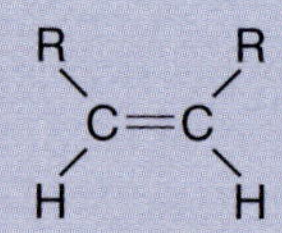

cis form

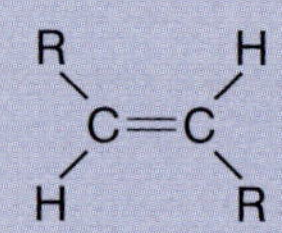

trans form

Before you go any further you will need to have some understanding of two areas of chemistry which may be new to you. **Geometric isomerism** (see box) is a phenomenon which arises in alkenes and a number of other types of compound. **Alkynes** are a homologous series of hydrocarbons in which there is a triple carbon to carbon bond. Ethyne, the simplest alkyne, has the structure H—C≡C—H.

You can read about alkynes in **Section 17.4**.

When molecules of ethyne polymerise, the product, poly(ethyne), is a molecule which contains alternating single and double bonds. There is more than one form of this polymer because geometric isomerism causes there to be two ways in which the double bonds can be arranged relative to one another. The structures of two forms of poly(ethyne) are shown in Figure 11; however, of course, mixed *cis* and *trans* arrangements are also possible.

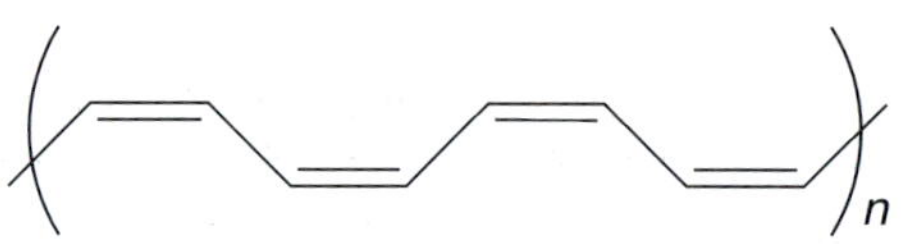

cis-poly(ethyne)

red

unstable; converts to the more stable *trans* form when warmed

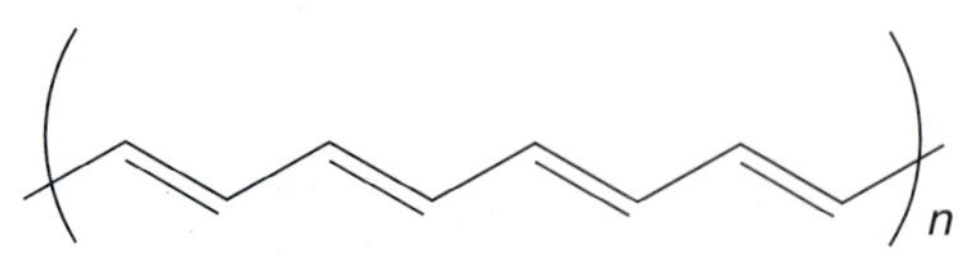

trans-poly(ethyne)

blue

the stable form of the polymer

Figure 11 The structures of two forms of poly(ethyne)

Research chemists in Natta's laboratory first made a mixture of these two forms in 1955. In 1971 Hideki Shirakawa and Sakuji Ikeda at the Tokyo Institute of Technology found that, when they directed a stream of ethyne onto the surface of a solution of a Ziegler–Natta catalyst at –78 °C, only the red *cis*-poly(ethyne) formed.

One of Shirakawa's graduate students then repeated the process, but with catalyst smeared over the inside walls of a glass reaction vessel. A red film of the *cis* polymer again formed, but this time when the film warmed up it turned silver instead of blue. The student had made a mistake and used one-thousand-times too much catalyst. Once again, a mistake was going to lead to the development of a new polymer.

The graduate student didn't realise the significance of his discovery. But Alan MacDiarmid did when he visited Shirakawa's laboratory in 1976. He had been trying to make metal substitutes at the University of Pennsylvania. The silvery film became a target for investigation.

If you look at the structure of poly(ethyne) you will see that from one end of the molecule to the other there are alternating single and double bonds. In systems like this it is best to regard the electrons in the double bonds as *delocalised* – spread out evenly along the molecule – rather than localised between adjacent carbon atoms.

Delocalised bonding alone does not make a polymer conduct. Before it can conduct, the polymer needs to have electrons added or removed. Electrons can then move along the polymer chain in the delocalised system, like a current flowing in a wire. Adding iodine is a way of removing some of the electrons from poly(ethyne). The method is called 'doping'.

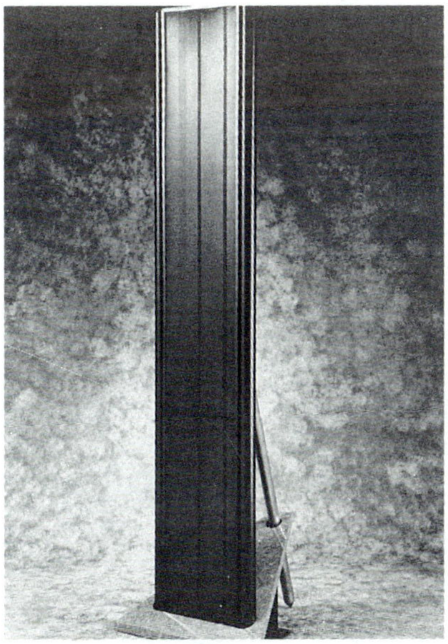

Figure 12 A conducting polymer is used as the membrane in this high-performance loudspeaker

Success came a year later. When the silvery material is 'doped' with iodine, a metallic golden sheet forms which conducts electricity 1 000 000 000 times better than previous forms of poly(ethyne).

5.5 *The invention of nylon*

Wallace Carothers

Carothers joined the US chemical company Du Pont in 1928. He led a team investigating the production of polymers that might be used as fibres. This was around the time when scientists were beginning to understand more about the structure of polymers, so Carothers had a scientific basis for his work.

Before you read this section, it will be helpful for you to find out how to recognise and name carboxylic acids, by reading **Sections 18.3 and 18.4**.

It was already known that wool and silk have protein structures and are polymers involving the linkage —CONH—. Chemists had also begun to discover that many natural fibres are composed of molecules which are very long and narrow – like the fibres themselves. Figure 13 shows part of a protein chain in a silk fibre. You will study the structure of proteins in more detail in **Section 5.9**.

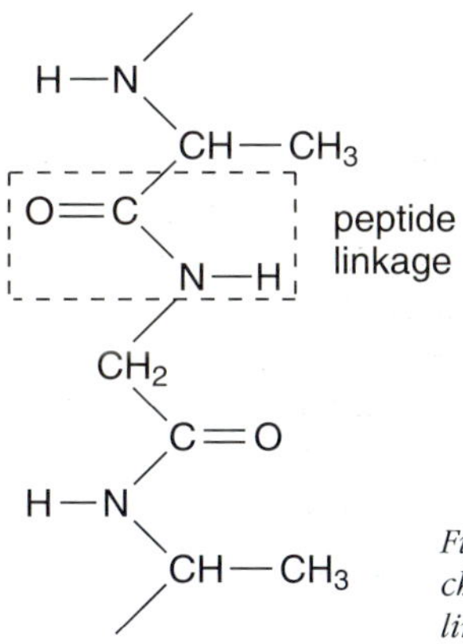

Figure 13 Part of a protein chain in silk, showing the linkage between monomers

Carothers did not make his discoveries by accident; he set about systematically trying to create new polymers. In one series of experiments he decided to try to make synthetic polymers in which the polymer molecules were built up in a similar way to the protein chains in silk and wool. Instead of using amino acids (the starting materials for proteins), Carothers began with amines and carboxylic acids.

Amines are organic compounds which contain the $—NH_2$ functional group. When an $—NH_2$ group reacts with the —COOH group in a carboxylic acid, an **amide** group —CONH— is formed. In the process, a molecule of water is eliminated. Reactions of this type are known as **condensation reactions**.

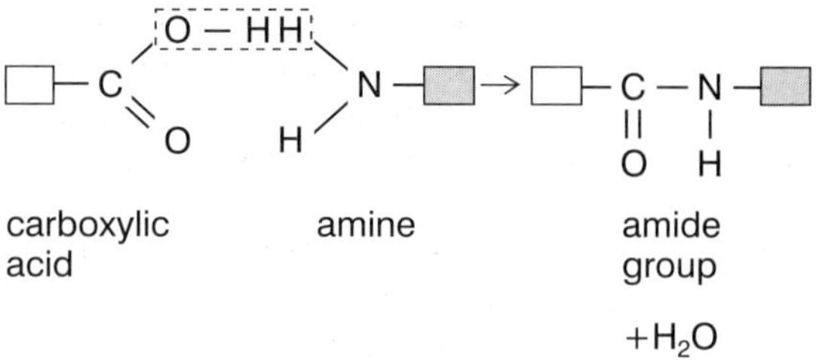

Carothers used *di*amines and *di*carboxylic acids which contained reactive groups in *two* places in their molecules, so they could link together to form a chain. In this way he was able to make polymers in which monomer units were linked together by amide groups. The process is called **condensation polymerisation** because the individual steps are condensation reactions.

Condensation polymerisation is also discussed in **Section 19.3**. You can find out about amines and amides in **Section 18.6**.

Figure 14 Nylon replaced silk as the fibre used for parachutes in the Second World War

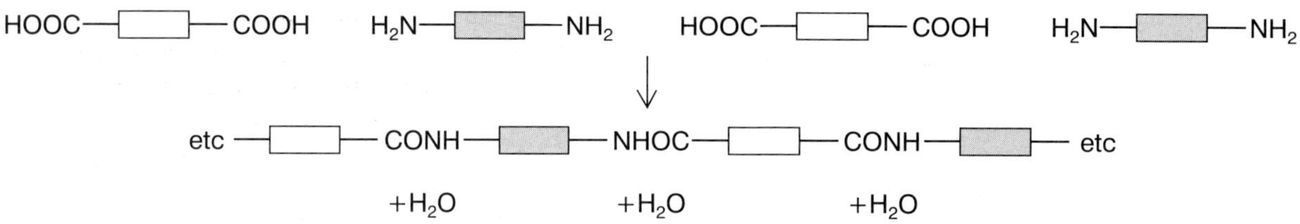

Because the group linking the monomer groups together is an amide group, these polymers are called **polyamides**. More usually, though, they are known as **nylons**.

Carothers discovered nylon in the spring of 1935. In 1938 the first product using nylon, 'Dr West's Miracle Toothbrush', appeared. Nylon stockings were seen for the first time in 1939. Most of the nylon produced at this time was used in place of silk for parachute material, so nylon stockings did not become generally available in Britain until the end of the Second World War.

Sadly, Wallace Carothers was not to see the development of his invention. He had been troubled with periods of mental depression since his youth. Despite his success with nylon, and other inventions such as neoprene (the first commercially successful synthetic rubber), he felt that he had not accomplished much and had run out of new ideas.

His unhappiness was compounded by the death of his favourite sister, and on 29 April 1937 he checked into a Philadelphia hotel room and died after drinking a cocktail of lemon juice laced with potassium cyanide.

He was 41. His daughter Jane was born 7 months later.

Plastic machine parts

You know that nylon is widely used as a fibre, but it is also a very important **engineering plastic** – a material which can be used in place of a metal in things like machine parts (Figure 15).

Its usefulness arises from its excellent combination of strength, toughness, rigidity and abrasion resistance, as well as its chemical unreactivity in many environments. As an engineering material it is far superior to poly(ethene) or poly(propene). The polymer chains in nylon need only be about half as long

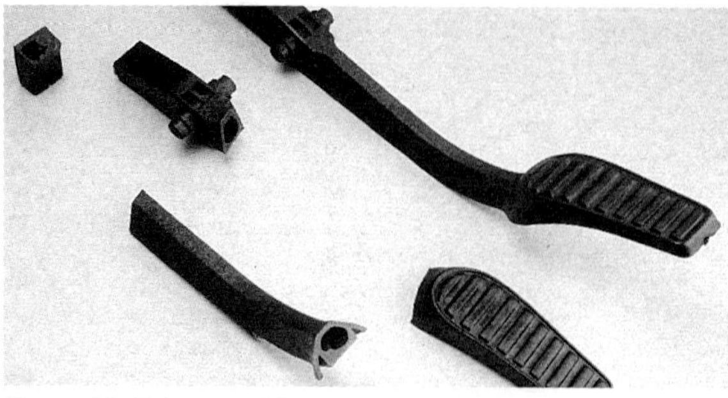

Figure 15 Nylon machine parts

as hdpe chains to show the same strength. The more powerful intermolecular forces which act between nylon chains are the source of this increased strength.

In **Unit 1, Section 16.4** you studied hydrogen bonds, the intermolecular forces responsible for nylon's strength.

In and out of fashion

Towards the end of the 1970s, people's tastes were moving away from nylon clothes and back towards the look and softer feel of natural fibres.

One of the problems with nylon fibres is that they are *hydrophobic* – they repel water. The nylon fabrics produced did not absorb moisture, and did not allow water vapour to escape through the weave. This made them rather sweaty and uncomfortable to wear (although, of course, the hydrophobicity of nylon could be used to advantage in such clothes as swimsuits and waterproof cagoules).

Chemical companies were facing a big downturn in the demand for their nylon, bringing major financial losses, but the high cost of developing a completely new polymer on the scale necessary to replace nylon was too high. To make things worse, the machines which had been specifically designed for making nylon fabrics could not be used for cotton or other natural fibres.

ICI's answer to this problem was to redevelop their nylon so that it bore a much closer resemblance to natural fibres. The first steps were to slim down the thickness of the nylon filament to the equivalent of cotton, and to add a delustrant to the fibre to reduce

Figure 16 Tactel fabrics 'breathe' but remain waterproof

the shiny appearance. The major breakthrough came when they developed a process for changing the shape and texture of the nylon yarn. They were able to create a large number of loops along the nylon filaments by blowing bundles of them apart with high-pressure air.

The new fibre was called *Tactel*. When the fibre is woven, the loops give the material a softness and texture similar to cotton.

Further refinement of the Tactel family of yarns has led to fabrics which are waterproof but which 'breathe'. For example, very fine yarns have been developed which allow water *vapour* to escape through the weave but do not allow *liquid* water in. Material made from these yarns is ideal for lightweight raincoats and ski-wear (Figure 16).

The solution to the problem of falling demand for nylon was not chemical – it did not involve the creation of a new polymer to suit the new fashion – but *technological*. It involved discovering new ways of handling the existing polymer to produce materials of the type people wanted.

5.6 *Polyesters*

The development of **polyesters** follows a similar story. A brilliant idea of UK chemist Rex Whinfield, in the 1940s, led to an incredible journey, developing materials undreamt of by their originators.

Whinfield knew that Carothers had tried to produce polyesters, but had not used 1,4-benzenedicarboxylic acid (terephthalic acid) and ethane-1,2-diol which Whinfield felt would give the necessary properties to the polyester. The structure of the polyester formed is

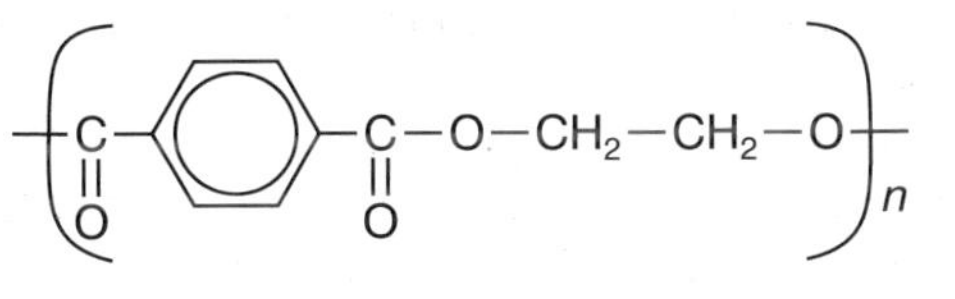

The 'old' name for this polyester is polyethyleneterephthalate, often known as PET.

You can read about the formation and reactions of esters in **Section 18.5**. You will need to know about alcohols and carboxylic acids to do this. You will find this information in **Sections 18.1** and **18.4**.

You can study the formation of polyesters in **Section 19.3**.

The polyester is produced as small granules, which are melted and squeezed through fine holes. The resulting filaments are spun to form a fibre (Figure 17a).

The fibre, known as Terylene and Dacron, is widely used in clothes, such as suits, shirts and skirts, and for filling anoraks and duvets, as it gives good heat insulation. The polyester can also be made into film for audio tapes and X-ray films (Figure 17b).

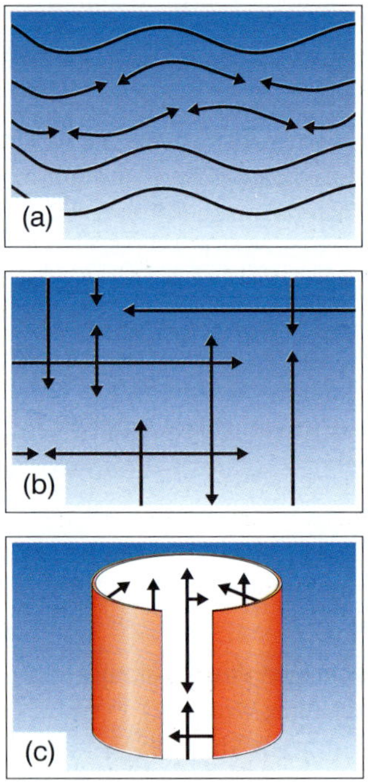

Figure 17 (a) Yarn: the molecules are mostly in one dimension (Terylene and Dacron). (b) Film: the molecules are in two dimensions. (c) For bottles: the polyester molecules are aligned in three dimensions

A newer use is for packaging. The granules of the polyester are heated to about 240 °C and further polymerisation takes place – a process known as *curing*. When the polymer is then stretched and moulded, the molecules are orientated in three dimensions. The plastic has great strength (Figure 17c) and is impermeable to gases. No wonder PET is widely used for bottling carbonated drinks.

5.7 Kevlar

The first aramids

After the invention of nylon, chemists began to make sense of the relationship between a polymer's structure and its properties. They were able to predict strengths for particular structures, and research was directed at inventing a 'super fibre'. In the early 1960s, Du Pont were looking for a fibre with the 'heat resistance of asbestos and the stiffness of glass'.

The aromatic polyamides seemed promising candidates: the planar aromatic rings should result in rigid polymer chains and, because the ratio of carbon to hydrogen is high, they require relatively large concentrations of oxygen before they burn.

You can read about aromatic hydrocarbons in **Section 17.5**.

The first polymeric aromatic amide – an **aramid** – was made from 3-aminobenzoic acid. The polymer could be made into fibres and was fire-resistant, but it was not particularly strong. The zig-zag nature of the chains prevented the molecules from aligning themselves properly.

A polymer was needed which had straighter chains, and which could be made from readily available and reasonably cheap starting materials. An example is shown in Figure 18.

This substance turned out to have all the right properties except one. The problem was its insolubility, which made it precipitate out of solution before long polymer chains had been able to grow.

The only suitable solvent seemed to be concentrated sulphuric acid. The company engineers were not impressed! However, the remarkable properties of the 'super fibre' were enough to encourage investment in a plant which uses concentrated sulphuric acid as a solvent. The polymeric material produced was called **Kevlar**.

ASSIGNMENT 3

a Draw out the structures of the two monomers needed to make Kevlar.

b What *type* of reaction is involved in the formation of Kevlar?

c The intermolecular forces in Kevlar are disrupted by concentrated sulphuric acid. That is why it dissolves. How do you think the forces are affected?

d Kevlar fibres are produced by squirting the solution in concentrated sulphuric acid into water. Suggest why the polymer precipitates out when the solution is diluted in this way.

Figure 18 The molecular structure of Kevlar

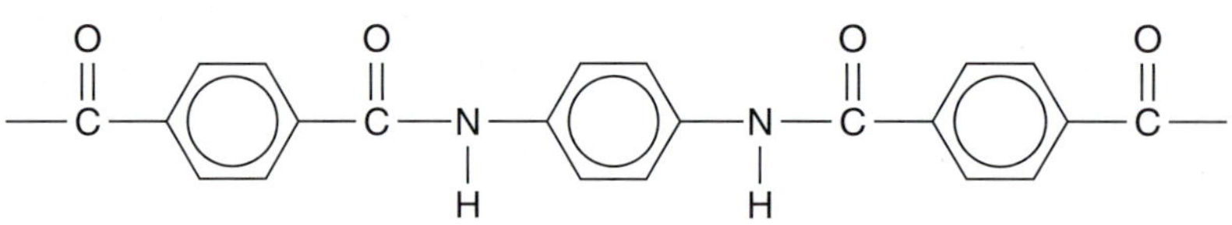

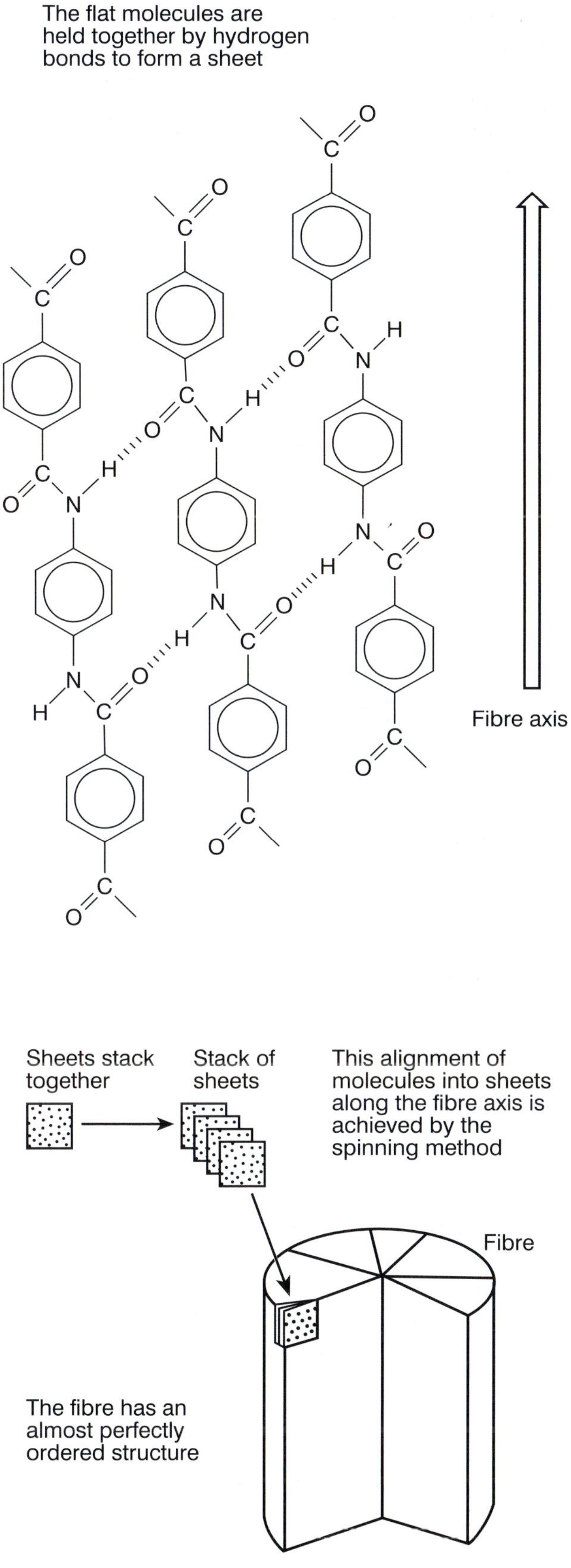

Figure 19 An illustration of the ordered structure of Kevlar

Why is Kevlar so strong?

Kevlar is a fibre which is fire-resistant, extremely strong and flexible. It also has a low density because it is made from light atoms: carbon, hydrogen, oxygen and nitrogen. Weight for weight, Kevlar is around five times stronger than steel! One of its early uses was to replace steel cords in car tyres: the Kevlar tyres are lighter and last longer than steel-reinforced tyres.

Kevlar is strong because of the way the rigid, linear molecules are packed together. The chains line up parallel to one another, held together by hydrogen bonds. This leads to sheets of molecules. The sheets then stack together regularly around the fibre axis to give an almost perfectly ordered structure. This is illustrated in Figure 19.

You studied hydrogen bonds in **Unit 1**, **Section 16.4**.

The important thing to realise about Kevlar is that it adopts this ordered structure because of the way the polymer is processed to produce the fibre. And this is a consequence of the work put in by the Du Pont team.

Developing a market for Kevlar

A full-scale commercial plant for the production of Kevlar required an investment of $400 million. It was therefore essential that there would be a market for the product. Although Kevlar has some remarkable properties it could not simply replace existing materials. Detailed market development had to take place alongside technical development. In other words, new uses had to be found for the new polymer.

Figure 20 Leading racing motorcyclists wear protective suits containing Kevlar (the small photo shows Kevlar fibres). The polymer's abrasion resistance is better than leather, the traditional material

Many uses have now been developed in addition to replacing steel in tyres. Kevlar ropes have 20 times the strength of steel ropes of the same weight. They last longer too. A stiffer form of Kevlar is used in aircraft wings, where its strength combined with its low density is important. And it's ideal for making bullet-proof vests and jackets for fencers.

5.8 *Throwing it away ... or not?*

Is there a problem?

Figure 21 shows some average figures for solid household waste in EC countries. Quantities are given as percentages by mass of total waste.

Much of our waste is disposed of by dumping in landfill sites, but sites are getting harder to find and waste disposal is becoming more expensive. Alternatives to dumping, such as recycling or incineration, and the use of degradable plastics, are beginning to look more satisfactory and more economical.

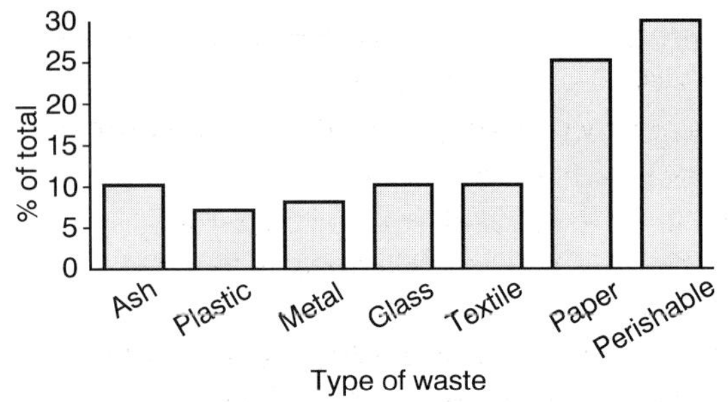

Figure 21 Types of household waste in EC countries

ASSIGNMENT 4

a Suggest why plastic waste is seen as a problem when it accounts for only about 7% of EC solid household waste by mass.

b The cost of disposing of solid waste in Tokyo is about £200 per tonne; in New York City it is about £80 per tonne. In a typical non-urban UK site the cost is around £10 per tonne. Suggest why it costs so much to get rid of waste in these two cities.

Degradable plastics

Most plastics are not degradable because decomposer organisms (bacteria and fungi) do not have the enzymes needed to break them down. There are two important categories of degradable plastic.

1 **Biopolymers.** These are polymers made by living organisms, which decomposers are able to break down.
2 **Synthetic plastics which are designed to break down naturally**. Three types are commonly used:

- polymers which break down in sunlight – **photodegradable** plastics
- **synthetic biodegradable** plastics which are broken down by bacteria
- **dissolving plastics** which dissolve in water.

Biopolymers

Poly(hydroxybutanoate), or PHB, is a natural polyester made by certain bacteria and used by them as a source of energy. Micro-organisms found in the soil, and in streams and the sea, are able to break the polymer down. PHB degradation in the environment is usually complete within 9 months. But there is a price to pay: PHB is about 15 times more expensive than poly(ethene) and production is being stopped. This also reflects a change in attitude from biodegradability to recyclability. The presence of biodegradable material in plastic waste makes it unsuitable for recycling.

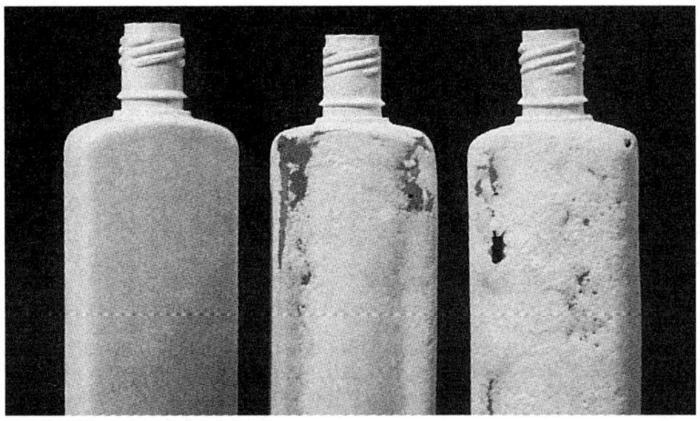

Figure 22 Bottles made from biodegradable plastic, in various stages of degradation

Photodegradability

Carbonyl groups (C=O) absorb sunlight in the near ultra-violet region of the spectrum. These groups can be incorporated into polymer chains to act as energy trappers. The trapped energy causes bonds to break in the neighbourhood of the carbonyl group, and the polymer chain breaks down into short fragments which can then biodegrade.

ASSIGNMENT 5

a Suggest some suitable uses for photodegradable polymers.

b Suggest some situations where their use would be unsuitable.

c What happens to a photodegradable polymer if it is disposed of by landfill?

Synthetic biodegradable plastics

Some plastic bags are made of poly(ethene) which has starch granules encapsulated in it. The starch is digested by micro-organisms in the soil when the plastic bag is buried. The bag then breaks up into very small pieces of leftover poly(ethene) which have a large surface area and which therefore biodegrade more quickly.

Dissolving plastics

If soiled laundry from a hospital is mishandled, there is a risk of infection. The risk can be avoided by making the laundry bags out of a dissolving plastic. The dirty linen is safely contained until the bag is placed in the wash – then the bag dissolves and the washing is let out.

The dissolving plastic which is used is **poly(ethenol)**. It is a new synthetic polymer, made from another polymer, poly(ethenyl ethanoate), by the process illustrated in Figure 23.

The extent of reaction can be controlled by adjusting either the temperature or the reaction time. The plastic's solubility depends on the percentage of ester groups on the polymer chain which have been converted to —OH groups. Table 1 shows how the two are related. Different solubilities give the plastic different uses.

% of OH groups	Solubility in water
100–99	insoluble
99–97	soluble in hot water
96–90	soluble in warm water
below 90	soluble in cold water

Table 1 Solubility of poly(ethenol)

Dissolving polymers have another medical use – as sutures or stitches in surgery.

Many materials can be used for the suture thread. Once the surgeon has chosen to use soluble thread, a decision is made about how quickly it should dissolve and the appropriate poly(ethenol) is selected.

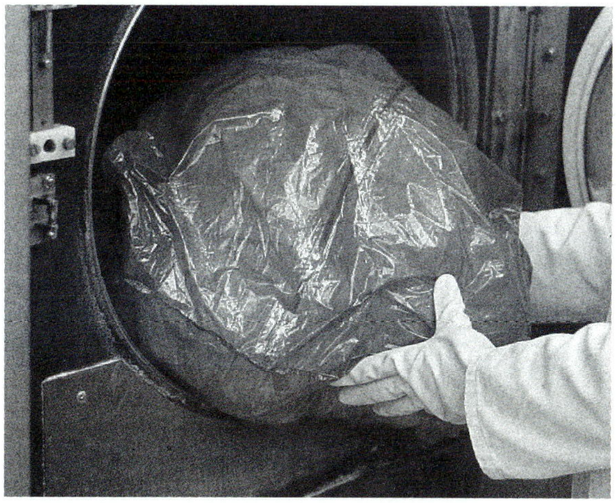

Figure 24 This hospital laundry bag has a section made of dissolving plastic, which dissolves in the washing machine to release the dirty washing

ASSIGNMENT 6

You may find it helpful to look back to **Unit 1**, **Sections 16.1** and **16.4** before you answer these questions.

a What type of intermolecular bonding will there be between the chains of poly(ethenol)?

b Explain why nearly pure poly(ethenol) is insoluble even in hot water.

c How is the intermolecular bonding in this polymer affected if ethanoate groups are still present?

d Explain the effect on solubility of increasing the number of ester groups in the polymer.

e Hospital laundry bags are made from the form of the polymer which is soluble in only hot water. Suggest why this form is chosen.

f Suggest some other uses for poly(ethenol) film.

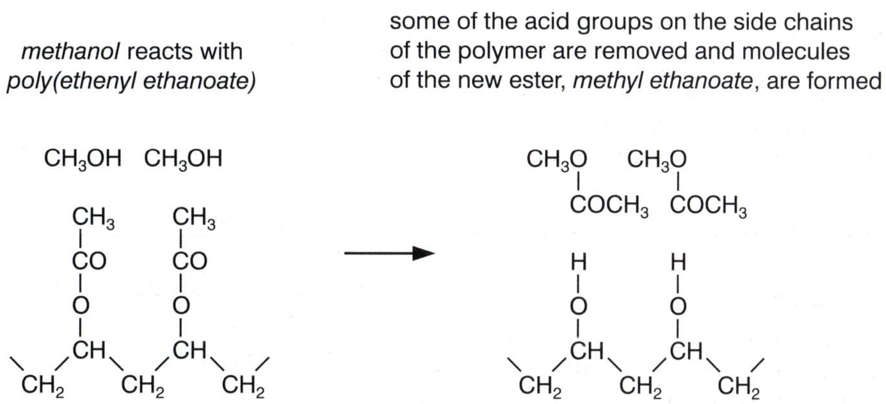

Figure 23 Formation of poly(ethenol) by ester exchange

Degrade or recycle?

Making any object, including a plastic one, requires energy. Some of the energy is needed to make the material – in this case a polymer – and the rest is required for fabrication (turning the polymer into the finished article).

Table 2 takes some familiar objects as examples, and shows how the energy used in production is divided between making the plastic and fabricating the object.

Object	Energy used/%	
	Plastic	Fabrication
5-litre container (pvc)	85	15
5-litre container (hdpe)	90	10
5-litre container (lpde)	94	6
Meat tray (polystyrene)	83	17

Table 2 Energy used in the production of some everyday objects

Most of the energy is used to make the plastics, so from these data alone you would conclude that recycling was worthwhile from the point of view of saving energy.

The situation is not as simple as recycling glass. Most glass is of only one type and comes in only three colours. Even so, it is often only economical to recycle glass if the disposer goes to the bottle bank as part of a normal journey. It is not economical if a special journey is made.

How much does it cost to collect and sort the plastic waste? And what can recycled plastic be used for? There are many applications for which its use is inappropriate. For example, most people do not like the idea of having their food wrapped in recycled plastic.

However, there are 60 companies in the UK which, between them, recycle about 150 000 tonnes of polymers per year. Two-thirds of the recycled plastic comes from industrial waste, and most of this is 'pure' plastic, eg unplasticised pvc from the makers of window frames, and old milk crates and car battery cases which are made from poly(propene).

It is not yet economical to recycle mixed plastics from domestic waste. When it is, it would be nice if we could go back to today's plastic waste to sort and recycle that too. That will not be possible if we spread our plastics around in landfill sites with all our other waste. So perhaps we should bury plastic waste

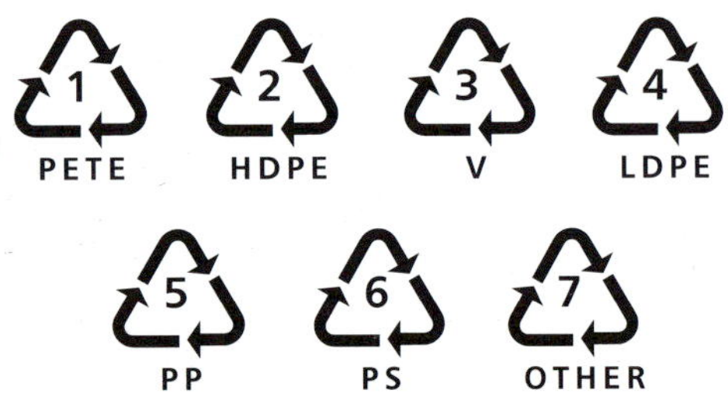

Figure 25 Recycling symbols help people to sort plastics into categories. These symbols are used throughout the US and are appearing on items within the EC

separately so that future generations have a resource which might be worth recovering.

ASSIGNMENT 7

a What problems might there be in operating a 'plastic bank' system for collecting household plastic waste, compared with a 'bottle bank'?

b Use the data given below for ldpe, hdpe and pvc to devise a scheme for separating these plastics. Present your solution in the form of a flow chart.

Polymer	Density/g cm^{-3}
ldpe	0.92
hdpe	0.96
pvc	1.40

As you have read this chapter you have learned about the reactions of many types of organic compounds. Reactions involving organic compounds rarely give *all* the product you would expect from the chemical equation (100% yield). **Section 18.8** shows you how to work out the percentage yield of a product.

5.9 *Proteins: natural polymers in action*

The final section of this chapter is about an important group of *natural* polymers called **proteins**. You have already met silk in **Section 5.5**. Silk and wool are both natural protein fibres. But proteins have many other functions – they are vital to all forms of life.

Christopher's story

Christopher is 11 years old. Just over 2 years ago he suddenly became very ill; he was taken into hospital where tests showed that he had developed *diabetes*. In

Figure 26 Christopher needs regular injections of insulin so that the cells in his body can absorb glucose

the UK, there are about 20 000 people under the age of 20 diagnosed with diabetes.

For some reason, possibly as a result of an earlier viral infection, Christopher's pancreas had stopped producing a *hormone:* **insulin**. Hormones are chemicals which regulate the rate at which we carry out our body activities. Insulin controls the uptake of glucose and some other sugars by our cells. Without insulin, Christopher's cells could not absorb glucose, and its concentration in his blood was building up. Untreated, he would have died within a few weeks.

Today, Christopher looks like any other 11-year-old boy. He goes to school; he rides his bike; he swims. In fact, he's very fit because he takes regular exercise and eats a healthy diet. However, regular insulin injections are essential.

More about insulin

Insulin is a hormone, and a **protein**. Many hormones are proteins (for example human growth hormone) but there are others, like the sex hormones, which are not.

The name *protein* (meaning 'first thing') was coined by Berzelius in 1838. He had little idea what proteins were, but he recognised their importance because they were so widespread in living things.

Proteins are big molecules with relative formula masses up to about 100 000. They play a key role in almost every structure and activity of a living organism. That's why they are regarded as among the most important constituents of our bodies (Figure 28).

Insulin was first isolated in 1921 by two young scientists, F G Banting and C H Best, working in Toronto. Only 1 year later it became available for the treatment of diabetes in the form of a preparation extracted from the pancreas of pigs or cows. Before then, diabetes was frequently fatal, and diabetics spent their numbered days living miserable lives on starvation diets which were designed to contain as little sugar or starch as possible. Vegetables which had been boiled three or four times and agar jelly (a protein) were standard fare.

Figure 28 Protein in the form of collagen is an important part of animal skins – leather is made of animal skin, so consists partly of protein

Injections of insulin have transformed the situation. Today we don't even need to use pig or beef insulin – synthetic human insulin is available from the biochemical industry. But the treatment still doesn't mimic natural insulin production as in non-diabetic people.

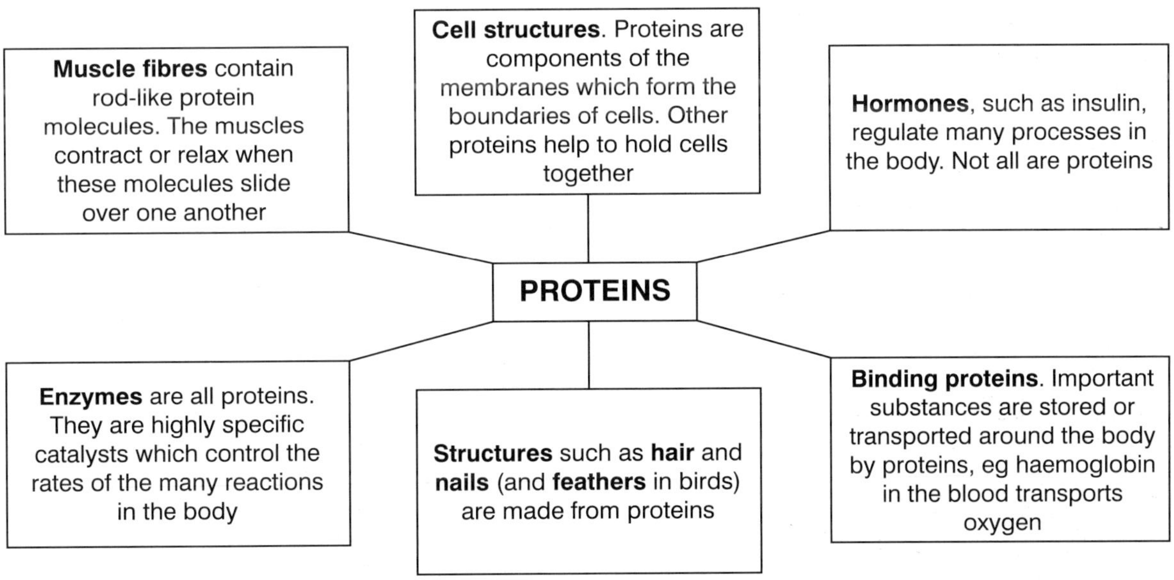

Figure 27 Proteins perform other functions in our bodies in addition to acting as hormones. ***Fibrous proteins*** *are the major structural materials.* ***Globular proteins*** *are involved in maintenance and regulation of processes, and include hormones and enzymes*

To help you understand what follows, you should read **Section 18.7** on amino acids.

You can remind yourself about the reactions of carboxylic acids by reading **Section 18.4**.

Splitting the six-pack

What happens to insulin levels when we eat? Figure 29 shows how the concentration of insulin in the blood of a non-diabetic person would be expected to behave after a meal.

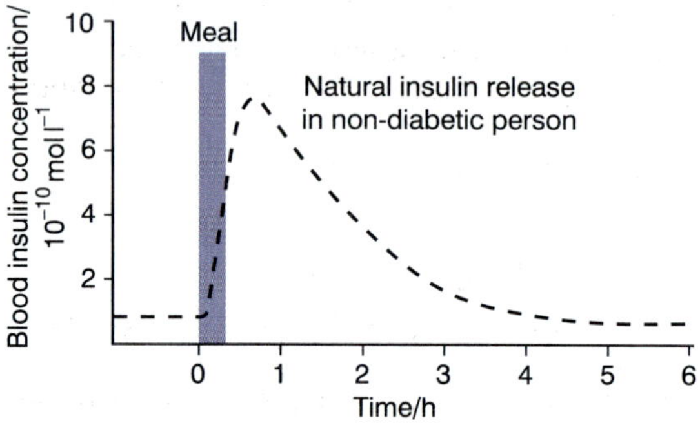

Figure 29 Insulin levels in the blood of a non-diabetic person

Notice four points about this graph:

- a concentration as low as 10^{-10} mol l^{-1} of insulin in the blood can have profound physiological effects
- there is a low-level, 'background' concentration of insulin in the body, even when food is not being digested
- insulin release follows rapidly after eating
- the insulin level peaks soon after eating and then falls off, eventually reaching the 'background' level.

We don't just make insulin when we eat: we make it all the time, and it is stored in special cells in the body. It is not stored in the form of individual molecules, but as *hexamers* – six molecules clustered together because of interactions between their surfaces.

When the concentration of glucose in the blood rises, insulin hexamers are released into the bloodstream. This makes the insulin solution very dilute, and the hexamers burst apart. The monomers are then quickly carried to where they are needed.

Christopher must wait for half an hour after his insulin injection before he eats his meal. Figure 30 shows how his blood insulin concentration might vary. (Also shown, for easy comparison, is the curve from Figure 29 for a non-diabetic person.)

Christopher doesn't have a store of insulin which special cells can release directly into his bloodstream. He has to inject hexamers just under his skin, and they

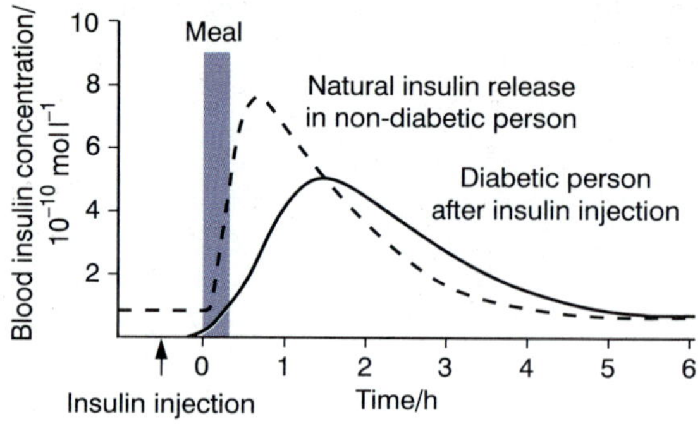

Figure 30 Insulin levels in the blood of a diabetic person following an insulin injection (injections also contain a slow-acting form of insulin which produces an effect for up to 12 hours; after that the insulin level falls to zero)

are too big to pass immediately through the capillary membranes into his blood. They have to spread out at the injection site and become diluted before they can break up into dimers and monomers which can diffuse away. This takes time. That's why Christopher has to inject his insulin half an hour before he starts to eat.

Christopher uses two bottles to prepare his dose because his injection is a mixture of two types of insulin. One has been prepared so that it breaks down easily and produces an effect close to the rapid response of normally produced insulin; the other breaks down slowly and gives him a low-level insulin concentration which is not permanent but which does last for up to 12 hours.

Christopher's injection is an improvement on the early preparations but, even so, the two curves in Figure 30 are clearly different. Because the insulin levels are different, diabetic people will still have unusual blood glucose levels, and this can lead to

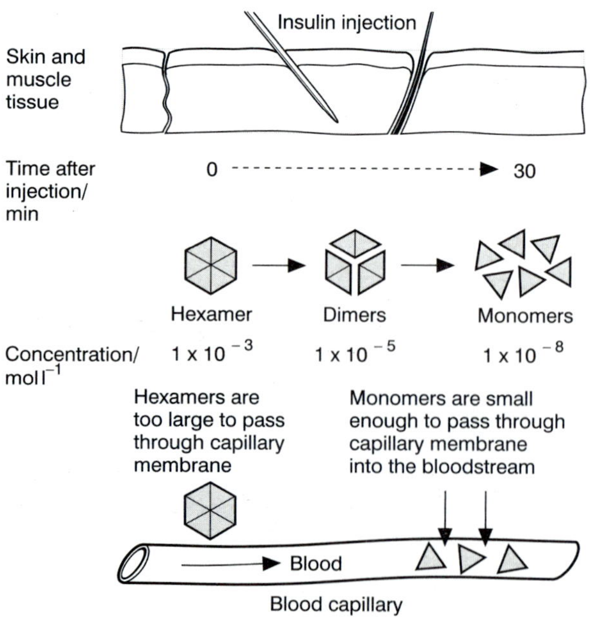

Figure 31 Injecting insulin

Figure 32 Christopher injects insulin hexamers into the tissues just below his skin; it takes about half an hour for the hexamers to diffuse apart, split into monomers and enter his bloodstream

complications and illness later in life. That's why Christopher also has to be so careful about his diet.

But the big difference, of course, is that normal insulin release switches on and off every time we eat – our insulin levels fit in with our food intake. It's the other way round for Christopher: his meals have to be planned around his levels of injected insulin.

To help Christopher further and make his injections mimic natural insulin release more closely, the half-hour time delay during which hexamers begin to break down needs to be eliminated. That would mean injecting insulin monomers which can get straight into the bloodstream: Christopher could then have injections along with his meal. Life would be closer to normal – and so would his blood-sugar levels.

Insulin monomers can be made. Clinical trials of the new form of the hormone began in 1987 and the results so far are encouraging.

Before monomeric insulin could be made, scientists needed a thorough understanding of the protein's structure and function. This required contributions over many years from many areas of science. In particular, scientists needed to know

- the composition and structure of insulin
- the shape of the monomer and the regions of the molecule which interact to form dimers and hexamers
- the types of intermolecular interaction which occur in the hexamer
- how to modify the monomer proteins to prevent them sticking together.

Modifying the structure of a protein is called **protein engineering**. One way of making and modifying proteins is to change the gene that codes for the production of the protein using **genetic engineering** – but first you need to know more about proteins and how they are made in the body.

The structure of insulin

Figure 33 illustrates the composition of a molecule of human insulin.

The abbreviations in circles represent the **α-amino acids** which have combined to form insulin. There are two short chains of these amino acid residues – the parts of the original molecules which are joined together to form the protein. This is called the **primary structure** of insulin. It is the *order* in which the amino acids are joined to one another.

See **Section 18.7** for information about amino acids.

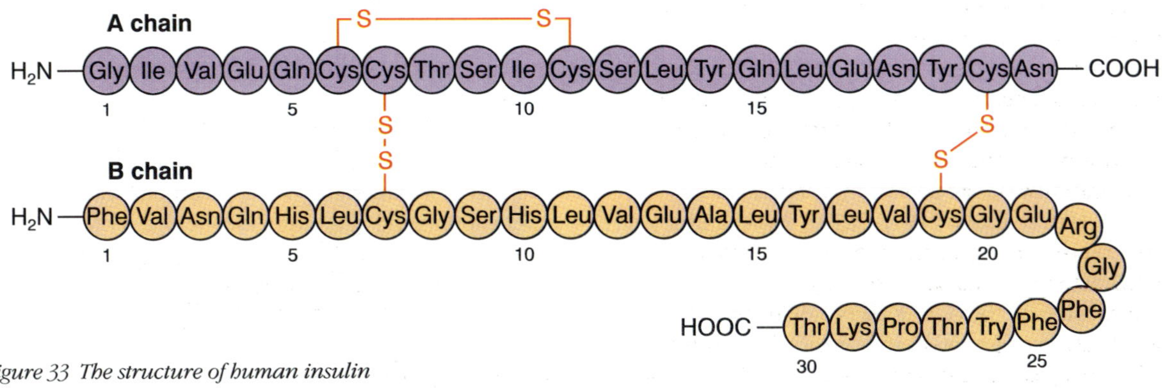

Figure 33 The structure of human insulin

Representing amino acid sequences

The dipeptide obtained by condensing the carboxylic acid group of glycine with the amino group of alanine has the structure

$$\mathrm{H_2N{-}\underset{\displaystyle H}{\overset{|}{CH}}{-}CO{-}NH{-}\underset{\displaystyle CH_3}{\overset{|}{CH}}{-}COOH}$$

This dipeptide would be abbreviated to **Gly Ala**. The convention of reading peptide groups in the direction with the free NH_2 group on the left is very important if the amino acid sequence is to be read unambiguously.

ASSIGNMENT 8

Read **Section 18.7** and then answer the following questions.

a Draw the structure of the dipeptide Ala Gly.

b In the tripeptide SerGlyAla, which amino acid has an unreacted

i NH_2 group?

ii COOH group?

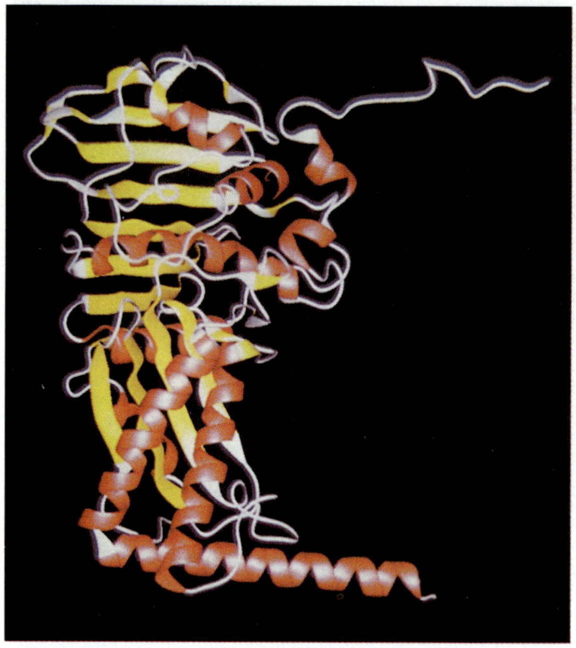

Figure 34 The tertiary structure of a protein; here the enzyme gyrase *is illustrated as a* ***ribbon diagram****. The helices (in red) and the sheets (in yellow) show up clearly in this representation. The thin white lines represent irregular regions. (Gyrase is involved in the DNA unwinding process)*

Folded chains

Beef, pig and human insulin all have different primary structures, but they are equally effective at treating diabetes. So scientists need to know about more than just primary structure before they understand how insulin works. One thing they now know is that insulin, like most proteins, has a precise shape which arises from the folding together of the chains. The action of insulin is critically dependent on this shape.

As long as different molecules fold to the same shape, they may have the same action. Chain folding gives proteins their three-dimensional shape; it also places chemical groupings in positions where they can interact most effectively.

Four types of interactions are important:

- **van der Waals forces** between non-polar side chains on amino acids like phenylalanine and leucine. The centres of protein molecules tend to contain amino acids like these so that the non-polar groups do not interfere with the hydrogen bonding between the surrounding water molecules.
- **hydrogen bonding** between polar side chains (eg $-CH_2OH$ in serine and $-CH_2CONH_2$ in asparagine), *and* between the peptide groups which link the chain together. If amino acids with these side chains are situated on the outside of proteins, hydrogen bonds can also form to water molecules surrounding the protein.
- **ionic attractions** between ionisable side chains, such as $-CH_2COO^-$ in aspartic acid and $-CH_2CH_2CH_2CH_2NH_3^+$ in lysine.
- **covalent bonding**. The –SH groups on neighbouring cysteine residues can be oxidised to form –S–S– links: look back at Figure 33. There are three such links in human insulin; two hold the A and B chains together.

The chains in a protein are often folded or twisted in a regular manner as a result of hydrogen bonding. Two arrangements of the protein chain are common:

- tightly *coiled* into a **helix** where the C=O group of one peptide link forms a hydrogen bond to an N–H group *four* peptide links along the chain
- stretched out into regions of *extended* chain, which lie alongside one another and hydrogen bond to form a **sheet**.

The two components of a protein's shape – helix and sheet – are sometimes referred to as its **secondary structure**. The chains may fold up further. The overall shape is stabilised by van der Waals forces, by hydrogen bonding, by ionic attractions and by covalent bonding. The overall shape of a protein is sometimes called its **tertiary structure**. Fibrous proteins are long and thin, whereas globular proteins have the spiral chains folded into a compact unit.

Compare the ribbon diagram of insulin in Figure 35 with Figure 33, which shows its primary structure.

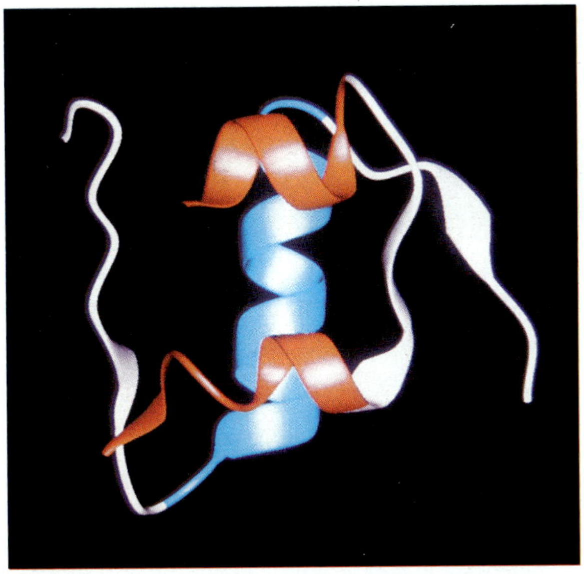

Figure 35 A ribbon diagram showing the two chains in insulin. There are two short helical sections (in red) in the A chain. The B chain contains a helical section (in blue) between amino acids 9 (Ser) and 19 (Cys)

You can read about van der Waals forces and hydrogen bonding in **Unit 1**, **Sections 16.1** and **16.4**.

Insulin monomers

Insulin monomers are better for diabetics than insulin hexamers because they work faster. So is there a way of injecting monomers rather than hexamers?

Once chemists had worked out the exact shape of human insulin (see box) and discovered how cells produce proteins, they were able to modify the structure of the monomers slightly in a process of genetic engineering to prevent them sticking together so readily. The modified insulin can be injected just before a meal so that people with diabetes do not have to be so careful about timing their injections and meals.

The modified insulin gets into the bloodstream more quickly than injected human insulin because it is monomeric. The hexamers needed time to breakdown into monomers. The release of injected insulin monomers more closely resembles natural insulin production.

Insulin hexamers

The same ideas about intermolecular interactions help to explain why insulin molecules stick together to form dimers and hexamers. The **space filling model** of the insulin monomer illustrated in Figure 36 shows that one side of its structure consists mainly of amino acids with non-polar side chains.

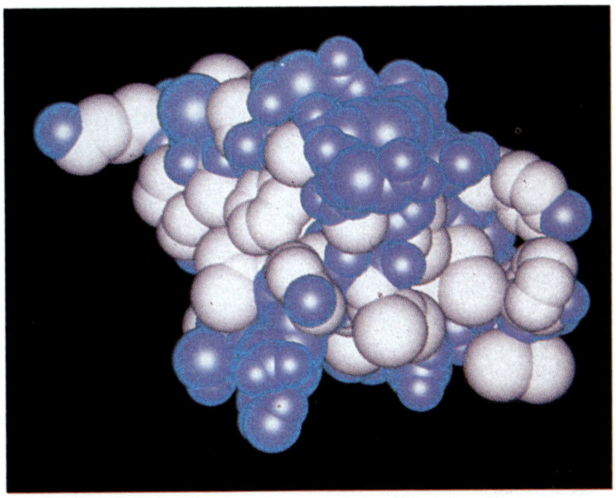

Figure 36 Space filling representation of an insulin monomer: the white spheres represent amino acid residues with non-polar side chains

Working out the structure of insulin

Insulin was recognised as a protein in 1928. Its primary structure was revealed in 1955 by Professor Fred Sanger and his team at Cambridge. They broke the protein into shorter bits by hydrolysis, then used chromatography to identify these bits. By piecing together the bits, they eventually worked out the amino acid sequence. Professor Sanger received a Nobel Prize for his work on the amino acid sequencing of proteins and a second Nobel Prize for more recent work on sequencing DNA.

The three-dimensional structure of insulin has been worked out from X-ray diffraction studies carried out by Professor Dorothy Hodgkin's group at Oxford. Professor Hodgkin received a Nobel Prize in 1964 for earlier work, particularly work on the structure of vitamin B_{12} and penicillin. At that time she was only the third woman, after Marie Curie and her daughter Irène Joliot-Curie, to receive a Nobel Prize.

Figure 37 Professor Dorothy Hodgkin with a model of insulin

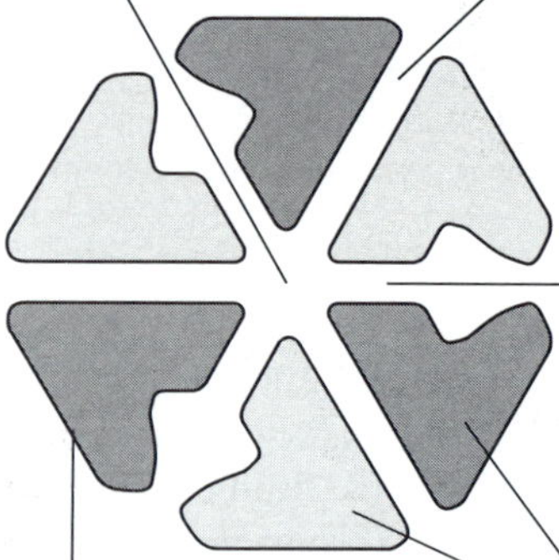

Figure 38 A diagram to illustrate an insulin hexamer

Left exposed to water, this non-polar side of the insulin monomer would disrupt water's hydrogen bonding. The non-polar sides of two insulin monomers come together when a dimer is formed, so the disruption is reduced. Figure 38 shows that a similar thing happens when hexamers form.

The arrangement of several protein sub-units into a bigger unit is sometimes referred to as **quaternary structure**.

So, for insulin, we have

- **primary structure**: the order of amino acid residues

Figure 39 A space filling diagram of the quaternary structure of an insulin hexamer: the white spheres represent amino acid residues with non-polar side chains

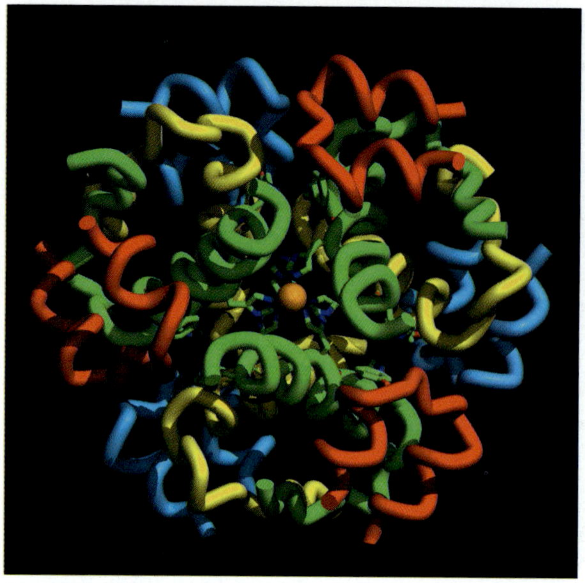

Figure 40 A ribbon diagram of the insulin hexamer

- **secondary structure**: coiling of parts of the chain into a helix or the formation of a region of sheet
- **tertiary structure**: folding of the chain into an insulin monomer
- **quaternary structure**: joining of monomers into a hexamer.

Figures 39 and 40 show a space filling diagram and a ribbon diagram of the insulin hexamer.

5.10 *Enzymes*

Testing glucose levels

There is a simple test you can do to check whether you are diabetic – although to be really certain you should go and see your GP. You can buy some reagent strips from a pharmacy and test for glucose in your urine. (When glucose builds up to a high level in the blood, it 'spills over' into the urine.)

The fresh reagent strip has a small pink square at one end which turns purple-blue in the presence of glucose.

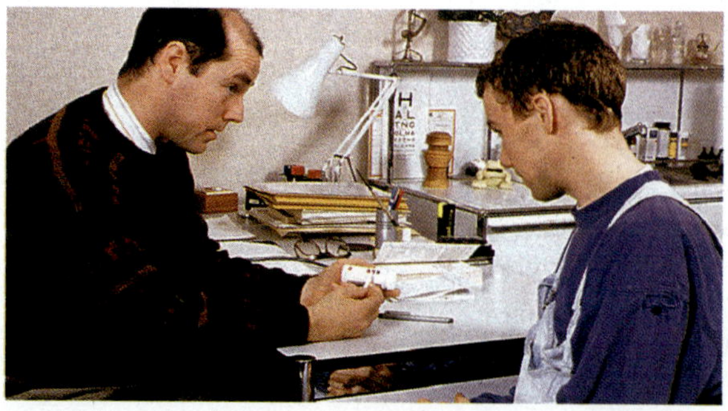

Figure 41 A simple urine test gives a measure of blood glucose concentration

The square is impregnated with four reagents:

- glucose oxidase, an enzyme which catalyses the reaction

  glucose + oxygen →
  gluconic acid + hydrogen peroxide

- an indicator (often orthotoluidine); this is present in its reduced form (XH_2) which is colourless, but it turns into a coloured form (X) when it is oxidised
- peroxidase, an enzyme which catalyses the oxidation of the indicator by hydrogen peroxide

  hydrogen peroxide + XH_2 → water + X

- a buffer: a mixture of chemicals which keeps the reagents at a fixed pH during the test.

If the manufacturer's instruction sheet is available, you will also see that it recommends storing the test strips below 30 °C but not in a refrigerator.

The test strips illustrate *four* important points about enzymes. Enzymes are

1. *catalysts*
2. *highly specific*: for example, the test strips only work with glucose – they give no response with other sugars
3. *sensitive to pH*: many work best at a particular pH and become inactive if the pH becomes too acidic or too alkaline
4. *sensitive to temperature*: many enzymes work best at temperatures close to body temperature; most are destroyed above 60 °C–70 °C.

The importance of shape

Enzymes are so *specific* because they have a precise tertiary structure which exactly matches the structure of the **substrate** – the molecule which is reacting.

The substrate molecule fits into the active site of the enzyme where reaction takes place. Within the active site are chemical groups – some of the side chains on the amino acid residues – which bind the substrate and possibly react with it.

The bonds which bind the substrate to the active site have to be weak so that the binding can be readily reversed when the products need to leave the active site after the reaction. The bonds are usually hydrogen bonds or interactions between ionic groups. The binding may cause other bonds within the substrate to weaken or it may alter the shape of the substrate so that atoms are brought into contact to help them to react.

If the enzyme's active site contains ionisable groups, the enzyme's action will be affected by a change in pH. For example, if there is a –COOH group which acts by donating an H^+ to the substrate, raising the pH will turn it into $–COO^-$ and the enzyme will not be able to work.

The enzyme will also become inactive if its shape is

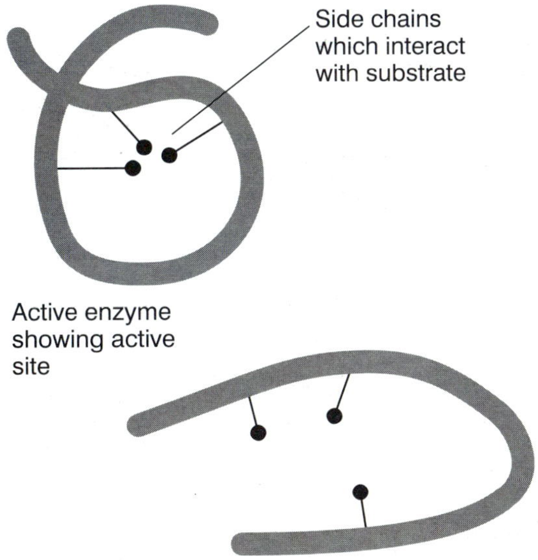

Figure 42 The active site is destroyed when an enzyme is denatured

destroyed. Figure 42 illustrates how the active site of an enzyme can be made up from the side chains of amino acids in different parts of the protein molecule. They are held close together by the enzyme's tertiary structure.

If the tertiary structure is broken, the enzyme loses its shape and the side chains are no longer close together. The active site is destroyed and the enzyme is said to be **denatured**.

The tertiary structure is held together by weak dipole–dipole bonds and hydrogen bonds. These can easily be broken by raising the temperature, which causes them to vibrate more vigorously. Ionic interactions holding the tertiary structure together can be broken by changing the pH.

So enzymes are *sensitive* to relatively small changes in *temperature* or *pH*.

You can read about the role of enzymes as catalysts in speeding up reactions in **Unit 1, Section 15.5**.

Figure 43 The proteins in egg white are denatured when the egg is cooked

5.11 *Summary*

This Chemical Storylines chapter is linked to Chemical Ideas Sections 17.4, 18.1, 18.3–18.8 and Chapter 19.

You have seen how our ideas about polymers and their structures have developed. As a result of this development, chemists have learned how to build up simple molecules into very large polymer chains. The polymers they have produced are new materials with unique sets of properties not possessed by any natural substances.

You have revisited earlier work on addition polymerisation and gone on to learn about condensation polymerisation, and how the properties of polymers depend on their structures and the intermolecular forces between the chains. Hydrogen bonds, dipole–dipole forces and van der Waals forces are important in this context.

Two classes of organic compounds – alkenes and alkynes – have been important to the development of this story. You have also learned about carboxylic acids, and looked at their reactions with amines to form amides (as in nylons) and with alcohols to form esters (as in polyesters). This led you to study oils and fats as examples of esters.

You then considered some environmental issues connected with the disposal and recycling of plastics.

You went on to look at one class of natural polymer, the proteins. It introduced you, through the story of insulin, to some of the things chemists have found out about proteins and their building blocks, amino acids. Proteins play a wide variety of roles in our bodies: they are molecules of almost infinite diversity, despite all being made by the same type of condensation reaction which results in the formation of a secondary amide or peptide group.

This diversity can be explained in terms of

- the sequencing of the different structural units of the 20 naturally occurring α-amino acids
- the interactions between these units which give proteins their precise three-dimensional shapes.

You ended with a more general look at one class of proteins – the enzymes – which control the rates at which many of the chemical reactions in our bodies take place. You learned why the shapes of enzymes are so important and why enzymes are so sensitive to changes in conditions.

6

OZONE DEPLETION IN THE STRATOSPHERE

6.1 *Screening the Sun*

The sunburn problem

Until the 1920s a suntan was something a white-skinned person could not avoid if they had to work outdoors in the Sun. Those who didn't, preferred to distinguish themselves by remaining pale. It was the clothes designer Coco Chanel who made sunbathing fashionable. She appeared with a golden tan after a cruise on the yacht belonging to the Duke of Westminster, who was one of the world's wealthiest men.

As we discover more about the effects of the Sun's radiation on the chemical bonds in living material, sunbathing seems less of a good idea.

Figure 1 The trend-setting clothes designer Coco Chanel, who set a new fashion for suntanned skin among white-skinned Europeans in the 1920s

ASSIGNMENT 1

a Look at the scatterplot in Figure 2. Suggest a reason for the variation of death rate from skin cancer in the US.

b Skin cancer is more common among office workers than among farmers. Suggest a reason why.

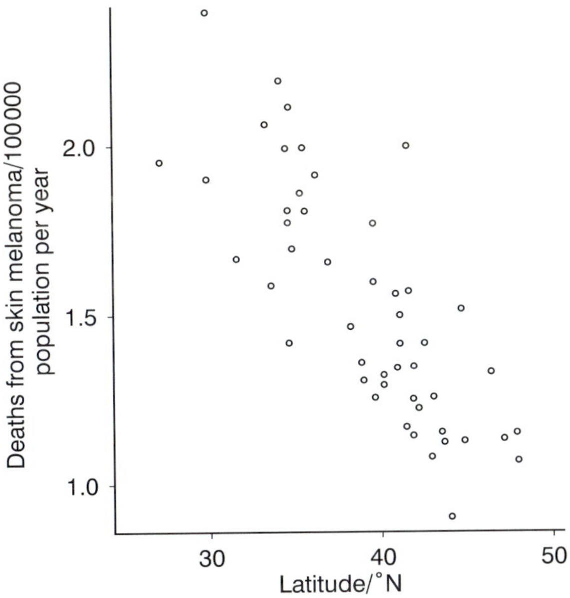

Figure 2 Deaths from skin melanoma in the US

The Sun radiates a wide spectrum of energy. Part of this spectrum corresponds to the energy required to break chemical bonds. Sunlight can therefore break bonds, including those in molecules such as DNA in living material. This can cause damage to genes and lead to skin cancer. On a less serious level, it can damage the proteins of the connective tissue beneath the skin, so that years of exposure to the Sun can make people look wrinkly and leathery. Even brief exposure to the Sun may cause irritation of the blood vessels in the skin, making it look red and sunburnt.

Chemical sunscreens

The diagram in Figure 3 shows the effect of different parts of the Sun's spectrum on the skin.

You can see that the most damaging region of this spectrum is in the ultra-violet. Fortunately, there are chemicals which absorb much of this radiation.

Have you ever wondered why people don't get sunburnt indoors? You can sit by a window for hours on a sunny day without burning. The glass in the window absorbs the damaging ultra-violet radiation, so it never reaches your skin. (Perspex does let through some ultra-violet, so it is possible to burn through Perspex.)

Chemists have found materials that do a similar job to glass, but can be spread onto the skin. They are called *sunscreens*, and tonnes of them are sold every summer.

But the best sunscreen of all is not made by chemists. It has always been with us. It is the atmosphere.

Why is the atmosphere such a good sunscreen?

Certain atmospheric gases absorb ultra-violet radiation strongly. They act as a global sunscreen, preventing

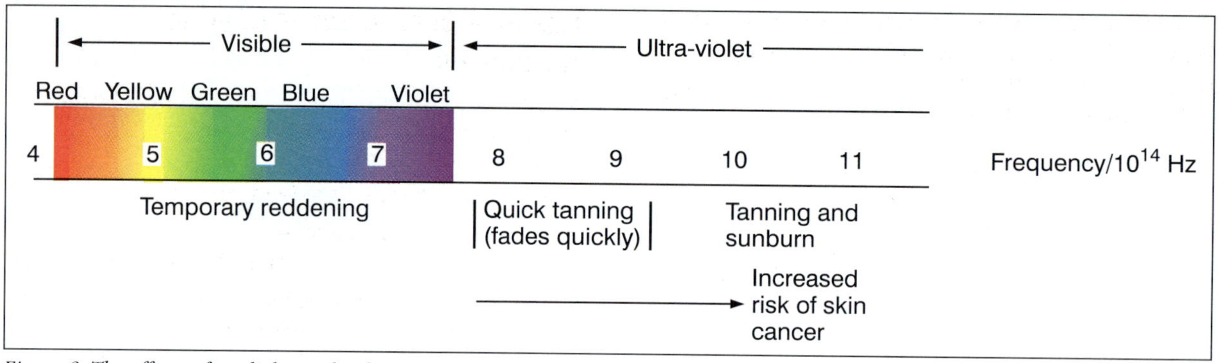

Figure 3 The effects of sunlight on the skin

Figure 4 A sunscreen's ability to protect skin is expressed as a 'Sun Protection Factor' or SPF number — SPFs indicate the time it will take for the Sun to produce a certain effect on your skin.

much of the Sun's harmful radiation from reaching the Earth.

Most of this absorption goes on in the upper part of the atmosphere, the **stratosphere**. Particularly important is the gas **ozone**, which absorbs ultra-violet radiation in the region which is most damaging to the skin. (Look back at Figure 3 which shows the effect on skin of radiation of different frequencies.) Much of the damaging ultra-violet radiation is absorbed by ozone in the stratosphere.

There is no life in the stratosphere, because the powerful ultra-violet radiation would break down the delicate molecules of living things. Indeed, even simple molecular substances get broken down in the stratosphere. Some of the covalent bonds break to give fragments of molecules (atoms or groups of atoms) called **radicals**.

6.2 *Ozone: a vital sunscreen*

Introducing ozone

Ozone is present in the atmosphere in only tiny amounts, dispersed among other atmospheric gases. If all the ozone in the atmosphere were collected and brought to the Earth's surface at atmospheric pressure, it would form a layer only 3 mm thick.

High up, in the stratosphere, ozone *protects* us by absorbing harmful ultra-violet radiation; however, lower down, in the troposphere, it can be a real nuisance.

In **Unit 1**, **Section 2.3** you read about the formation of ozone in photochemical smogs. Ozone is a very reactive gas and is a powerful oxidising agent. It reacts with many synthetic materials such as plastics, paints and dyes, causing, for example, rubber car tyres to crack.

It isn't really surprising that there is so little ozone in the atmosphere. It reacts so quickly with other substances and gets destroyed. In fact, we might ask why the ozone in the atmosphere hasn't run out. Some reactions must be producing it too.

How is ozone formed in the atmosphere?

Ozone is formed when an oxygen atom (an example of a radical) reacts with a dioxygen molecule:

$$O + O_2 \rightarrow \underset{\textit{ozone}}{O_3}$$

One way to make oxygen atoms is by splitting up dioxygen molecules. This requires quite a lot of energy. In this case the energy can be provided by ultra-violet radiation, or by an electric discharge.

As soon as oxygen atoms have been produced they react with the dioxygen molecules which are always present in the air. You can often smell the sharp odour of ozone near electric motors or photocopiers. The electric discharges happening inside the machine make some of the dioxygen molecules in the air split up into atoms. Have you ever noticed the ultra-violet lamps used to kill bacteria in food shops? You can often smell ozone near them too.

Some of the ozone in the troposphere is formed in the complex series of reactions taking place in photochemical smogs. These develop in bright sunlight over large cities which are heavily polluted by motor vehicle exhaust fumes. In this case, oxygen atoms are produced by the action of sunlight on the pollutant gas, nitrogen dioxide.

In the stratosphere, oxygen atoms are formed when dioxygen absorbs ultra-violet radiation of the right frequency to break the covalent bond.

The reaction can be summarised as:

$$O_2 \xrightarrow[\text{radiation}]{\text{u.v.}} O + O \quad \text{(reaction 1)}$$

The oxygen atoms produced can do a number of things when they meet another particle and collide with it. The least interesting of these is that the particles just bounce apart again. But even when the oxygen atom collides with a particle it can react with, not every collision results in a reaction.

Unit 1, **Chapter 15** discusses rates of reactions.

More interesting is when the oxygen atom sticks onto the particle it collides with. This could be one of three things: O_2, another O or O_3. The three possible outcomes can be represented by reactions 2–4 below:

$$O + O_2 \rightarrow O_3 \quad \text{(reaction 2)}$$
$$O + O \rightarrow O_2 \quad \text{(reaction 3)}$$
$$O + O_3 \rightarrow O_2 + O_2 \quad \text{(reaction 4)}$$

Reaction 2 is of course the one which produces ozone.

When the ozone absorbs radiation of the right frequency, some molecules split up again:

$$O_3 \xrightarrow[\text{radiation}]{\text{u.v.}} O_2 + O \quad \text{(reaction 5)}$$

It is reaction 5 which is responsible for the vital screening effect of ozone, since it absorbs the radiation which is responsible for sunburn.

ASSIGNMENT 2

Look at reactions 1–5 above.

a Which reaction or reactions *remove* ozone from the atmosphere?

b Which reaction or reactions absorb ultra-violet radiation?

c What relationship is there between

i reactions 1 and 3?

ii reactions 2 and 5?

Reactions 1 and 5 are examples of **photochemical reactions**. You met other examples of photochemical reactions in **Unit 1**, **Section 15.4**.

Ozone – here today and gone tomorrow

You can see from reactions 1–5 that ozone is being made and destroyed all the time. Left to themselves, these reactions would reach a point where ozone was being made as fast as it was being used up:

rate of producing ozone = rate of destroying ozone

At this point, the concentration of ozone would remain constant. This is called a **steady state**.

It's like the situation in Figure 5 when you are running water into a basin with the plug out of the waste pipe. Before long you get to the point where water is running out as fast as it's running in, and the level of water in the basin stays constant. If you turned the tap on more, the level of the water would rise – but that would make the water run out faster, because of the higher pressure. Before long you would get to a steady state again, but this time with more water in the basin. What would happen if you made the waste pipe larger?

The water coming out of the tap is like the reactions producing ozone, and the water going down the waste pipe is like the reactions that destroy it.

To estimate the concentration of ozone in the stratosphere, you need to know the **rates** of the reactions that produce and destroy ozone.

Chemists have studied these reactions in the laboratory and are able to write mathematical equations giving the rates of all the reactions involved in producing and destroying ozone. So, for example, taking reactions 1–5, the rate of producing ozone will be the rate of reaction 2. If there is a steady state, this will be balanced by the rate of destruction of ozone: the rate of reaction 4 *plus* that of reaction 5. From these relations chemists can work out what the concentration of ozone *should* be at different altitudes, at different times of day and at different times during the year.

However, when chemists compared their *calculated* concentrations of ozone with *measured* values, they found that the actual concentration of ozone was in fact a good deal *less* than expected.

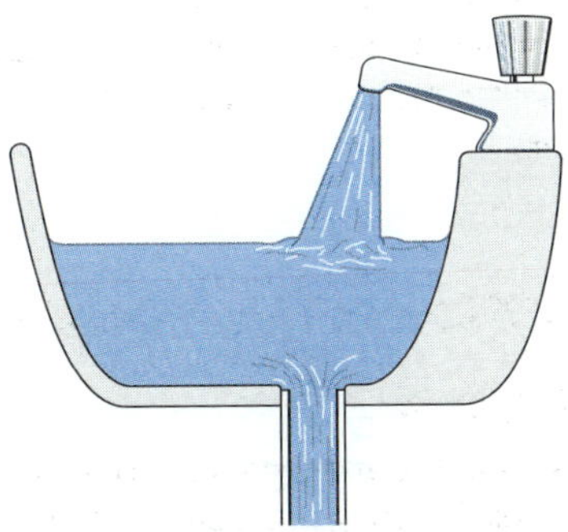

Figure 5 One example of a steady state situation

ASSIGNMENT 3

A number of factors can affect the rate of ozone production and destruction. Remembering that the reactions are taking place in the gas phase and in strong sunlight, suggest *three* factors which could affect the rates of these reactions. (You may find it helpful to look back to **Unit 1**, **Chapter 15**.)

Explain how altering each factor would affect the rate of reaction.

This suggests that the ozone is being removed faster than expected. Going back to the analogy of the basin and the running tap, it's as if the waste pipe had been made larger. But by what?

What is removing the ozone?

We have seen that ozone is very reactive and reacts with oxygen atoms. But oxygen atoms aren't the only radicals to be found in the stratosphere. There are other radicals which can remove ozone by reacting with it.

6.3 *The CFC story*

For many years, chemists have been concerned that substances put into the atmosphere by human activities may be destroying the ozone layer. In the early 1970s there was concern about high-flying jet aircraft. Jet engines release nitrogen oxides in their exhaust gases: could this make a significant difference to the amount of NO_x ($NO + NO_2$) in the stratosphere and so damage the ozone layer? In the end, it turned out that this wasn't a significant problem – the number of aircraft concerned was then too small to make much difference. (The problem, however, may reappear because large aircraft are being designed in the US which will fly at supersonic speeds in the stratosphere.)

In 1974 another concern was raised.

Halogenoalkanes

Organic halogen compounds have one or more halogen atoms (F, Cl, Br or I) attached to a hydrocarbon chain. The simplest examples are the **halogenoalkanes**, which have the halogen atoms attached to an alkane chain.

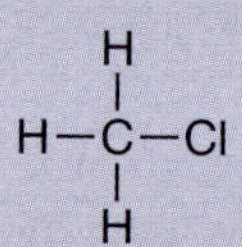

chloromethane

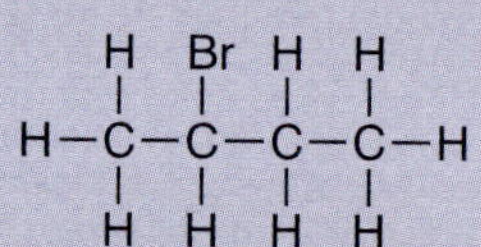

2-bromobutane

CFCs (chlorofluorocarbons) are halogenoalkanes. Examples of the uses of CFCs are given in Figure 13.

Sherry Rowland's predictions

Figure 6 Professor Sherwood ('Sherry') F Rowland

Professor Sherry Rowland is the American scientist who, with **Dr Mario Molina**, predicted back in the early 1970s that chlorofluorocarbons (CFCs) would damage the ozone layer. We talked to him when he visited London, and he described how he made his discovery.

"I originally started looking at the CFC compounds as an interesting *chemical* problem in an environmental setting: whether we could predict from our laboratory knowledge what the fate of the CFCs would be in the Earth's atmosphere. We were not at the beginning thinking of the destruction of the CFCs as an environmental *problem*, but rather as just something that would be happening there.

I've always been attracted to chemical problems – my first research experiment after graduate school involved putting a powdered mixture of lithium carbonate and ordinary glucose in the neutron flux of a nuclear reactor. A nuclear reaction in lithium produces tritium, a radioactive isotope of hydrogen, and I wanted to see if the tritium atoms could replace hydrogen atoms in glucose. (They did – and this led to many further interesting experiments in a field called *hot atom* chemistry.)

With CFCs, we wanted to find out how quickly they would break down in the atmosphere, and by what chemical process. I was working with a young postdoctoral research associate called Mario Molina – this was in 1973 at the University of California at Irvine – and we knew that CFCs are very stable compounds. But how stable? How long could they hang around in the atmosphere? One year or 100 years?

Mario and I looked at all of the processes that could conceivably affect CFCs in the troposphere, and calculated how rapidly such reactions could occur. The answer was very slowly indeed. CFCs remain unreacted for *many decades* – even centuries.

So now we knew that CFCs survive unchanged for a very long time. But we also knew that when they eventually reach the stratosphere they must be broken down by the fierce ultra-violet radiation there – everything is! The CFCs contain atoms of

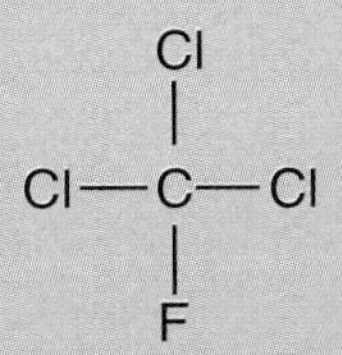

CFC-11

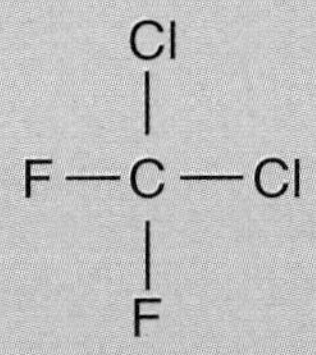

CFC-12

chlorine, fluorine and carbon, and their ultra-violet breakdown releases free chlorine atoms.

For example

$CCl_3F \rightarrow CCl_2F + Cl$

So we did some calculations to find out *how many* chlorine atoms would be formed now and in the future, and then asked what would happen to the chlorine atoms. In the stratosphere, we found that chlorine atoms are about a thousand times more likely to react with ozone than with anything else, leaving still another chlorine-containing chemical, chlorine oxide (ClO)."

[See the next section for details of how Cl atoms destroy ozone.]

"So, once more we asked the same kind of question: what was going to happen to chlorine oxide in the stratosphere? And we found that it would react with oxygen atoms, releasing atomic chlorine again. The two reactions seem to go around in circles – chlorine atoms form chlorine oxide; chlorine oxide forms chlorine atoms – a seemingly endless chain. But with ozone being destroyed at every step! We then calculated how much ozone could be destroyed – each chlorine atom on average destroys about 100 000 molecules of ozone, and mankind has been putting about 1 million tonnes of CFCs into the atmosphere every year since the 1970s.

We couldn't believe the answer! The calculated ozone loss was so high that we thought we must have moved a decimal point by mistake! But we checked very carefully and couldn't find any errors – there really was that much chlorine up there, with much more expected in the future, and the ozone losses would eventually be enormous!

We had started on this problem at the beginning of October 1973, and by mid-December we realised we were onto something very important. When a scientist makes a discovery, the first instinct is to publish it so that other scientists can learn about it, and test the ideas with their own experiments. But the calculated ozone loss was so large that we wanted to be extra certain that no mistake had been made.

So we visited Hal Johnson at the University of California at Berkeley, because he had played a major role in showing how nitrogen oxides from high-flying aircraft – such as Concorde – could affect stratospheric ozone. From Hal we learned that the chlorine chain reaction with ozone had just been discovered, but without a source for chlorine. And we had discovered that the CFCs would be an enormous source of chlorine released directly into the stratosphere!

Now, we were sure that we had discovered something really significant – now it was a major environmental problem. We published our results and conclusions in the scientific journal *Nature*, and waited to see if others could pick holes in them. Many tried – because that's the way science works – but none succeeded. The issue of CFCs and stratospheric ozone had really arrived.

The state of Oregon banned CFCs as propellant gases in aerosols in 1975, and the whole of the US, as well as Canada, Norway and Sweden, followed in the next 2 or 3 years. Unfortunately, the rest of Europe and Japan did not ban aerosol propellant use, and the other major applications – refrigeration, insulation, cleaning electronics, etc – continued to grow until the appearance of the hole in the Antarctic ozone layer resulted in international action in 1988–1990."

Ozone destruction by Cl atoms

When chlorine-containing compounds reach the stratosphere they absorb high-energy solar radiation and break down to give chlorine atoms. Chlorine atoms can react with ozone like this:

$$Cl + O_3 \rightarrow ClO + O_2 \qquad \text{(reaction 6)}$$

The ClO formed is reactive and can react with oxygen atoms:

$$ClO + O \rightarrow Cl + O_2 \qquad \text{(reaction 7)}$$

So now we have two reactions competing with one another to remove ozone from the stratosphere:

$$O + O_3 \rightarrow O_2 + O_2 \qquad \text{(reaction 4)}$$

and

$$Cl + O_3 \rightarrow ClO + O_2 \qquad \text{(reaction 6)}$$

The concentration of Cl atoms in the stratosphere is much less than the concentration of O atoms. So how significant is reaction 6?

Chemists measured the rates of these two reactions in the laboratory under different conditions. They showed that, at temperatures and pressures similar to those in the stratosphere, the reaction of O_3 with Cl atoms takes place more than 1500 times faster than the reaction of O_3 with O atoms.

Even when they took into account the fact that Cl atoms have a much lower concentration than O atoms in the stratosphere, the chemists still found that the reaction of Cl atoms takes place sufficiently quickly to make a very large contribution to removal of ozone.

Competing reactions

Activation energies are very important when *comparing* the rates of two competing reactions taking place under similar conditions. If the activation energy of a reaction is large, only a small proportion of colliding particles will have enough energy to react, so the reaction proceeds slowly. If, however, the activation energy is very small, most of the colliding particles will have sufficient energy to react and the reaction occurs very quickly.

Figure 7 shows the activation energies for the reactions of O atoms and Cl atoms with ozone. The reaction with the lower activation energy (ie O_3 + Cl) will proceed more quickly.

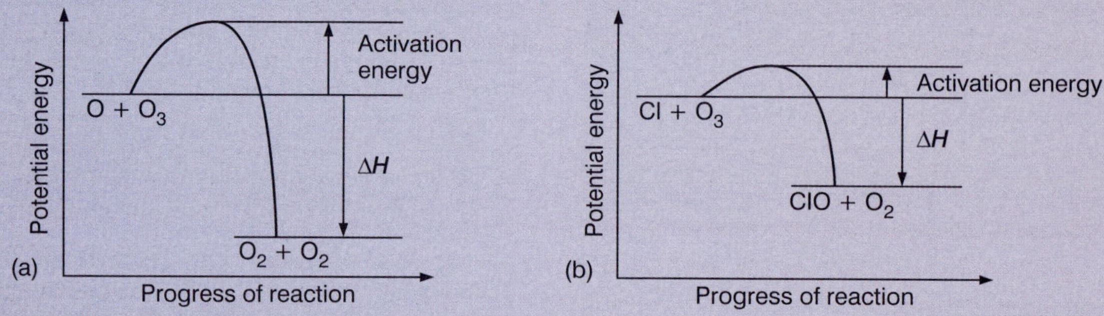

Figure 7 Potential energy diagrams for the reactions of ozone with (a) O atoms and (b) Cl atoms taking place under identical conditions

What's more, the Cl atoms are regenerated in a **catalytic cycle** (reactions 6 and 7) and can go on to react with more O_3. So you can see why their effect could be so devastating.

Chlorine atoms are not the only substances which can destroy ozone in the stratosphere through a catalytic cycle. Hydroxyl radicals, OH, from the reaction of water with oxygen atoms, nitrogen monoxide, NO, and bromine atoms, formed when bromomethane reaches the stratosphere, also destroy ozone.

Mario Molina and Sherwood Roland, along with Paul Crutzen in Germany, were awarded the 1995 Nobel Prize in Chemistry for their work in atmospheric chemistry, particulary concerning the formation and decomposition of ozone.

ASSIGNMENT 4

Chlorine atoms are particularly effective at removing ozone. A single Cl atom can remove about 1 million ozone molecules.

Add equations 6 and 7 together to produce the equation for the *overall* reaction caused by chlorine atoms.

Comment on the result. What role are Cl atoms playing in the overall reaction?

You have already met the reaction of chlorine atoms with ozone when you were learning to draw energy pathways in **Unit 1, Section 14.1** and when you were studying homogeneous catalysis in **Unit 1, Section 15.5**.

The predictions come true

Science is all about making predictions, then testing them experimentally. The problem with Sherry Rowland's predictions is that they involve a long time scale. What is more, they need a large laboratory to test them: in fact they need the largest laboratory in the world – the Earth's atmosphere.

But in 1985, scientists examining the atmosphere above the Antarctic made a momentous discovery.

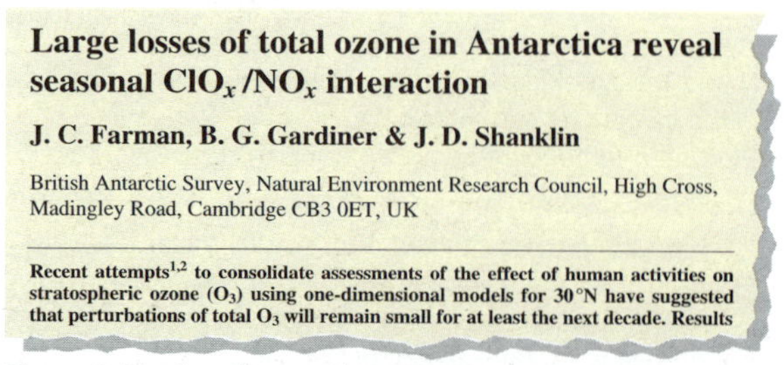

Large losses of total ozone in Antarctica reveal seasonal ClO_x/NO_x interaction

J. C. Farman, B. G. Gardiner & J. D. Shanklin

British Antarctic Survey, Natural Environment Research Council, High Cross, Madingley Road, Cambridge CB3 0ET, UK

Recent attempts[1,2] to consolidate assessments of the effect of human activities on stratospheric ozone (O_3) using one-dimensional models for 30°N have suggested that perturbations of total O_3 will remain small for at least the next decade. Results

Figure 8 The headline-making paper which appeared in the scientific journal Nature *in May 1985, reporting the discovery of a 'hole' in the ozone layer over Antarctica*

Joe Farman's story

Figure 9 Dr Joe Farman

Dr Joe Farman is the British scientist whose group first discovered the 'hole' in the ozone layer. We went to Cambridge to talk to him, and this is how he described their discovery.

"To stand up and make a fuss you need to have your background secure.

We were measuring ozone concentrations over the Antarctic as part of our work with the British Antarctic Survey. We use ultra-violet spectroscopy: ozone absorbs ultra-violet radiation of a particular frequency. If you measure how strongly the atmosphere is absorbing ultra-violet of that frequency, you can work out the concentration of ozone in that part of the atmosphere. Our measurements started in 1981, and over 2 or 3 years I became convinced that there was something seriously wrong. The concentrations of ozone were much lower than expected, particularly in October, which is the Antarctic spring. In 1984 we put in a new instrument to check our readings, and that gave us confirmation. There really was a 'hole' in the ozone layer.

Figure 10 The British Antarctic Survey's Halley Base

We were making our measurements from *below*, by looking up into the atmosphere from the Earth's surface at our base in Antarctica. The Americans were making similar measurements, but from *above*: NASA satellites were using ultra-violet spectroscopy to measure ozone concentrations by looking down from above.

So why didn't NASA spot the 'hole' first? The trouble is, NASA satellites make *too many* measurements. They collect enormous quantities of data on all sorts of things, not just ozone concentrations.

They can't possibly process all the data, so their computers are programmed to ignore any data that seem impossibly inaccurate. In the case of the ozone measurements, this meant that they ignored most of them, because they were so far out from what was expected. About 80% of the data for October were discarded by the computer because no-one believed ozone concentrations could get that low.

Later, when our own measurements showed the concentration of ozone really was that low, NASA went back and re-examined the discarded data, which confirmed our own measurements.

In 1985 we published our findings in the scientific journal *Nature*. Then it was a matter of convincing the world how serious the problem is. Once a scientist has made a momentous discovery of this kind, there is a duty to tell everyone about it.

To me the real horror is the sheer speed with which it has happened. In 1985 the United Nations published a report saying there was plenty of time to study the ozone problem. At about the same time we published our own results which show that the ozone gets turned over in a period of about *5 weeks*! We still don't know exactly what the effects of ozone depletion will be, but when you are affecting a system that fast you have to be aware that almost anything could happen.

We may think we understand a lot about the way that nature works, but to my mind we're so ignorant that anything could happen. However good your models are, you have to *keep making measurements* to make sure that nature really does work the way you think.

It's easy to see now, with the benefit of hindsight, where we went wrong with CFCs. The original idea was to look for something *very stable* which would not be flammable, poisonous or corrosive. But the trouble with very stable substances is that they stay around for a long time – and when they do eventually break down they form something *very unstable* and reactive, which is likely to cause trouble."

ASSIGNMENT 5

a Explain how concentrations of ozone in the atmosphere are measured.

b Why did NASA discard 80% of their October 1984 ozone concentration measurements?

c Joe Farman's group were first aware of the low concentrations of ozone around 1983, but did not publish their findings until 1985. Why did they delay publication?

d Joe Farman gives two reasons why the great stability of CFCs makes them an environmental problem. What are the two reasons?

Figure 11 Helium-filled balloons are sent up into the stratosphere to measure ozone concentrations. A cord almost 10 miles in length attaches the balloon to measuring instruments on the launch vehicle; the cord can be reeled in and out to obtain measurements at different altitudes

Since the mid-1980s, monitoring of ozone concentrations has continued with even greater urgency. Satellite readings and measurements from balloons and high-altitude planes have supported the measurements taken from the ground, and have confirmed the presence of the 'hole' over the Antarctic.

Satellite measurements have also shown that ozone concentrations have gone down in other parts of the globe too. The effect is particularly dramatic in the Antarctic spring (Figure 12) because of the special weather conditions existing then – but there is evidence of a smaller 'ozone hole' appearing above the Arctic and a general depletion of ozone stretching across northern latitudes, including Canada, the US and Europe.

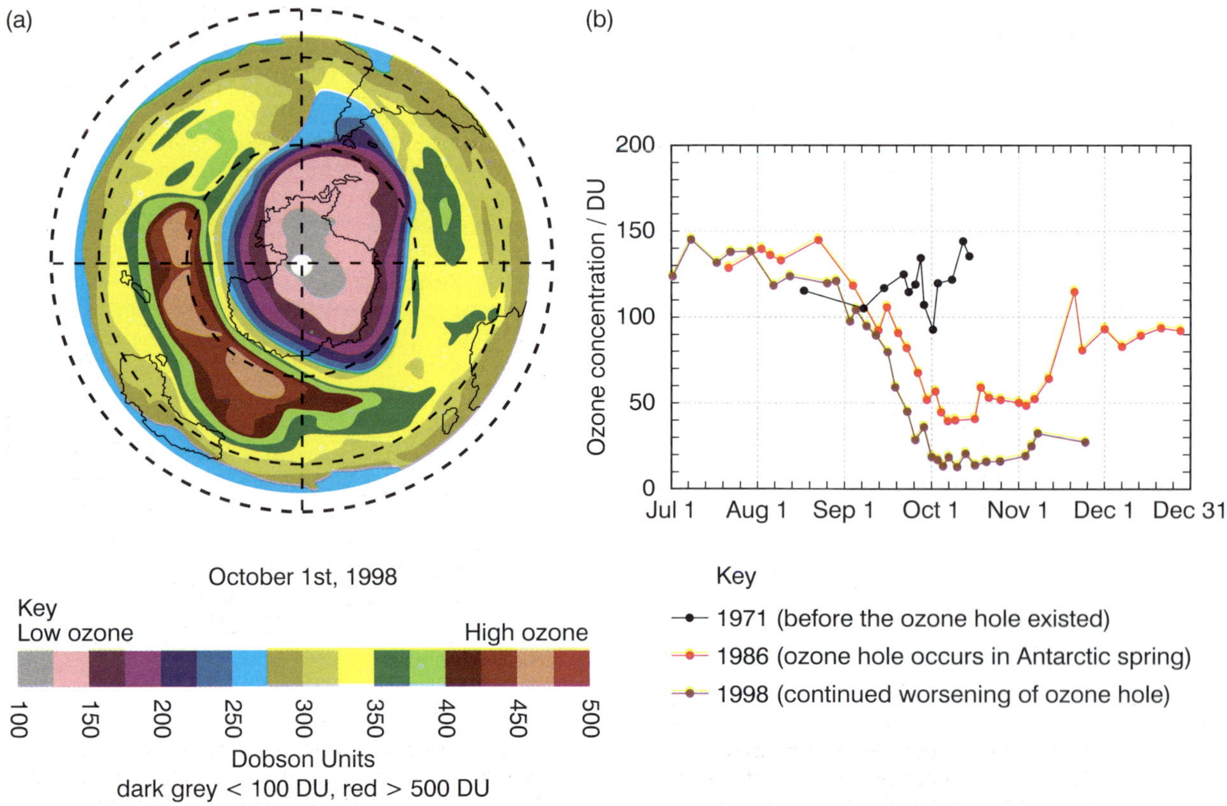

Figure 12 (a) Maps like this are obtained every day by satellite and can be seen on the web (http://toms.gsfc.nasa.gov/). (b) The graph shows how the ozone concentration varies over Antarctica during the year

CFCs: very handy compounds

In 1930 the American engineer Thomas Midgley demonstrated a new refrigerant to the American Chemical Society. He inhaled a lungful of dichlorodifluoromethane (CCl_2F_2) and used it to blow out a candle.

Midgley was flamboyantly demonstrating two important properties of CCl_2F_2: its lack of toxicity and its lack of flammability. Up to that time, ammonia had been the main refrigerant in use. Ammonia has a convenient boiling point, –33 °C, which means it can easily be liquefied by compression. Unfortunately, it is also toxic and very smelly.

Midgley had been asked to find a safe replacement for ammonia, and that meant finding a compound with a similar boiling point but which was very unreactive and therefore safe. He came up with the compound CCl_2F_2.

CCl_2F_2 belongs to a family of compounds called chlorofluorocarbons (CFCs), which contain chlorine, fluorine and carbon. There are several members of the family, all with different boiling points. That's one of

As propellants for aerosols

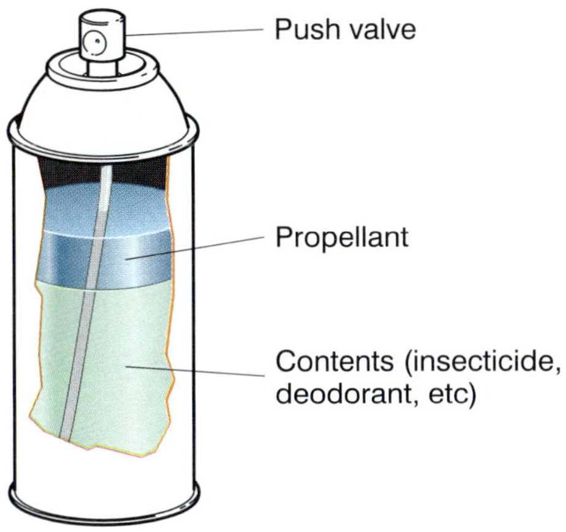

When the valve opens, the pressure falls inside the can causing the propellant to vaporise. The propellant escapes into the atmosphere along with the other contents of the can.

As blowing agents for making expanded plastics

A volatile CFC is incorporated in the plastic when it is made. The heat given off during the polymerisation reaction vaporises the CFC so it 'blows' tiny bubbles in the plastic making a foam. Inevitably, some of the CFC escapes into the atmosphere during the blowing process, and more escapes when the plastic is finally disposed of.

As refrigerants

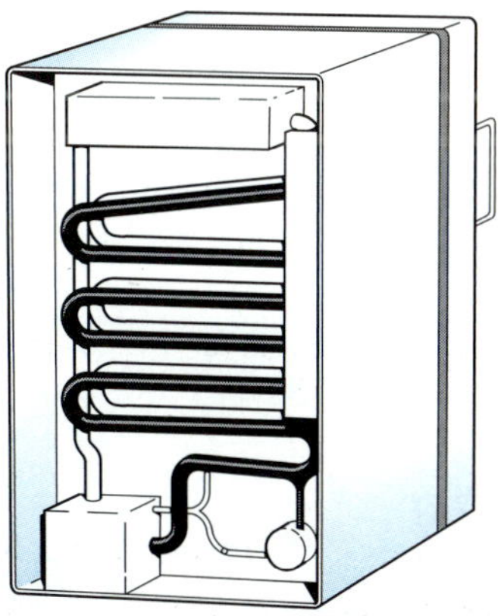

CFCs are used as refrigerants in food refrigerators and air-conditioning units. A total of 55 000 tonnes of CFCs went into the air conditioners of cars in the US in 1987. Eventually some of these refrigerants end up in the atmosphere, through leakage or when the refrigerator is scrapped.

As cleaning solvents

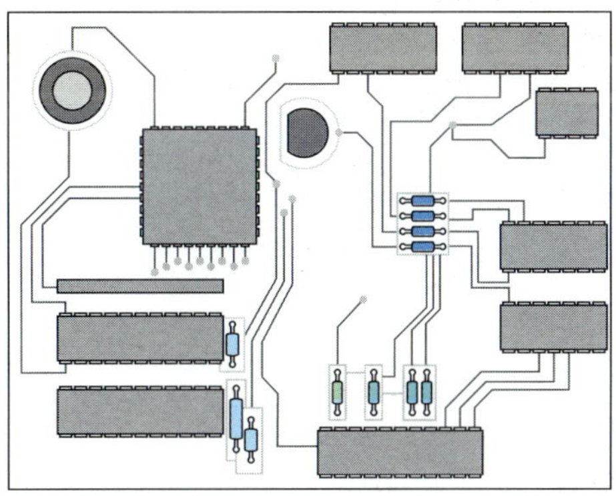

CFCs dissolve grease and are used as solvents in dry-cleaning, cleaning electronic circuits, etc. Some of the solvent escapes into the atmosphere during use.

Figure 13 Everyday applications of CFCs

the things that makes them so useful: you can find CFCs with boiling points to suit different applications.

Another useful thing about CFCs is that they are very unreactive, and have low flammability and low toxicity.

You can see the main uses of CFCs in Figure 13.

The trouble with CFCs

When Thomas Midgley and other chemists developed CFCs, they did their job too well. They found a family of compounds that are very unreactive, and this makes them excellent for the jobs just described. The trouble is, they are *too* unreactive.

The estimated lifetimes for CFCs in the troposphere are about 100 years. This gives plenty of time for them to be transported up into the stratosphere – where they are no longer unreactive.

As more became known about the effect of CFCs on ozone in the stratosphere, people began to press for their replacement. This meant that the chemical industry had an important job to do: to find compounds that would replace CFCs, but have no significant damaging effect on the ozone layer.

Progress has been very rapid. The replacement compounds are hydrochlorofluorocarbons (HCFCs) or hydrofluorocarbons (HFCs). The difference is that these molecules contain H–C bonds and are broken down in the troposphere.

Sadly, they are not a perfect solution. Both CFCs and their replacements are greenhouse gases and contribute to global warming (a problem that is discussed in **Unit 3**, **Chapter 7**). There are fears too that some of the decomposition products may be toxic.

Methane to the rescue

Cl atoms are obviously bad news in the stratosphere. They attack ozone and destroy it. Indeed, they are so reactive that, left to themselves, they would quickly destroy most of the ozone there. Fortunately, they have not done that yet, because there are other molecules in the stratosphere that react with Cl atoms.

Methane, CH_4, is an important example of such a molecule. It is produced on Earth in large quantities by living organisms. Most of the methane released is oxidised in the troposphere, but significant amounts of it are eventually carried up into the stratosphere.

Once in the stratosphere, methane molecules remove chlorine atoms by reacting with them like this:

$$CH_4 + Cl \rightarrow CH_3 + HCl$$

The hydrogen chloride made in this reaction may eventually be carried down into the troposphere where it can be removed in raindrops. This gets rid of some of the troublesome Cl atoms from the stratosphere, although there is not enough methane to stop levels of chlorine rising overall.

Figure 14 Rice paddy fields cover large areas of land and are one of the biggest sources of methane

6.4 *How bad is the ozone crisis?*

Ozone is a vital sunscreen gas which protects us from ultra-violet radiation. Removing it from the stratosphere may have serious consequences for the Earth. The trouble is, we do not know just what these will be.

One thing that seems clear is that cases of skin cancer and eye cataracts will increase as ozone is destroyed. It has been estimated that reducing ozone by 10% could cause a 30%–50% increase in skin cancer cases.

But what about species other than humans? Increased ultra-violet radiation could affect species such as plankton in the oceans. That in turn could affect other organisms involved in the food chain.

And what about the weather? Changes in the amount of radiation reaching the Earth will affect the

Figure 15 Children in New Zealand have to wear hats on school outings in summer, to protect them from harmful ultra-violet radiation

World unites on ozone deal

Paul Brown
Environment Correspondent

A worldwide agreement to phase out CFCs and other ozone depleting chemicals by the year 2000 was reached in London last night.

A group of environmentally advanced countries pushed to have the date brought forward to 1997, but were blocked by the United States, the Soviet Union, and Japan. The compromise solution was a 50 per cent reduction by 1995, 85 per cent by 1997, and 100 per cent by 2000. A sub-clause agreed the position would be reviewed in 1992, to see if the timetable could be improved.

The agreement provided for the establishment of a new global front to help the Third World adapt to the changes.

Chris Patten, the Environment Secretary, said: "This is a major step forward in environmental diplomacy. It is a unique agreement bringing together, as it does, the establishment of environmental objectives with provision of funds and the transfer of technology."

It boded well for future global environmental agreements.

Answering criticisms that the agreement still did not go far enough, he said: "We would all like to have stopped CFCs production tomorrow but this was the best agreement that was possible, taking all considerations into account".

Negotiations had run well over the time yesterday as detailed timetables for phasing out chemicals were hammered out for inclusion in the agreement.

There was personal success for Mr Patten, who, as chairman of the conference, redrafted the final document so that both China and India felt they could pledge to sign the Montreal Protocol.

By yesterday, 59 nations had signed and most of the other 39 at the conference were expected to ratify soon.

In spite of ministerial joy at the agreement, there were doubts that the timetable would be quick enough to prevent ozone depletion being a serious problem.

Yesterday's deal meant that CFCs, the main ozone depleter, used in fridges and air conditioning, will go by 2000. Methyl chloroform, a metal cleaning agent, will be banned by 2005.

Much of the argument centred on how quickly individual chemicals could be phased out.

One of the triumphs was getting India and China, with more than one third of the world's population between them, to join. Maneka Gandhi, the Indian Environment Minister, had held out for two days for the transfer of technology from the West to be included in the agreement so India could manufacture CFCs substitutes itself.

Mrs Gandhi said yesterday the agreement now said that if technology was not transferred, that India did not have to stop the manufacture of CFCs. That placed the onus on the West to keep its promises, and on that basis she was prepared to recommend her government to ratify the protocol.

Joe Farman, the British scientist who discovered the hole in the ozone layer, said he feared the agreement was still not stringent enough.

Mr Farman, head of atmospheric dynamics at the British Antarctic Survey, said the ozone hole would go on getting bigger for some time.

Chlorine is currently 3.6 parts per billion in the atmosphere. This would grow to 4.8 parts per billion in 10 years, he said. He calculated this could lead to 18 per cent depletion in ozone in the northern hemisphere during the winter and spring by the year 2000.

Under the agreement, he calculated that it would be 2030 before the chlorine level went down to the 1986 level when the hole was first announced.

Wednesday 30 September 1998

Biggest hole in ozone forecast

THIS YEAR'S ozone hole over the Antarctic is shaping up to be the biggest on record, capable of burning skin protected by factor 30 sunscreen in just 20 minutes, a British scientist warned yesterday.

Jonathan Shanklin, a meteorologist at the British Antarctic Survey, which first detected the development of the hole in the 1980s, also said that the northern hemisphere will suffer equally bad ozone depletion in the next 20 years, raising the incidence of skin cancer in northern Europe.

Preliminary satellite data collated by New Zealand teams in the Antarctic suggest that the hole in the layer of the upper atmosphere, which protects the

Figure 16 The destruction of ozone in the stratosphere is a global problem. The two articles appeared in June 1990 (after the London meeting) and in September 1998

temperature of the Earth itself, which of course affects the weather.

Governments have gradually realised that it's not worth risking the global experiment that is needed to find the answers to these questions. In 1987, at an international meeting in Montreal, a procedure was agreed for restricting the production and release of CFCs into the atmosphere.

Since then, new global observations of ozone depletion have prompted amendments to strengthen the treaty: in London in 1990, Copenhagen in 1992, and in Montreal in 1997. The aim is to reduce and eventually eliminate emissions of ozone-depleting substances as a result of human activity. These now include bromomethane and other halogenoalkanes as well as CFCs.

By 1998 the developed nations had almost phased out their use of CFCs. A special fund was set up to help developing countries move away from CFC use, and many are committed to stopping their use by 2000. Yet CFC smuggling is now reckoned to be one of the world's biggest undercover businesses.

CFCs are very stable, and they take a long time to reach the stratosphere. If the Montreal protocol *is* observed, the amounts of chlorine and bromine atoms in the stratosphere should reach a maximum early in the 21st century, and then decline slowly. The ozone layer could return to normal by the middle of the 21st century. But stratospheric ozone depletion will still get worse before we begin to notice any improvement.

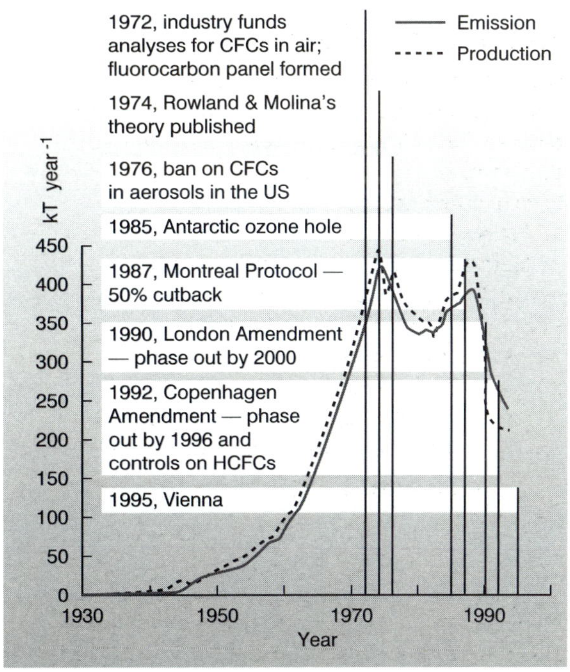

Figure 17 CFC-12 global production and emission

ASSIGNMENT 6

The following is an extract from a letter to a newspaper:

"… Ozone is a gas which occurs in the atmosphere in very small quantities. So small, that the total amount, if it were separated from the air and brought to a condition of normal temperature and pressure, would form a layer only a few millimetres high over the whole Earth. The entire atmosphere under the same conditions would reach 8 km.

This tiny amount of ozone absorbs ultra-violet solar energy but is under threat from the CFCs we hear so much about.

If so small a quantity of ozone can produce the protection we require, why can we not manufacture ozone and release it into the atmosphere? …"

What would be your reply to this? Do you think this could be a possible solution?

6.5 *Summary*

This Chemical Storylines chapter is not linked to any specific Chemical Ideas chapters, though parts of it revise work you did in Unit 1, Chapter 15 on rates of reactions.

In this chapter you have looked at one of the problems in atmospheric chemistry: the destruction of the ozone layer. Ozone depletion is a *stratospheric problem*. At its centre is the idea of the absorption of radiation by gas molecules. If ultra-violet radiation is not absorbed in the stratosphere, dangerous effects arise from the increase in ultra-violet radiation at the surface of the Earth.

The details of many chemical processes in the stratosphere require an understanding of radicals and their reactions. Chlorine atoms (radicals) produced from CFCs are involved in ozone destruction in a catalytic cycle. CFCs have properties which made them useful in a variety of consumer products, but their use is now severly restricted by international agreement.

UNIT 3 CHEMICAL REACTIONS
OVERVIEW

Chemical Storyline chapters

Chapter 7 Equilibria in the Atmosphere and the Oceans

This chapter begins by explaining the greenhouse effect caused by some gases in the atmosphere, such as carbon dioxide, absorbing infra-red radiation form the Earth. It looks at the build up of carbon dioxide in the atmosphere as a result of human activities and how this could lead to global warming. Central to this is a study of the equilibria concerning carbon dioxide in the atmosphere and the oceans, since much of the carbon dioxide released is taken up by the oceans.

Chapter 8 The Chemical Industry

The chemical industry is a major contributor to our economy. This chapter outlines the general principles on which the chemical industry works. It begins by looking at the factors involved in setting up a new chemical plant and at the people employed to operate it. It considers safety and environmental issues, the choice of feedstock and how these factors are influenced by the need to make a profit.

Chapter 9 Case Studies from the Chemical Industry

This chapter takes examples of some processes in the chemical industry and shows how basic chemical principles are applied to industrial processes. The case studies include the Haber process for making ammonia, extracting bromine from the ocean, extracting and refining copper and several applications involving radioisotopes.

Chemical Ideas chapters

Each Chemical Storyline is linked to relevant sections in the Chemical Ideas:

Chapter 20 *Equilibrium in Chemistry*
Chapter 21 *Acids and Bases*
Chapter 22 *Hess's Law and Enthalpy Cycles*
Chapter 23 *Managing a Chemical Process*
Chapter 24 *Redox*
Chapter 25 *Nuclear Chemistry*

Overview of chemical principles

From previous work, you should:
- know that some chemical reactions are reversible
- understand what is meant by the enthalpy change for a reaction
- understand the role of a catalyst in a chemical reaction
- be able to classify substances as acids or bases
- understand what is meant by oxidation and reduction.

At the end of this unit you will have covered all the content statements for Unit 3 in the Arrangements for Higher Chemistry. The table below shows where each topic is covered. Remember that the *order* in which the content statements are done is governed by the Chemical Storyline and so is different from the way they are listed in the Arrangements.

UNIT 3 CHEMICAL REACTIONS		
Areas in the content statements for Higher Chemistry	**Chemical Storylines**	**Chemical Ideas**
(a) The chemical industry	8.1–8.9	23.1–23.5
(b) Hess's law	7.2	22.1
(c) Equilibrium		
The concept of dynamic equilibrium	7.2, 7.3	20.1
Shifting the equilibrium position	7.2, 7.3, 9.1	20.2 (15.5)
(d) Acids and bases		
The pH scale	7.2, 7.3	21.1
The concept of strong and weak		21.2, 21.3
The pH of salt solutions		21.4 (18.5)
(e) Redox reactions		
Oxidising and reducing agents	9.2	24.1
Redox titrations		24.2
Electrolysis	9.3	24.3
(f) Nuclear chemistry		
Types of radiation		25.1
Half-lives		25.1
Radioisotopes	9.4	25.1
Nuclear of fossil fuels?	9.4	
Nuclear fusion	(1.1)	25.2

(The entries in brackets refer to previous use of these ideas in Units 1 and 2.)

7

EQUILIBRIA IN THE ATMOSPHERE AND THE OCEANS

7.1 *Keeping the Earth warm*

The composition of our atmosphere is determined by biological processes acting together with physical and chemical changes and is essential to the support of life on Earth as we know it.

Of all the planets in the solar system, Earth has the largest relative abundance of water vapour in its atmosphere. Water is unique in that it can exist as a solid, a liquid or a gas at the temperatures found on or near the Earth's surface. Most of the Earth's water is liquid, with only 0.001 per cent in the atmosphere at any one time. Snow and ice sheets make up 2.4 per cent. The rest is present in oceans, rivers and lakes.

In **Unit 1**, **Chapter 3** you learned how the global water cycle acted to transfer energy, and how currents in the oceans themselves redistribute heat. But the oceans and atmosphere interact in another important way. Oceans can hold gases in solution and provide a supply of gases to the atmosphere to moderate against changes in atmospheric composition.

Together, the atmosphere and oceans have a very important role in controlling the climate of the Earth, a climate which we know has varied considerably over the Earth's lifetime. To understand more we will start by looking at the way the Sun keeps the Earth warm.

Radiation in, radiation out

When things get hot, they send out electromagnetic radiation. The hotter the object, the higher the energy of the radiation.

The surface of the Sun has a temperature of about 6000 °C, and this means that it radiates energy in the ultra-violet, visible and infra-red regions.

The Earth is heated by the Sun's radiation. Its average surface temperature is about 12 °C – a lot cooler than the Sun, but still hot enough to radiate electromagnetic radiation. But at this lower temperature, the energy radiated is mainly in the infra-red region. The situation is illustrated in Figure 1.

The Sun radiates energy mainly around the visible and ultra-violet region. Part of this energy is absorbed by the Earth and its atmosphere, and part is reflected back into space. The part that gets absorbed helps to heat the Earth, and the Earth in turn radiates energy back into space. A steady state is reached, where the Earth is radiating energy as fast as it absorbs it. Under

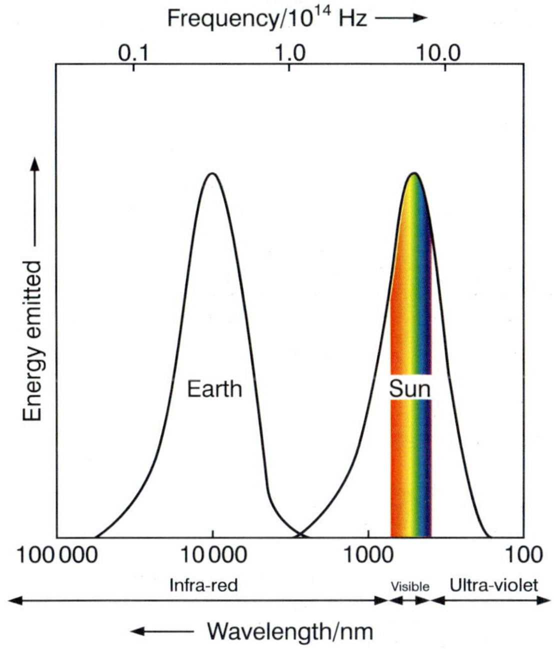

Figure 1 The radiation from the Sun which reaches the outer limits of the atmosphere and the radiation given off from the surface of the Earth (the frequencies and wavelengths are plotted here on a logarithmic scale, so each division is a factor of 10 greater than the one before)

such conditions, the average temperature of the Earth remains constant.

As in all steady states, the delicate balance can be disturbed by changes to the system – in particular by changes to the quantities of various atmospheric gases.

Several atmospheric gases – carbon dioxide and methane are important examples – do not absorb visible or ultra-violet radiation from the Sun, but do absorb infra-red radiation. They stop some of the energy radiated by the Earth from escaping and the effect is to warm up the troposphere. The interactions are summarised in Figure 2. Gases which act like this are called **greenhouse gases**, because their effect resembles that of the glass of a greenhouse which lets visible light in but prevents infra-red from escaping and so makes the greenhouse warmer.

Some gases have a more powerful greenhouse effect than others. Table 1 lists some gases, their abundance in the atmosphere and their **greenhouse factors**. The greenhouse factor is a measure of the greenhouse effect caused by the gas, relative to the same amount of carbon dioxide, which is assigned a value of 1. One molecule of methane, for example, has the same effect as about 30 molecules of carbon dioxide.

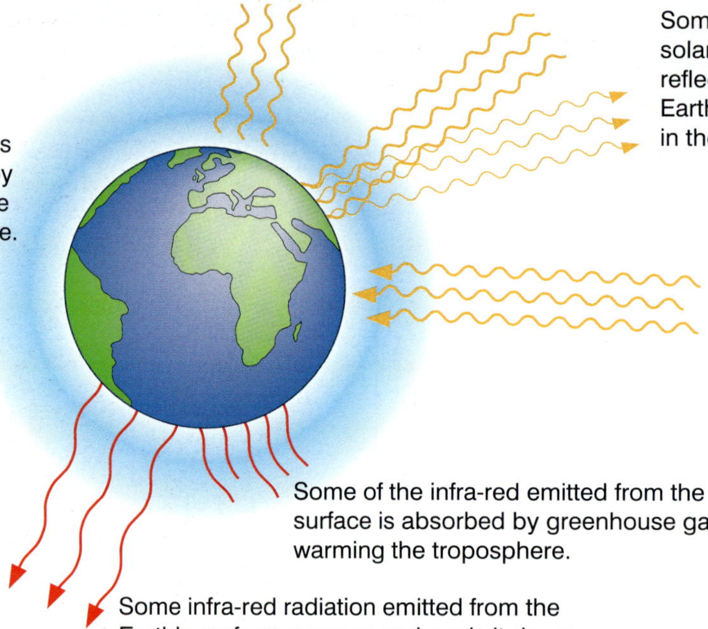

Figure 2 The Earth – input and output of energy

Gas	Tropospheric abundance (by volume)/%	Greenhouse factor
N_2	78	negligible
O_2	21	negligible
Ar	1	negligible
$H_2O(g)$	1*	0.1
CO_2	3.7×10^{-2}	1
CH_4	1.8×10^{-4}	20
N_2O	3.1×10^{-5}	310
CCl_2F_2	5.3×10^{-8}	8100
CCl_3F	2.6×10^{-8}	3800

Table 1 Relative contributions to the greenhouse effect of various gases in the atmosphere (averaged figure)*

ASSIGNMENT I

Use the data in Table 1 to help with the following questions.

a Bearing in mind both their abundance and the greenhouse factor, list the gases in order of how much they contribute to the total greenhouse effect on Earth.

b Which of these gases are produced in significant amounts by human activities?

c Which gases in part **b** would it be most fruitful to try to control in order to control the greenhouse effect?

d Human activities produce a lot of water vapour, but this has very little effect on the net amount of water vapour in the atmosphere. Why?

The greenhouse effect is good for you

Without the greenhouse effect, we would not be here. It keeps the average temperature of the Earth high enough to support life.

If we had no atmosphere on Earth, it would be like the Moon – barren and lifeless. The surface of the Moon gets very hot in the day, but is bitterly cold at night. If, on the other hand, the composition of our atmosphere resembled that of our neighbouring planet Venus, with 96.4% carbon dioxide, the greenhouse effect would make it so hot that life-forms would find it impossible to survive.

The carbon dioxide atmosphere on Venus is thick and the greenhouse effect is extreme. The warming due to the greenhouse effect on Venus is 446 °C, compared to 33 °C on Earth.

Our other neighbouring planet, Mars, has an atmosphere which is mostly carbon dioxide, like that on Venus. But on Mars the atmosphere is very thin. There is only a small greenhouse effect, and so Mars is colder. The warming due to the greenhouse effect on Mars is only 3 °C.

We are used to the stable, comfortable temperature on Earth and small changes in that temperature could have a dramatic effect on life.

Is the Earth getting hotter?

About 100 years ago, the Swedish chemist Arrhenius predicted that increasing amounts of carbon dioxide could lead to warming of the Earth. Average temperatures did indeed rise from 1880 to 1940, by about 0.25 °C. But then between 1940 and 1970 they fell again, by 0.2 °C.

So why are we worried today? During the 1970s, measurements of carbon dioxide in the atmosphere began to show a significant increase, and new predictions began to be made about the effect of the carbon dioxide increase on the Earth's climate.

Making predictions about the climate is very difficult because it involves so many variable factors, many of which are still poorly understood.

The information available is fed into computers to give an overall **mathematical model** of the Earth's climate.

The amount of information needed is enormous. It includes, for example, the concentrations and distribution of the gases in the atmosphere, their predicted lifetimes, and details of how the concentrations are expected to change. Add to this information about variations in the intensity of the Sun's radiation, meteorological data such as air circulation patterns and cloud cover, as well as ocean temperatures and currents, and you can see why powerful computers are needed.

Models are constantly being improved as more reliable information becomes available. Figure 4 shows the global pattern of temperature rises as predicted by a computer model for a doubling in carbon dioxide concentration.

Figure 3 The eruption of Mt Pinatubo in the Philippines in June 1991 released enormous clouds of dust and sulphur dioxide, which for a few years would have contributed to global cooling; such events make predictions about global change even more difficult. It is also likely that the sulphur dioxide released would have caused some additional ozone depletion in the stratosphere

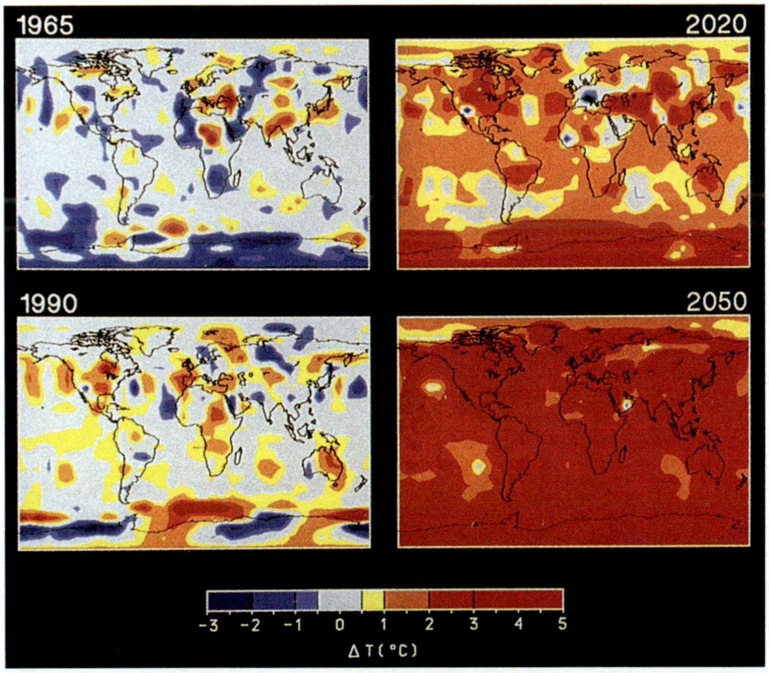

Figure 4 The global pattern of summer temperature rises as predicted by a NASA climate model, on the basis that the emission of greenhouse gases will continue to increase at current rates

At the time predictions like these were being made, field measurements of the Earth's temperature began to show increases too. Atmospheric scientists use an average global temperature, taken between the years 1961 and 1990, as a baseline. They then calculate an average temperature for any particular year from measurements from around the world, and calculate the difference between the temperature that year and the 1961–1990 average.

Over the last 100 years, the five hottest years, in global terms, are all in the 1990s: in descending order 1998, 1997, 1995, 1990 and 1991.

In 1998, the average global temperature was nearly 0.6 °C above the 1961–1990 baseline.

So, is **global warming** really taking place? The climate displays so much natural variation that spotting trends is very difficult. But there is now a great deal of scientific evidence that the Earth is indeed getting warmer, and that this warming is due to human-made emissions of greenhouse gases.

The two most significant greenhouse gases are carbon dioxide and water. Because they are so abundant in the atmosphere, they absorb a lot of the infra-red radiated by the Earth. Water vapour makes the larger contribution – simply because more of it is present.

Gases produced by human activities can increase the natural greenhouse effect of the atmosphere. There are two types:

- Gases already naturally present in the atmosphere, which are increased in amount by human activities. Carbon dioxide is an important example. Humans burn fossil fuels, and this increases the amount of carbon dioxide in the atmosphere. This in turn increases the greenhouse effect.
- Gases which are not naturally present in the atmosphere. CFCs are an example. Although CFCs are only present in the atmosphere in small amounts, they are important because they have a very large greenhouse factor – so each molecule has a big effect (see Table 1).

ASSIGNMENT 2

Suppose the average temperature of the Earth increases due to an increased greenhouse effect.

a What effect will this have on the amount of water vapour in the atmosphere?

b What effect will your answer to part **a** have on further global warming?

c Explain why the role of water vapour in global warming is an example of **positive feedback**.

Water is different from other greenhouse gases (see **Assignment 2**). Under most conditions on Earth, water is a liquid, with some vapour associated with it. $H_2O(l)$ and $H_2O(g)$ are quickly interconverted. If human activities such as burning fuels put $H_2O(g)$ into the atmosphere, most of it will condense to $H_2O(l)$ and

Figure 5 It is difficult to predict accurately the effect of atmospheric water on global warming; water is a greenhouse gas, but water droplets in low clouds tend to block out the Sun and cool the surface of the Earth

eventually return to Earth. So in that sense $H_2O(g)$ isn't nearly as much of a greenhouse problem as $CO_2(g)$.

But there are two more things to consider. First, if the Earth *does* get warmer, more $H_2O(g)$ will evaporate from the oceans. That would tend to increase global warming. Second, although $H_2O(g)$ is a greenhouse gas, the droplets of $H_2O(l)$ in clouds tend to block out the Sun, as people living in the UK know well. So more water in the atmosphere *could* work in either direction – to increase or to decrease warming. This is one reason why climate modellers find it so hard to predict what will happen to the Earth's climate in the future.

At least half of the expected *increase* in the greenhouse effect due to human activities is likely to be caused by carbon dioxide. So control of the greenhouse effect must focus mainly on control of the amount of carbon dioxide we produce.

7.2 *Focus on carbon dioxide*

In this section we look more closely at the role of carbon dioxide in the greenhouse effect.

It is possible to get an accurate picture of the way carbon dioxide concentration has changed over the years. The graph in Figure 6 shows measurements of atmospheric carbon dioxide concentrations made at Mauna Loa Observatory, situated at an altitude of about 3500 m in Hawaii.

ASSIGNMENT 3

a The graph in Figure 6 shows a zig-zag pattern. What time of year corresponds to
  i the peaks?
  ii the troughs?

b Suggest a reason for this zig-zag pattern.

c What was the average percentage increase in carbon dioxide concentration at Mauna Loa between 1960 and 1997?

d What problems would be involved in making a similar record of carbon dioxide concentration in London?

Where does the carbon dioxide come from – and go to?

The increase in concentration of carbon dioxide in the Earth's atmosphere is due to the increasing use of fossil fuels. For the 100 years prior to 1990, the amount of fossil fuels burned increased by about 4% every year. Because of a fall in demand in the former Soviet Union, the consumption levelled off during the early 1990s but is now increasing again. In 1997 the world consumption of fossil fuels for energy was almost 8.5 Gt.

1 Gt = 1 gigatonne = 1×10^9 tonnes

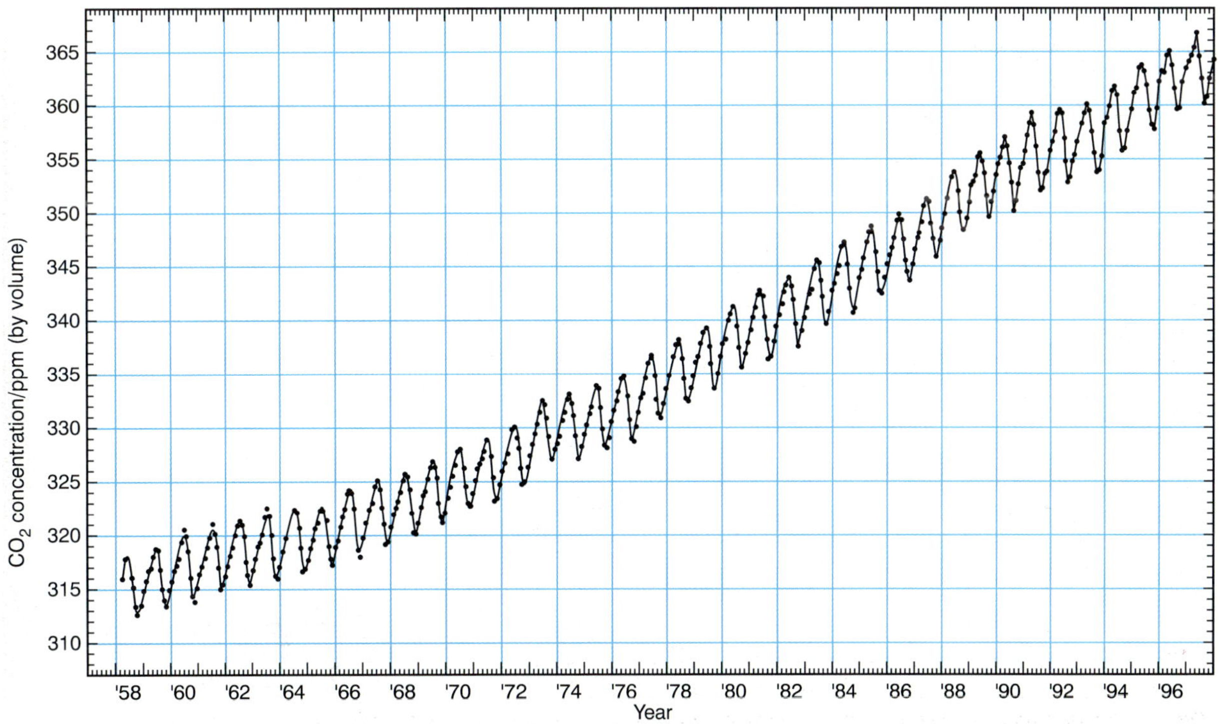

Figure 6 The build up of CO_2 in the atmosphere as recorded at Mauna Loa Observatory, Hawaii

ASSIGNMENT 4

In 1994, the concentration of carbon dioxide in the atmosphere at Mauna Loa (see Figure 6) was 359 ppm (by volume). The *total mass* of carbon dioxide in the atmosphere was estimated to be about 2650 Gt at that time.

Through the four years, 1994 to the end of 1997, it is estimated that 33 Gt of fossil fuel were burned throughout the world.

In 1997 the measured percentage of carbon dioxide in the atmosphere at Mauna Loa was found to be 365 ppm (by volume).

Now answer these questions.

a Assume that the 33 Gt of fossil fuel burned in the four years were all carbon (this is a reasonable approximation since the other main element in fossil fuels, hydrogen, has a very low relative atomic mass). Calculate the mass of carbon dioxide formed by burning this mass of fuel.

b Calculate the new total mass of carbon dioxide in the atmosphere at the end of 1997, assuming that the extra carbon dioxide came only from burning fossil fuels and that it stayed in the atmosphere.

c By what percentage would you expect the concentration of carbon dioxide in the atmosphere to have increased in the four years?

d By what percentage did the concentration of carbon dioxide in the atmosphere actually increase in the four years?

e Comment on your answers to parts **c** and **d**.

Calculations of the kind you have just done suggest that – judging by the quantity of fossil fuels being burned – the increase in carbon dioxide concentration in the atmosphere *should* have been about *twice* as much as it actually has been. Atmospheric carbon dioxide concentration is not increasing as fast as we might expect. Good news – but where is all the carbon dioxide going?

Oceans soak up carbon dioxide

Oceans cover almost three-quarters of the Earth's surface. Carbon dioxide is fairly soluble in water, so large amounts of atmospheric $CO_2(g)$ dissolve in the oceans.

When carbon dioxide dissolves in water, it forms hydrated CO_2 molecules:

$$CO_2(g) \rightleftharpoons CO_2(aq)$$

To understand the chemistry in this section you need an idea of dynamic equilibrium. **Section 20.1** will help you with this.

This is a reversible reaction, as you will know if you have watched what happens when you take the top off a bottle of fizzy drink. However, it is a fairly slow reaction (which is a good thing from the point of view of fizzy drink consumers), and it takes quite a long time for equilibrium to be reached.

Storing carbon dioxide

Imagine you are planning a couple of hours on the beach in the middle of your summer holiday. You have a choice – to take cans of fizzy drink with you, or to leave them in the fridge to drink when you return. From experience you decide to leave the cans in the fridge. You know that more of the fizz stays in a fizzy drink if you keep it cool. If you leave a can standing in the Sun, the gas will come frothing out explosively when you open the top.

The carbon dioxide in the drink is like all gases: its solubility *decreases* as temperature increases. Table 2 shows this for carbon dioxide, oxygen and nitrogen.

Temperature/°C	Solubility/mg per 100 g water CO_2	O_2	N_2
0	338	7.01	2.88
10	235	5.47	2.28
20	173	4.48	1.89
30	131	3.82	1.65
40	105	3.35	1.46
50	86	3.02	1.35

Table 2 Variation of the solubilities of some gases with temperature (at atmospheric pressure)

Carbon dioxide is put into fizzy drinks under considerable pressure – about 14 atmospheres in most cases. That's why there is such a 'pop' when you open the can. The high pressure makes more gas dissolve, as Figure 7 shows.

Section 20.2 looks at how the position of equilibrium is shifted by changes in concentration, temperature and pressure.

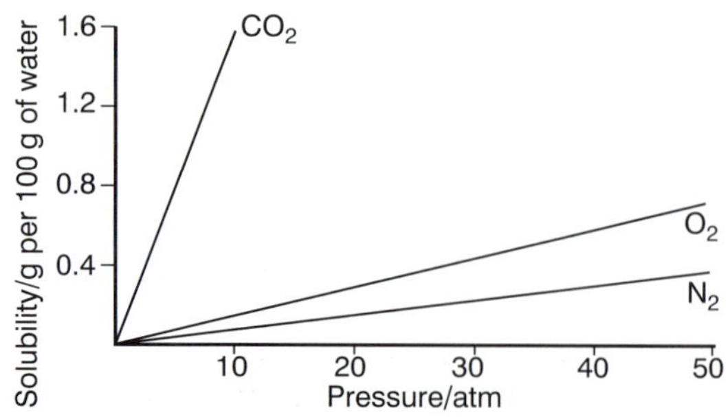

Figure 7 Variation of the solubilities of some gases with pressure, at 25 °C

A high pressure of gas and a low temperature help to keep your drink fizzy. They also encourage carbon dioxide to dissolve in the oceans, which takes the gas out of the atmosphere and helps to maintain a stable environment on Earth.

The exchange of carbon dioxide between the atmosphere and oceans takes place quickly – more quickly than if you left some water in a beaker to come to equilibrium. Like so many environmental processes, we cannot explain it in terms of physical factors alone. Uptake of carbon dioxide by the oceans is speeded up by the action of marine life, as shown in Figure 8.

Table 2 and Figure 7 show that carbon dioxide is more soluble in water than oxygen or nitrogen.

$$CO_2(g) \rightleftharpoons CO_2(aq) \qquad \text{(reaction 1)}$$

In addition, some of the carbon dioxide molecules react chemically with water and are removed from the equilibrium. More carbon dioxide therefore dissolves to maintain the equilibrium position:

$$CO_2(aq) + H_2O(l) \rightleftharpoons H^+(aq) + HCO_3^-(aq) \qquad \text{(reaction 2)}$$

$$HCO_3^-(aq) \rightleftharpoons H^+(aq) + CO_3^{2-}(aq) \qquad \text{(reaction 3)}$$

The reaction with water produces a mixture which contains mainly hydrogencarbonate ions (HCO_3^-) and H^+ ions together with some carbonate ions (CO_3^{2-}).

Much of the excess carbon dioxide we release into the atmosphere from the combustion of fuels is absorbed by the oceans. Estimates vary, but it seems likely that 35%–50% is removed this way. The oceans continue to soak up carbon dioxide because surface water, rich in CO_2, is constantly being removed and stored away for hundreds of years in the ocean deeps. The maximum amount of carbon dioxide is removed because this takes place in cold regions where CO_2 is most soluble. So currents, chemistry and marine life together make up a very efficient CO_2 removal system.

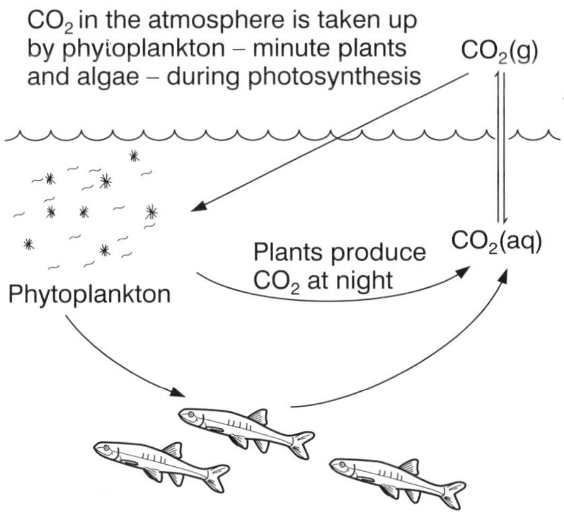

Figure 8 Phytoplankton increase the rate at which carbon dioxide dissolves in the oceans

Sinking shells

Shells are also involved in the reactions that influence the solubility of carbon dioxide in the oceans. The three reactions that have been discussed so far are linked together – the product of reaction 1 is the reactant in 2, and so on. The three equations can be added together to produce just one equation which shows how the reactants in reaction 1 lead to the products of reaction 3:

$$CO_2(g) + H_2O(l) \rightleftharpoons 2H^+(aq) + CO_3^{2-}(aq) \qquad \text{(reaction 4)}$$

Remember, the reaction does not happen as simply as this, but the equation should make the next part of the story clearer.

Marine organisms build protective shells composed of insoluble calcium carbonate using CO_3^{2-} ions in the sea water. The shells are often very beautiful. They provide a way of mopping up carbon dioxide – perhaps from someone's factory chimney – and keeping the composition of our atmosphere constant.

Figure 9 Mussels have protective shells made of calcium carbonate, made from carbon dioxide dissolved in the oceans

Billions of years ago, the Earth's atmosphere contained very much more carbon dioxide than it does now – probably about 35% carbon dioxide by volume. Once the process of photosynthesis had evolved, marine life had plenty of raw materials to work on in the form of carbon dioxide and water. Shell production flourished. Limestone and chalk rocks are the remains of shells of marine organisms which lived at that time and changed carbon dioxide from the atmosphere into solid calcium carbonate. When you go down into a cave you are quite literally making a journey into the past. When you walk on limestone hills or chalk downland, you are treading on Earth's prehistoric atmosphere.

Calcium carbonate is a good material for shellfish to use for protection at the surface of the oceans. It does not dissolve in sea water. But it does dissolve, slightly,

Figure 10 The chalk cliffs of the Seven Sisters are the legacy of marine organisms which lived billions of years ago

Figure 11 Machair, North Uist. Machair sand is 80%–90% carbonate from crushed shells. It is driven by waves and wind over grass and peatland, neutralising acid from the peat. Machair is one of the rarest habitats in Europe

in pure water. It is an example of a sparingly soluble solid. The dissolving of sparingly soluble solids is controlled by equilibria such as

$$CaCO_3(s) \rightleftharpoons Ca^{2+}(aq) + CO_3^{2-}(aq) \qquad \text{(reaction 5)}$$

in which the ions in the saturated solution are in dynamic equilibrium with the undissolved solid present.

Calcium carbonate is a safe material from which to build sea shells because $[Ca^{2+}(aq)]$ and $[CO_3^{2-}(aq)]$ are already high enough at the surface of the sea for the calcium carbonate in the shells to be effectively insoluble. Remember, though, that the shells are in equilibrium with the ions in sea water, and there will be a constant exchange of Ca^{2+} and CO_3^{2-} ions between the two.

When studying reactions, chemists often want to know the enthalpy changes involved. However, for reaction 5 the enthalpy change is so small it is difficult to obtain an accurate value by direct experimental measurement. Fortunately, it is possible to determine enthalpy changes indirectly.

You will need to find out how to calculate enthalpy changes. **Section 22.1** will help you with this.

ASSIGNMENT 5

a Use Hess's law to calculate the standard enthalpy change for the process

$$CaCO_3(s) \rightarrow Ca^{2+}(aq) + CO_3^{2-}(aq)$$

Use the following data:

$$Ca(s) + C(s) + 1\tfrac{1}{2}O_2(g) \rightarrow CaCO_3(s) \qquad \Delta H = -1207\ \text{kJ mol}^{-1}$$

$$Ca(s) + aq \rightarrow Ca^{2+}(aq) + 2e^- \qquad \Delta H = -543\ \text{kJ mol}^{-1}$$

$$C(s) + 1\tfrac{1}{2}O_2(g) + aq + 2e^- \rightarrow CO_3^{2-}(aq) \qquad \Delta H = -677\ \text{kJ mol}^{-1}$$

b Explain in terms of equilibria (reactions 1–5) how calcium carbonate production encourages more carbon dioxide to dissolve from the atmosphere.

But things are different deeper in the ocean where the pressure is higher and the temperature is lower. Under these conditions calcium carbonate is more soluble. There is also a continuous downward drift of material from above. It's like a perpetual snowstorm. In fact the falling material is called *marine snow*. It contains the remains of dead organisms and the waste products from live creatures. Most of the organic material, such as tissue, is consumed or decomposed higher up, but some reaches the deeper water where bacteria break it down to produce carbon dioxide. The shells fall down intact, but then react with the extra carbon dioxide and dissolve. These processes are summarised in Figure 12.

There are no shells on the deep ocean floor: they've all dissolved. The creatures which live there cannot use calcium carbonate for a protective coating.

The calcium carbonate deposits which built up to form our limestone hills could not have formed in deep water. They must have been laid down when our landmass was in shallower seas. The abundance of life suggests also that it was warm, tropical water. Evidence like this helps scientists piece together the distant history of the Earth, and helps us explain how the continents have drifted and how the climate has changed throughout time.

7.3 *Controlling acidity*

The reactions associated with dissolving carbon dioxide in the oceans also help the oceans to control their acidity. This ability of the oceans to withstand external changes has been essential for the unbroken evolution of life. The pH of the oceans has remained close to 8 for millions of years. Why were the oceans not very much more acidic when the atmosphere contained 35% CO_2 – 1000 times its present level?

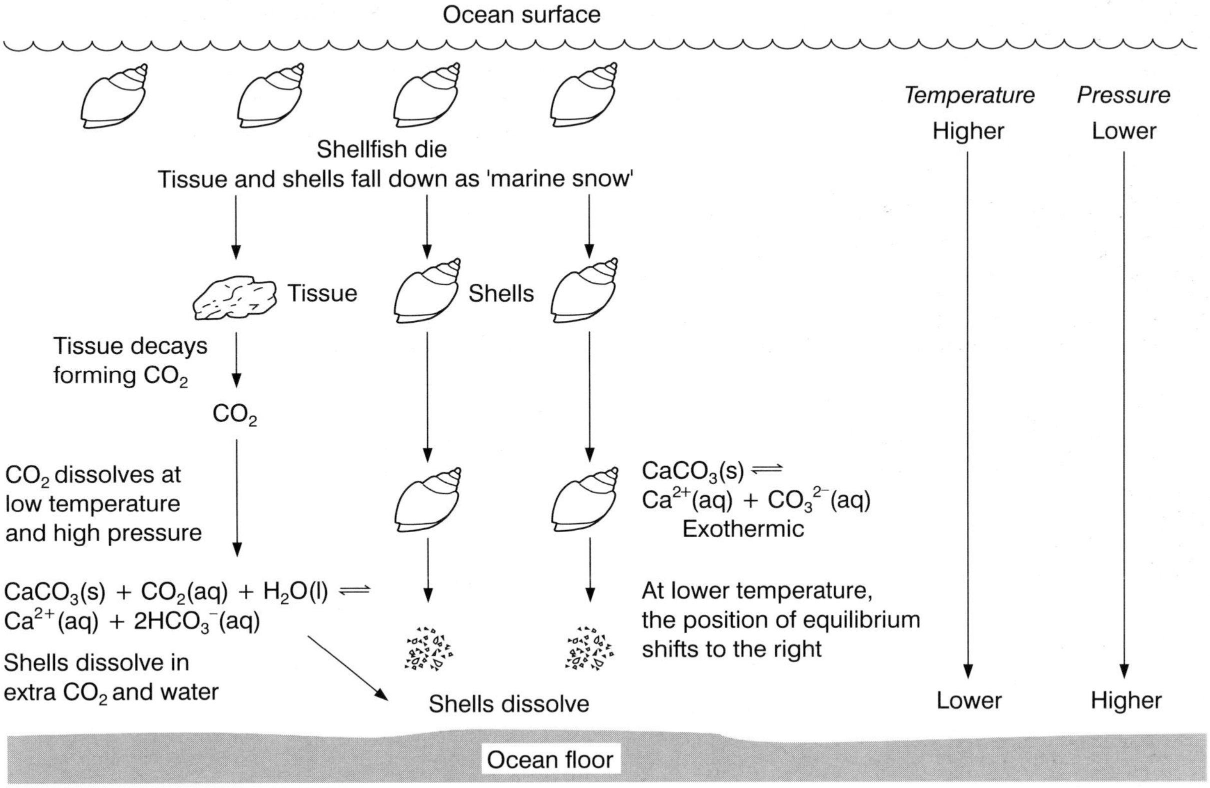

Figure 12 The dissolving of shells on the deep ocean floor

Sections 21.1–21.4 will tell you about acids and bases.

One reason is that a solution of carbon dioxide in water is a **weak acid**. A weak acid reacts incompletely with water, so in this case only a small *proportion* of the carbon dioxide molecules react to form hydrogen ions and hydrogencarbonate ions, as shown in reaction 2.

Figure 13 River waters provide a constant supply of HCO_3^- ions to the oceans from the weathering of limestone rocks. If $[HCO_3^-(aq)]$ falls, the equilibrium shown in reaction 2 will move to the right

There are also $HCO_3^-(aq)$ ions which flow into the sea from material dissolved in river water (Figure 13). But there are two other processes which can supply an almost limitless amount of hydrogencarbonate ions. There are the reactions involving carbonate ions from calcium carbonate. Figure 14 illustrates the processes which prevent the oceans from becoming acidic.

The shells, chalk and limestone in the sea are the reservoir of the anions needed to prevent changes in acidity.

What would happen if the carbon dioxide in the atmosphere rose to the high level of several billion years ago? Solid calcium carbonate would dissolve to produce the carbonate and hydrogencarbonate ions needed to remove the extra H^+ ions. Few CO_3^{2-} ions would remain. Most of the carbon would be in the form of HCO_3^- and dissolved CO_2. The sea would be like a mixture of Perrier water and bicarbonate of soda; the shells and white cliffs would disappear!

Limestone deposits could not form while the atmosphere contained 35% CO_2. The earliest sedimentary rocks are *silicaceous* (they consist predominantly of silicon dioxide). The first significant calcium carbonate deposits formed only about 2 billion years ago, over halfway through the Earth's lifetime. Much of the atmospheric carbon dioxide had been used up by then – so much so that its greenhouse effect had diminished and a major ice age had set in.

The carbon dioxide/calcium carbonate system is made up of linked equilibria which respond rapidly to

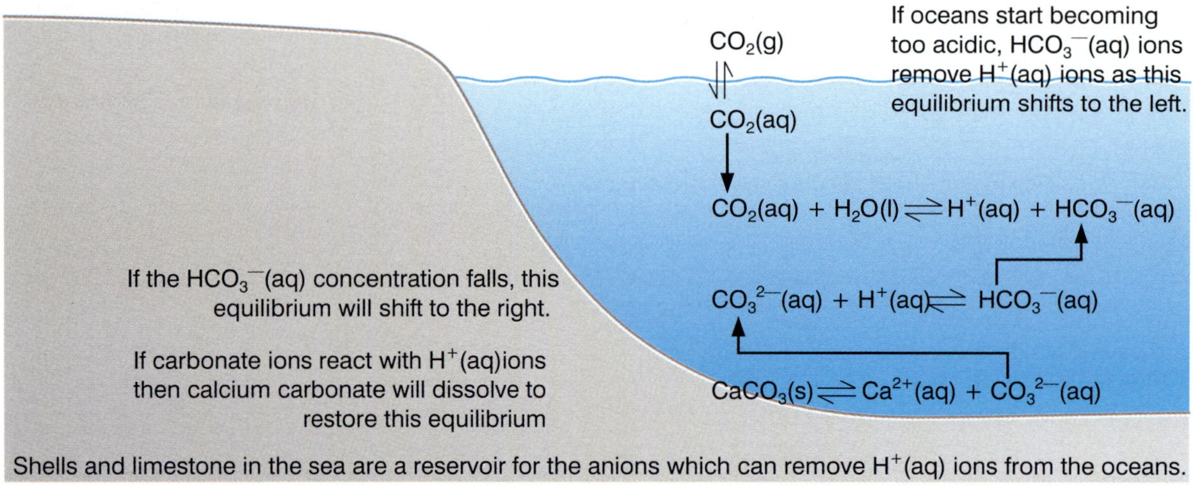

Figure 14 Why the oceans do not become more acidic

change. In the longer term, ion exchange between H^+ ions in the water and Na^+ or K^+ ions in clay sediments provides another very powerful pH control mechanism.

This process can only take place at the bottom of the ocean where sea water and sediment are in contact. Deep ocean water circulates slowly, perhaps taking 1000 years to complete one cycle. The ion-exchange equilibria may be important over millions of years, but in the surface water and throughout the oceans on a shorter time scale it is the carbon dioxide/calcium carbonate system which keeps the pH of the ocean stable.

Coping with carbon

It is estimated that in the middle of the 19th century, before the Industrial Revolution had really started pumping carbon dioxide into the atmosphere, the atmospheric concentration of carbon dioxide was 270 ppm. By 1997 it was 365 ppm. By the 2080s it may have doubled from its pre-Industrial Revolution value to about 540 ppm.

The increase in atmospheric carbon dioxide is a serious problem which will have far-reaching effects on the world's climate.

It is thought that the doubling of carbon dioxide concentration will cause an average temperature increase of about 2 °C. If you take into account the other greenhouse gases such as methane and CFCs, which are also increasing in concentration, this 2 °C rise will probably be with us by the 2030s – within the lifetime of most people reading this. Although 2 °C may not sound much, it will be enough to have a dramatic effect on the global climate.

The science of the effects of carbon dioxide in an atmosphere, which is what has been presented in this chapter, is fairly well known. We are far less certain about the extent to which these effects will occur, the role of other processes and the overall balance which will emerge.

One of the predicted results of global warming is a rise in the mean sea level because of the thermal expansion of water and the melting of some land ice.

Modelling our future climate is extremely difficult. There are people who believe we are heading for disaster from accelerating global warming fuelled by positive feedback. Others believe the Earth will develop ways of compensating for any serious departure from equilibrium.

7.4 *Summary*

This Chemical Storylines chapter is linked to Chemical Ideas Chapters 20, 21 and 22.

In this chapter you looked at the problem of an increase in the greenhouse effect leading to global warming. This is a problem that affects the whole world.

The greenhouse effect is a *tropospheric phenomenon.* It arises from the absorption of radiation by gas molecules. You have focussed on the role of carbon dioxide and the need to monitor and detect its concentration. This led you to the key idea of chemical equilibrium and the need to find out more about how the position of equilibria can be changed.

You looked at the role of the oceans in absorbing much of the added carbon dioxide gas, and at how equilibria involving carbon dioxide and carbonate ions allow the oceans to absorb carbon dioxide without changing in acidity. To understand these processes, you needed to read about acids and bases, the nature of weak acids and how to calculate enthalpy changes which are not easy to measure directly.

One of the big challenges scientists face is to understand how human activities are likely to affect the global environment. What is becoming clear is that global conditions are determined by the way in which the oceans, the atmosphere, the land and life interact with one another. This is an enormous task involving all branches of science. As these chapters have shown, chemistry has a major part to play.

8 THE CHEMICAL INDUSTRY

8.1 *Introduction*

The UK chemical industry is a major contributor to both the quality of our life and our national economy. It is the only major sector of UK manufacturing industry which makes a positive contribution to the UK's balance of trade with the rest of the world.

As well as chemists, chemical companies employ people of most other disciplines to fulfil their activities, including accountants, lawyers, linguists, physicists, biologists, mathematicians, computer programmers, chemical engineers and other engineers of many kinds (mechanical, civil, electronic and so on).

The UK chemical industry is the nation's third largest manufacturing industry as measured by the value added, or money generated, in converting basic raw materials to more sophisticated, advanced chemicals and end products. The largest sectors are: food, drink and tobacco; electrical and instrument engineering; and equal third chemicals and paper, printing and publishing. The chemical industry is the largest export earner among the manufacturing industries in the UK and, in global terms, is the fifth largest chemical industry behind the US, Japan, Germany and France in terms of sales (Table 1).

This study of the industry enables you to see the practical importance of some of the chemical ideas you have met.

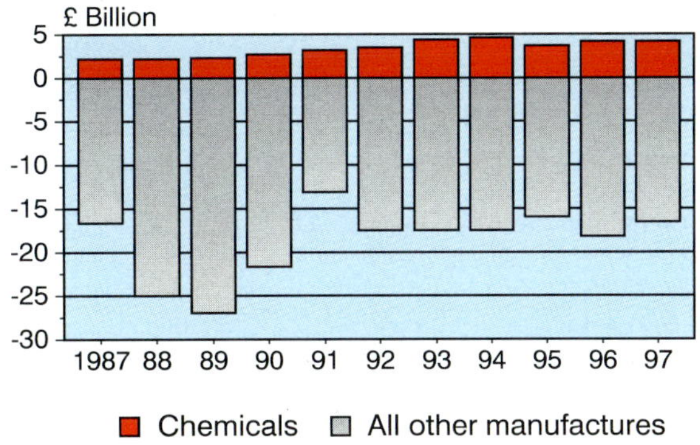

Figure 1 UK trade balance in chemicals and all other manufactured products, 1987–1997

Country	Sales/billion US$
US	405.4
Japan	209.7
Germany	112.9
France	81.0
UK	55.4
Italy	52.8
Belgium	38.4
Spain	31.0
The Netherlands	30.2

Table 1 World chemical sales in 1997 (excluding oil and synthetic fibres)

Figure 2 The steam cracker for ethene production at BP Grangemouth

Table 2 shows the ten largest chemical companies in the world. Although only one company has its head office in the UK, most have manufacturing plants here. Among the large installations in Scotland are Exxon Chemicals at Mossmorran and BP Chemicals in Grangemouth.

Company	Sales/ billion US$	Location of head office
BASF	32.2	Germany
Bayer	31.7	Germany
Hoechst	30.0	Germany
Du Pont	24.1	US
Merck	23.6	US
Novartis	21.5	Switzerland
Dow	20.0	US
ICI	18.1	UK
Rhône-Poulenc	15.4	France
Mitsubishi Chem	14.3	Japan

Table 2 The ten largest chemical companies (sales in 1997)

The chemical business

The chemical industry earns its money by carrying out chemical conversions and selling products either for further chemical reactions or for formulation into final products. These final products cover a very wide range of uses, such as cleaners, paints, inks, medicines, pesticides, polymers, fertilisers, fuels, fuel additives, dyes, and soaps.

ASSIGNMENT I

Consider a typical school day. Make a list of 10 manufactured items that you usually come into contact with during the day between waking up and going to sleep.

Next to each item write down what contribution you think the chemical industry has made to its production. You may prefer to do this with a group of two or three other students.

Now sort your list of items into different categories of chemical industry, for example the polymer industry.

How many items do you have left which have not been dependent on a contribution from the chemical industry?

8.2 *Research and development*

The phrase 'optimum conditions' is much used when discussing industrial processes (as indeed it is in the laboratory).

The phrase 'optimum conditions' is discussed in detail in **Section 23.3**.

When you have prepared a compound from some reactants, either you or your teacher will have decided beforehand what conditions to use. For example:

- what reagents to use
- how much of each reagent
- what temperature
- whether or not to use a catalyst
- how long to allow the reaction to proceed.

You probably did not have time to check whether or not you had the optimum conditions before you did the experiment. However, in industry, much research is carried out to determine optimum conditions. A number of stages can be identified in ironing out difficulties that might occur in transferring the process from the research laboratory to the industrial scale (Figure 3).

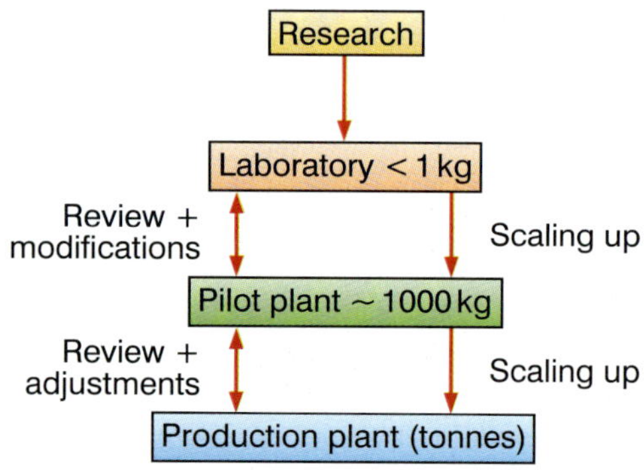

Figure 3 The search for optimum conditions requires several stages before the chemical plant is built

- In the laboratory, small-scale experiments are performed to see whether the manufacture of the compound will be feasible.
- Increasingly larger scale tests are carried out to learn as much as possible about the reaction and the effects of changing conditions such as the catalyst or the concentrations of reagents (Figure 4).

Figure 4 Experiments in the laboratory

- At this point, experiments are carried out on a pilot scale. Here a significant quantity of product will be made. It is usually necessary to change from glass to steel vessels at this stage and the equipment resembles that used on a chemical plant rather than in the laboratory (Figure 5).

Figure 5 Experiments in a pilot plant

- Finally, the fateful decision (possibly involving the investment of hundreds of millions of pounds) to build the plant is taken (Figure 6).

Figure 6 A full scale production plant

At each of these stages, the information obtained is reviewed and modifications are made to the process. When scaling up experiments, it is necessary not only to calculate the masses of reactants to be used, but also to calculate the likely temperature changes. On a larger scale, a large and rapid temperature rise can cause a reaction to run out of control, causing a fire or even an explosion.

8.3 *Building the chemical plant*

The equipment needed to carry out the chemical conversions, the chemical plant, is specifically built or adapted to manufacture a required product. Plants are constructed according to the chemical process being operated and their size depends on the demand for the product. The process may be operated in continuous or batch sequences.

Section 23.1 provides details of continuous and batch processes.

There are many important issues to address in deciding which product to make. For example:

- what is the size of the potential market?
- are there any competitors and, if so, can the company produce the chemical more profitably?
- will the product sell for a price which generates a profit for the company?
- what investment is required to manufacture the product?
- how does the product 'fit in' with the company's other activities?

The plant is the major resource allocated to the manufacture of a product. The economic viability of the product will depend on the returns generated by the capital committed to the plant. The marketing department will generate the necessary information to answer the above questions.

When the first set of questions has been answered, others arise, such as:

- what feedstock should be used?
- does the company have sufficient expertise to develop the process and run the plant?
- how will the plant be made safe for employees and those living nearby?
- how will the environment be protected?
- where should the plant be built?

After answering all of these questions, the company has the information needed to find what are the optimum conditions for making the chemical.

These questions do not simply apply to new plants. Very often, a company will be considering improvements to their existing plants. All the same questions apply and need answering. Further, a company may be manufacturing a chemical and be looking for an alternative process – one for example that is more environmentally friendly, perhaps by using less energy or producing harmless effluent.

The following considerations will be discussed in the rest of this chapter:

- the people needed to run a chemical plant
- safety
- environmental issues
- the choice of location
- the choice of feedstock
- making a profit.

8.4 *People*

The UK chemical industry employs around 230 000 people from many disciplines with many different kinds of expertise and skills. It is necessary, because of the nature of chemical processes, that those directly responsible (chemists, chemical engineers, process operators) are highly qualified and extremely well trained. They work together as a team, with each individual contributing according to his or her different discipline.

The chemical industry is, by and large, capital rather than labour intensive and the relative size of the workforce reflects this. Indeed a large plant, such as one making sulphuric acid, may be operated by only four or five people at a time. Computers and automation are used intensively (Figure 7).

The chemical industry invests heavily in research and development, particularly in pharmaceuticals. Thus, its research and development expenditure was about 7% of sales in 1996, compared to an average of 1.5% for other manufacturing industries.

ASSIGNMENT 2

The list below shows some of the types of employee who might be required in a chemical plant.

Choose *four* from the list and suggest why they might be needed:

accountants
lawyers
linguists
physicists
biologists
computer programmers
engineers
– mechanical
– civil
– electrical
– production
– quantity
biochemists
analytical chemists
safety advisors
cleaners
caterers
transport workers
construction workers
fitters
clerical workers
laboratory technicians
market researchers
economists
librarians.

Figure 7 Computers play a major part in the chemical industry, for example in controlling the temperature, pressure and other conditions for all the reactions in the plant

8.5 *Safety matters*

Safety is a major consideration in all operations in the chemical industry. All aspects of safety are affected by national and European Community legislation. In particular, the Health and Safety at Work Act in the UK places responsibility for health and safety with the employer.

ASSIGNMENT 3

a What other legislation do you know about that affects the safe handling of chemicals?

b Does this legislation apply to the work which you do in your study of chemistry?

Figure 8 Protective clothing in the pharmaceutical industry

An analysis of all possible hazards and an examination of safety will be applied to any proposed project to build new or modified chemical plant. A Hazard and Operability Study (HAZOP) is a systematic procedure frequently used to carry out such an analysis. Every valve, pipe, vessel, pump, etc, is examined and the risk associated with failure is assessed and minimised by design.

Safety considerations play a crucial part in deciding the plant layout. Legislation requires specific design features to minimise risks of uncontrollable reactions and undesirable emissions. The designs of new plants have to demonstrate that the possible risks have been minimised and that emissions or leaks will be prevented.

On a site, there will be people who are responsible for considering specific aspects of safety and for ensuring that every person who works on the site or visits it constantly acts safely and is aware of all possible hazards. Training in the safe operation of the plant is very important.

Figure 9 Safety training at BP Grangemouth

Personal safety is rated very highly and on a typical visit to a production site you may see eye-baths, showers, toxic gas refuges, breathing apparatus, emergency control rooms and, on larger sites, the company's fire brigade, ambulance service and a well-equipped medical centre with its own qualified doctors and nurses.

Some reagents used by chemical companies could become a hazard to people living near the plant in the event of an accident. This particularly applies to poisonous gases or volatile liquids. Where a plant requires the use of a reagent such as chlorine, it will work with the local authority and emergency services. Emergency procedures will be carefully planned and rehearsed at regular intervals. Leaflets and, possibly, a video will brief people what to do if the plant siren sounds continuously. The siren is tested at monthly intervals and this test ensures everyone recognises the warning. The usual procedure for an emergency is for everyone to go indoors and close all windows and doors until the all clear is given.

8.6 *Environmental issues*

The environment is a major factor for chemical companies. It is not acceptable to allow harmful substances to escape into the environment. The resultant pollution which occurs if there is an accidental escape of materials from a plant or disposal of untreated waste products not only damages the environment but also jeopardises the future of the company itself.

There is an increasing quantity of legislation (local, national, European and international) which regulates the performance of chemical operations with respect to environmental issues. The worldwide voluntary 'Responsible Care' scheme has set new standards for chemical operations, leading to fundamental improvements in environmental matters.

Waste generation, treatment and disposal are major issues in process development. Minimising waste by getting the right conditions has already been discussed and the adoption of new clean technologies is helping in this area. However, chemical processes will always produce some waste which has to be dealt with.

Figure 10 Manufacturing chemicals in Widnes, Lancashire, in the 1890s

Figure 11 Widnes, Lancashire, in the 1990s

Figure 12 To remove waste products from the effluent water, the effluent is treated very carefully before it is discharged into a river, a canal or the sea

In the past, waste has been dumped into the nearest convenient place (the atmosphere, old quarry, river, lake, sea). Alternatively, it was contained in purpose-built ponds or tips which have caused problems with toxic materials leaching out into nearby streams and waterways. These methods are no longer acceptable and increasingly are becoming illegal.

Waste must be treated and can only be disposed of when in a state which is not harmful to the environment. Liquid waste from chemical works has to meet legal requirements on such things as pH and metal ion content before being released into natural waterways or sewage systems. It must be treated appropriately, eg to neutralise any acid. Water containing organic waste cannot be discharged into rivers or canals if it would significantly reduce the oxygen content of the water, causing fish to die from lack of oxygen.

Gases which contain contaminants are purified by bubbling them through neutralising solutions to remove soluble contaminants; particles of dust can be removed by filtration or other methods.

Sulphur dioxide, for example, provides two of the most successful examples of the removal of a potential pollutant – thus improving our environment.

In the manufacture of many metals, for example, copper, lead and zinc, the ores are sulphides. They are roasted in air to form the oxide as one of the first stages in their manufacture. For example:

$$2PbS(s) + 3O_2(g) \rightarrow 2PbO(s) + 2SO_2(g)$$

The sulphur dioxide is not allowed to escape but is led away from the furnace to a nearby plant which converts it into sulphuric acid by the Contact process (see **Section 23.3**).

$$2SO_2(g) + O_2(g) \rightarrow 2SO_3(g)$$
$$SO_3(g) + H_2O(l) \rightarrow H_2SO_4(aq)$$

High grade sulphuric acid is produced.

Sulphur is a contaminant in many coals, so when coal is burnt to generate electricity, large amounts of sulphur dioxide are produced.

Longannet Power Station, which is the second largest power station in the UK and one of the largest in Europe, burns Scottish coal, which has some of the lowest sulphur levels in the UK, at less than 1% sulphur. However, even when this coal is burnt, many tonnes of sulphur dioxide are put into the atmosphere and there have been many experiments to reduce this amount.

Figure 13 Longannet Power Station

Scientists at ABB Environmental, working with scientists at the power station, have developed a remarkable process in which the flue gases are washed with sea water. Sea water is naturally alkaline. Sulphur dioxide reacts to produce sulphite ions:

$$SO_2(g) + H_2O(l) \rightleftharpoons SO_3^{2-}(aq) + 2H^+(aq)$$
$$HCO_3^-(aq) + H^+(aq) \rightleftharpoons CO_2(g) + H_2O(l)$$

Air is then passed through the solution to oxidise the sulphite ions to sulphate ions:

$$2SO_3^{2-}(aq) + O_2(g) \rightarrow 2SO_4^{2-}(aq)$$

At the end of the process, the water, containing a very dilute solution of sulphate ions, is discharged into the nearby river and hence to the sea. The carbon dioxide is discharged into the atmosphere, but, in relative terms, these are small amounts compared to those discharged by combustion of the coal and discharging carbon dioxide is much more environmentally favourable than discharging sulphur dioxide.

Other gases which cause great problems in the atmosphere are the oxides of nitrogen, NO and NO_2, known collectively as NO_x. They are discharged by cars, lorries, factories, power stations and anywhere where air is heated to high temperatures and nitrogen and oxygen react:

$$N_2(g) + O_2(g) \rightarrow 2NO(g)$$

Subsequently nitric oxide, NO, is oxidised to nitrogen dioxide, NO_2, in the cool atmosphere.

The oxides are removed catalytically in car exhausts (see **Unit 1**, **Section 2.4**). However, this process is too expensive for power stations, as very large amounts of gases must be treated. An ingenious process is being tried at Longannet. Natural gas is fed into the flue gases, which contain nitrogen oxides, above the flames of the burning coal. The methane is oxidised by the oxides of nitrogen to form nitrogen. The overall reaction can be represented as:

$$CH_4(g) + 4NO(g) \rightarrow 2N_2(g) + CO_2(g) + 2H_2O(g)$$

Excess natural gas is then burnt off, producing more heat which is used in the overall production of electricity. The technique, known as *gas reburn*, is being tested on the largest scale ever attempted.

8.7 *Location*

The chemical industry in the UK grew up around the source of its major raw materials or the customer industry which it served or in a location with good transport links. With increasing specialisation in the UK and a much improved communication network, these factors are no longer the primary considerations in locating manufacturing sites.

New chemical plants are often built near to existing works. This may be because a district specialising in the manufacture of a particular product is more likely to be able to provide the skilled labour needed for new

Figure 14 An aerial view of BP Grangemouth – the visible emissions are water vapour from cooling towers

developments. That's why there are specialist steel producers in Sheffield, a traditional centre for steelmaking.

It may be that the feedstock for the new process is already being produced at the existing site. There may be less opposition to building a new plant at an existing works than to the development of a new site. In addition to these considerations, it may save money if facilities such as a canteen, medical centre and administration block are shared between existing and new plants.

However, many well-established locations are in built-up areas where the increasing concerns of the local population give rise to problems in extending plant or building new facilities. Increasingly, chemical companies are seeking more remote locations for their operations. Then another problem arises for they are then accused of spoiling the countryside.

The actual site chosen should be served by a good communications network, as building this from scratch would add significantly to the cost of opening the plant. Road, rail and water are all important means of transport, and deep sea access is useful for the import and export of bulk materials, such as oil. Ideally the site would be level, free from danger of subsidence and have scope for further expansion. Waste treatment and disposal are always concerns to be borne in mind.

The story of Paraffin Young

Just north of Edinburgh, about 20 minutes by train or car, lies the site of one of the world's most successful chemical industries, Grangemouth. As you get nearer, you will see a huge collection of hardware which is the subject of countless photographs (Figure 14).

Grangemouth looks the perfect location. It is well away from any houses, just in case of serious accidents, yet near enough to cities and towns to allow the many thousands of people who work there to live within a comfortable distance. The site is convenient to a major road network, to allow supplies to be delivered and for its products to be transported away. It is near the edge of the North Sea to allow natural gas and oil to be piped ashore from one of the world's most profitable sources. A triumph of planning, an example for the textbook? Well, the answer is equivocal. It is yes, and no.

Given its proximity to the North Sea oil and gas fields, and that the plant is as modern as anywhere in the world, one might imagine that Grangemouth was chosen as the site at the start of the growth of the chemical industry based on oil – the petrochemical industry. But it can be traced back much further – to 1851 when Dr James Young found oil seeping from coal in the area (Figures 15 and 16). He set up a company based on heating the coal gently to collect the crude oil. He then distilled the oil into several fractions: one was 'lighting oil', which we would now call paraffin, another was an oil for lubrication, and another was a solid wax. Known as 'Paraffin' Young, his work was copied by many others and within 15 years there were 120 similar companies in the area. However, their market was then taken away by the oil industry in the United States, which had large reserves of oil found in states such as Pennsylvania and Texas.

Figure 15 Dr James 'Paraffin' Young (1811–1883)

Figure 16 'Paraffin' Young's company

By 1920, only six companies had survived so, to make themselves competitive, they formed a single company 'Scottish Oils'. Eventually this became part of one of the largest oil companies in the world, BP Amoco. It was Scottish Oils that saw the potential of the Grangemouth site as a refinery, distilling crude oil imported from the Middle East (principally Iran, then called Persia) into its various fractions including petrol, lubricating oil, heating oils, bitumen and waxes. Grangemouth is near to the coast and the smaller tankers could discharge the oil nearby. Later, large tankers had to use Finnart, a port on the west coast of Scotland which has a deep water anchorage, and a pipeline was laid to Grangemouth. The sea, road and rail links ensured that the products could be sent away easily.

In the Second World War, it was impossible to get enough oil from the Middle East, and Grangemouth closed down. It reopened again after the war, but political uncertainty in the Middle East in the 1950s reinforced the obvious conclusion that other sources of oil were needed. Oil and gas is now supplied by the great oilfields in the North Sea (Figure 17).

Not only did coal supply Paraffin Young with oil, it was the basis of the British chemical industry until the Second World War. Natural gas and oil began to supersede coal but it wasn't until the late 1950s and 1960s that the huge growth in the petrochemical industry began. With such a good site as Grangemouth, it was natural to build the chemical plants adjacent to the refinery. Similar developments in the UK occurred in Cheshire (Shell) and near Southampton (Esso).

Figure 17 Bringing the oil and gas ashore

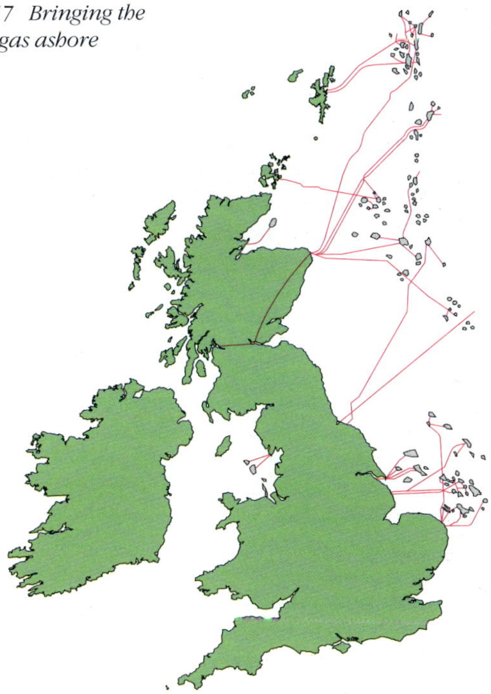

ASSIGNMENT 4

Use a road map (preferably scale 1:250,000) to sketch the area around Grangemouth, showing Edinburgh, Dunfermline, Stirling and Glasgow. Draw in major roads, railways and pipelines.

Use your map, together with information in this section and in **Section 23.2**, to illustrate that the Grangemouth site really does have adequate road, rail and sea connections, and has the ability to manufacture and distribute chemicals.

8.8 *Choosing a feedstock*

The reactants which go into a chemical process are called the **feedstock**. Feedstock is produced from the raw materials. Sometimes there may be more than one possible raw material to choose from. The choice will normally be decided by cost and suitability for the process.

You can find out more about raw materials and feedstocks in **Section 23.2**.

Ethanoic acid is an interesting example. Over the last 100 years, the feedstock has changed several times. The main source of ethanoic acid used to be from ethanol (with the ethanol coming from the fermentation of natural products) by air oxidation. An example of this is the air oxidation of wine into vinegar. By the 1950s, in the UK, the main feedstock was produced from oil (in particular butane and naphtha). Butane or naphtha were dissolved in a solvent and heated under pressure at about 200 °C. A range of products was produced including methanoic, ethanoic and propanoic acids. Although the yield of ethanoic acid was small, it was an economic process as the feedstock was relatively cheap. However, it became increasingly difficult to dispose of the by-products.

Now, ethanoic acid is made from methanol and carbon monoxide, using a rhodium/iodine catalyst:

$$CH_3OH(l) + CO(g) \rightarrow CH_3CO_2H(l)$$

It produces a pure product from a relatively cheap feedstock. (The methanol is produced from carbon monoxide and hydrogen.)

ASSIGNMENT 5

There are five ways of manufacturing phenol (C_6H_5OH). Four of them start from benzene (C_6H_6) and one from methylbenzene ($C_6H_5CH_3$).

Methylbenzene is cheaper per tonne than benzene and this would seem to give an advantage for the route starting from methylbenzene.

a Calculate the mass of
  i benzene
  ii methylbenzene
  needed to manufacture 1 tonne of phenol. (The relative atomic masses you need are H = 1, C = 12, O = 16.)

b Explain how your answer to part **a** shows that judging the most economical raw material is not always a straightforward matter of comparing price per tonne.

Changing the feedstock allows for products to be tailor-made. An example is the production of poly(ethene). High density poly(ethene) (hdpe) is produced using a Ziegler–Natta catalyst (for example titanium(IV) chloride and an aluminium alkyl, see **Unit 2, Section 5.3**). If, instead of using pure ethene, small amounts of another alkene, for example, hex-1-ene, are added, small side-chains are produced along the chain:

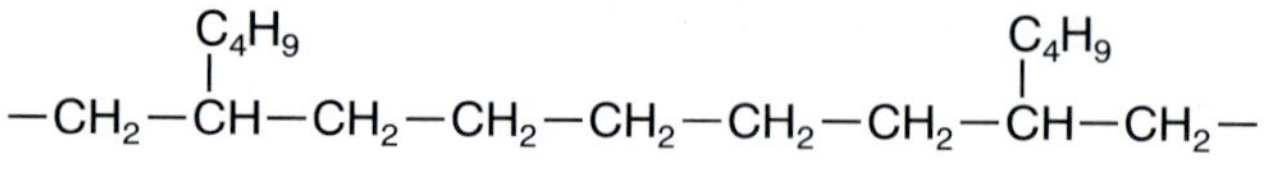

Different alkenes and different relative amounts of them change the structure and hence the properties of the polymer. As there are now side-chains, the polymer molecules do not pack together so regularly as in the linear high density poly(ethene). Consequently, the new polymer, known as *linear low density poly(ethene)*, lldpe, has a lower melting point (the intermolecular bonding is weaker) and is more flexible. The polymers are widely used and are in great demand. It is not necessary to shut down the plant when changing from making high density to low density poly(ethene). As the mixture of reactants is changed, so the end product changes and can be collected separately. This is an example of how versatile modern chemical plants are becoming.

Changing the feedstock provides ingenious ways in which flexibility can be incorporated in the chemical industry. For example, both ethene and propene are needed in large quantities to make a variety of polymers, such as poly(ethene), poly(propene), poly(chloroethene) (pvc) and poly(phenylethene) (polystyrene). The demand for each of these can change suddenly and thus a company needs to be able to alter the proportion of ethene to propene that it produces.

Ethene and propene are made by heating alkanes in the presence of steam at about 800 °C–900 °C, a process known as **steam cracking** (see Figure 2). By changing the feedstock as illustrated in Table 3, the proportions of the different products are changed. It is also possible to change the proportion of different products from one feedstock by changing the conditions under which the feedstock is reacted.

See **Section 23.3** for a discussion on the best conditions for a chemical process.

Product	Feedstock				
	Ethane	Propane	Butanes	Naphtha	Gas oil
hydrogen	5–6	3	1	1	1
methane	10–12	26–28	15–25	13–18	10–12
ethene	75–80	40–45	20–30	25–37	22–26
propene	2–3	14–18	15–25	12–16	14–16
C_4 alkenes	4–6	4–6	15–24	6–10	6–11
others	1	4–6	5–14	22–36	35–44

Table 3 The proportion of products (%) on cracking various feedstocks

ASSIGNMENT 6

Your company is using naphtha as a feedstock. There is a sudden increase in demand for poly(ethene). Which feedstocks could be used instead of naphtha to help you meet this increase in demand? (Use the data in Table 3.)

8.9 *Making a profit*

Chemicals have to be sold at a high enough price to make a profit. This profit benefits society in a number of ways:

- in the form of taxes used for public expenditure on, for example, roads, schools or hospitals
- dividents to shareholders who have lent the company money by investing in it
- money for the company to invest in the future, through new ideas, new research, new developments and new plant.

Without this cycle of investment, followed by research and development of new or improved products, the company would gradually cease to be competitive and would eventually go out of business. This would be bad news for all of us, the shareholders and especially for the company employees.

It is of paramount importance to determine the optimum conditions for the chemical process to obtain the best yield at minimum cost. The factors which are likely to affect the yield are:

- temperature
- concentration (or pressure for a gas-phase reaction)
- catalyst.

These factors will also affect the rate of reaction. Often a compromise has to be reached between the yield and the rate at which the product is obtained.

The Haber process for the manufacture of ammonia from nitrogen and hydrogen illustrates such a compromise – see **Sections 9.1** and **20.2**.

It is also important to obtain feedstocks as cheaply as possible. This can cause a conflict between a company's interest and what is perceived as a national or social interest. One example, often debated, is whether to use gas or oil to generate electricity in the chemical plant or to use coal, which may be more expensive. Coal has the advantage of being more plentiful in the long term and keeps the coal mining industry alive.

And how should we deal with the countryside? Should we quarry limestone from our beautiful countryside, thereby scarring the landscape and creating traffic, or buy it from abroad? If it can be obtained elsewhere with less damage to the landscape and transported by sea, it may be more expensive but our landscape will be preserved. Costs will rise and we may well price ourselves out of a market, thereby creating unemployment. There are simply no easy answers.

Industry is increasingly trying to save materials and energy, thereby improving the environment and cutting costs. An increasingly important issue is to use any other products of the main reaction (known as co-products) and by-products (from side reactions) as effectively as possible. The overall efficiency of a process is eventually seen in its costs.

You can read more about 'Becoming even more efficient' in **Section 23.4**, where you will also find examples of the use of co-products and by-products.

An introduction to the consideration of costs in the chemical industry is given in **Section 23.5**.

ASSIGNMENT 7

Deeside Titanium

Titanium is an important strategic metal whose high strength-to-weight ratio makes it invaluable in the aerospace industry. There is a £25 million titanium manufacturing plant on Deeside, near Chester (see Figure 18). Why Deeside? The plant was originally planned for Hartlepool, but local residents objected and the Deeside location was chosen, attracted by government subsidies. The site was near to a steelworks which was in the process of closing down part of its plant. The site had a number of advantages, including access to a suitable means of waste disposal.

Making titanium

Titanium is manufactured from titanium(IV) chloride, $TiCl_4$, by reducing it with sodium. Deeside has two main suppliers for each of these two raw materials. These four suppliers are situated in France, Germany and the UK.

The raw materials are brought to the plant by road tanker or rail tanker. Both sodium and $TiCl_4$ react violently with water, and sodium reacts with air.

The sodium metal is melted on delivery and is kept molten throughout the process. A precisely measured quantity of sodium is added slowly to liquid $TiCl_4$ in a large pot filled with argon. A highly exothermic reaction occurs, and when it is complete the products, a mixture of titanium and sodium chloride, are allowed to solidify.

The contents of the pot are then broken into lumps by a small explosion so they can be removed. The lumps are washed with dilute hydrochloric acid which dissolves away the sodium chloride. The resulting solution is the main waste product of the process. It is neutralised and filtered before being discharged into the sea.

The Deeside plant can produce up to 5000 tonnes of titanium per year and guarantees a supply of high-quality titanium and titanium alloys to the British aerospace industry.

a List the factors which made Deeside a favourable location for the new plant.

b Suggest why Deeside Titanium has contracts with two suppliers for $TiCl_4$ and two for sodium.

c Why is the sodium kept in a liquid state during the process?

d Why is the sodium added in a precisely measured quantity?

e Write an equation, including state symbols, for the reaction in which $TiCl_4$ is reduced to titanium.

f Why is the reaction vessel filled with argon?

g What would be an appropriate material for making the reaction vessel?

h Which type of process is this, batch or continuous? Why is this the more appropriate type for the process?

Figure 18 The location and transport connections of Deeside

i **i** What is titanium's co-product in this process?
ii What by-products would you expect the process to have?

j Name a chemical that could be used to neutralise the waste solution before pouring it into the sea. Give the reason for your choice.

k It is considered environmentally acceptable to discharge the waste from this process into the sea.
i Explain why.
ii Do you foresee environmental problems that may arise from this waste discharge, in spite of your previous answer?

l For this process, identify
i the major fixed costs
ii the major variable costs.

m What particular safety considerations would arise in the Deeside Titanium plant?

8.10 *Summary*

This Chemical Storylines chapter is linked to Chemical Ideas Chapter 23 and uses ideas first met in Chapters 14, 15 and 20.

These chapters demonstrate the importance of the chemical industry to the many manufactured items we use in our daily lives as well as to our national income. The process of research and development of a chemical from the laboratory stage, through pilot plant to the full size plant, is described. These stages are needed in order to establish the optimum conditions for manufacturing the desired product.

To appreciate what is involved in the design of industrial processes you need to understand ideas about enthalpy changes, factors that affect rates of reactions and factors that affect the positions of equilibria. You also need to understand the characteristics of batch and continuous processes and the problems posed by scaling-up processes.

When it comes to building the plant, other issues have to be addressed – such as the capital investment required, potential market and competition and likely profit. A new plant is very expensive and mistakes could lead to the collapse of the company.

Other decisions required include:

- what people are required to run the plant
- how the plant will be made safe and friendly to the environment
- where to build the plant
- which feedstock to use.

You will find case studies on the extraction of bromine from sea water, the manufacture of ammonia and the refining of copper in **Chapter 9**.

9

CASE STUDIES FROM THE CHEMICAL INDUSTRY

In this chapter you will look at a number of ways in which chemicals and the chemical industry benefit our everyday lives. The examples chosen illustrate aspects of chemistry that you have already met in the course, such as chemical equilibria and rates of reactions, and also introduce you to some new chemical ideas. You will find out about the chemistry of reactions involving oxidation and reduction – **redox reactions** – and about reactions in which the atomic nucleus is involved – **nuclear chemistry**.

You will have opportunities to apply some of the basic principles of chemistry to real-life situations as well as gaining a deeper insight into aspects of the chemical industry.

The case studies discussed are:

- the Haber process for making ammonia
- extracting bromine from sea water
- from minerals to elements: obtaining copper
- using radioisotopes.

Figure 1 Harvesting wheat in Oxfordshire

9.1 *The Haber process for making ammonia*

Why do we need to make ammonia?

How can we grow crops again and again on the same soil without decreasing its fertility?

The fertility of a soil depends on many things, such as the clay and humus content, the moisture content and temperature. It can be damaged by destroying the arrangement of particles and pore spaces in the soil, and by removing too many nutrients.

Nutrient cycling

Apart from carbon dioxide from the air, most plants get all their nutrients from the soil (Figure 2). These nutrients are drawn from an **inorganic store** in the soil, and an **organic store** partly on top of and partly in the soil. Elements are cycled continuously between living systems, the organic store and the inorganic store. The general routes are shown in Figure 3.

The organic store is replenished by animal excretions, and by death and decay of living organisms. Micro-organisms act on organic matter and convert it into *humus*. They also convert it into inorganic ions (*mineralisation*), producing ammonium, nitrate, phosphate and sulphate ions. Weathering of soil and rock minerals releases more ions into the inorganic store (see **Unit 1**, **Section 3.2**).

Nutrients are lost by being leached out of the top layers of soil by rainwater, and nitrogen can be lost by conversion into gases, such as N_2 and N_2O, which disperse into the atmosphere. Figure 4 shows typical values for the total quantities of some elements in different soils.

Many crops make severe demands on the ability of the soil to supply nutrients. This makes the rates at which nutrients are interconverted in the cycles very important. The rate of supply of ions to plant roots has to be rapid enough during the peak growing period to meet the demands of the crops. Crop yields are reduced if there is a shortage of even one nutrient (Figure 5).

In the next section, you will look at one nutrient cycle in more detail.

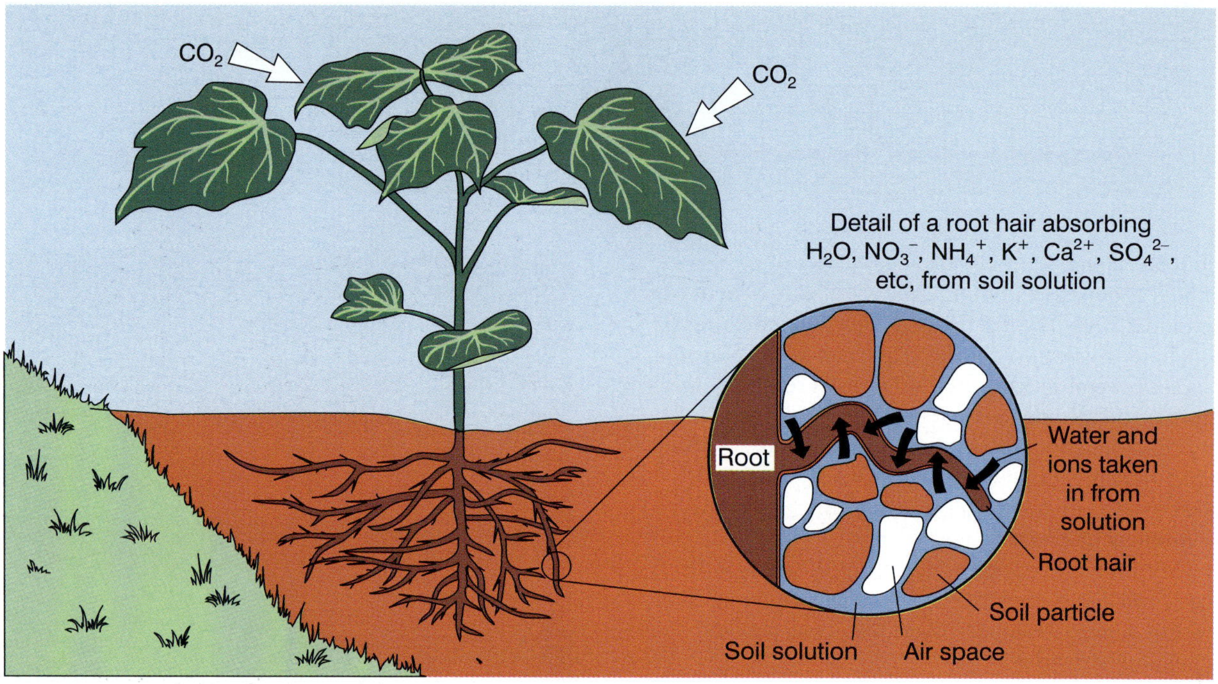

Figure 2 Sources of plant nutrients

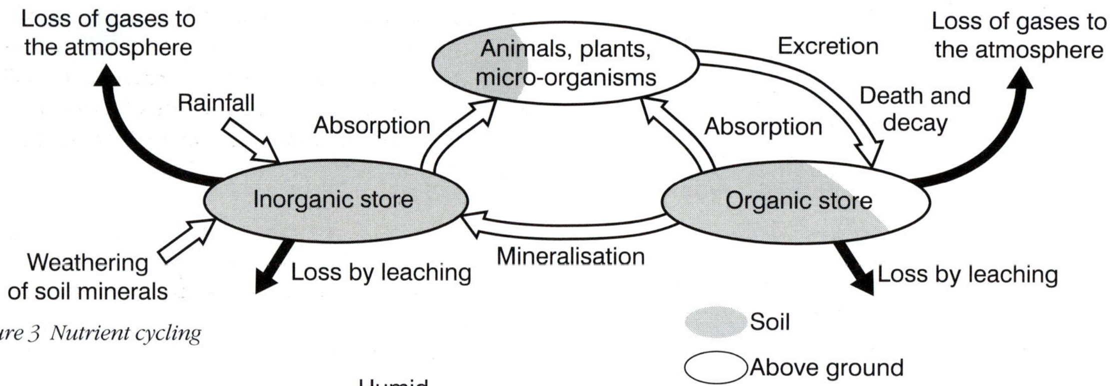

Figure 3 Nutrient cycling

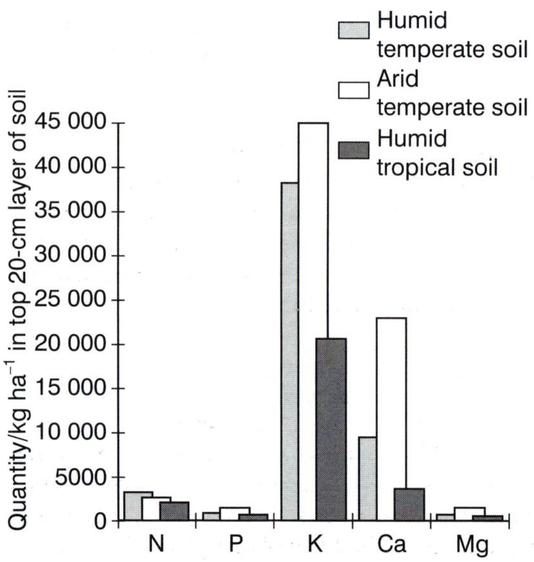

Figure 4 Quantities of some elements present in different soils

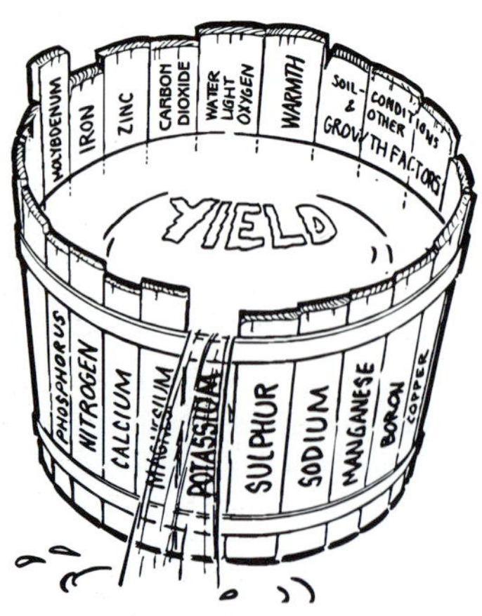

Figure 5 A German chemist, Justus von Liebig, used a barrel to illustrate to his students that a deficiency of any single plant nutrient is enough to limit growth – in the diagram, potassium is the nutrient limiting the yield

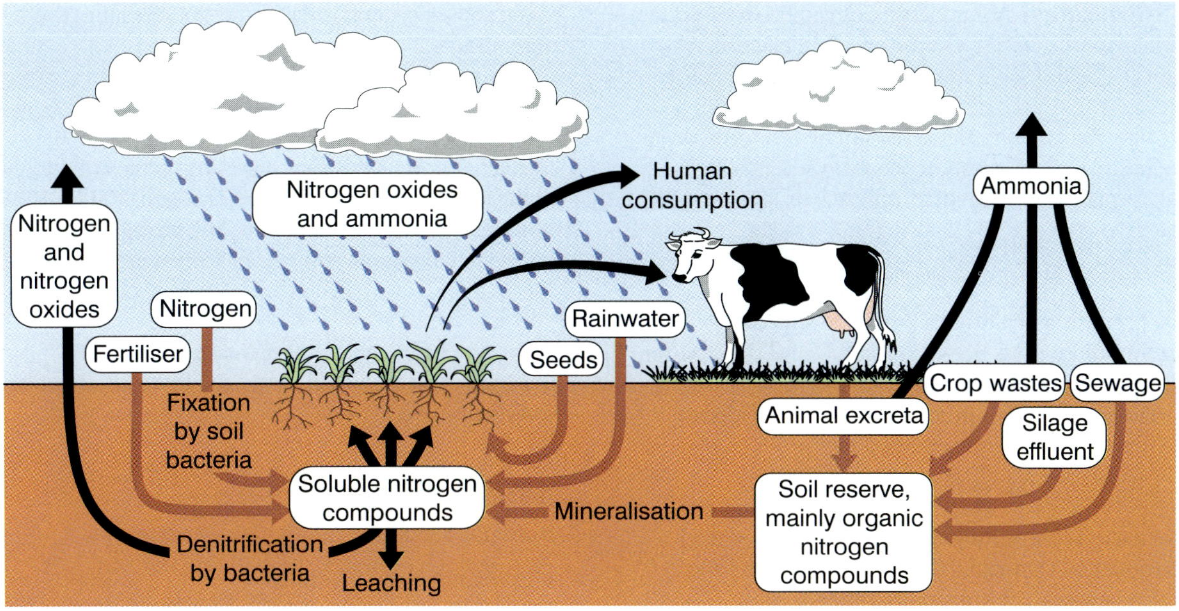

Figure 6 The nitrogen cycle

The nitrogen cycle

Almost all the nitrogen in soil is present in complex organic compounds and so is not readily available to plants. Various processes convert gaseous nitrogen and organic nitrogen compounds into the soluble ammonium and nitrate ions which plants *can* use. The main processes in the nitrogen cycle are listed below. As you read through, refer to the diagram of the nitrogen cycle in Figure 6.

Gains of nitrogen to the soil

Biological fixation Some kinds of bacteria in soil, and in root nodules in legumes such as peas and beans, can convert nitrogen gas to ammonium ions. A reducing agent is needed to provide the electrons. The ion-electron equation for the reduction is

$$N_2(g) + 8H^+(aq) + 6e^- \rightarrow 2NH_4^+(aq)$$

Reduction reactions are discussed in **Section 24.1**.

Figure 7 Lightning generates enough energy to convert nitrogen and oxygen in the air to nitrogen monoxide (nitric oxide)

Other additions to soil nitrogen Lightning, burning of hydrocarbon fuels and natural fires all cause some oxidation of nitrogen in the atmosphere to nitrogen oxides which are carried down to the soil in rainwater. In Europe about 20 kg–40 kg nitrogen are deposited from the air onto each hectare of the soil each year. Some of this is in the form of nitrogen oxides and some as ammonium ions which have come from ammonia emitted by animal excreta.

Hectares

Land areas are conveniently measured in **hectares** (ha).

$1\ \text{ha} = 1 \times 10^4\ \text{m}^2$ (= 2.47 acres)

A football pitch is about 0.5 hectares. Fields range from about 6 hectares to about 60 hectares.

Transformations in the soil

Mineralisation Soil bacteria and other micro-organisms break down organic nitrogen compounds into simpler molecules and ions. Any nitrogen not needed by the organisms themselves is released into the soil as ammonium ions:

$$\text{organic N} \xrightarrow{\text{several steps}} \rightarrow \rightarrow NH_4^+(aq)$$

The NH_4^+ ions are held by clay minerals as exchangeable positive ions (see **Unit 1**, **Section 3.2**).

A well-drained soil is best for mineralisation, and the reaction is much faster at the higher temperatures of tropical soils. Radioactive labelling with ^{15}N (see **Section 9.4**) shows that only 1%–3% of soil nitrogen is mineralised each year.

Nitrification Ammonium ions can be oxidised by certain aerobic bacteria in the soil. The bacteria carry out the reactions as a means of obtaining respiratory energy. The overall process is called nitrification because the end product is the nitrate ion formed via the nitrite ion. Nitrification occurs in several stages. The formation of the nitrite ion can be represented by:

$$NH_4^+(aq) + 1\tfrac{1}{2}O_2(g) \rightarrow NO_2^-(aq) + 2H^+(aq) + H_2O(l)$$

The bacteria which do this are called *Nitrosomonas*. The optimum pH is between 7 and 9, and the reaction stops in dry conditions.

The nitrite, NO_2^-, produced is rapidly oxidised further. This can be represented by:

$$NO_2^-(aq) + \tfrac{1}{2}O_2(g) \rightarrow NO_3^-(aq)$$

The bacteria converting nitrite to nitrate are called *Nitrobacter*. They can tolerate dry conditions and higher acidity than *Nitrosomonas*.

Some nitrification occurs in soils at temperatures down to 0 °C, but it ceases in waterlogged soils where the oxygen content is too low.

Losses of nitrogen from the soil

Denitrification Where oxygen content is low, anaerobic bacteria reduce nitrate ions in the sequence:

$$NO_3^-(aq) \rightarrow NO_2^-(aq) \rightarrow NO(g) \rightarrow N_2O(g) \rightarrow N_2(g)$$

In flooded soils, like those used in rice cultivation, losses by denitrification can be high.

Leaching Nitrogen is lost by leaching mainly as the nitrate ion, which is not held by clays or humus in temperate soils. The quantities lost depend on the soil structure and the amount of rainfall, as well as the nitrate level in the soil.

Loss of ammonia gas Ammonium ions are converted into ammonia under alkaline conditions in some soils. The ammonia then evaporates into the atmosphere.

Uptake by plants In natural systems the quantity of nitrogen removed per year is relatively small. For example, a coniferous forest takes up 25 kg–80 kg nitrogen per hectare each year. Crops cultivated to give high yields take up much more nitrogen, 100 kg–500 kg nitrogen per hectare each year. Soil nitrogen cannot be converted into ammonium ions fast enough to meet this demand, even when the reserves of organic nitrogen are high.

In order to maintain or increase crop yields, nitrogen must be added to the soil in a form that plants can readily use.

Figure 8 Famous fields: researchers at Rothamsted Experimental Station study plots set up more than a century ago

Adding nutrients

Nutrients are added to soil in three ways:

- as manure from livestock
- as inorganic compounds found in nature, such as phosphate rock
- as inorganic compounds specially manufactured, such as ammonium nitrate.

Long-term experiments have been conducted at the Rothamsted Experimental Station in Hertfordshire since 1843, investigating the effects of adding nutrients to the soil (Figure 8). They have shown that yields can be increased, and high productivity maintained, by adding nutrients to the soil.

After a century, the application of farmyard manure to plots of land has more than doubled the organic nitrogen content of soil, increasing its nitrogen reserve. In addition, farmyard manure has encouraged flourishing populations of small animals, such as earthworms, and soil micro-organisms. These assist in converting organic nitrogen into forms usable by plants.

The water content of farmyard manure is high (around 90%–95%), so large quantities of manure have to be used to supply adequate nutrients (see Table 1).

Figure 9 Manure is a good source of nitrogen and other nutrients

	Manure produced/ $t\,yr^{-1}$	Nutrient content/$kg\,t^{-1}$ N	P	K
1 dairy cow	23	4.7	0.6	4.4
10 pigs	21	6.3	1.5	2.9

Table 1 The nutrient content of some farmyard manures

Because of the bulk, long-distance transport of manure is uneconomical. Drying processes have been developed, but processing costs are high.

ASSIGNMENT I

A farmer plans to add 144 kg nitrogen per hectare to a small field of wheat. The area of the field is 7 ha.

a Calculate the total mass of nitrogen needed.

b Calculate the mass of cow manure needed to supply the nitrogen for the field.

While manure is a resource to be used wherever possible, it only returns some of the nutrients removed by livestock back to the soil. It cannot *completely replace* nutrients removed in crops and exported from a farm. Another source of nutrients is needed in order to maintain the productivity of the soil.

Long-term experiments as at Rothamsted show that high levels of productivity can also be sustained using inorganic fertilisers.

You can see the increasing importance of inorganic fertilisers over the last 50 years in Figure 10. During this time average grain yields have almost tripled. Today, inorganic fertilisers support about half the world's cereal production. Without them we cannot hope to feed the world.

How is inorganic fertiliser produced?

The raw materials for fertiliser manufacture are

- water
- air
- natural gas
- phosphate rock (calcium phosphate)
- minerals containing potassium chloride
- sulphur.

The range of fertilisers produced includes *ammonium nitrate*, *ammonium sulphate*, *ammonium phosphate*, *urea* ($CO(NH_2)_2$) and *potassium nitrate*. The compounds are sold individually, or mixed to produce a range of products with different N:P:K ratios to meet farmers' needs.

Fritz Haber's discovery

At the heart of fertiliser production is the synthesis of ammonia from nitrogen and hydrogen. The apparatus shown in Figure 11 was used by the German chemist Fritz Haber to develop the method of synthesis, and with it he managed to produce 100 g of ammonia in 1909.

The process was scaled up by Carl Bosch, a chemical engineer employed by BASF (Badische Anilin und Soda Fabrik) near Mannheim in Germany. In 1913 the first industrial plant went into production with a capacity of 30 tonnes of ammonia per day. Modern plants use the same basic design principles, but with capacities of about 1500 tonnes per day.

About 6500 experiments were carried out between 1910 and 1912 to discover an effective catalyst.

Both Haber and Bosch were awarded Nobel Prizes in Chemistry: Haber in 1918 for his academic work, and Bosch in 1931 for inventing and developing the high-pressure technology.

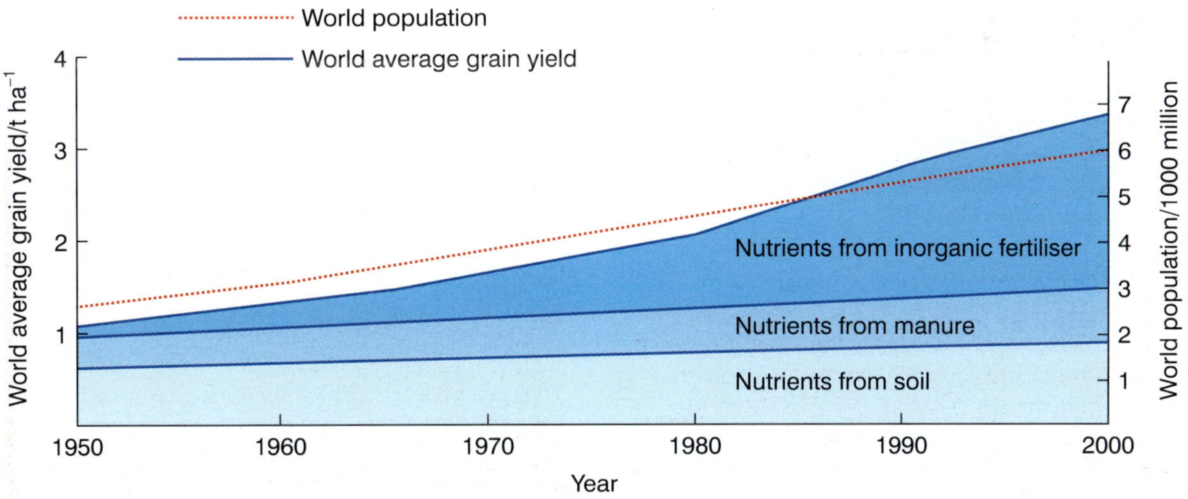

Figure 10 World population figures shown against world trends in grain yields; the shaded areas show estimates of the relative contributions of the different sources of plant nutrients

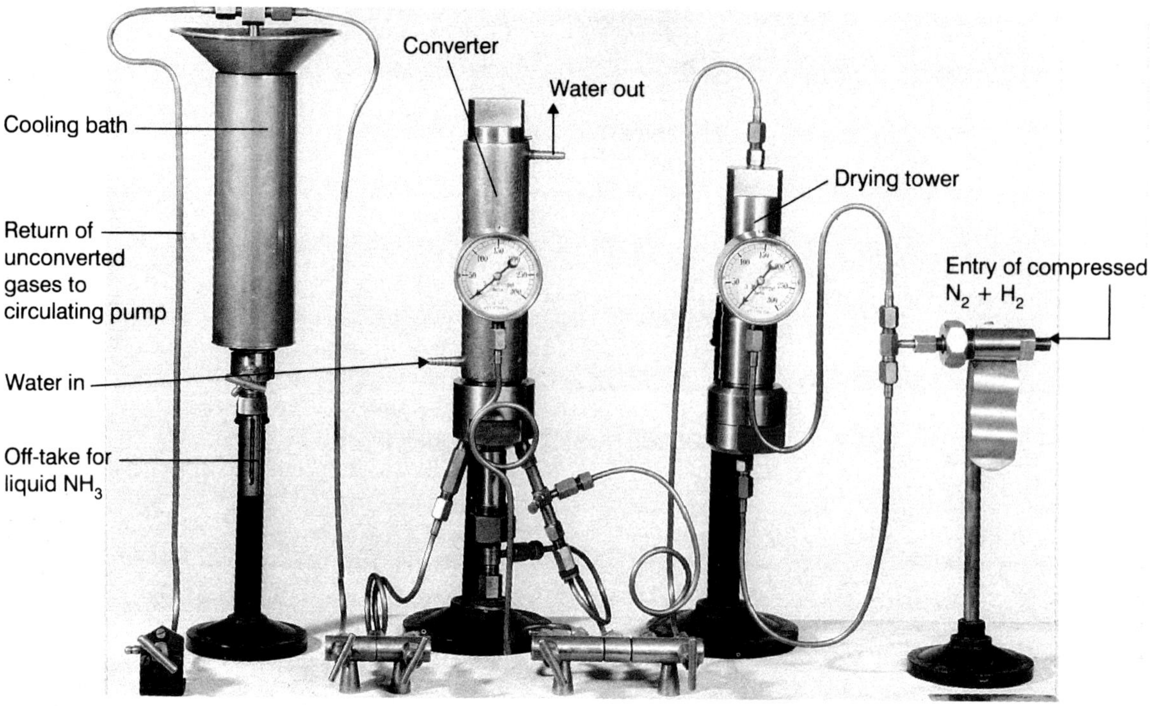

Figure 11 Haber's apparatus for the synthesis of ammonia

Figure 12 Fritz Haber (1868–1934)

The modern Haber process plant

In modern chemical plants in Scotland, the hydrogen needed to make ammonia is made from methane (obtained from natural gas) and steam. Nitrogen is obtained from the air.

The mixture of nitrogen and hydrogen is heated and passes over the catalyst where the gases react to form ammonia:

$$N_2(g) + 3H_2(g) \rightleftharpoons 2NH_3(g) \quad \Delta H = -92\,kJ\,mol^{-1}$$

The reaction is reversible. Ammonia produced in the forward reaction can decompose to nitrogen and hydrogen in the reverse reaction. If a mixture of nitrogen and hydrogen is placed in contact with the catalyst and left for a sufficient time then an equilibrium mixture containing all three gases will be obtained.

Making nitrogen and hydrogen today

In stage 1, methane and steam are passed over a solid catalyst (nickel oxide) contained in tubes (sometimes 350 in parallel) is a furnace at about 800 °C and under pressure of about 30 atmospheres.

$$CH_4(g) + H_2O(g) \rightleftharpoons CO(g) + 3H_2(g)$$
$$\Delta H = +210\,kJ\,mol^{-1}$$

About 90% of the methane is converted into a mixture of carbon monoxide and hydrogen, known as *synthesis gas*.

Air is added in stage 2. Oxygen reacts with some of the hydrogen, producing heat

$$2H_2(g) + O_2(g) \rightarrow 2H_2O(g) \quad \Delta H = -484\,kJ\,mol^{-1}$$

and this heat and steam ensures that the remaining methane forms synthesis gas. Now the gas contains mainly hydrogen, carbon monoxide and nitrogen.

In stage 3, the gases are mixed with more steam and passed over an iron(III) oxide catalyst at about 400 °C.

$$CO(g) + H_2O(g) \rightleftharpoons CO_2(g) + H_2(g)$$

The carbon dioxide is then washed out (with a solution of a base) and the composition of the gas emerging is about 75% hydrogen and 25% nitrogen – ready for the actual synthesis of ammonia.

Pressure/ atm	NH$_3$ present at equilibrium/% 100 °C	200 °C	300 °C	400 °C	500 °C	700 °C
10	–	50.7	14.7	3.9	1.2	0.2
25	91.7	63.6	27.4	8.7	2.9	–
50	94.5	74.0	39.5	15.3	5.6	1.1
100	96.7	81.7	52.5	25.2	10.6	2.2
200	98.4	89.0	66.7	38.8	18.3	–
400	99.4	94.6	79.7	55.4	31.9	–
1000	–	98.3	92.6	79.8	57.5	12.9

Table 2 Volume percentage of NH_3 in equilibrium mixtures in the reaction $N_2(g) + 3H_2(g) \rightleftharpoons 2NH_3(g)$

To investigate how the highest yields can be obtained, nitrogen and hydrogen in the volume ratio 1:3 were mixed and held at different temperatures and pressures. The equilibrium yield of ammonia was recorded. Some results are shown in Table 2.

Section 20.2 explains how variations in pressure or temperature affect the equilibrium position.

ASSIGNMENT 2

To answer this assignment refer to Table 2 and the equilibrium equation for the Haber process.

Explain the observed change in the percentage of ammonia at equilibrium on increasing

a pressure

b temperature.

The equilibrium position depends on the temperature and pressure chosen for the reaction. As you can see from the data in Table 2, the percentage of ammonia in the mixture increases with increasing pressure and falling temperature. Maximum conversion to ammonia is obtained at high pressures and low temperatures.

However, high pressures are very expensive to achieve, and at low temperatures ammonia is produced only very slowly. The reaction conditions chosen are those which give the most economic production of ammonia. It is better to get moderate yields of ammonia rapidly than to wait a long time for a higher yield.

Ammonia is separated from unreacted nitrogen and hydrogen before equilibrium is reached. The unreacted gases are recycled over the catalyst. Thus overall the conversion is always proceeding at a satisfactory rate, leading to a higher yield of ammonia. To separate the ammonia, the gases from the catalytic reactor are cooled. As the pressure is maintained, the ammonia liquefies and can be run off. It is stored in pressurised tanks (Figure 13).

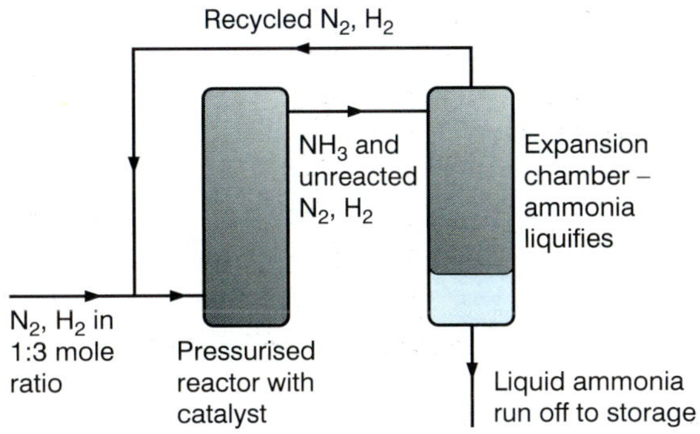

Figure 13 The Haber process

Most reactors now operate at pressures between 25 atm and 150 atm, and at temperatures between 400 °C and 500 °C. Energy consumption in modern factories is about 35 MJ for every kilogram of nitrogen converted to ammonia.

The catalyst used in the Haber process is finely divided iron. Small amounts of potassium, aluminium, silicon and magnesium oxides are added to improve its activity. A modern ammonia plant is shown in Figure 14.

You can read about catalysts in **Unit 1**, **Section 15.5** and about the effect of catalysts on the rate of reaction in **Section 23.3**.

ASSIGNMENT 3

a Explain the difference between homogeneous and heterogeneous catalysis. Classify the catalyst in the Haber process.

b The feedstock for making ammonia must be purified. For example, carbon monoxide should not be present.

  i What is the effect of impurities on the catalyst?

  ii Why may impurities prove costly to the company?

c Draw a fully labelled potential energy diagram to explain why a catalyst increases the rate of a reaction.

Figure 14 A high-technology plant for making ammonia. It uses less energy and produces fewer emissions than more traditional plants

The variety of products made from ammonia are shown in Figure 15. By far the largest use is to convert ammonia to fertilisers such as urea and ammonium nitrate. To make ammonium nitrate, some ammonia is oxidised by a catalytic process with oxygen to nitric acid. Ammonia is then reacted with nitric acid to produce ammonium nitrate.

ASSIGNMENT 4

a Write a balanced equation for the reaction of ammonia with nitric acid.

b What type of reaction is taking place?

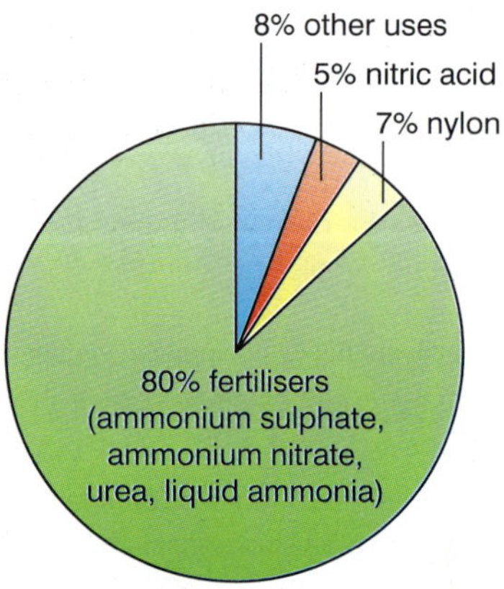

Figure 15 Uses of ammonia

However, ammonia is used to make nitric acid for other purposes, for example, to produce explosives such as nitroglycerine. Ammonia is also used in the manufacture of nylon and its characteristic smell will be most familiar in the kitchen and bathroom where it is used in cleaning agents.

9.2 *Extracting bromine from sea water*

The lowest point on Earth

The Dead Sea is the lowest point on Earth, almost 400 m below sea level in the rift valley which runs from East Africa to Syria. It is like a vast evaporating basin: water flows in at the north end from the River Jordan but there is no outflow. The countryside around it is desert and in the scorching heat so much water evaporates that the air is thick with haze, making it hard to see across to the mountains a few kilometres away on the other side. Steady evaporation of the water for thousands of years has resulted in huge accumulations of salts so that the water in the Dead Sea is much denser than usual.

The Dead Sea is such a natural curiosity that surveys of the salt concentration were conducted as early as the 17th century, even though many of the elements in the salts were then unknown. The water contains about $350\,g\,l^{-1}$ of salts compared with $40\,g\,l^{-1}$ in water from the oceans. A comparison of the most abundant ions present in Dead Sea water and typical ocean water is shown in Table 3.

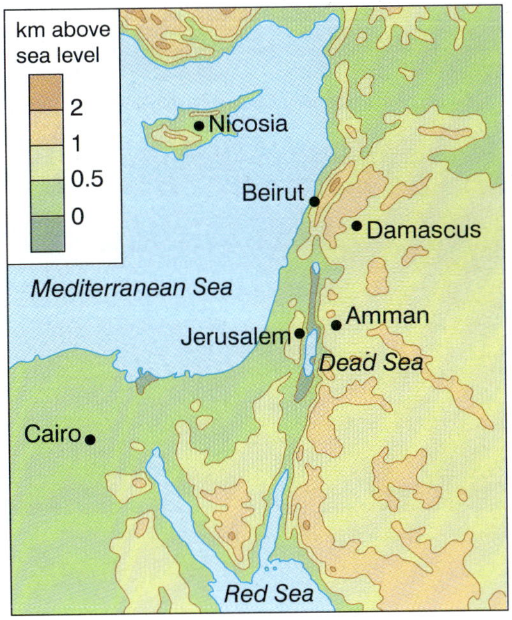

Figure 16 The region around the Dead Sea in relief

Ocean water		Dead Sea water	
Ion	Mass in 1 litre of water/g	Ion	Mass in 1 litre of water/g
Na^+	10.70	Na^+	31.5
K^+	0.39	K^+	6.8
Mg^{2+}	1.29	Mg^{2+}	36.2
Ca^{2+}	0.40	Ca^{2+}	13.4
Cl^-	19.22	Cl^-	183.0
Br^-	0.07	Br^-	5.2
HCO_3^-	0.14	HCO_3^-	trace
SO_4^{2-}	2.51	SO_4^{2-}	0.6

Table 3 Composition of samples of typical ocean water and Dead Sea water

Estimates suggest that there are 43 billion tonnes of salts in the Dead Sea and a particular feature is the relatively high proportion of bromides.

The sea is the major source of minerals in the region. A chemical industry has grown up around the Dead Sea in Israel and it has become the largest exporter of bromine compounds in the world. The annual production of bromine compounds in Israel exceeds 100 000 tonnes.

Figure 17 The Dead Sea – a natural curiosity

Figure 18 Despite the low concentration of bromide ion in the ocean, it is economic to extract the element from sea water

ASSIGNMENT 5

Dead Sea water is certainly more salty than ocean water, but Table 3 shows that there are differences in ionic composition between the two.

a What do you notice about the abundances of the ions of Group 1 and Group 2 in the two samples?

b Compared with ocean water, Dead Sea water has a particularly high proportion of one ion. Which ion is this? How many times more abundant is this ion in Dead Sea water than in ocean water?

The UK bromine industry

In the UK, bromine is extracted from ocean water by Great Lakes (Europe) Ltd, at Amlwch, Anglesey (Figure 18). Currently, just under 40 000 tonnes of bromine are produced annually and this figure is rising as uses for new bromine compounds are found.

ASSIGNMENT 6

a Use the data in Table 3 to calculate the mass in tonnes of ocean water that contain 1 tonne of bromide ion. (Assume that 1 litre of ocean water has a mass of 1 kg.)

b At Amlwch approximately 22 000 tonnes of ocean water are used to produce 1 tonne of bromine. Using your answer in part **a**, calculate the percentage yield of the extraction process at Amlwch.

Turning bromide ions into bromine is simple chemistry: you can do it in the laboratory by adding chlorine water to a solution containing bromide ions.

This is an example of a **redox reaction**. You can find out more about such reactions in **Sections 24.1** and **24.2**.

Industrially, it is more complicated than that and involves some clever chemical engineering, as you will see later. Because the concentration of bromide ions in the ocean is so low, it is necessary to use a four-stage process to extract bromine:

1 oxidation of bromide ions to bromine
2 removal of bromine vapour
3 reduction of bromine to hydrobromic acid
4 oxidation of hydrobromic acid to bromine.

Figure 19 The main pump house at Amlwch – a third of a million gallons of sea water are pumped every minute

Figure 20 The sea water main at Amlwch

1 Oxidation of bromide ions to bromine

At Amlwch, water is pumped from the ocean (Figure 19). After marine debris is removed, the water passes through a large pipeline (Figure 20) where it is acidified with sulphuric acid. The pH of the ocean water is changed from 8 to 3.5. Acidification is necessary because, at higher pH, bromine and chlorine react with water.

$$Br_2(aq) + H_2O(l) \rightarrow HBr(aq) + HBrO(aq)$$
$$Cl_2(aq) + H_2O(l) \rightarrow HCl(aq) + HClO(aq)$$

The first reaction would reduce the overall yield of the process considerably and chlorine would be wasted by the second reaction, adding to the overall costs.

A slight excess of chlorine is injected into the acidified ocean water in the sea water main to displace the bromine by a redox reaction:

$$Cl_2(aq) + 2Br^-(aq) \rightarrow 2Cl^-(aq) + Br_2(aq)$$

Representing this reaction by two ion-electron equations helps to make clear what has been oxidised and what has been reduced.

$$2Br^-(aq) \rightarrow Br_2(aq) + 2e^-$$
$$Cl_2(aq) + 2e^- \rightarrow 2Cl^-(aq)$$

Electrons appear on the product side of the first ion-electron equation so it is clear that bromide ions have been oxidised to bromine because they have lost electrons. Electrons appear on the reactant side of the second ion-electron equation. Chlorine has been reduced because it has gained electrons to form chloride ions.

$Cl_2(aq)$ is the oxidising agent because it has removed electrons from the bromide ions. $Br^-(aq)$ is the reducing agent because it has given electrons to the chlorine.

2 Removal of bromine vapour

The treated ocean water then passes into a 'blowing-out tower' (Figures 21a and 21b). A strong current of air passes up the tower as the water is sprinkled over the packing. Bromine is volatile so the air removes most of the bromine and some chlorine from the water. At this point the bromine concentration in the air is too low for effective separation of liquid bromine.

Figure 21a The blowing-out towers – here bromine is blown from the treated water by air and then reduced to hydrobromic acid

Figure 21b Inside a blowing-out tower

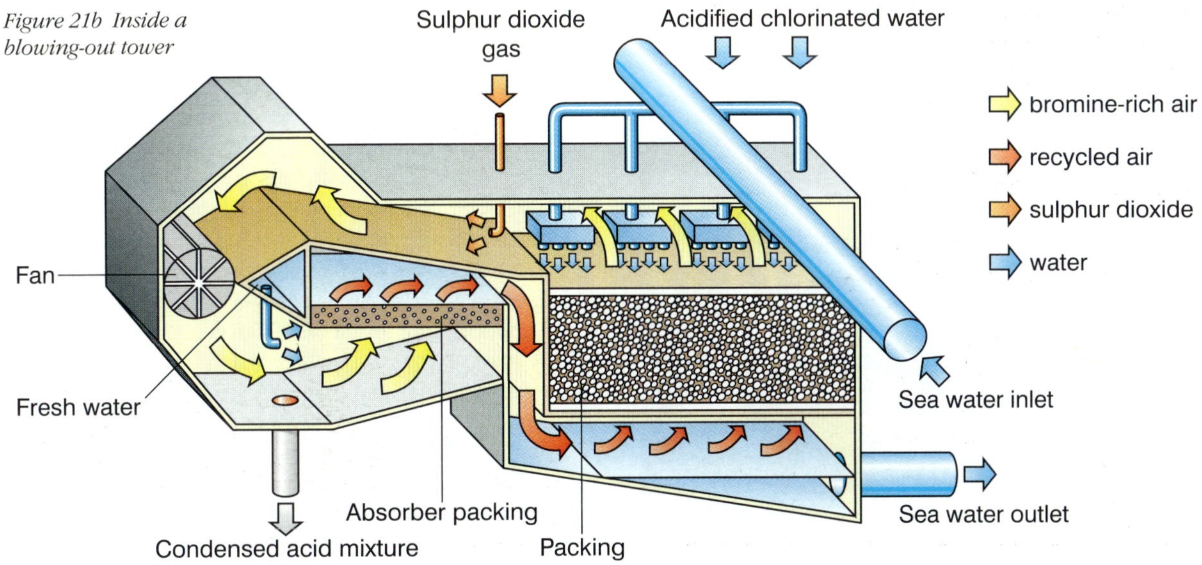

3 Reduction of bromine to hydrobromic acid

To raise the concentration of bromine further the element is first reduced to hydrobromic acid by injecting sulphur dioxide gas into the blowing-out tower (Figure 21b). The reaction also produces sulphuric acid:

$$Br_2(aq) + SO_2(g) + 2H_2O(l) \rightarrow 2HBr(aq) + H_2SO_4(aq)$$

Injection of fresh water into the blowing-out tower produces a fine mist of the acids which is condensed in the absorber section of the tower. The acid mixture at the end of this stage contains about 13% by mass of bromine as hydrobromic acid.

4 Oxidation of hydrobromic acid to bromine

The collected acid is passed into the 'steaming-out tower'. Here, chlorine regenerates the bromine, which is then distilled out of the tower with steam (Figure 22).

$$2HBr(aq) + Cl_2(g) \rightarrow Br_2(g) + 2HCl(aq)$$

The hot vapour mixture is condensed to form an aqueous layer and a lower layer of bromine. The resulting damp bromine is dried with concentrated sulphuric acid.

The spent sea water is returned to the Irish Sea. Any chlorine or bromine present is destroyed by the presence of a small excess of sulphur dioxide before the water leaves the blowing-out tower. On discharge it has a pH of 3.5 but this rapidly falls to pH 8 within 100 m of the outfall.

Sulphur dioxide, sulphuric acid and hydrochloric acid are all produced at Amlwch. Chlorine is produced in Cheshire by the electrolysis of brine and transported to Amlwch by road or rail.

Figure 22 After leaving the steaming-out tower, bromine condenses to a deep red liquid

ASSIGNMENT 7

Write ion-electron equations to represent the oxidation and reduction processes occurring when sulphur dioxide is used to convert bromine to hydrobromic acid.

What is the oxidising agent and what is the reducing agent in this redox reaction?

Dark, dense and dangerous

Bromine was discovered in 1826 by a French chemist, Jérôme Balard, who later became a professor at the Sorbonne in Paris. His discovery helped Johann Döbereiner spot the idea of 'triads', which was a step on the way to the development of the Periodic Table.

Bromine has a dense and choking vapour – hence its name which is based on a Greek word, *bromos*, meaning *stench*. The liquid produces painful sores if spilled on the skin.

The safety of employees and the surrounding population of Amlwch is of paramount concern. The company continually makes checks on the quality of air and the fresh and ocean water around the site. Other independent bodies such as the Health and Safety Executive, the Environment Agency and the Pollution Inspectorate also make regular checks.

The extraction process is a continuous process and takes place in sealed equipment. Such sealed systems ensure a high level of safety and protection. When maintenance or repair is required, a section is isolated by valves. A continuous process enables close monitoring of the whole plant. This is done by a system of sensors connected to a computer in a control room. Signs of trouble can be spotted long before they become serious. A system of automatic alarms and shutdowns provides further protection.

As a result of this care, the plant is regarded as a very clean plant. Nevertheless, one can never be complacent when handling harmful material. If a major emergency did occur, well-rehearsed plans would come into effect. These plans are a regular feature of staff training with the local emergency services. A siren on the plant also alerts the local population. The public are briefed to go indoors immediately and close all windows and doors. They must stay indoors until the all clear is given.

Great care has to be taken when transporting bromine. Most of it is carried in lead-lined steel tanks supported in strong metal frames (Figure 23). Each tank holds several tonnes of the element. International regulations control the design and construction of road and rail tankers. The industry's safety record is good but there have been some accidents.

Figure 23 A road tanker for the transport of bromine – note the protective clothing worn by the operator

The need to transport chemicals safely cannot be overemphasised. This is shown by the report in Figure 24, which describes an accident in Israel.

ASSIGNMENT 8

According to *The Jerusalem Post* report (Figure 24), a safety expert recommended more careful monitoring of the transportation of dangerous chemicals. In the light of the report, what do you think the chemical industry does to reduce the risk of injury or death when chemicals such as bromine are transported?

Figure 24 Report from The Jerusalem Post

CHEMICAL SCARE SHAKES NEGEV

The driver of the truck that overturned near here early yesterday morning, sending an orange cloud of poisonous bromine gas into the sky over the Arava, said before he died that the accident occurred because he fell asleep at the wheel. The cloud of gas formed by the evaporating bromine, which had a diameter of some 10 kilometres later in the morning, forced scores of residents to leave their homes for several hours. Negev police closed off the road for the day.

According to a senior member of the Dead Sea Works rescue crew, the driver, Yisrael Taib, 32, of Dimona, was trapped when his flatbed truck flipped over on its side at about 5.15 am after rounding a bend some 20 kilometres south of the Arava junction.

Eight people were injured as a result of the accident, most of them rescue workers who tried frantically but in vain to extricate Taib, who was crushed by the steering wheel and the engine block in the mangled cabin.

A senior police officer on the scene praised the rescue crew for the speed with which it reached the spot.

The rescuers arrived shortly after 6 am and worked for several hours, even after a wind change sent the poisonous gas straight towards them.

A safety expert from a major industrial concern in the Negev told *The Jerusalem Post* that the bromine had apparently been packed according to international safety standards, but that yesterday's accident highlighted the need for more careful monitoring of the transportation of dangerous chemicals.

The expert, who sits on a national safety committee, told *The Post* that repeated efforts over the years to have 'black boxes' installed in the cabins of trucks carrying dangerous materials had been thwarted by trucking officials. The expert, who asked not to be identified, conceded that the black boxes wouldn't have prevented yesterday's accident, but would have told safety officials how fast the driver was travelling and how many hours he had been working.

The truck, on its way from the Dead Sea bromine plant south of Beersheva to the port in Eilat, was carrying two containers with over 20 tons of liquid bromine.

They said it was the first accident ever in Israel involving bromine, a deep rust-coloured liquid used in industry, agriculture and medicine. The bromine evaporated quickly in the desert heat. Bromine's boiling point is 59 °C.

Bromine may once have been just a laboratory curiosity but it is now an important industrial commodity which is manufactured on a large scale. An important use is in the production of flame retardants (Figure 25). It is also used in water purification and the manufacture of dyes. Bromomethane, CH_3Br, is one of the most important chemicals used against insect pests. The problem is that bromomethane has a strong ozone depletion potential (see **Unit 2**, **Chapter 6**) and there is a very large effort to find a replacement for it as a fumigant. Silver bromide is **photosensitive** – an aspect of bromine chemistry which makes modern techniques of photography possible.

Leaded petrol contains antiknock agents such as tetraethyllead and tetramethyllead to improve its ignition properties. Until the advent of unleaded petrol, 1,2-dibromoethane was manufactured in considerable quantities as it is added to leaded petrol as a scavenger for the lead. It prevents lead oxides being deposited in the engine by converting them to volatile lead bromides which are removed in the exhaust gases. As much less leaded petrol is now used, the relative proportions of organic bromine compounds produced has changed.

At the Amlwch site, there is a batch reactor (Figure 26) which can be used for the manufacture of different organic bromine compounds. Many of the organic bromine products are specialised chemicals in the sense that they are required by other companies in relatively small amounts and at irregular intervals. The company is able to manufacture one product and then switch the reactor to a different product. It may be a few years before there is a need to produce more of any one product.

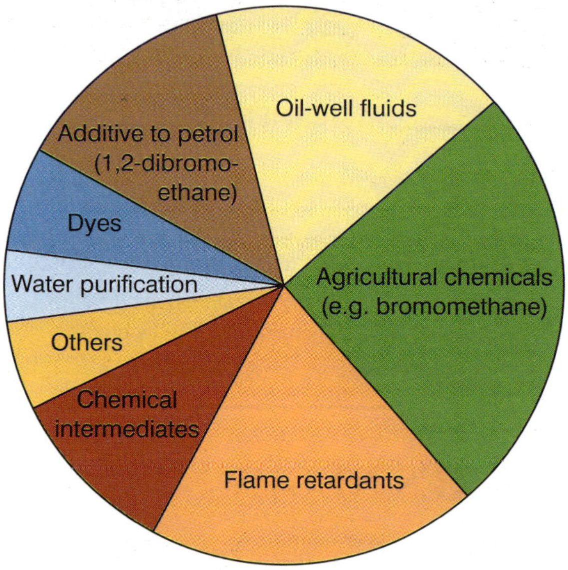

Figure 25 A pie chart showing the uses of organic bromine compounds

Figure 26 The multi-product bromination reactor

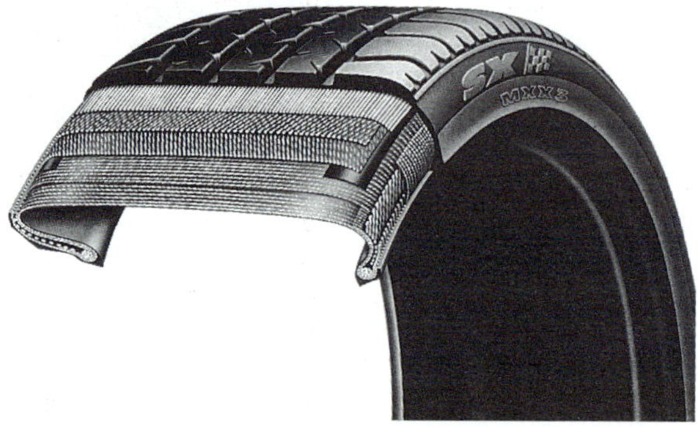

Figure 27 A cut-away section of a radial car tyre: some bromine is used in the manufacture of bromobutyl rubber which forms the airtight lining on the inside of the tyre

ASSIGNMENT 9

Explain why a batch reactor is more suitable than a continuous reactor for producing a range of chemicals.

With the help of the sketch map (Figure 28), suggest factors which make Amlwch on Anglesey an appropriate location for a bromine plant.

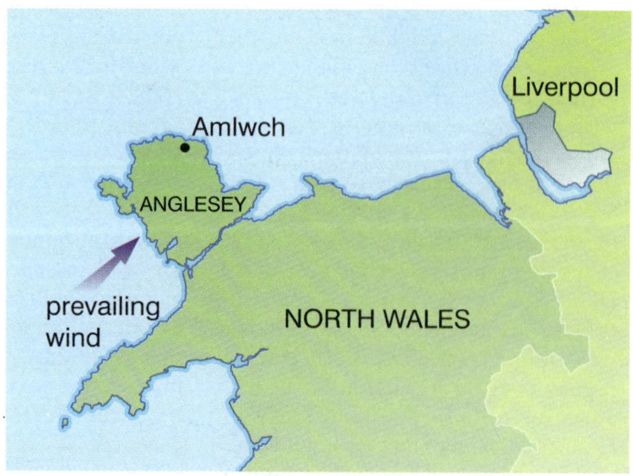

Figure 28 The location of the bromine plant at Amlwch, Anglesey

9.3 From minerals to elements: obtaining copper

A few unreactive elements, such as the precious metals gold and silver, occur in the Earth's crust in the free state. Most elements, however, occur combined with other elements in compounds.

Although native copper can be found in some areas, most of the copper we use today has been extracted from copper-containing minerals. This case study describes how pure copper is obtained from the mineral chalcopyrite ($CuFeS_2$).

The biggest hole in the world

Towards the end of the 19th century, the demand for copper rose rapidly in response to the need of 'the age of electricity' for wires and other articles. There were not enough rich copper deposits to meet this demand. Deposits with only 0.5%–1% copper in the rock had to be used. One place where this was profitable was at Bingham Canyon in the US (Figure 29).

Bingham Canyon is near Salt Lake City in Utah. Copper has been extracted since 1896. Although the percentage of the mineral chalcopyrite in the rock is small, the total quantity of the mineral is enormous. So it was worth building up a process which could handle the low-grade material and pick out the small proportion of chalcopyrite present. Over 5 000 000 000 tonnes of material have been removed so far.

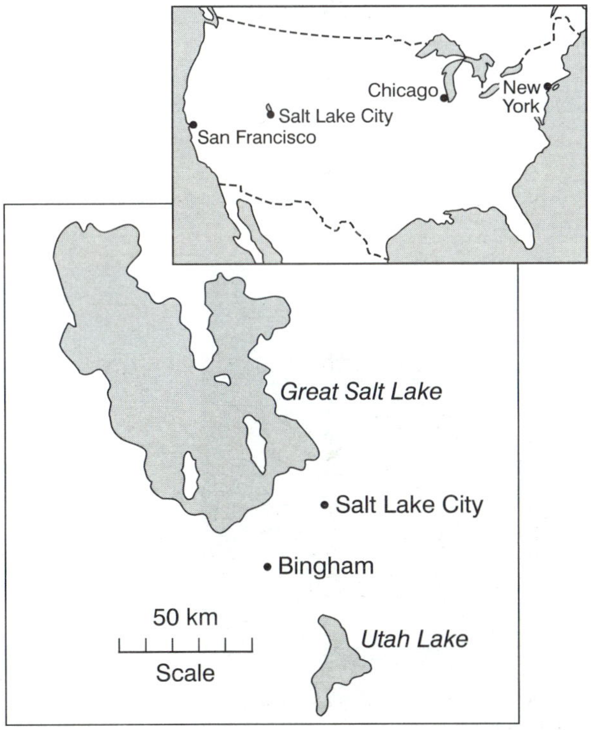

Figure 29 Location of Bingham Canyon, Utah, US

Development of the technique of **froth flotation** made it possible to separate the chalcopyrite 'needles' from the rock 'haystack' and made mining at Bingham Canyon profitable. The crushed ore is mixed with water and detergent in a large tank. Oil is added and when air is blown through the mixture, a froth forms on the top which contains most of the mineral. The waste rock sinks to the bottom. The operation was shut down from 1985 to 1988 while it was modernised to fall in line with tighter environmental legislation.

Figure 30 Bingham Canyon mine

Getting the copper

Figure 31 describes the Bingham Canyon process in outline. There are four main stages: **mining**, **concentration**, **smelting** and **electrolytic refining**. The scale is huge.

During the smelting stage, the concentrated ore is mixed with sand (silica) and roasted with a controlled amount of oxygen. Sulphur is converted to sulphur dioxide, the iron forms a silicate slag and impure liquid copper is run off from the smelter. The copper is cast into positive electrodes ready for the **electrolytic refining**, which is explained in detail later.

Smelting produces large quantities of sulphur dioxide which leads to serious pollution if its release into the atmosphere is not strictly controlled. Breathing air which contains sulphur dioxide causes respiratory damage, and sulphur dioxide is also one of the gases responsible for acid deposition (or 'acid rain' as it is sometimes called). Acid deposition is known to have damaged the ecology of many lakes in parts of Europe, Scandinavia and North America, and to be partly to blame for disease and death among trees in large areas of forest in the same regions.

ASSIGNMENT 10

Make a list of some of the environmental hazards which may be associated with mining and mineral processing.

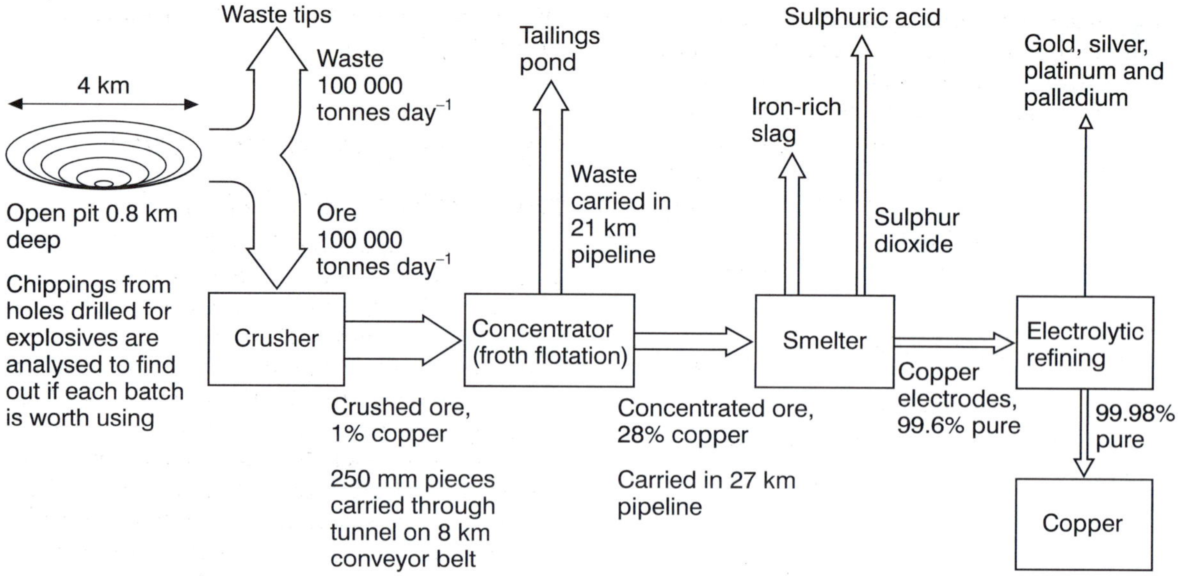

Figure 31 Outline of the Bingham Canyon process

ASSIGNMENT 11

This assignment makes you think about the waste created at Bingham Canyon each day. The company mines 200 000 tonnes of rock of average density 2.5 tonnes m^{-3} each day. The rock contains 0.5% copper by mass on average.

a What volume of rock is removed each day?

b What is the approximate volume of waste which has to be disposed of each day?

c Which of the stages – mining, concentration or smelting – contribute significantly to the production of waste?

d A large office block built recently in Glasgow has a ground floor area of 1200 m^2 and is 50 m high. Compare the volume of the building with the volume of waste produced each day at Bingham Canyon.

You can read about electrolysis in **Section 24.3**.

Electrolytic refining

The copper positive electrodes which are produced from the smelter are 99.6% pure, but this still isn't pure enough for the copper which will be used to conduct electricity. This has to be at least 99.98% pure.

The final purification is done using electrolysis. Figure 32 shows you an illustration of how this is done.

The ion-electron equation for the reaction occurring at the negative electrode is

$Cu^{2+}(aq) + 2e^- \rightarrow Cu(s)$ reduction

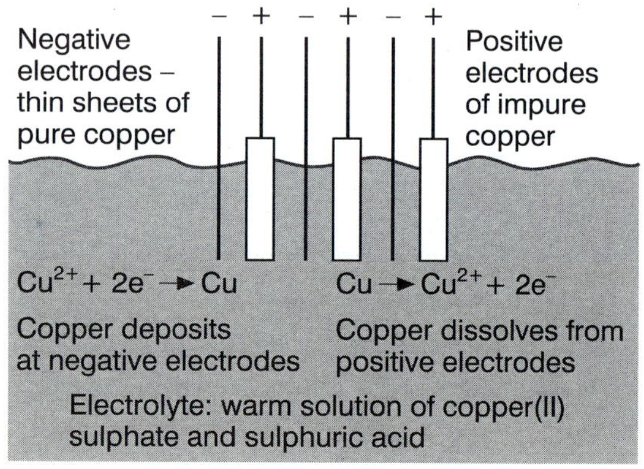

Figure 32 Copper refining

and at the positive electrode is

$Cu(s) \rightarrow Cu^{2+}(aq) + 2e^-$ oxidation

Electrolysis takes about 2 weeks and during this time the negative electrodes increase in mass from as little as

Figure 33 Electrolytic refining of copper

5 kg to 120 kg. Some modern cells use thin sheets of stainless steel rather than pure copper for the negative electrodes. The negative electrodes are removed and washed. The copper is stripped from the stainless steel sheets which can be re-used many times. Some of the insoluble impurities are valuable (eg silver and gold) and these are extracted from the sludge which forms. Other impurities are soluble and contaminate the solution, which has to be replaced as electrolysis goes on.

ASSIGNMENT 12

In one copper refinery, the current is passed for 11 days. The voltage is 0.5 V and the current density is 220 A m^{-2}.

a Under these conditions, what is the mass of copper deposited on a negative electrode of area 1 m^2?

The refinery employs 190 people and produces 280 000 tonnes of copper annually. From the impurities around 15.5 tonnes of gold and 124 tonnes of silver are obtained annually.

b Use these figures to estimate the approximate percentage by mass of gold and silver in the impure copper.

c If the market price for each metal, in £ $tonne^{-1}$, is as follows, Cu 900, Ag 125 000, Au 6 400 000, calculate the value of sales for each of these metals in one year.

Using copper

Copper chemistry also plays an essential role in the processes of life, but the everyday uses of copper which are most familiar to us rely on copper's *lack* of chemical reactivity!

Because it does not react with water, even when heated, copper is used for domestic water pipes and hot water cylinders. Copper reacts very slowly with air and retains its untarnished, attractive, metallic appearance when used for objects of art or decoration. Often it is combined with other metals to form **alloys**, for example *bronze* (95% Cu, 5% Sn) and *brass* (70% Cu, 30% Zn).

Our coinage is based on copper: 'copper' coins contain 97% copper mixed with tin and zinc, although since 1992 1p and 2p coins have been made of copper-plated steel; 'silver' coins are really made from *cupronickel* (75% Cu, 25% Ni).

Pure copper has a high electrical conductivity and is tough – it can be bent and twisted without breaking. That is why it has to be electrolytically refined: tiny quantities of impurity cause it to snap easily or to get dangerously hot when current is passed through it.

Almost 60% of all the copper produced is used in electrical applications.

9.4 *Using radioisotopes*

Radioactivity and Marie Curie

Radioactivity was discovered in 1896 by the French physicist, Antoine Henri Becquerel. He left an unexposed photographic plate in the dark near a sample of a uranium salt. When the plate was developed it was found to be fogged, just as it might if it had been exposed to light. The fogging of the plate was caused by a form of radiation from the uranium.

The term **radioactivity** was coined by Marie Curie and her husband Pierre Curie. They worked together and showed that radioactivity was an atomic property not a chemical change. The discovery of radioactivity won the Curies and Becquerel the Nobel Prize for Physics in 1903.

Marie Curie continued her work on radioactive elements and isolated radium (Figure 34). For this work she won her second Nobel Prize, this time for Chemistry, in 1911. To isolate radium, she started with several tonnes of uranium ores and after very many repeated fractional crystallisations obtained the salt radium chloride. Her work with radioactive elements eventually caused Marie Curie's death, in 1934, from leukaemia.

You can read more about radioactivity and radioisotopes in **Section 25.1**.

Beneficial uses of radioisotopes and radioactivity

Although radioactivity can be exceedingly harmful, there are now very many beneficial uses of radioisotopes. In medicine, they may help diagnose illness as well as

Figure 34 Marie Curie at work in her laboratory

effecting cures. In industry, radioactivity, in the form of γ-rays, may help with diagnosing invisible faults in metal constructions. In chemical and biochemical research, radioisotopes help us to identify reaction pathways. In biochemistry, this understanding may help us to develop new ways to prevent or treat diseases. Carbon dating has enabled archaeologists to date historical items with more certainty. We have also learnt how to harness energy from the fission of certain radioisotopes to produce electricity.

Radioisotopes in medicine

An iodine-deficient diet leads to swelling of the thyroid gland, a condition known as goitre. This condition can be relieved simply by increasing iodine consumption.

Iodine is taken into the body as either iodide ions, sometimes added in small quantities to ordinary salt and also present in very small quantities in drinking water, or as iodate ions, (IO_3^-) present in marine plants and animals. Iodide ions are also present in cows' milk, particularly in winter when cows are given iodised salt licks to supplement their diet.

Almost all the iodine stored in the body can be found in the thyroid gland, in the neck. This store of iodine is referred to as the *iodine pool*. If the amount of iodine taken into the body is in excess of the body's requirements, the surplus iodine is excreted.

Ingested iodine enters the bloodstream and, if required by the iodine pool, is quickly taken up by the thyroid gland. Here, a sequence of reactions takes place to produce two iodine-containing hormones, tetraiodothyronine and triiodothyronine (Figure 35).

An underactive thyroid gland produces insufficient quantities of these hormones. In children this causes retarded mental and physical development and a characteristic appearance known as cretinism. Cretinism can be cured through treatment with tetraiodothryonine and triiodothyronine.

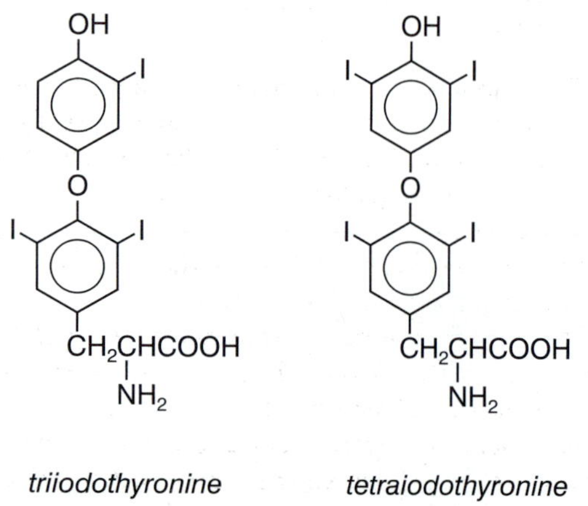

Figure 35 Iodine-containing thyroid hormones

ASSIGNMENT 13

Identify the functional groups present in the iodine-containing hormones.

An over-active thyroid gland produces too much of these hormones, which results in an increase in heart rate and other body functions. Diagnosis of this condition, or other thyroid gland problems, may involve the use of iodine-123. This isotope of iodine decays with a half-life of 13 hours, releasing γ-radiation. In the past, iodine-131 was used for such diagnostic studies. Iodine-131 decays with a half-life of 8 days and releases β-particles as well as γ-radiation with higher energy than that from iodine-123. The radioactive iodine-123 is taken in with food and incorporated into the thyroid gland in exactly the same way as normal iodine.

When atoms of a radioactive isotope decay, dangerous ionising radiation is released. Some ionising radiation is around us all the time because of radioactive isotopes in the natural environment. This type of radiation is called **background radiation**. It is very unlikely to affect us as it is very low level radiation. Higher levels of ionising radiation can damage or even destroy cells. Hence the dose of radioactive isotopes used in medicine must be very carefully controlled.

Radioactive iodine atoms will concentrate in the thyroid gland and in the hormone molecules. The γ-radiation from iodine-123 or iodine-131 can be detected from outside the body using a scintillation camera. The resulting scan enables doctors to make an accurate diagnosis and recommend appropriate treatment. The use of iodine-131 has been discontinued for diagnostic purposes due to the large number of β-emissions. The absorbed radiation dose to the thyroid gland was correspondingly high. Such a radiation dose could result in the development of thyroid cancer much later in the life of a patient.

Nowadays, iodine-131 is only recommended for the actual treatment of cancerous thyroid cells in elderly patients for whom other treatments have failed.

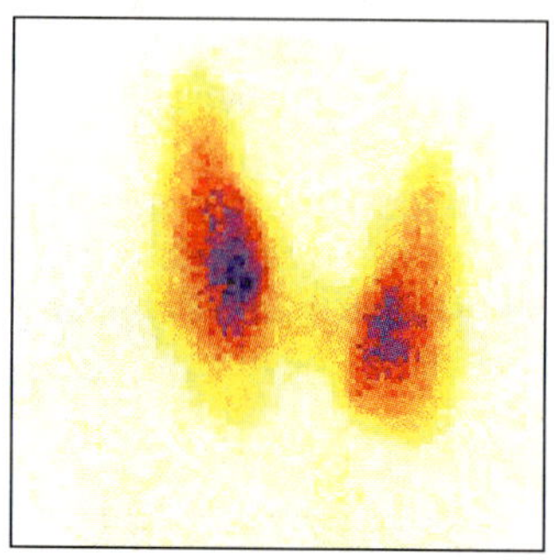

Figure 36 A technetium-99 scan of the two lobes and isthmus of the thyroid gland. Technetium-99 is now a preferred alternative to iodine isotopes

Large doses of iodine-131 can be fatal. In normal circumstances this would be unlikely to happen, but iodine-131 is a common product of nuclear fission reactions. In the event of a nuclear accident, such as that which occurred at Chernobyl in 1986, iodine-131 would be released into the atmosphere. From there it can enter the food chain or be ingested directly. Once incorporated into the thyroid gland, the iodine-131 will start to destroy the gland.

The body can be protected from iodine-131 if it has a plentiful supply of stable iodine in the iodine pool. Immediately after the accident at Chernobyl, people in areas likely to be affected were given potassium iodate tablets to ensure their bodies had a plentiful supply of stable iodine. Any extra iodine (like iodine-131) which then entered the body was excreted. Countries now hold supplies of potassium iodate tablets for use in the event of a nuclear accident.

Other radioactive isotopes have also been successfully used for the treatment of cancers. Such treatment is very tricky as the radiation kills both healthy and cancerous cells. The aim is to destroy a tumour but cause minimal damage to neighbouring healthy cells. The dose of radiation has to be very carefully calculated and a narrow beam is applied. By applying brief exposures from different directions on several occasions, the tumour receives a much higher dose than surrounding tissue. Cobalt-60 is a frequently used source of γ-radiation for such treatments (Figure 37).

Radioisotopes in industry

Radioactive isotopes have found a number of uses in industry, often in detecting unseen faults in metal castings or for measuring the rate of wear in machinery.

Measuring engine wear Wear occurs in machinery when two surfaces rub together. In vehicle engines, such rubbing occurs very frequently and car manufacturers and oil producers need to be able to measure the rate of this wear. One method is to dismantle an engine after a period of use and take accurate measurements to compare with the original dimensions. Clearly this is very time consuming and expensive as much labour is involved.

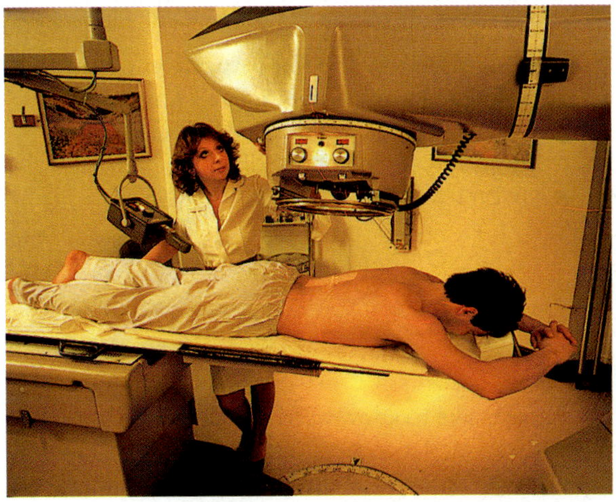

Figure 37 A patient receiving treatment for a tumour using γ-radiation from cobalt-60

A much quicker method is to make use of a radioactive tracer. This involves using very small amounts of material. For example, a thin radioactive layer can be produced on the inside of a cylinder in an engine or on the chromium-plated piston rings which press against the wall of the cylinder. This is done by bombarding a small area with a beam of charged particles. The particles, called deuterons, are produced in a particle accelerator and consist of a proton and a neutron. The chromium on the piston ring or the cylinder wall can be made radioactive to a depth of 0.1 mm (Figure 38).

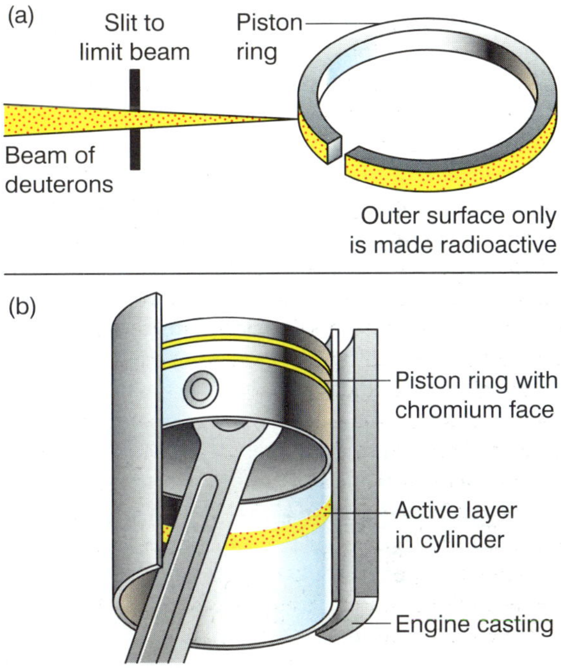

Figure 38 Using a radioactive tracer to measure engine wear. Either (a) the outer surface of the piston ring is made radioactive, or (b) a thin layer of the cylinder wall is made radioactive

The detection of wear inside the cylinder is now simple and quick (Figure 39). The sparking plug for the cylinder is removed and a radiation detector inserted through the sparking plug hole into the cylinder. Repeated measurements can be made to determine the rate of wear. Alternatively, debris in the oil can be monitored for radioactive material.

Other tests can be made to measure the wear of the different parts of an engine. The same technique is used to make the part under test radioactive. Such measurements will enable a car manufacturer to improve the materials used in the engine, and so prolong the life of the engine. The measurements can also help oil companies to improve the quality of their lubricating oils.

Figure 39 Measuring radiation levels inside an engine cylinder

As only small surface layers are made radioactive, hazards in handling the parts are much reduced.

Checking jet engines for cracks γ-Radiation from iridium-192 is used in the detection of cracks in the turbine blades of jet engines (Figure 40). A section of the turbine is first covered by photographic film. The iridium-192 source is then placed in the centre of the engine. After exposure the radiation source is removed. The iridium-192 source is handled throughout by remote control as the strength of radiation is very high. When the film is developed, any cracks show up as dark lines.

Figure 40 Safety check on a jet engine using γ-radiation

Radioisotopes in scientific research

Radioactive isotopes have provided scientists with a versatile tool which has been used in a variety of ingenious ways. We shall look at two applications:

- determining the age of archaeological specimens
- finding the pathway of a chemical reaction.

Carbon dating Probably one of the most widely known useful applications of a radioisotope is for carbon dating. This method is frequently used in archaeological research and is suitable for use with any specimen containing material which was once living. The ratio of carbon-14 to carbon-12 in the specimen is measured. By comparing this ratio with that in a living specimen, a calculation can be made of the age of the object based on the half-life of carbon-14. This is possible for the following reasons:

- In the upper atmosphere neutrons are produced by the action of cosmic rays on atoms. These neutrons react with nitrogen atoms to form carbon-14:

  $${}^{14}_{7}N + {}^{1}_{0}n \rightarrow {}^{14}_{6}C + {}^{1}_{1}H$$

  Carbon-14 is a radioactive isotope which decays with a half-life of 5730 years. This means that carbon-14 is continually forming and decaying, thus resulting in a fairly constant proportion of carbon-14 to carbon-12 in the atmosphere.
- The carbon that finds its way into the food chain via photosynthesis in green plants will also be made up of a fairly constant proportion of carbon-14 to carbon-12.
- The proportion of the isotope in living organisms (plants or animals) remains fairly constant until they die. In the dead organism, the level of carbon-14 begins a slow decline as the isotope decays and is no longer replaced.

ASSIGNMENT 14

The ratio of ^{14}C to ^{12}C in living organisms is $1{:}1 \times 10^{12}$. The half-life of carbon-14 is 5730 years.

a What is the ratio of ^{14}C to ^{12}C after 5730 years?

b How many years will have elapsed since the death of a specimen when the ratio of ^{14}C to ^{12}C has fallen to $1{:}8 \times 10^{12}$?

Reaction pathway for the formation of an ester In **Unit 2, Section 18.5**, you learnt how an alcohol reacts with a carboxylic acid to form an ester and water. For example, methanol forms methyl ethanoate on treatment with ethanoic acid:

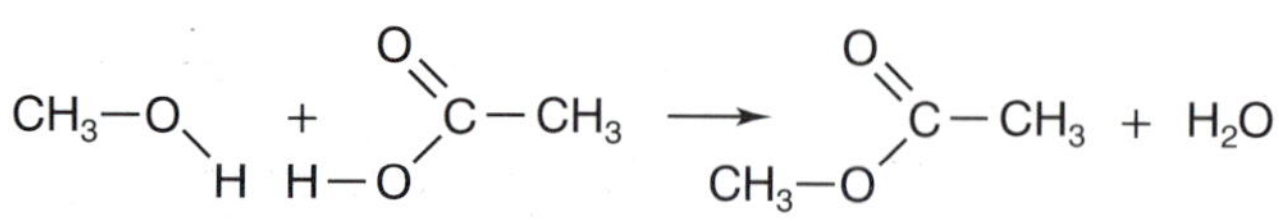

Chemists like to know which bonds are broken and which are formed in a chemical reaction. If you look closely at the above reaction, you will see that there are two possibilities for bonds to be broken to form the ester:

- the C—O bond in the acid together with O—H bond in the alcohol
- the O—H bond in the acid together with C—O bond in the alcohol.

By using methanol in which the oxygen atom is the radioisotope oxygen-18, rather than the usual oxygen-16, chemists have been able to show that the oxygen from methanol becomes incorporated in the ester. The conclusion is that the O—H bond in the alcohol is broken:

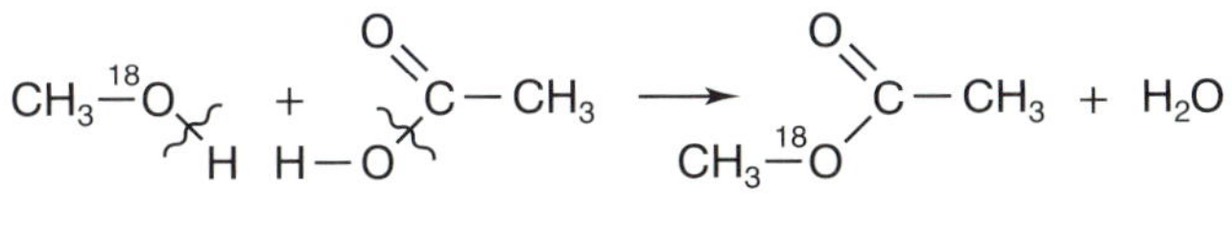

This technique, called **isotopic labelling**, has proved to be useful in helping to unravel many reaction pathways in living organisms.

Energy production

Both **fusion** and **fission** of nuclei provide potential for generating large amounts of energy.

Fission When nuclei of radioactive atoms split, energy is emitted. Uranium-235 is normally used as the fuel in a nuclear reactor. The fission occurs when a slow neutron collides with the nucleus of a uranium-235 atom. Two smaller nuclei of different elements are formed and two or three neutrons are released; for example:

$$ {}^{1}_{0}n + {}^{235}_{92}U \rightarrow {}^{143}_{56}Ba + {}^{90}_{36}Xe + {}^{1}_{0}n + {}^{1}_{0}n + {}^{1}_{0}n $$

If the neutrons released during the fission hit other uranium-235 nuclei, further fissions will occur, releasing more neutrons, which in turn cause more fissions. There is then the possibility of a chain reaction

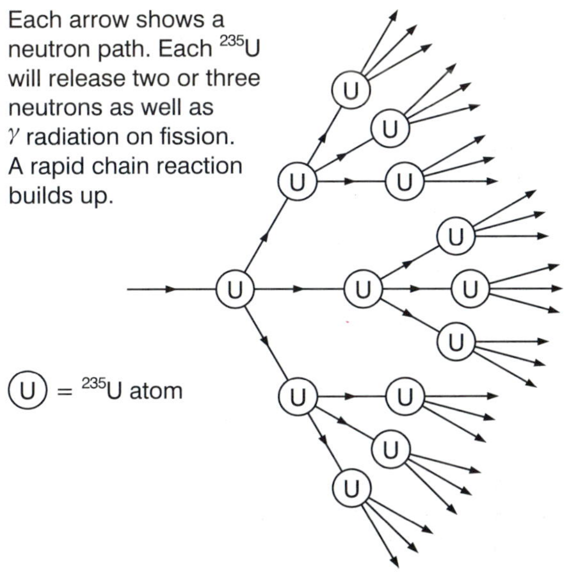

Figure 41 Chain reaction in uranium-235

occurring (Figure 41). In a nuclear reactor the uranium fuel rods are surrounded by graphite which acts as a moderator and slows down the neutrons so that they are more effective.

Above a critical mass of uranium-235, a nuclear explosion would occur, However, inside a nuclear reactor the amount of uranium-235 in each fuel rod is small and the process is controlled by absorbing neutrons by lowering boron rods into the reactor.

Nuclear power stations usually use the energy released by the fission process to convert water to steam to drive turbines which generate electricity.

Fusion When two nuclei fuse, energy is released. Fusion occurs in our Sun and other stars and is the way that heavier elements are formed in the Universe.

You can find out more about the formation of elements in the stars in **Unit 1**, **Section 1.1** and in **Section 25.2**.

Scientists have been endeavouring to build fusion reactors which can produce energy at an economic rate (Figure 42). Fusion of the hydrogen isotopes deuterium and tritium releases huge amounts of energy:

$$ {}^{2}_{1}H + {}^{3}_{1}H \rightarrow {}^{4}_{2}He + {}^{1}_{0}n $$

These isotopes of hydrogen are derived from water, so there is an almost unlimited supply. Unfortunately, such reactors are extremely expensive. The dream of fusion reactors for generating electricity has yet to be realised.

Figure 42 The JET experimental fusion reactor at Culham, Oxfordshire

Fuel used	1990	1997
coal	65	38
nuclear	21	30
gas	1	27
oil	11	2
hydro	1	1
other	1	2

Table 4 Percentage of UK electricity generated from different fuels

Nuclear or fossil fuels?

Energy can be produced from both nuclear and fossil fuels. The proportion of the electricity generated in the UK using each fuel is compared in Table 4. The most noticeable changes between 1990 and 1997 were the fall in the use of coal and the increase in the use of gas. There are issues associated with these changes but for many the major issue is whether future demands for electricity should be met by the use of fossil or nuclear fuels.

When deciding which fuel should used, the efficiency of the fuel, the size of known fuel reserves and any safety hazards or environmental implications associated with its use must be taken into account.

Although the environment is so important to us, there is still very little use of renewable resources to generate electricity in the UK. However, one of the few major hydroelectric generating companies is in Scotland, Hydro-Electric. You will see from Table 4 how little hydroelectric power is generated overall in the UK.

Energy efficiency

It has been calculated that the energy obtained by fission of 100 g of uranium-235 would require about 30 tonnes of high grade coal to be burned. The burning of this amount of coal would also result in about 100 tonnes of carbon dioxide being released into the atmosphere.

Fuel reserves

The reserves of fuel for nuclear power and fossil fuels are finite in the sense that they are non-renewable. But, bearing in mind its greater efficiency, this is less of a problem for nuclear fuel than it is for fossil fuels. Natural uranium is mostly uranium-238, with only a small proportion of it being uranium-235. Although uranium-238 is non-fissionable it can be converted into plutonium-239, which is fissionable.

Pollution and safety

Fossil fuels produce carbon dioxide, which is a greenhouse gas. The proportion of carbon dioxide in the atmosphere is increasing slowly and it seems likely that carbon dioxide produced by human activity will contribute to global warming (see **Sections 7.1** and **7.2**). Fossil fuels can also produce sulphur dioxide and oxides of nitrogen, which contribute to acid rain. Nuclear fuels do not produce such pollutants, but major disadvantages are the potentially catastrophic consequences of a serious accident at a nuclear plant and the difficulties of disposing of radioactive waste fission products. Throughout the life of a nuclear reactor, nuclear waste needs to be made safe by reprocessing or by long-term containment. When a nuclear reactor comes to the end of its working life, it requires many years of expensive work for decommissioning. Further, highly radioactive materials from the decommissioning require reprocessing or containment for many years until the radioactivity falls to safe levels.

Over 20% of our energy needs are supplied by nuclear reactors. The longer term consequence of not replacing nuclear reactors will be that we become increasingly reliant on fossil fuel power stations. To replace the nuclear component of our electricity production with fossil fuel power stations will effectively prevent the UK from meeting the targets set for reducing atmospheric carbon dioxide levels in the early decades of the 21st century.

Gas-fired power stations do produce less carbon dioxide per kilowatt of electricity than oil- or coal-fired power stations. However, replacing nuclear power stations with gas-fired power stations will still produce an overall rise in carbon dioxide emissions. Also, known reserves of gas and oil are more limited than those of coal. We clearly need to control our energy use and manage our energy resources more carefully if we are going to reduce carbon dioxide emissions.

9.5 *Summary*

This Chemical Storylines chapter is linked to Chemical Ideas Chapters 20, 24 and 25.

This chapter has used selected case studies to show you how chemistry is used in the real world. You have been able to apply what you have learned about rates of reactions (from Unit 1, Section 15) and chemical equilibria to understand why a compromise is necessary when choosing the conditions for making ammonia.

You then saw how redox reactions are put to good use in extracting bromine from sea water and you met an example of industrial electrolysis in the refining of copper.

Finally, you studied both fission and fusion nuclear reactions, and encountered a wide range of applications for radioisotopes in medicine, in industry and for scientific research.

10 MEASURING AMOUNTS OF SUBSTANCE

10.1 *Amount of substance*

Relative atomic mass

Imagine you have a bag containing 10 golf balls and 20 table tennis balls. The golf balls would make up most of the mass, but the table tennis balls would make up most of the contents.

Chemists encounter a comparable situation when they try to work out the composition of water, or any other substance. Oxygen makes up nearly 90% of the mass of water, but two thirds of the atoms are hydrogen atoms. The difference between the two situations is that, whereas golf balls and table tennis balls can be picked up and counted, the chemist is unable to pick up and count the atoms in water molecules – they are far too small.

You know the formula of water is H_2O, and this means there are *two* hydrogen atoms combined with *one* oxygen atom in *each* water molecule. The link between the mass of an element and the number of atoms it contains is the **relative atomic mass** of the element. It is this link which allows chemists to work out chemical formulae.

The relative atomic mass scale is used to compare the masses of different atoms. The hydrogen atom, the lightest atom, was originally assigned a relative atomic mass of 1, and relative atomic masses of other atoms were compared with this. The reference that is used now is the carbon-12 isotope (^{12}C) which is assigned a relative atomic mass of exactly 12. In chemistry, approximate relative atomic masses are used most of the time, and you can simply think of carbon atoms, set at relative atomic mass = 12, as being the reference point for the relative atomic mass scale. The approximate relative atomic masses of several elements are listed in Table 1.

Table 1 Some approximate relative atomic masses

Element	Symbol	Relative atomic mass
hydrogen	H	1
carbon	C	**12**
nitrogen	N	14
oxygen	O	16
magnesium	Mg	24
sulphur	S	32
calcium	Ca	40
iron	Fe	56
copper	Cu	64
mercury	Hg	200

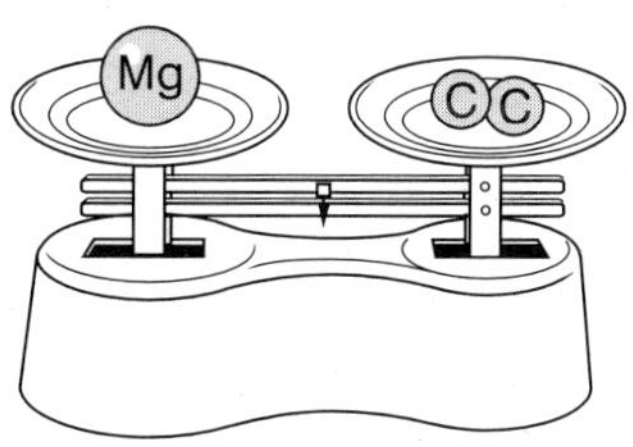

Figure 1 Two ^{12}C atoms have the same mass as one Mg atom. So the relative atomic mass of Mg is $2 \times 12 = 24$

Notice that relative atomic masses have no units. Copper atoms *do not* have a mass of 64 g or 64 'anythings'. They are just 64 times heavier than hydrogen atoms, or 4 times heavier than oxygen atoms, or twice as heavy as sulphur atoms, and so on.

Chemical quantities

Suppose you have two bottles containing equal masses of copper (relative atomic mass = 64) and sulphur (relative atomic mass = 32). You would know that you had twice as many sulphur atoms as copper atoms because sulphur atoms have only half the mass of copper atoms. If you had a bottle of mercury (relative atomic mass = 200) that was five times heavier than a bottle of calcium (relative atomic mass = 40), you would know that both bottles contained equal numbers of atoms because each mercury atom has five times the mass of each calcium atom.

12 g of carbon, 1 g of hydrogen and 16 g of oxygen all contain equal numbers of atoms because these masses are in the same ratio as the relative atomic masses. This amount of each of these three elements has a special significance in chemistry. It is called a **mole** (or **mol** for short). The mole is the unit that measures **amount of substance** in such a way that equal amounts of elements consist of equal numbers of atoms. The mole is the unit that measures amount in the same way as the kilogram is the unit that measures mass.

Chemical amounts are defined so that the mass of one mole (the **molar mass**) is equal to the relative atomic mass in grams. Thus, the molar mass of carbon is 12 g mol^{-1}.

If you had 6 g of carbon you would have 0.5 mol of carbon atoms; 4 g of oxygen would contain 0.25 mol of oxygen atoms and 3 g of hydrogen would contain 3 mol of hydrogen atoms. You can work out these amounts like this:

6 g of carbon contains 6 g $\div$ 12 g mol^{-1} = 0.5 mol of carbon atoms

4 g of oxygen contains 4 g $\div$ 16 g mol^{-1} = 0.25 mol of oxygen atoms

3 g of hydrogen contains 3 g $\div$ 1 g mol^{-1} = 3 mol of hydrogen atoms

Each of these calculations uses the relationship:
mass in grams ÷ molar mass = amount in moles of atoms

Relative formula mass

You can use the mole to deal with all substances, compounds as well as elements. A molecule of methane, CH_4, for example, is formed when one carbon atom combines with four hydrogen atoms: this also means that one *mole* of methane is formed when one *mole* of carbon atoms combines with four *moles* of hydrogen atoms.

Just as there are relative atomic masses for atoms, so chemists use **relative formula masses** to compare other substances. The relative formula mass of a substance can be worked out by first writing the formula of the substance and then adding together the relative atomic masses of each of the atoms in the formula.

For example
for methane, CH_4
relative formula mass $= (1 \times 12) + (4 \times 1)$
$= 16$
for calcium nitrate, $Ca(NO_3)_2$
relative formula mass $= (1 \times 40) + (2 \times 14) + (6 \times 16)$
$= 164$

Relative formula masses have no units and are on the same scale as relative atomic masses, which has the relative atomic mass of ^{12}C set at 12.

Formula units

Just as computer memory is made up of individual bits and a box of chalk is made up of individual pieces, the 'bits and pieces' that make up a substance are called **formula units**. They are the basic units or building blocks of substances and, not surprisingly, they match the formulae of the substances. Formula units can be single atoms, molecules or groups of ions.

Take copper, for example. The formula unit in the metal is simply a Cu atom. Likewise, the formula unit in carbon, a non-metal, is the carbon atom. In fact, the formula unit in most elements is a single atom and so their relative formula mass is identical to their relative atomic mass. There are some exceptions, though. In oxygen gas, the formula unit is the O_2 molecule rather than the O atom and so the relative formula mass is *twice* the relative atomic mass of oxygen.

In many covalent compounds, the formula unit is a molecule. In methane, for example, the formula unit is the CH_4 molecule. In ionic compounds, however, the formula unit is a *group of ions*. In calcium nitrate, for example, the formula unit is $Ca^{2+}(NO_3^-)_2$ and contains a group of three ions. These groups of ions are not labelled with any special name and so we just use the general name 'formula unit' when referring to ionic compounds.

Some definitions

The **relative atomic mass** of an element is the mass of its atom relative to $^{12}C = 12$.

The **relative formula mass** of a substance is the mass of its formula unit relative to $^{12}C = 12$.

A **mole** (1 **mol**) of a substance is the **amount of substance** which contains as many elementary entities (atoms, molecules, formula units, etc) as there are atoms in 12 g of ^{12}C.

The **molar mass** of a substance is the mass of substance which contains 1 mol.

Moles of formula units

The relative formula mass in grams is equal to the molar mass, so the molar mass of methane is 16 g and that of calcium nitrate is 164 g. If you had 8 g of methane you would have 0.5 moles of CH_4 formula units; 41 g of calcium nitrate contains 0.25 moles of $Ca^{2+}(NO_3^-)_2$ formula units. In general:

mass in grams ÷ molar mass = amount in moles of formula units

As well as describing 16 g of methane as containing 1 mole of formula units (or molecules) of CH_4, we can also say that it contains 1 mole of C atoms and $(1 \times 4) = 4$ moles of H atoms. Similarly, 164 g of calcium nitrate contains 1 mole of formula units of $Ca^{2+}(NO_3^-)_2$; it also contains 1 mole of Ca^{2+} ions and 2 moles of NO_3^- ions.

For elements that exist as diatomic gases, such as oxygen, you must be especially careful. For example, 32 g of oxygen gas contains 1 mole of O_2 molecules, but 2 moles of O atoms. To avoid ambiguity when using moles, it is essential to give the formula unit you are referring to (eg moles of O atoms or moles of O_2 molecules).

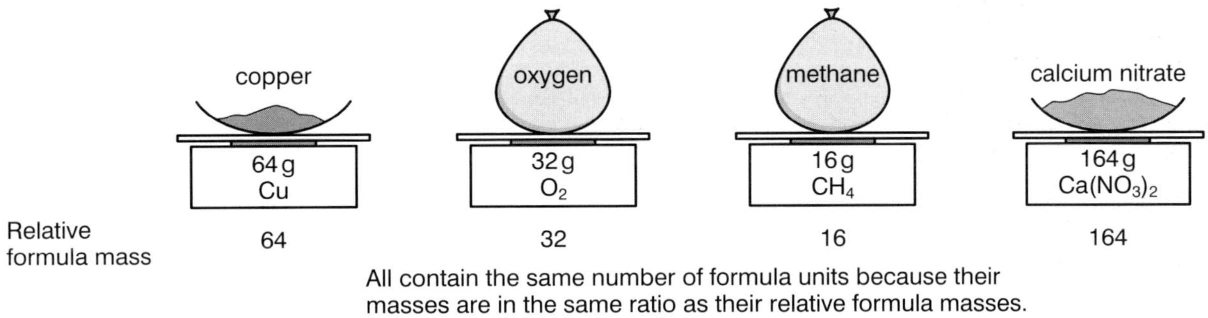

Figure 2 The important thing to remember is that equimolar amounts of substances contain equal numbers of formula units

The Avogadro constant

The number of formula units in a mole of a substance is a constant. It is called the **Avogadro constant** after the Italian scientist, Amedeo Avogadro (1776–1856) who did some work which was fundamental to this method of counting formula units. The value of the Avogadro constant (symbol L) is 6.02×10^{23} formula units mol^{-1}. If you are counting atoms, it is 6.02×10^{23} atoms mol^{-1}; for molecules it is 6.02×10^{23} molecules mol^{-1}; for electrons it would be 6.02×10^{23} electrons mol^{-1}. The number is so huge that it is difficult to comprehend. Figure 3 shows one way to illustrate how big it is.

The Avogadro constant

One mole of any substance contains 6.02×10^{23} (L) formula units.

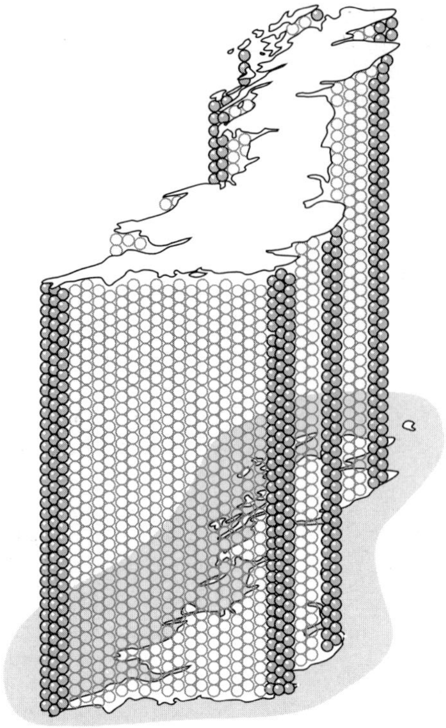

Figure 3 If the 6.02×10^{23} atoms in 12 g of carbon were turned into marbles, the marbles could cover Great Britain to a depth of 1500 km!

If you want to make sure you have the same number of formula units of different substances, you don't have to count out incredibly small formula units, neither do you have to work with numbers as big as the Avogadro constant. You just have to measure out a mole. This is easily done by weighing out the molar mass, and is the reason why chemists find the mole so useful.

Chemical formulae

We use moles when we work out chemical formulae. If you did a laboratory experiment to determine the formula of magnesium oxide, you would burn a known mass of magnesium and find out what mass of oxygen combined with it. Here are some specimen results and, from them, a calculation of the formula:

mass of magnesium = 0.84 g   mass of oxygen = 0.56 g

amount of Mg atoms
$= 0.84\,g \div 24\,g\,mol^{-1} = 0.035\,mol$

amount of O atoms
$= 0.56\,g \div 16\,g\,mol^{-1} = 0.035\,mol$

ratio of moles of atoms of Mg : O in magnesium oxide = 1 : 1

formula of magnesium oxide is MgO

Another way of analysing a compound is to find the *percentage mass* of each element it contains. Here are some specimen results for methane:

% mass of carbon = 75%   % mass of hydrogen = 25%

mass of carbon in
100 g of methane = 75 g

mass of hydrogen in
100 g of methane = 25 g

amount of C atoms
$= 75\,g \div 12\,g\,mol^{-1} = 6.25\,mol$

amount of H atoms
$= 25\,g \div 1\,g\,mol^{-1} = 25\,mol$

ratio of moles of atoms of C : H in methane = 1 : 4

formula of methane is CH_4

In a methane molecule there is a central carbon atom surrounded by four hydrogen atoms. The simple ratio of atoms from which it is formed is the same as in the formula of the molecules.

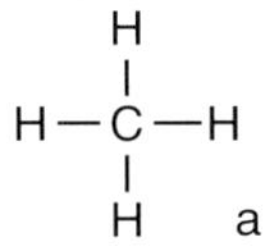

a *methane* molecule

But this isn't the case for all substances. Ethane molecules contain eight atoms: two carbon atoms and six hydrogen atoms.

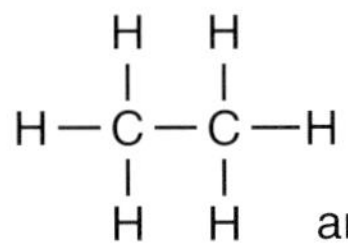

an *ethane* molecule

The **molecular formula** of ethane is C_2H_6, but the simplest ratio for the moles of atoms of C : H is 1 : 3. So a calculation from percentage masses would lead you to a formula CH_3. Chemists call this type of formula an **empirical formula**. Table 2 shows you some more examples of the two types of formulae.

The **empirical formula** of a substance tells you the *ratio* of the numbers of different types of atom in the substance.

The **molecular formula** tells you the *actual numbers* of different types of atom.

Substance	Molecular formula	Empirical formula
ethene	C_2H_4	CH_2
benzene	C_6H_6	CH
butane	C_4H_{10}	C_2H_5
phosphorus(V) oxide	P_4O_{10}	P_2O_5
oxygen	O_2	O
bromine	Br_2	Br

Table 2 Some molecular formulae and empirical formulae

PROBLEMS FOR 10.1

1 How many times heavier are
- a Mg atoms than H atoms?
- b Mg atoms than C atoms?
- c H_2 molecules than H atoms?
- d O_2 molecules than O atoms?
- e CO_2 molecules than He atoms?
- f SO_2 molecules than O_2 molecules?
- g $Mg^{2+}O^{2-}$ formula units than He atoms?
- h $Cu^{2+}O^{2-}$ formula units than $Mg^{2+}O^{2-}$ formula units?
- i $Ca^{2+}(Br^-)_2$ formula units than CH_4 molecules?

2 Which contains the most atoms?
- a 36 g of carbon
- b 24 g of magnesium
- c 20 g of calcium
- d 64 g of sulphur

3 Which contains the most formula units?
- a 8 g of CH_4
- b 17 g of NH_3
- c 9 g of H_2O
- d 11 g of CO_2

4 Which contains the fewest formula units?
- a 10 g of $CaCO_3$
- b 10 g of MgO
- c 10 g of NaCl
- d 10 g of $CuSO_4$

5 Which contains the fewest hydrogen atoms?
- a 16 g of CH_4
- b 17 g of NH_3
- c 18 g of H_2O
- d 4 g of H_2

6 Calculate the molar mass in g mol^{-1} of
- a phosphorus, P
- b iodine, I_2
- c benzene, C_6H_6
- d copper(II) carbonate, $CuCO_3$
- e sodium thiosulphate pentahydrate, $Na_2S_2O_3.5H_2O$

7 How many moles of formula units are contained in the following masses?
- a 32 g of S
- b 28 g of Fe
- c 64 g of O_2
- d 1.8 g of H_2O
- e 11 g of CO_2
- f 9 g of $C_6H_{12}O_6$ (glucose)
- g 82 g of $Ca(NO_3)_2$
- h 190 kg of $MgCl_2$

8 a How many moles of atoms are contained in the following masses?
  i 54 g of aluminium
  ii 10 g of neon
 b How many moles of molecules are contained in the following masses?
  i 12.7 g of iodine, I_2
  ii 1.00 tonne of water
 c How many moles of formula units are contained in the following masses?
  i 2.0 kg of sodium chloride
  ii 0.1 g of calcium carbonate

9 Copy and complete the following table.

Substance	Molar mass/g mol^{-1}	Mass of sample/g	Amount of sample/mol	Number of formula units
carbon	12	12.00	1.00	6.02×10^{23}
hydrogen	2	2.00		
oxygen	32	64.00		12.04×10^{23}
water	18		1.00	
glucose	180		0.50	
sodium hydroxide	40			3.01×10^{23}
calcium nitrate	164	1.64		

10 a How many moles of atoms are contained in the following amounts of substance?
  i 2.0 mol of O_2
  ii 0.1 mol of CO_2
  iii 10 mol of CH_4
  iv 0.5 mol of CCl_4
 b How many moles of ions are contained in the following amounts of substance?
  i 1.0 mol of Al_2O_3
  ii 6.0 mol of MgO
  iii 0.1 mol of $CaCO_3$
  iv 0.2 mol of $(NH_4)_2SO_4$

11 a Work out the empirical formulae of the compounds formed from the following masses and elements.
  i 32 g of copper and 8 g of oxygen
  ii 10 g of calcium and 4 g of oxygen
  iii 8 g of sulphur and 8 g of oxygen
 b Work out the empirical formulae of the compounds with the following percentage composition by mass
  i 92.3% carbon and 7.7% hydrogen
  ii 52.2% carbon, 13.0% hydrogen and 34.8% oxygen

12 What are the empirical formulae of the following substances?
 a C_3H_6
 b $C_{10}H_8$
 c P_4O_6
 d Al_2Cl_6

13 You are given the empirical formulae and the relative formula masses of three compounds, I, II and III. Write down their molecular formulae.

Compound	Empirical formula	Relative formula mass
I	HO	34
II	CO	28
III	CH	26

10.2 *Balanced equations*

A balanced chemical equation tells you the reactants and products in a reaction, and the relative amounts involved. The equation is balanced so that there are equal numbers of each type of atom on both sides. For example, the equation

$$CH_4(g) + 2O_2(g) \rightarrow CO_2(g) + 2H_2O(l)$$

tells you that one molecule of methane reacts with two molecules of oxygen to form one molecule of carbon dioxide and two molecules of water. These are also the *amounts in moles* of substances involved in the reaction.

The number written in front of each formula in a balanced equation tells you the number of formula units involved in the reaction. A formula unit may be a molecule, or another species such as an atom or an ion. The small subscript numbers are part of the formulae, and cannot be changed.

> **State symbols**
>
> State symbols are included in chemical equations to show the *physical state* of the reactants and products:
>
> (g) gas
> (l) liquid
> (s) solid
> (aq) aqueous solution

Writing balanced equations

The only way to be sure of the balanced equation for a reaction is to do experiments to find out what is formed in the reaction and what quantities are involved. But chemists use equations a lot, and it isn't possible to do experiments every time. Fortunately, if we know the reactants and products, we can usually find their formulae and *predict* the balanced equation.

The steps for predicting balanced equations

We will illustrate the steps by using the reaction that occurs when propane, C_3H_8, burns in air.

Step 1 Decide what the reactants and products are.
propane + oxygen → carbon dioxide + water

Step 2 Write formulae for the substances involved. State symbols should be included.
$C_3H_8(g) + O_2(g) \rightarrow CO_2(g) + H_2O(l)$

This equation is *unbalanced* because there are different numbers of each type of atom on each side.

Step 3 Balance the equation so that there are the same numbers of each type of atom on each side.
$C_3H_8(g) + 5O_2(g) \rightarrow 3CO_2(g) + 4H_2O(l)$

Equations can only be balanced by putting numbers in front of the formulae. You cannot balance them by altering the formulae themselves, because that would create different substances.

> Atoms are not created or destroyed in chemical reactions. They are simply rearranged ...
>
> ... so equations *must* balance.

10.3 *Using equations to work out reacting masses*

A balanced equation tells you the *amount in moles* of each substance involved in the reaction. For example, the equation

$$CH_4(g) + 2O_2(g) \rightarrow CO_2(g) + 2H_2O(l)$$

tells you that 1 mole of CH_4 reacts with 2 moles of O_2 to give 1 mole of CO_2 and 2 moles of H_2O.

The mass of 1 mole of CH_4	is	$12\,g + (4 \times 1)\,g$	$= 16\,g$
The mass of 1 mole of O_2	is	$(2 \times 16)\,g$	$= 32\,g$
The mass of 1 mole of CO_2	is	$12\,g + (2 \times 16)\,g$	$= 44\,g$
The mass of 1 mole of H_2O	is	$(2 \times 1)\,g + 16\,g$	$= 18\,g$

So, 16 g CH_4 react with 64 g O_2 to give 44 g CO_2 and 36 g H_2O.

(Note that the *total* mass on each side of the equation must always be the same.)

This means that chemists can use equations to work out the masses of reactants and products involved in a reaction, without having to do an experiment.

The steps for working out reacting masses

Step 1 Write a balanced equation.

Step 2 In words, state what the equation tells you about the amount in moles of the substances you are interested in.

Step 3 Change amounts in moles to masses in grams.

Step 4 Scale the masses to the ones in the question.

Example

What mass of carbon dioxide is produced when 64 g of methane is burned in a plentiful supply of air?

Step 1	$CH_4(g)$ + $2O_2(g)$	$\rightarrow$	$CO_2(g)$	+	$2H_2O(l)$
Step 2	1 mole		1 mole		
Step 3	16 g		44 g		
Step 4	64 g		(44×4) g = 176 g		

So, 176 g CO_2 are produced when 64 g CH_4 is burned.

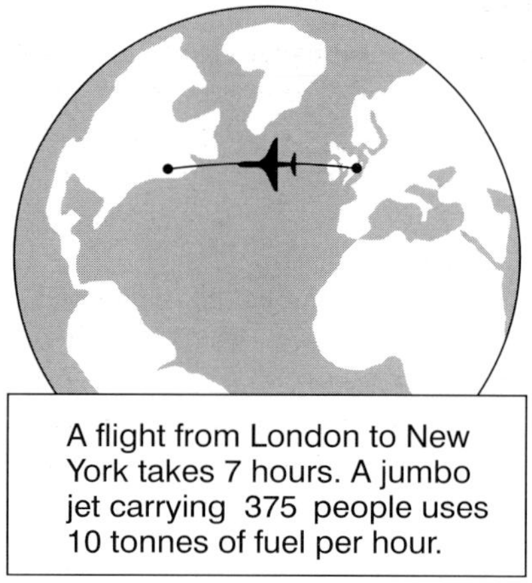

Figure 4 What mass of CO_2 is produced during one transatlantic flight?

PROBLEMS FOR 10.2 AND 10.3

You will need to consult a table of relative atomic masses when you do these questions.

1 Write balanced equations for the following reactions:
- **a** ethane, C_2H_6, burning in oxygen to form carbon dioxide and water
- **b** ethane burning in oxygen to form carbon monoxide and water
- **c** hexane, C_6H_{14}, burning in oxygen to form carbon dioxide and water
- **d** ethanol, C_2H_5OH, burning in oxygen to form carbon dioxide and water
- **e** nitrogen reacting with oxygen to form nitrogen monoxide, NO
- **f** nitrogen reacting with oxygen to form nitrogen dioxide, NO_2

2 1 kg of charcoal is burned in a plentiful supply of air. Assuming charcoal is pure carbon, what mass of CO_2 will be formed?

3
- **a** Write an equation for the complete combustion of octane.
- **b** What mass of oxygen would be needed for the complete combustion of 50 kg of octane?
- **c** What mass of carbon dioxide would be produced by the complete combustion of 50 kg of octane?

4 Calcium carbonate is decomposed by heating to give calcium oxide and carbon dioxide,

$$CaCO_3(s) \rightarrow CaO(s) + CO_2(g)$$

What mass of calcium carbonate would be needed to produce 1.4 g of calcium oxide?

5 A coal-burning power station burns coal which contains 1% sulphur by mass. If the power station burns 5600 tonnes of coal a day, what mass of sulphur dioxide gas is released into the air each day?

6 Use the information in Figure 4 to work out, for one transatlantic flight,
- **a** the total mass of CO_2 produced
- **b** the mass of CO_2 produced per person.

(Jet fuel is a mixture of hydrocarbons, but you can assume it is typically $C_{12}H_{26}$.)

7 When iron ore is reduced by carbon monoxide in a blast furnace, the equation for the reaction is

$$Fe_2O_3(s) + 3CO(g) \rightarrow 2Fe(l) + 3CO_2(g)$$

- **a** What is the relative formula mass of Fe_2O_3?
- **b** Calculate the mass of iron produced from 1 mole of Fe_2O_3.
- **c** Calculate the mass of iron produced from 1000 tonnes of Fe_2O_3.
- **d** How many tonnes of Fe_2O_3 would be needed to produce 1 tonne of iron?
- **e** If the iron ore contains 12% of iron(III) oxide, how many tonnes of ore are needed to produce 1 tonne of iron?

10.4 *Concentrations of solutions*

Chemists often carry out reactions in solution. When you are using a solution of a substance, it is important to know how much of the substance is dissolved in a particular volume of solution.

Concentrations are sometimes measured in grams per litre. A solution containing 80 g of sodium hydroxide in 1 litre of solution has a concentration of $80\,g\,l^{-1}$.

However, chemists usually prefer to measure out quantities in moles rather than in grams, because working in moles tells you about the number of particles present. So the preferred units for measuring concentrations in chemistry are moles per litre, or $mol\,l^{-1}$.

To convert grams per litre to moles per litre, you need to know the molar mass of the substance involved. For example, the molar mass of sodium hydroxide, NaOH, is $40\,g\,mol^{-1}$. So a solution containing $80\,g\,l^{-1}$ has a concentration of

$$\frac{80\,g\,l^{-1}}{40\,g\,mol^{-1}} = 2\,mol\,l^{-1}$$

In general:

$$\text{concentration/mol}\,l^{-1} = \frac{\text{concentration/g}\,l^{-1}}{\text{molar mass/g}\,mol^{-1}}$$

In some books you may see '$mol\,l^{-1}$' abbreviated to 'M'. This is quick to write, and was once widely used, but nowadays $mol\,l^{-1}$ is preferred.

When we make a solution, its concentration will depend on:

- the amount of solute
- the volume of solution it is dissolved in (Figure 5).

If you know the concentration of a solution, you can work out the amount of solute in a particular volume.

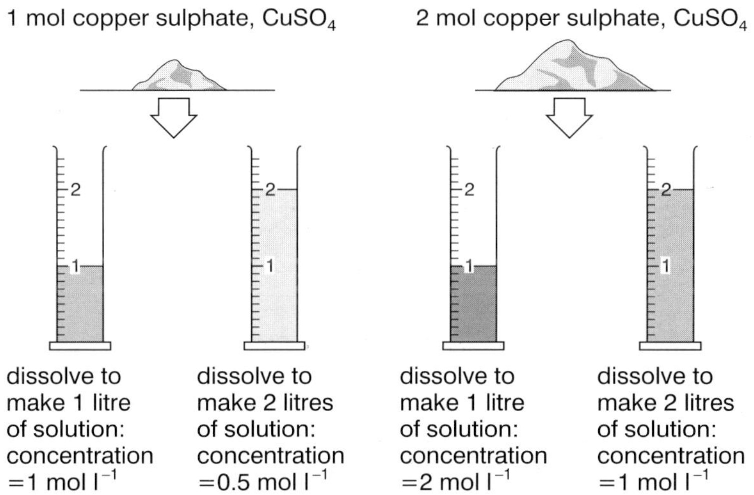

Figure 5 The concentration of a solution depends on the amount of solute and the volume of solvent

Example

Suppose we have 250 cm^3 of a solution of sodium hydroxide whose concentration is $2\,mol\,l^{-1}$. How many moles of NaOH do we have?

Answer $250\,cm^3$ is 0.25 litres, and there are 2 mol in each litre. So in 0.25 litres we must have ($2\,mol\,l^{-1} \times 0.25$ litres)
= 0.5 mol of NaOH

In general:

amount/mol = (concentration of solution/$mol\,l^{-1}$) × (volume of solution/l)

Measuring volumes

The units you will use to measure volume will depend on how big the volume is. Large volumes are measured in litres (l); smaller volumes are measured in cubic centimetres (cm^3).

You may also come across other names for these units:

1 litre = $1\,dm^3$
$1\,cm^3$ = 1 ml

Using concentrations in calculations

If you are carrying out a chemical reaction in solution, and you know the equation for the reaction, you can use the concentrations of the reacting solutions to predict the volumes you will need.

Example

Consider the reaction of sodium hydroxide with hydrochloric acid. Suppose the concentrations of both solutions are $2\,\text{mol}\,l^{-1}$. If you have 0.25 litres of sodium hydroxide solution, what volume of hydrochloric acid would be needed to neutralise it?

Answer The equation for the reaction is

$$NaOH(aq) + HCl(aq) \rightarrow NaCl(aq) + H_2O(l)$$

The equation shows that 1 mole of NaOH reacts with 1 mole of HCl.

Amount of sodium hydroxide $= 2\,\text{mol}\,l^{-1} \times 0.25$ litres $= 0.5\,\text{mol}$

$\Rightarrow$ amount of HCl needed $= 0.5\,\text{mol}$

$\Rightarrow$ volume of HCl needed $= \dfrac{0.5\,\text{mol}}{2\,\text{mol}\,l^{-1}} = 0.25$ litres

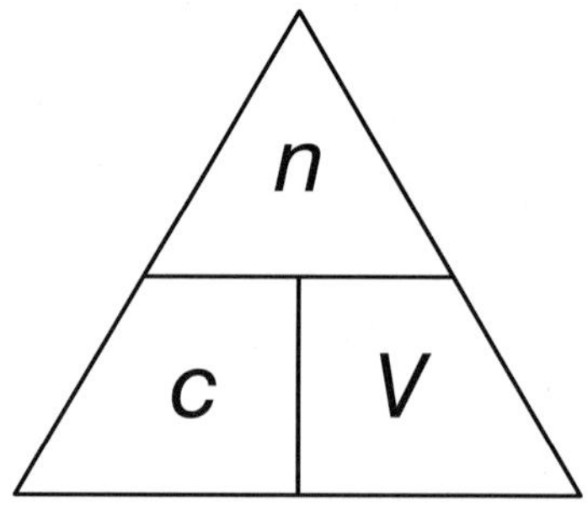

Figure 6 A memory aid showing the relationship between concentration (c) in mol l^{-1}, the amount in moles (n) and the volume of solution (V) in litres

$c = \dfrac{n}{V}$, $n = c \times V$, $V = \dfrac{n}{c}$

You could probably have worked out this simple example without going through these stages, but it isn't so easy when the solutions are of differing concentrations, and when they don't react in a 1 : 1 ratio. In general, the steps for working out the reacting volumes of solutions are as follows (compare these steps with those used for calculating reacting masses in **Section 10.3**).

Step 1 Write a balanced equation.
Step 2 Say what the equation tells you about the amounts in moles of the substances you are interested in.
Step 3 Use the known concentrations of the solutions to change amounts in moles to volumes of solutions.
Step 4 Scale the volumes of solutions to the ones in the question.

Reactants in excess

At Standard Grade you were introduced to calculations involving concentrations, where the amounts of the reactants were in the correct proportions to react completely. If this was not the case, it was made clear which one of the reactants was in excess. An example of each type of question is shown below.

Example 1

Calculate the mass of sodium chloride produced when a sodium hydroxide solution is completely neutralised by $20\,\text{cm}^3$ of $1.25\,\text{mol}\,l^{-1}$ hydrochloric acid.

Step 1 Write the balanced equation and the mole ratios.

NaOH(aq) +	HCl(aq) →	NaCl(aq) +	H_2O(l)
1 mol	1 mol	1 mol	1 mol

Step 2 Calculate the amount of acid reacting and hence find the amount NaCl produced.

Amount of HCl = concentration × volume (in litres)
$= 1.25\,\text{mol}\,l^{-1} \times 0.02\,l = 0.025\,\text{mol}$

From the equation,

$NaOH(aq) + HCl(aq) \rightarrow NaCl(aq) + H_2O(l)$

1 mol 1 mol

0.025 mol 0.025 mol

Step 3 Calculate the mass of product from the amount in moles.

Relative formula mass (NaCl) = 58.5

Mass = amount in moles × molar mass

$= 0.025\ mol \times 58.5\ g\ mol^{-1}$

$= 1.46\ g$ NaCl

Example 2

Calculate the mass of zinc chloride produced when 6.5 g of zinc are added to excess hydrochloric acid.

Step 1 Write the balanced equation and the mole ratios.

$Zn(s) + 2HCl(aq) \rightarrow ZnCl_2(aq) + H_2(g)$

1 mol 2 mol 1 mol 1 mol

Step 2 Calculate the amount of zinc reacting and hence find the amount of $ZnCl_2$ produced.

$$\text{Amount of Zn} = \frac{\text{mass}}{\text{molar mass}}$$

$$= \frac{6.5\ g}{65\ g\ mol^{-1}} = 0.1\ mol$$

From the equation,

$Zn(s) + 2HCl(aq) \rightarrow ZnCl_2(aq) + H_2(g)$

1 mol 1 mol

0.1 mol 0.1 mol

Step 3 Calculate the mass of product from the amount in moles.

Relative formula mass ($ZnCl_2$) = 136

Mass = amount in moles × molar mass

$= 0.1\ mol \times 136\ g\ mol^{-1}$

$= 13.6\ g\ ZnCl_2$

Normally, however, we cannot assume that reactants will be present in the exact amounts required for complete reaction. Often there will be less of one reactant than is needed to allow all the reactants to combine. As soon as one of the reactants is used up, the reaction will stop. This reactant is called the **limiting reactant** because it limits the extent of the reaction. Any reactant left over at the end is said to be **in excess**.

Calculations must be based on the amount of the limiting reagent. Examples 3 and 4 are similar to Examples 1 and 2, except that you first have to work out which is the limiting reactant.

Example 3

Calculate the mass of sodium chloride produced when $25\ cm^3$ of $0.1\ mol\ l^{-1}$ sodium hydroxide solution reacts with $10\ cm^3$ of $0.2\ mol\ l^{-1}$ hydrochloric acid.

Step 1 Write the balanced equation and the mole ratios.

$NaOH(aq) + HCl(aq) \rightarrow NaCl(aq) + H_2O(l)$

1 mol 1 mol 1 mol 1 mol

Step 2 Calculate the amount of each reactant and decide which is the limiting reactant.

Amount in moles = concentration × volume (in litres)

Amount NaOH = $0.1\ \text{mol l}^{-1} \times 0.025\,\text{l} = 0.0025\ \text{mol}$

Amount HCl = $0.2\ \text{mol l}^{-1} \times 0.01\,\text{l} = 0.002\ \text{mol}$

Limiting reactant is HCl (0.002 mol)

Excess reactant is NaOH (0.0025 mol)

Step 3 Using the amount of the limiting reactant, calculate the amount of NaCl produced.

From the equation,

$NaOH(aq) + HCl(aq) \rightarrow NaCl(aq) + H_2O(l)$

1 mol 1 mol

0.002 mol 0.002 mol

Step 4 Calculate the mass of product from the amount in moles.

Relative formula mass (NaCl) = 58.5

Mass = amount in moles × molar mass

= $0.002\ \text{mol} \times 58.5\,\text{g mol}^{-1}$

= 0.117 g NaCl

Example 4

Calculate the mass of zinc chloride produced when 6.5 g of zinc reacts with $500\,\text{cm}^3$ of $0.1\ \text{mol l}^{-1}$ hydrochloric acid.

Step 1 Write the balanced equation and the mole ratios.

$Zn(s) + 2HCl(aq) \rightarrow ZnCl_2(aq) + H_2(g)$

1 mol 2 mol 1 mol 1 mol

Step 2 Calculate the amount of each reactant and decide which is the limiting reactant.

Amount in moles = concentration × volume (in litres)

$$\text{Amount of Zn} = \frac{\text{mass}}{\text{molar mass}}$$

$$= \frac{6.5\,\text{g}}{65\,\text{g mol}^{-1}} = 0.1\,\text{mol}$$

Amount of HCl = $0.1\ \text{mol l}^{-1} \times 0.5\,\text{l} = 0.05\ \text{mol}$

Limiting reactant is HCl (0.05 mol)

Excess reactant is Zn (0.1 mol)

Step 3 Using the amount of the limiting reactant, calculate the amount of $ZnCl_2$ produced.

From the equation,

$Zn(s) + 2HCl(aq) \rightarrow ZnCl_2(aq) + H_2(g)$

2 mol 1 mol

0.05 mol 0.025 mol

Step 4 Calculate the mass of product from the amount in moles.

Relative formula mass ($ZnCl_2$) = 136

Mass = amount in moles × molar mass

= $0.025\ \text{mol} \times 136\,\text{g mol}^{-1}$

= 3.4 g $ZnCl_2$

PROBLEMS FOR 10.4

1 How many moles of solute are dissolved in each of the following solutions?
  a 1 litre of 0.5 mol l^{-1} KCl(aq)
  b 2 litres of 2 mol l^{-1} NaOH(aq)
  c 250 cm^3 of 4 mol l^{-1} HCl(aq)
  d 100 cm^3 of 2 mol l^{-1} $Na_2CO_3(aq)$
  e 250 cm^3 of 0.2 mol l^{-1} $H_2SO_4(aq)$
  f 50 cm^3 of 0.04 mol l^{-1} NaI(aq)

2 What is the concentration (in mol l^{-1}) of the solutions with the following compositions?
  a 2 mol of KOH in 1 litre
  b 1 mol of NaCl in 500 cm^3
  c 0.5 mol of HCl in 100 cm^3
  d 2 mol of HCl in 10 litres
  e 0.1 mol of HNO_3 in 25 cm^3
  f 0.05 mol of $CaCl_2$ in 250 cm^3
  g 10 g of NaOH in 1 litre
  h 4 g of NaOH in 500 cm^3
  i 100 g of $CaBr_2$ in 100 cm^3
  j 2 g of $CaBr_2$ in 25 cm^3
  k 8 g of $CuSO_4$ in 100 cm^3
  l 40 g of $CuSO_4$ in 2 litres

3 Complete the following table for the concentrations of ions in Dead Sea water.

Ion	Concentration/ g l^{-1}	Concentration/ mol l^{-1}
Cl^-	183.0	
Mg^{2+}	36.2	
Na^+		1.37
Ca^{2+}		0.335
K^+	6.8	
Br^-	5.2	
SO_4^{2-}		0.00625

4 For each of the following examples show, by calculation, which reactant is in excess and find the mass of the salt produced in each case.
  a 10 g of copper(II) oxide were added to 50 cm^3 of 1.0 mol l^{-1} sulphuric acid.
  $CuO(s) + H_2SO_4(aq) \rightarrow CuSO_4(aq) + H_2O(l)$
  salt
  b 2.7 g of lead(II) carbonate were added to 80 cm^3 of 0.5 mol l^{-1} nitric acid.
  $PbCO_3(s) + 2HNO_3(aq) \rightarrow$
  $Pb(NO_3)_2(aq) + H_2O(l) + CO_2(g)$
  salt

5 Calculate the volume of 0.100 mol l^{-1} $AgNO_3$ required to precipitate all the chloride ions in 5 cm^3 of 0.5 mol l^{-1} NaCl. The reaction involved is

$Ag^+(aq) + Cl^-(aq) \rightarrow AgCl(s)$

6 10.0 cm^3 of a sample of domestic bleach was added to an excess of acidified aqueous potassium iodide. The bleach contains sodium chlorate(I). In the presence of acid, the chlorate(I) ion, $ClO^-(aq)$, oxidises the iodide ion to iodine and the reaction mixture turns yellow orange.

$$ClO^-(aq) + 2I^-(aq) + 2H^+(aq) \rightarrow Cl^-(aq) + I_2(aq) + H_2O(l)$$

The iodine formed was titrated against 0.500 mol l^{-1} sodium thiosulphate. 20.0 cm^3 of the sodium thiosulphate was required. The thiosulphate ion, $S_2O_3^{2-}(aq)$, reduces the iodine back to iodide ions.

$$2S_2O_3^{2-}(aq) + I_2(aq) \rightarrow S_4O_6^{2-}(aq) + 2I^-(aq)$$

  a Calculate the amount in moles of thiosulphate ion required to reduce the iodine and hence, using the second equation, the amount in moles of iodine present.
  b What is the amount in moles of chlorate(I) needed to produce this amount of iodine?
  c This amount of chlorate(I) was present in the 10.0 cm^3 sample of bleach. What is the concentration of chlorate(I) in the domestic bleach?

10.5 Calculations involving gases

Chemists use chemical equations to work out the masses of reactants and products involved in a reaction. If one or more of these is a gas, it is sometimes more useful to know its volume rather than it mass.

Avogadro's law

Equal volumes of all gases at the same temperature and pressure contain an equal number of molecules.

Molar volume

The number of molecules in 1 mole of any gas is always 6.02×10^{23}. This quantity is known as **Avogadro's constant**, ***L***. The molecules in a gas are very far apart so that the actual size of each molecule has negligible effect on the total volume the gas occupies. So, 1 mole of any gas always occupies the same volume, no matter which gas it is. Avogadro realised this as long ago as 1811, when he put forward his famous law (sometimes called Avogadro's hypothesis).

The volume occupied by 1 mole of any gas at a particular temperature and pressure is called the **molar volume**. At *standard temperature and pressure* (stp), the molar volume of a gas is 22.4 litres. Standard temperature and pressure is a temperature of 0°C and a pressure of 1 atmosphere.

At room temperature, around 25 °C, and 1 atmosphere pressure, the volume of 1 mole of any gas is about 24 litres (Figure 7).

The idea of molar volume allows you to calculate the amount in moles from the volume of a gas, and vice versa, provided you know the temperature and pressure of the gas. Here are two examples. You will need to assume that the molar volume of a gas at room temperature and pressure (rtp) is $24\,l\,mol^{-1}$.

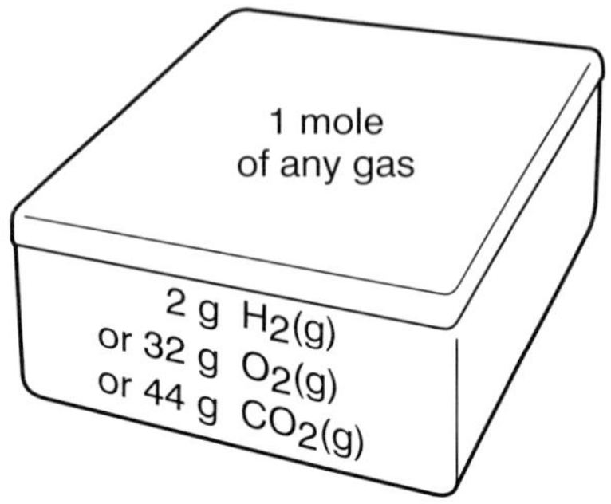

Figure 7 1 mole of any gas at room temperature and pressure occupies 24 litres (about the volume of a large biscuit tin)

Example 1

Calculate the volume occupied by 4.4 g of carbon dioxide at room temperature and pressure.

Step 1 Calculate the amount in moles from the mass.

Relative formula mass $(CO_2) = 44$

$$\text{Amount of } CO_2 = \frac{\text{mass}}{\text{molar mass}}$$

$$= \frac{4.4\,g}{44\,g\,mol^{-1}} = 0.1\text{ mol}$$

Step 2 Calculate the volume.

1 mol CO_2 at rtp has a volume of 24 litres
0.1 mol CO_2 at rtp has a volume of 2.4 litres

Example 2

Calculate the mass of 1.2 litres of methane gas (CH_4) at room temperature and pressure.

Step 1 Calculate the amount in moles.

$$\text{Amount of } CH_4 = \frac{\text{volume}}{\text{molar volume}}$$

$$= \frac{1.2\text{ litres}}{24\text{ litres}} = 0.05\text{ mol}$$

Step 2 Calculate the mass.

Relative formula mass $(CH_4) = 16$

$$\text{Mass of } CH_4 = \text{amount in moles} \times \text{molar mass}$$
$$= 0.05\text{ mol} \times 16\,g\,mol^{-1} = 0.8\,g$$

Reacting volumes of gases

We can use molar volumes to work out the volumes of gases involved in a reaction.

For example, consider the manufacture of ammonia from nitrogen and hydrogen:

$$N_2(g) + 3H_2(g) \rightarrow 2NH_3(g)$$

From the equation:

1 mole N_2	+	3 moles H_2	$\rightarrow$	2 moles NH_3

Using the idea that 1 mole of each gas occupies 24 litres at rtp, we can write:

	24 litres N_2	+	(3×24) litres H_2	$\rightarrow$	(2×24) litres NH_3
or	1 litre N_2	+	3 litres H_2	$\rightarrow$	2 litres NH_3
or	1 volume N_2	+	3 volumes H_2	$\rightarrow$	2 volumes NH_3

So if we had 10 cm^3 nitrogen it would react with 30 cm^3 hydrogen to form 20 cm^3 ammonia, *provided all the measurements were taken at the same temperature and pressure.*

For a reaction *involving only gases*, we can convert a statement about the numbers of *moles* of each substance involved to the same statement about *volumes*.

$N_2(g)$	+	$3H_2(g)$	$\rightarrow$	$2NH_3(g)$
1 mole	+	3 moles	$\rightarrow$	2 moles
1 volume	+	3 volumes	$\rightarrow$	2 volumes

Reactants in excess

In **Section 10.4** you found out how to calculate which substance is the limiting reactant in a reaction and which reactant is in excess. The same type of calculation can be carried out for reactions involving gases.

Example

A mixture of 20 cm^3 ethane and 120 cm^3 oxygen was ignited and then allowed to cool. Calculate the volume and composition of the gas mixture which would be left at the end assuming that all volumes are measured at room temperature and pressure.

Step 1 Write the balanced equation and the mole ratios.

$$C_2H_6(g) + \tfrac{7}{2}\,O_2(g) \rightarrow 2CO_2(g) + 3H_2O(l)$$
$$1\text{ mol} + \tfrac{7}{2}\text{ mol} \rightarrow 2\text{ mol} + 3\text{ mol}$$

Step 2 Using Avogadro's law

$$1\text{ vol} + \tfrac{7}{2}\text{ vol} \rightarrow 2\text{ vol}$$
$$20\,cm^3 + 70\,cm^3 \rightarrow 40\,cm^3$$

Step 3 Decide which is the limiting reactant and work out the volume and composition of the resulting gas mixture.

Ethane:	limiting reactant and will react completely
Oxygen:	120 cm^3 at start; 70 cm^3 will react, so $(120\,cm^3 - 70\,cm^3) = 50\,cm^3$ will be left at the end
Carbon dioxide:	40 cm^3 produced

The resulting gas mixture will contain 50 cm^3 oxygen and 40 cm^3 carbon dioxide.

PROBLEMS FOR 10.5

1 a Calculate the volume occupied by 0.30 g of ethane gas (C_2H_6) at room temperature and pressure.

b Calculate the mass of 0.60 litres of nitrogen dioxide gas (NO_2). The volume was measured at room temperature and pressure.

2 Which of the following gases will have the largest volume? (All the volumes are measured at the same temperature and pressure.)

a 44 g CO_2 c 56 g N_2
b 48 g CH_4 d 71 g Cl_2

3 Methane burns in oxygen to produce carbon dioxide and water:

$$CH_4(g) + 2O_2(g) \rightarrow CO_2(g) + 2H_2O(g)$$

a What are the relative volumes of the reactants and products involved (assuming all measurements are made at 500 °C and 1 atmosphere pressure)?

b What volume of oxygen would be needed for the complete combustion of 10 cm^3 of methane?

c What would be the *total* volume of the products formed?

d What would be the *change* in volume as the reaction is carried out?

e How would your answer to part **d** differ if the initial and final volume measurements were made at room temperature?

4 a Write an equation for the complete combustion of pentane, C_5H_{12}, in air.

b Suppose 18 g of pentane are burned completely in a car engine, forming carbon dioxide and water. Calculate

i the amount in moles of pentane that is burned.
ii the amount in moles of oxygen that is needed to burn the pentane.
iii the *volume* of oxygen that is needed.
iv the volume of *air* that is needed (assume air is 20% oxygen by volume).
v the volume of carbon dioxide that is produced.

5 What volume of oxygen is required to react completely with a mixture of 10 cm^3 hydrogen and 20 cm^3 carbon monoxide? (Assume all volumes are measured at the same temperature and pressure.)

6 10 cm^3 of a gaseous hydrocarbon react completely with exactly 40 cm^3 of oxygen to produce 30 cm^3 of carbon dioxide. What is the empirical formula of the hydrocarbon? (All volumes are measured at the same temperature and pressure.)

7 In each of the following examples, calculate

i which reactant is in excess,
ii the volume and composition of the resultant gas mixture (measured at room temperature and pressure).

a A mixture of 10 cm^3 methane and 30 cm^3 oxygen was burned.
$CH_4(g) + 2O_2(g) \rightarrow CO_2(g) + 2H_2O(l)$

b A mixture of 20 cm^3 propane and 25 cm^3 oxygen was burned.
$C_3H_8(g) + 5O_2(g) \rightarrow 3CO_2(g) + 4H_2O(l)$

8 Unlike gases, the molar volumes of liquids and solids are far from constant. Why do you think this is so?

11 ATOMIC STRUCTURE

11.1 *A simple model of the atom*

What is inside atoms?

No one has yet been able to look directly inside atoms to see what they are really like, but a considerable amount of experimental evidence has given us a good working model of atomic structure. Scientific models should not be considered as the 'truth' or the 'right answer'. They are useful because they help to explain our observations of nature and they guide our thinking in productive directions.

Sometimes we can explain things by using a simplified version of a model. For example, regarding atoms as tiny snooker balls is sufficient to explain the states of matter, but not detailed enough to describe how chemical bonds are formed.

Many chemical and nuclear processes can be explained by a simple model of atomic structure in which atoms are thought to be made from three types of *sub-atomic particle*: **protons**, **neutrons** and **electrons**. This is not the whole story, but it's enough for present purposes. Protons and neutrons form the dense nucleus (or centre) of atoms. Electrons are much more diffuse and move around the nucleus. Figure 1 shows a very simple model of the atom. It is not to scale. The nucleus is tiny compared with the volume occupied by the electrons. If you imagined the atom to be the size of the stadium at Hampden Park, the nucleus would be the size of a pea on the centre spot! Since most atoms have a radius of 0.1 nm–0.2 nm (1×10^{-10} m–2×10^{-10} m), the nucleus must be very, very small – about 1×10^{-15} m radius, in fact.

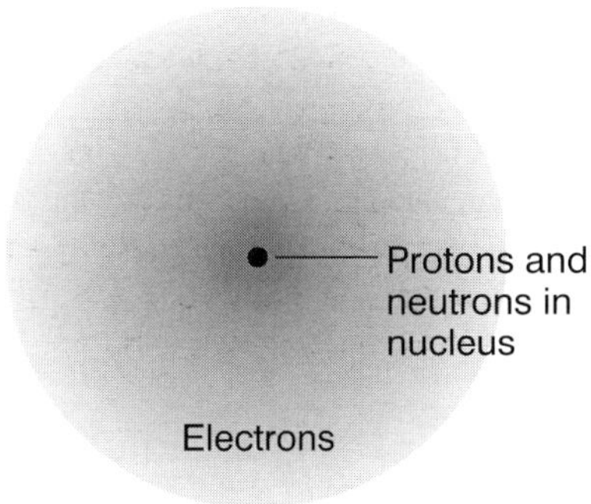

Figure 1 A simple model of the atom – not to scale

Sub-atomic particles

Some properties of protons, neutrons and electrons are summarised in Table 1.

Particle	Mass on relative atomic mass scale	Charge (relative to proton)	Location in atom
proton	1	+1	in nucleus
neutron	1	0	in nucleus
electron	0.00055	−1	around nucleus

Table 1 Some properties of sub-atomic particles

Protons and electrons have equal but opposite electrical charges. Neutrons have no charge. Protons and neutrons have almost equal masses, and are much more massive than electrons. The nucleus accounts for almost all the mass of the atom and hardly any of its volume. Most of the atom is empty space.

It is the outer parts of atoms which interact together in chemical reactions, so for chemists, electrons are the most important particles.

Nuclear symbols

The nucleus can be described by just two numbers – the **atomic number** and the **mass number**.

The atomic number is the number of protons in the nucleus. It is also numerically equal to the charge on the nucleus. The atomic number is the same for every atom of an element: for example, all carbon atoms have an atomic number of 6.

The mass number is the number of protons plus neutrons in the nucleus.

Nuclear symbols identify the mass number and the atomic number as well as the symbol for the element. Figure 2 gives an example.

Sometimes the atomic number is omitted since the chemical symbol tells us which element it is anyway. Thus $^{12}_{6}C$ is simplified to ^{12}C.

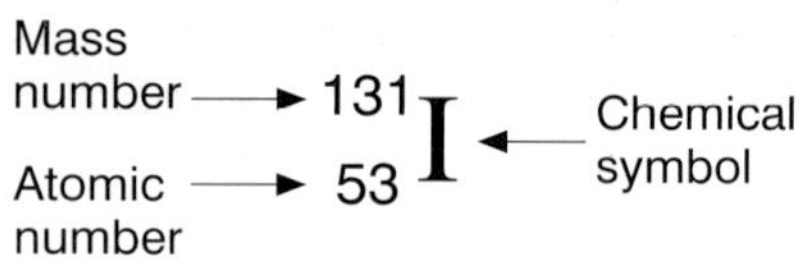

Figure 2 The nuclear symbol for iodine-131

What are isotopes?

Atoms of the same element which have different mass numbers are called **isotopes**. Since the number of protons is the same for all atoms of a particular element, the differences in mass must arise from different numbers of neutrons.

Most elements exist naturally as a mixture of isotopes. The relative atomic mass is an average of the masses of the isotopes which also takes account of their abundances. Isotopes of some elements are given in Table 2, together with their percentage abundances in a naturally occurring sample. From the figures for chlorine, we can work out that the relative atomic mass of chlorine is $[(35 \times 75) + (37 \times 25)]/100 = 35.5$.

Element	Isotope	Abundance
chlorine	^{35}Cl	75%
	^{37}Cl	25%
bromine	^{79}Br	50%
	^{81}Br	50%
calcium	^{40}Ca	96.9%
	^{42}Ca	0.7%
	^{43}Ca	0.1%
	^{44}Ca	2.1%
	^{48}Ca	0.2%
iron	^{54}Fe	5.8%
	^{56}Fe	91.7%
	^{57}Fe	2.2%
	^{58}Fe	0.3%

Table 2 Isotopes of some elements

PROBLEMS FOR II.I

1 Copy and complete the following table.

Isotope	Symbol	Atomic number	Mass number	Number of neutrons
carbon-12	$^{12}_{6}C$	6	12	6
carbon-13			13	7
oxygen-16	$^{16}_{8}O$	8	16	
strontium-90	$^{90}_{38}Sr$	38		52
iodine-131	$^{131}_{53}I$			

2 How many protons, neutrons and electrons are present in the following atoms?

a $^{79}_{35}Br$

b $^{81}_{35}Br$

c $^{35}_{17}Cl$

d $^{37}_{17}Cl$

3 Calculate the relative atomic masses of Br and Fe from the data in Table 2.

4 The relative atomic mass of iridium is 192.2. Iridium occurs naturally as a mixture of iridium-191 and iridium-193. What is the percentage of each isotope in naturally occurring iridium?

II.2 *Electron arrangement*

In our simple model of the atom, the electrons are arranged in a series of concentric **shells** surrounding the nucleus. These shells are labelled according to how close they are to the nucleus. The first shell is nearest the nucleus, the second shell is a bit further away, and so on. The electrons in the shells have different energies – the further a shell is from the nucleus, the higher the energy associated with the shell. Although each shell can hold more than one electron, there is a limit. The maximum numbers of electrons which can be held in the first three shells are

first shell 2 electrons
second shell 8 electrons
third shell 18 electrons

A shell which contains its maximum number of electrons is called a **filled shell**. Electrons are arranged so that the lowest energy shells are filled first. Table 3 shows the **electron shell arrangement** for elements 1 to 20. The electron shell arrangement of sodium, for example, is written as 2.8.1.

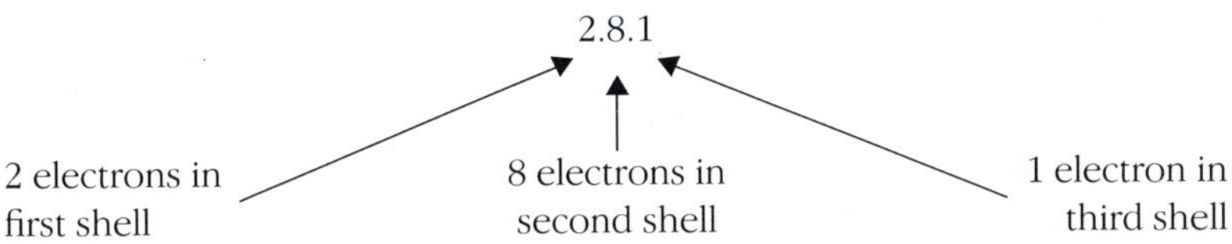

Table 3 Electron shell configurations for the first 20 elements

Element	Atomic number	1st shell	2nd shell	3rd shell	4th shell
hydrogen	1	1			
helium	2	2			
lithium	3	2	1		
beryllium	4	2	2		
boron	5	2	3		
carbon	6	2	4		
nitrogen	7	2	5		
oxygen	8	2	6		
fluorine	9	2	7		
neon	10	2	8		
sodium	11	2	8	1	
magnesium	12	2	8	2	
aluminium	13	2	8	3	
silicon	14	2	8	4	
phosphorus	15	2	8	5	
sulphur	16	2	8	6	
chlorine	17	2	8	7	
argon	18	2	8	8	
potassium	19	2	8	8	1
calcium	20	2	8	8	2

Chemists can explain many of the properties of atoms without needing to use a detailed theory of atomic structure. Much chemistry is decided only by the outer shell electrons, and one very useful model treats the atom as being composed of a **core** of nucleus plus inner shells, surrounded by an **outer shell**. Figure 3 shows how a sodium atom would be thought of using this model.

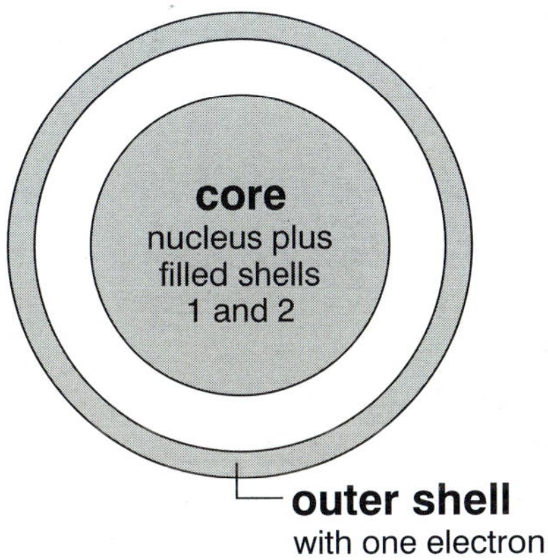

Figure 3 The core and outer shell model of a sodium atom (2.8.1)

PROBLEMS FOR II.2

1 Isotopes of elements have different numbers of neutrons. Do they have different electron arrangements? Explain your answer.

2 Look at Figure 3 and draw a similar labelled diagram for the core and outer shell model of a fluorine atom.

3 The electron arrangement for sodium can be written as 2.8.1. Use this notation to write down the electron arrangements for

a lithium
b phosphorus
c calcium.

4 The noble gases, helium, neon and argon, are very unreactive.

a Write down the electron arrangements for helium, neon and argon.
b Suggest a reason for the stability of these elements.

12 THE PERIODIC TABLE

12.1 *The modern Periodic Table*

Each cell gives: atomic number, symbol, name, relative atomic mass.

Period	Group 1	2											3	4	5	6	7	0
			Transition elements															
1	1 H hydrogen 1.0																	2 He helium 4.0
2	3 Li lithium 6.9	4 Be beryllium 9.0											5 B boron 10.8	6 C carbon 12.0	7 N nitrogen 14.0	8 O oxygen 16.0	9 F fluorine 19.0	10 Ne neon 20.2
3	11 Na sodium 23.0	12 Mg magnesium 24.3											13 Al aluminium 27.0	14 Si silicon 28.1	15 P phosphorus 31.0	16 S sulphur 32.1	17 Cl chlorine 35.5	18 Ar argon 39.9
4	19 K potassium 39.1	20 Ca calcium 40.1	21 Sc scandium 45.0	22 Ti titanium 47.9	23 V vanadium 50.9	24 Cr chromium 52.0	25 Mn manganese 54.9	26 Fe iron 55.9	27 Co cobalt 58.9	28 Ni nickel 58.7	29 Cu copper 63.5	30 Zn zinc 65.4	31 Ga gallium 69.7	32 Ge germanium 72.6	33 As arsenic 74.9	34 Se selenium 79.0	35 Br bromine 79.9	36 Kr krypton 83.8
5	37 Rb rubidium 85.5	38 Sr strontium 87.6	39 Y yttrium 88.9	40 Zr zirconium 91.2	41 Nb niobium 92.9	42 Mo molybdenum 95.9	43 Tc technetium (99)	44 Ru ruthenium 101.1	45 Rh rhodium 102.9	46 Pd palladium 106.4	47 Ag silver 107.9	48 Cd cadmium 112.4	49 In indium 114.8	50 Sn tin 118.7	51 Sb antimony 121.8	52 Te tellurium 127.6	53 I iodine 126.9	54 Xe xenon 131.3
6	55 Cs caesium 132.9	56 Ba barium 137.3	57 La lanthanum 138.9	72 Hf hafnium 178.5	73 Ta tantalum 181.0	74 W tungsten 183.9	75 Re rhenium 186.2	76 Os osmium 190.2	77 Ir iridium 192.2	78 Pt platinum 195.1	79 Au gold 197.0	80 Hg mercury 200.6	81 Tl thallium 204.4	82 Pb lead 207.2	83 Bi bismuth 209.0	84 Po polonium (210)	85 At astatine (210)	86 Rn radon (222)
7	87 Fr francium (223)	88 Ra radium (226)	89 Ac actinium (227)	104 Rf rutherfordium (261)	105 Db dubnium (262)	106 Sg seaborgium (263)	107 Bh bohrium (262)	108 Hs hassium (265)	109 Mt meitnerium (266)	110 Uun ununnilium	111 Uuu unununium	112 Uub ununbium						

metals | non-metals

Lanthanide elements

58 Ce cerium 140.1	59 Pr praseodymium 140.9	60 Nd neodymium 144.2	61 Pm promethium (147)	62 Sm samarium 150.4	63 Eu europium 152.0	64 Gd gadolinium 157.3	65 Tb terbium 158.9	66 Dy dysprosium 162.5	67 Ho holmium 164.9	68 Er erbium 167.3	69 Tm thulium 168.9	70 Yb ytterbium 173.0	71 Lu lutetium 175.0

Actinide elements

90 Th thorium 232.0	91 Pa protactinium (231)	92 U uranium 238.1	93 Np neptunium (237)	94 Pu plutonium (242)	95 Am americium (243)	96 Cm curium (247)	97 Bk berkelium (245)	98 Cf californium (251)	99 Es einsteinium (254)	100 Fm fermium (253)	101 Md mendelevium (256)	102 No nobelium (254)	103 Lr lawrencium (257)

KEY

6.9 — relative atomic mass
Li — symbol
3 — atomic number
lithium

Figure 1 *The modern Periodic Table*

Groups and periods

The modern Periodic Table (Figure 1) is based on one proposed by the Russian chemist, Mendeleev, in 1869. He organised the 61 elements known at that time into a table made up of horizontal rows and vertical columns. Elements with similar properties were placed underneath each other in the columns. Mendeleev left gaps in his Periodic Table. Not only did he predict that new elements would be discovered to fill the gaps, he also predicted the properties of these yet undiscovered elements.

You can read more about the history of the Periodic Table in **Section 1.4**.

In Mendeleev's Periodic Table the elements were arranged in order of increasing relative atomic mass. At first glance, the same seems to be true of the modern version – but not quite. Some pairs of elements are 'out of order', eg Te (relative atomic mass = 128) comes before I (relative atomic mass = 127). It is **atomic number**, the number of protons in the nucleus, which really determines the place of an element in the Periodic Table.

Vertical columns in the Periodic Table are called **groups**. The elements in a group show specific similarities. You may already know some of the common features of the elements of Group 1 (the alkali metals), Group 7 (the halogens) and Group 0 (the noble gases).

Horizontal rows in the Periodic Table are called **periods**. Since they cut across the groups, there are fewer common features among the elements of a period. After hydrogen and helium, there are two *short periods* (Li to Ne and Na to Ar) and four *long periods* (K to Kr etc). There are 8 elements in a short period, but 18 in the first long period because of the inclusion of 10 transition elements.

Elements change from metallic to non-metallic across a period. They become increasingly metallic *down* a group – and increasingly non-metallic *up* a group.

Electronic structure and the Periodic Table

The arrangement of elements by rows and columns in the Periodic Table is a direct result of the electronic structure of atoms.

The *number of outer shell* electrons determines the **group** number. So, for example, Li, Na, K all have one outer shell electrons and are in Group 1; C and Si have four outer shell electrons and are in Group 4. The noble gases all have filled outer shells and are in Group 0 (the zero refers to the shell *beyond* the filled outer one, which does not have any electrons in it yet).

The number of the *shell* which is being filled determines the **period** to which an element belongs. Thus from Li to Ne, the outer electrons are being placed into shell 2, so these elements belong to Period 2.

PROBLEMS FOR 12.1

1 State the group and period to which each of the following elements belong.
   a sulphur, S
   b radium, Ra
   c antimony, Sb
   d gallium, Ga

2 For each of the following electron arrangements state the group and period to which you would expect the element to belong.
   a 2.8.7
   b 2.8.8.2
   c 2.1
   d 2.8.5

3 Name *three* alkali metals and give a distinctive chemical property shown by these metals.

4 Astatine (At) is an unstable radioactive element with atomic number 85. Would you expect astatine to show metallic or non-metallic properties? Explain your answer.

5 The Periodic Table drawn up by Mendeleev was missing a whole group of elements. Which elements are these? Suggest why Mendeleev did not leave a space for them.

12.2 *Patterns in the Periodic Table*

When we arrange the elements in order of atomic number, we see the patterns in physical properties which Mendeleev noted. There are *trends* in properties as you go down a group but also many properties vary in a fairly regular way as you move across a period from left to right. The pattern is then repeated as you go across the next period. The occurrence of periodic patterns is called **periodicity**.

Density

The **density** of an element is its mass per unit volume. Elements with high densities contain heavy particles which are closely packed together so that they occupy a relatively small volume. The opposite, of course, is true for low density elements. Figure 2 shows the variation in the densities of the elements across Periods 2 and 3. Across Period 2 (Li to Ne) the densities of the elements increase to a maximum at boron (B) in Group 3 and then fall. A similar pattern is observed in Period 3 (Na to Ar), with the maximum at the Group 3 element, aluminium (Al).

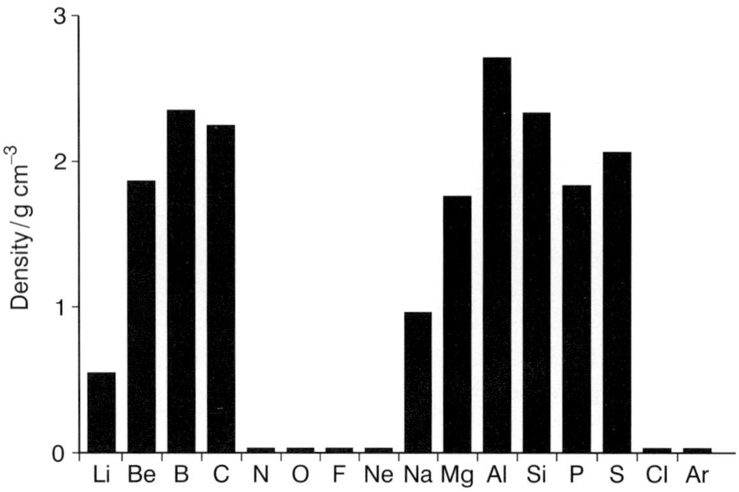

Figure 2 Densities of elements in Periods 2 and 3

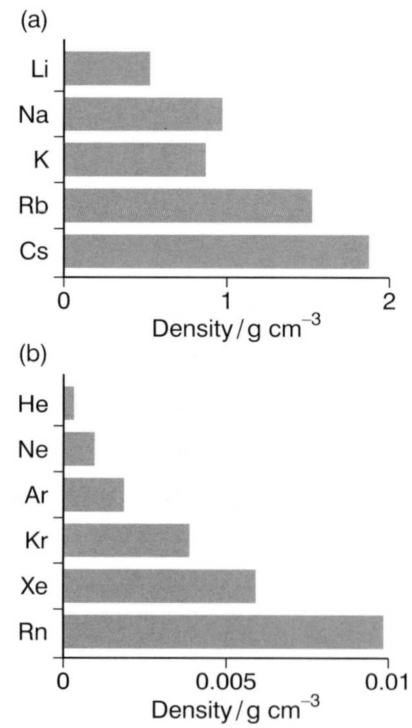

Figure 3 (a) Densities of Group 1 elements (b) Densities of Group 0 elements

On going down Group 1 (the alkali metals) (Figure 3a) there is an overall increase in density. Although the densities of the noble gases (Group 0) (Figure 3b) are very much smaller than those for the alkali metals, there is a similar but more regular increase as you go down the group.

Melting points and boiling points

When elements are melted and then boiled, the bonds between their constituent particles are broken. It is the strength of these bonds which determines whether an element has a high or low melting point or boiling point.

Figure 4 shows the variation in melting and boiling points of the elements in Period 3 (Na to Ar). Both the melting and boiling points initially increase as you go across the period and then fall dramatically from silicon (Si) to phosphorus (P). This means that the bonds between the particles in phosphorus must be very much weaker and easier to break than those between the particles in silicon. Similar patterns in melting and boiling points are observed for the elements in other periods.

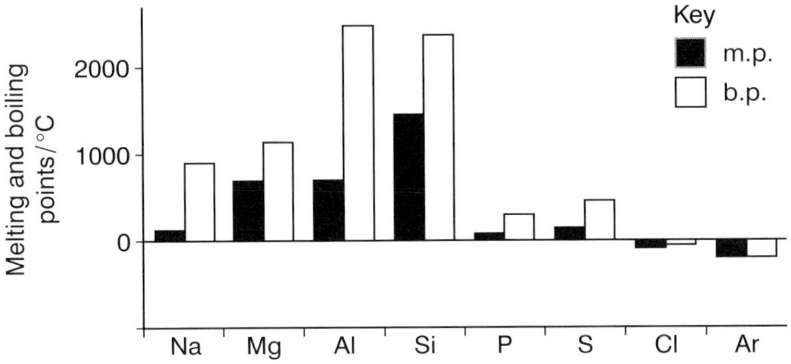

Figure 4 Melting and boiling points of elements in Period 3

On going down a group, the melting and boiling points of the elements decrease. Figure 5 shows the trend in melting and boiling points for the alkali metals in Group 1. For the halogens (Group 7), fluorine and chlorine at the top of the group are gases at room temperature and pressure, whereas the next element, bromine, is a liquid and iodine is a solid.

You will find out more about the types of bonds which are broken on melting and boiling later in this unit (**Chapter 16**).

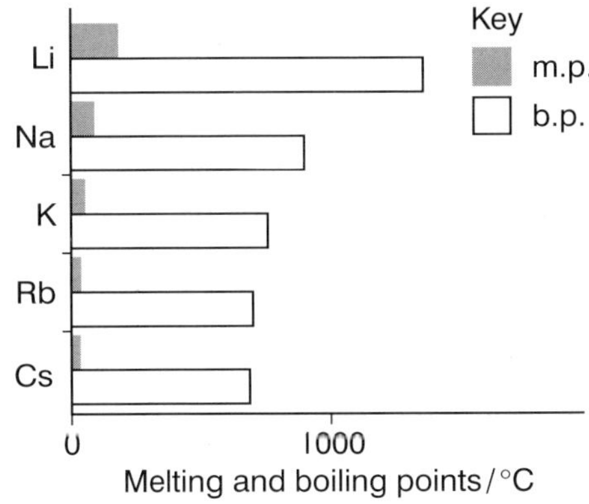

Figure 5 Melting and boiling points of elements in Group 1

Atomic size

The size of an atom is determined by the space its electrons occupy. So you might expect that, as the number of electrons in an atom increases, so will its atomic size. However, you can see from Figure 6 that, on crossing a period, atoms become smaller despite the increasing number of electrons. This is because the successive electrons added on crossing a period enter the same electron shell. In addition, the number of protons in the nucleus steadily increases across a period and the increasing positive nuclear charge attracts the electrons more and more strongly. As a result, the atoms get smaller even though they have more electrons.

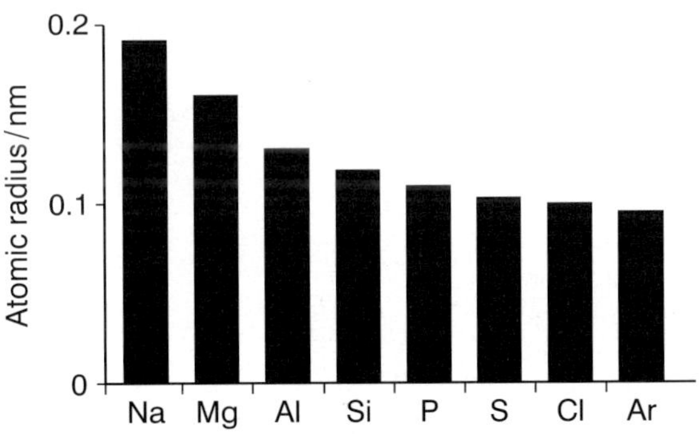

Figure 6 Variation in atomic size across Period 3

On going down a group, the atomic size increases (Figure 7). The reason for this is that another electron shell is added as we move from one element to the next one immediately below. Each new electron shell is further from the nucleus than the previous one and so the atoms get bigger.

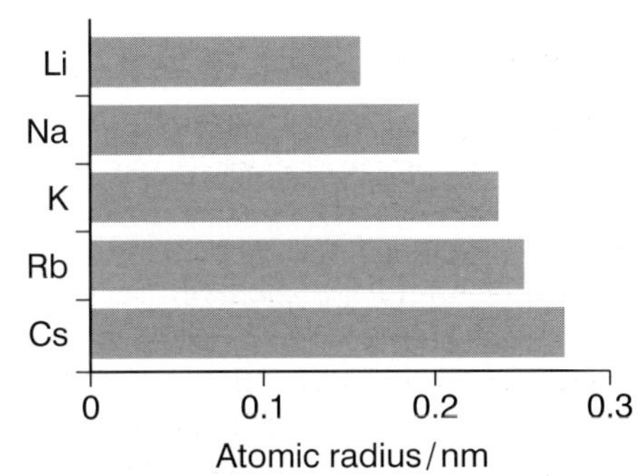

Figure 7 Variation in atomic size down Group 1

Ionisation energy

If sufficient energy is given to an atom, an electron is lost and the atom becomes a positive ion. **Ionisation** has taken place. Because electrons are attracted to the nucleus, an input of energy is *always* needed to remove them.

First ionisation energies

When one electron is pulled from an atom, the energy required is called the **first ionisation energy**. We define the first ionisation energy of an element as the energy needed to remove one electron from every atom in 1 mole of isolated gaseous atoms of the element. 1 mole of gaseous ions with one positive charge is formed. The general equation for the first ionisation process is:

$$E(g) \rightarrow E^+(g) + e^-$$

where E represents the symbol for the element. Let's consider oxygen. Its first ionisation energy is +1320 kJ mol^{-1}. This means that an input of 1320 kJ of energy is necessary to bring about the process:

$$O(g) \rightarrow O^+(g) + e^-$$

It is the most loosely held electron that is removed on first ionisation. This will be one of the outer shell electrons since they are furthest from the nucleus.

The first ionisation energies of elements 1–20 are plotted in Figure 8. They show a clearly repeating pattern. There is an overall increase in first ionisation energy across each period, peaking at the noble gases (He, Ne and Ar) and then falling sharply as you move on to the alkali metals (Li, Na and K).

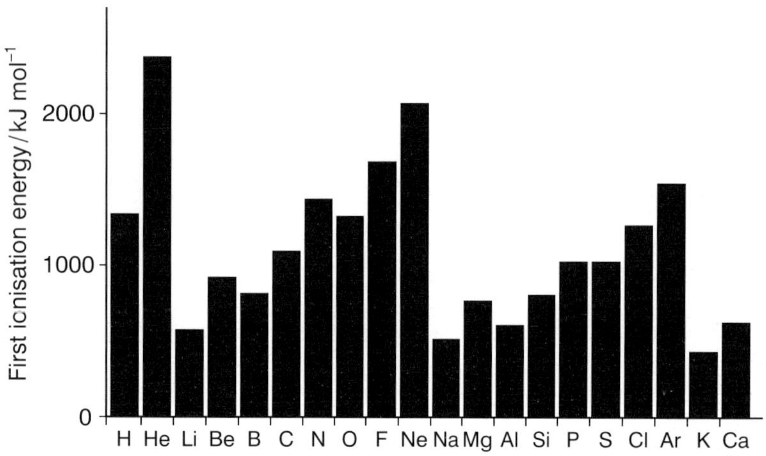

Figure 8 First ionisation energies of elements 1–20

Why does it become more difficult to remove an electron from an atom as you go across a period in the Periodic Table? Across a period, electrons are being added to the same shell but, at the same time, protons are being added to the nucleus. As the nuclear charge becomes more positive, the electrons will be held more tightly and so it gets harder to pull one from the outer shell.

On going down a group, the first ionisation energy decreases; it becomes easier to remove an electron. Figure 9 shows this trend for the alkali metals (Group 1). In this case, extra shells of electrons are added on going down the group. This has the effect of increasing the distance between the outer electrons and the nucleus. It also means an increase in the number of inner shells. Electrons in inner shells *shield* the outer ones from the full nuclear charge and the more inner electrons there are the more effective will be the shielding. Both factors, the increasing separation between the outer electrons and the nucleus and their increased shielding from the nucleus, weakens the attraction between the outer electrons and the nucleus. This makes it easier to remove one of the outer electrons.

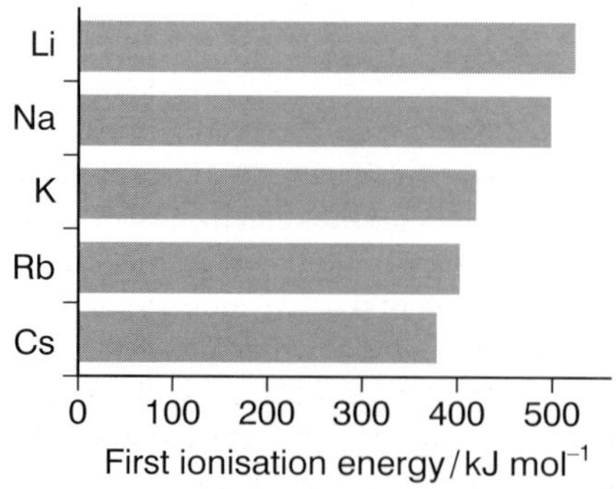

Figure 9 First ionisation energies for the Group 1 elements

To summarise, the ease with which an electron is removed from an atom depends on:

- the distance of the electron from the nucleus
- the nuclear charge
- the screening effect of inner shell electrons.

Successive ionisation energies

More than one electron can be removed from an atom (apart from a hydrogen atom, of course). So, as well as first ionisation energies, there are second, third and fourth ionisation energies, etc, which refer to the energies required to remove further moles of electrons. The general equations for these ionisation processes are:

first ionisation: $E(g) \rightarrow E^{+}(g) + e^{-}$
second ionisation: $E^{+}(g) \rightarrow E^{2+}(g) + e^{-}$
third ionisation: $E^{2+}(g) \rightarrow E^{3+}(g) + e^{-}$
fourth ionisation: $E^{3+}(g) \rightarrow E^{4+}(g) + e^{-}$

and so on.

Notice that each of the second and subsequent ionisation processes involve the removal of an electron from a *positive ion*. The second ionisation energy, for example, is the energy needed to remove one electron from an $E^{+}(g)$ ion (and *not* the energy required to remove two electrons from an E(g) atom).

Figure 10 shows eight successive ionisation energies for aluminium. The striking feature here is that ionisation energies increase as successive electrons are removed. It is more difficult to remove the second electron than the first; more difficult to remove the third than the second, and so on. This is true for all elements, not just aluminium. It makes sense because once an electron has been removed from an atom or ion, the remaining electrons will be attracted more strongly to the nucleus.

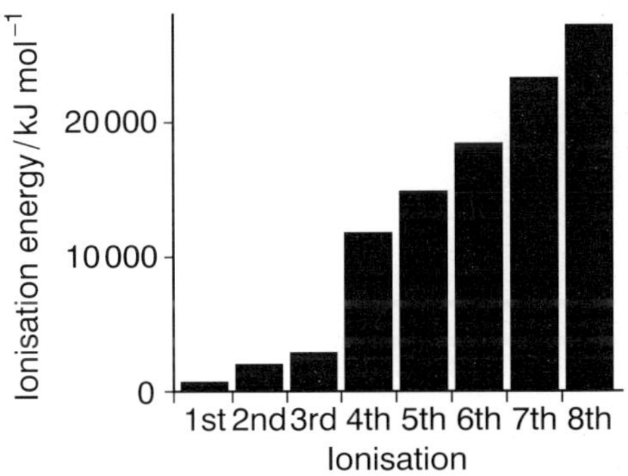

Figure 10 Successive ionisation energies for aluminium

Another important feature of Figure 10 is the sharp increase between the third and fourth ionisation energies of aluminium. For an explanation we need to look at the electron arrangement, 2.8.3, of the aluminium atom. The first, second and third ionisations correspond to the removal of the three electrons from the outer third shell. The fourth ionisation, on the other hand, involves pulling an electron from a filled inner shell. Because the second shell is closer to the nucleus, it is much more difficult to remove an electron from this shell than from the third shell. Other Group 3 elements also show 'big jumps' between the third and fourth ionisation energies.

If this explanation is correct, then Group 1 elements will have sharp rises between the first and second ionisation energies, Group 2 elements between the second and third ionisation energies, and so on. So you would predict that the 'big jump' for phosphorus, a Group 5 element, would come between its fifth and sixth ionisation energies. And it is clear from Figure 11 that this indeed is the case.

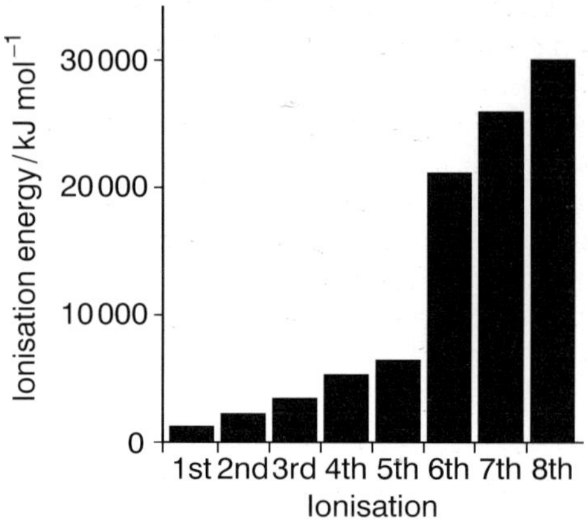

Figure 11 Successive ionisation energies for phosphorus

PROBLEMS FOR 12.2

1 Molar atomic volume is the volume occupied by 1 mole of a solid or liquid element. The molar atomic volumes for elements 1–56 are plotted below.

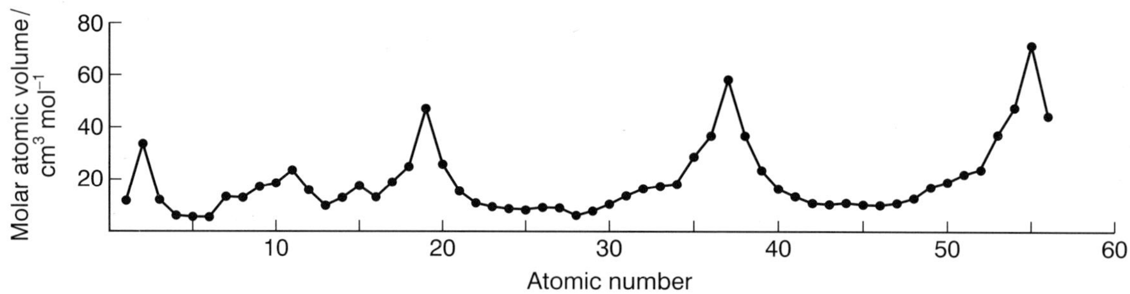

a What do the elements at the peaks of the molar atomic volume graph have in common?

b Explain why this graph provides evidence for periodicity.

2 **a** Which element in Period 5 contains the largest atoms?

b Which element in Group 7 contains the smallest atoms?

3 The graph below shows the variation in the first ionisation energies with atomic number for 16 elements in the Periodic Table. The element at which the graph starts is not specified.

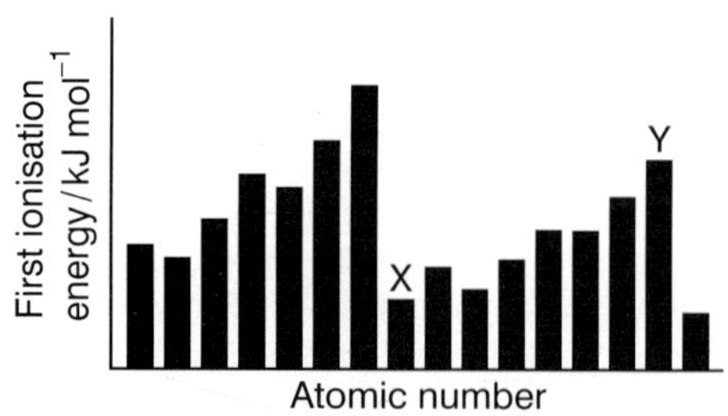

In which group of the Periodic Table is

a element X? **b** element Y?

Explain your answers.

4 The first ionisation energies of sulphur, chlorine and selenium are shown below.

Element	First ionisation energy /kJ mol^{-1}
sulphur	+1010
chlorine	+1260
selenium	+941

a Write an equation corresponding to the first ionisation of sulphur.

b Explain why the first ionisation energy of sulphur is smaller than that of chlorine.

c Explain why the first ionisation energy of sulphur is larger than that of selenium.

5 The first, second and third ionisation energies of calcium are +596 kJ mol^{-1}, +1160 kJ mol^{-1} and +4930 kJ mol^{-1}, respectively.

a Write equations corresponding to each of these three ionisation energies.

b Explain why the second ionisation energy of calcium is larger than its first ionisation energy.

c Explain why there is a very sharp rise between the second and third ionisation energies of calcium.

6 Explain why the second ionisation energy of sodium (+4560 kJ mol^{-1}) is much larger than the second ionisation energy of magnesium (+1460 kJ mol^{-1}).

13 CHEMICAL BONDING

13.1 *Ionic bonding*

Noble gas electron arrangements

In 1916, W Kossel and G N Lewis independently realised that, with the exception of helium, all the noble gases had eight electrons in their outer shells. They linked the chemical stability of the noble gases to this outer shell electron arrangement. They suggested that other elements try to achieve eight outer shell electrons by losing or gaining electrons when they react to form compounds. Some light elements are able to achieve stability by reaching the helium electron arrangement of two outer shell electrons.

Formation of ions

Metal atoms in Groups 1 and 2 of the Periodic Table have only one or two outer shell electrons. What makes them metals is that a noble gas electron arrangement is reached if these electrons are *lost* to form positively charged ions.

Most non-metal atoms have more than three outer shell electrons. One way they can reach a noble gas electron arrangement is by *gaining* electrons to form negatively charged ions.

There are limits to how many electrons an atom can pick up. One electron leads to an ion with a single negative charge. A second electron is repelled by this ion because their charges are the same, so making an ion with a 2– charge is much harder. Getting a third electron to stick, by the process: $A^{2-} + e^- \rightarrow A^{3-}$, is very difficult and does not often happen. Ions with a charge of 3– are unusual.

It is also hard to *remove* three or more electrons from atoms. Ions with a 4+ charge are almost unknown.

When metals react with non-metals, electrons are *transferred* from the metal atoms to the non-metal atoms. This usually gives both metal and non-metal a stable electronic structure like a noble gas. The positive and negative ions which are formed are held together by **electrostatic attraction** because of their opposite charges.

We use **electron dot-cross diagrams** to represent the way atoms bond together. In these, the outer shell electrons of one atom are represented by dots, with crosses for the other. Figure 1 shows dot-cross diagrams for the formation of sodium chloride and magnesium fluoride. The dots and crosses represent *only the outer shell electrons*. The numbers underneath the chemical symbols show the arrangement of electrons in the shells.

Each sodium atom loses one electron and each chlorine atom gains one electron, so the compound formed has the formula NaCl. Each magnesium atom loses two electrons but each fluorine atom gains only one electron, so the formula for magnesium fluoride is MgF_2.

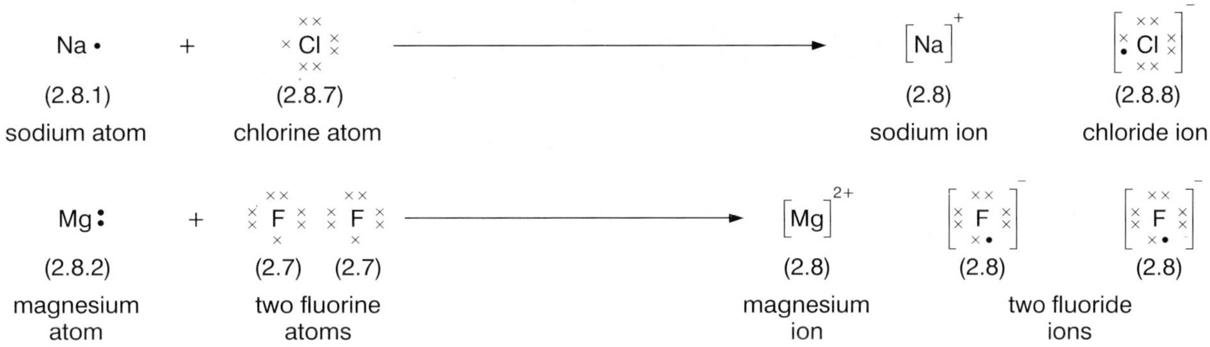

Figure 1 Dot-cross diagrams for two ionic compounds

The oppositely charged ions attract each other strongly: this electrostatic attraction is an **ionic bond**. In the solid compound, each ion attracts many others of opposite charge, and the ions build up into a giant lattice like the one shown in Figure 2. There is more about this in **Section 16.3**.

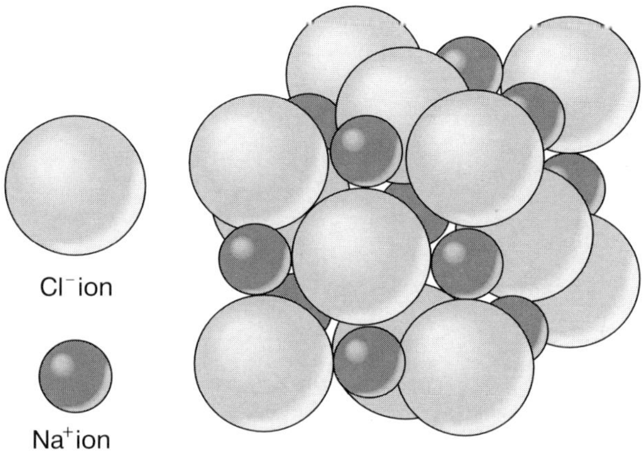

Figure 2 A model of the sodium chloride lattice, built up from oppositely charged sodium ions and chloride ions

PROBLEMS FOR 13.1

1 Use electron arrangements to suggest the formulae of the ions commonly formed from each of the following atoms:
  - **a** Li
  - **b** Al
  - **c** I
  - **d** S
  - **e** Ba

2 Draw electron dot-cross diagrams for the following ionic compounds:
  - **a** potassium fluoride, KF
  - **b** magnesium oxide, MgO
  - **c** calcium chloride, $CaCl_2$
  - **d** sodium sulphide, Na_2S

3 Suggest why the carbon atom does not normally form ions.

13.2 *Covalent bonding*

Sharing electrons

Ionic compounds are usually formed when metals react with non-metals. But there are many compounds containing only non-metallic elements. These cannot bond ionically because some of the atoms would have to lose too many electrons. Instead, noble gas electron arrangements are achieved by *sharing* electrons. Shared electrons count as part of the outer shell of *both* atoms in the bond. The dot-cross diagram for the H_2 molecule is shown in Figure 3.

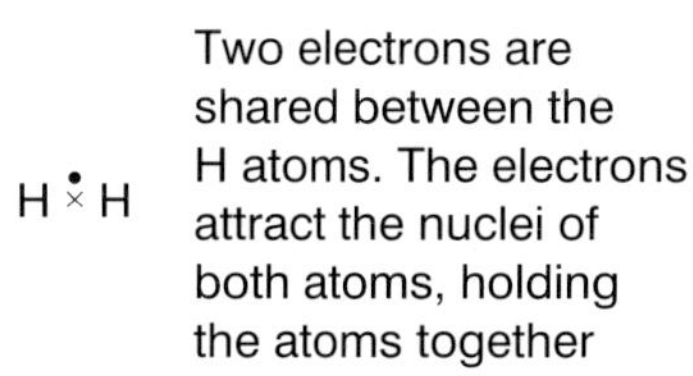

Figure 3 Electron sharing in the hydrogen molecule

Bonds formed by shared electrons are called **covalent bonds**. The two atoms are held together because their positively charged nuclei are simultaneously attracted to the negatively charged shared electrons. If a *pair* of electrons is involved, the bond is called a *single* covalent bond, or more simply a **single bond**.

Examples of dot-cross diagrams for two more covalent compounds are shown in Figure 4. Notice that, by sharing electrons in this way, the atoms all achieve stable electron structures like noble gases.

ammonia, NH_3 *water*, H_2O

Figure 4 Dot-cross diagrams for NH_3 and H_2O

Electron pairs which form bonds are called **bonding pairs**. Pairs of electrons not involved in bonding are called **lone pairs**. Both water (H_2O) and ammonia (NH_3) have lone pair electrons.

When two pairs of electrons form a covalent bond, it is called a **double bond**. The bonds in molecular oxygen and carbon dioxide are double covalent bonds (Figure 5).

oxygen, O_2 *carbon dioxide*, CO_2

Figure 5 Dot-cross diagrams for O_2 and CO_2

When three pairs of electrons form a bond, it is called a **triple bond**. The bonds in nitrogen (N_2) and between carbon and nitrogen in hydrogen cyanide (HCN) are examples (Figure 6).

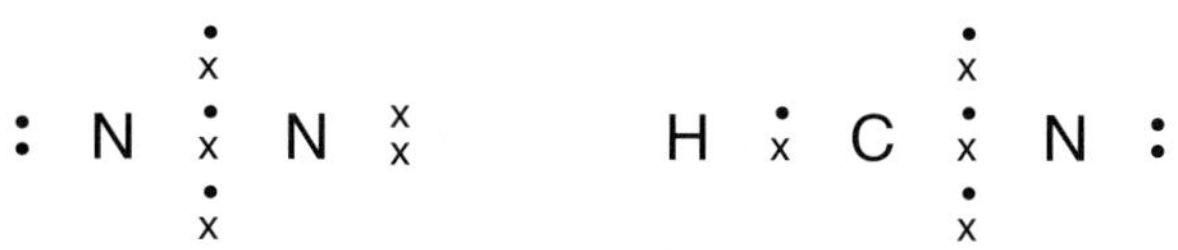

nitrogen, N_2 *hydrogen cyanide*, HCN

Figure 6 Dot-cross diagrams for N_2 and HCN

Electron dot-cross diagrams are useful for representing individual electrons in chemical bonds. Chemists often draw structures in a simpler way by using lines to represent a pair of electrons shared between two atoms. A single line represents a single covalent bond. Double and triple lines represent double and triple covalent bonds.

Some of the molecules used as examples earlier are shown in this way in Figure 7.

H—H *hydrogen*

O=C=O *carbon dioxide*

H—C≡N *hydrogen cyanide*

O=O *oxygen*

N≡N *nitrogen*

Figure 7 Covalently bonded molecules

Why are bonds like bears?

Figure 8 shows a more detailed diagram of the way the hydrogen molecule is bonded.

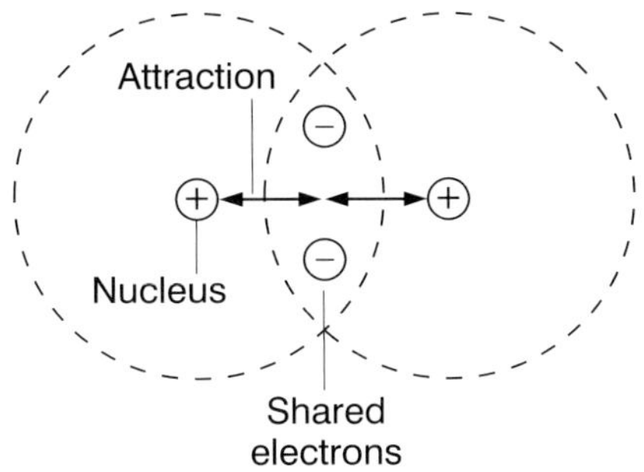

Figure 8 In a hydrogen molecule, the atoms are held together because their nuclei are both attracted to the shared electrons

The atoms are held together because their nuclei are both attracted to the electron pair which is shared between them. Both atoms are identical so the electrons are shared equally.

With atoms more complicated than hydrogen, it is the atomic cores which are attracted to the shared electron pairs. The atomic core is made up of everything except the outer shell electrons (see **Section 11.2**). A fluorine atom, for example, has an electron arrangement of 2.7 and its core is made up from the nucleus with a charge of 9+ and the two electrons in the innermost shell with a charge of 2–. This gives a core with a charge of 7+. The F_2 molecule is held together because the 7+ core charges are attracted to the negative charges on the shared electrons.

Very often the two atoms bonded together have different sizes. In this case, the core of the smaller atom will be closer to the shared electrons and will exert a stronger pull on them (Figure 9a). A similar situation arises when the atoms are from different groups and have different core charges (Figure 9b).

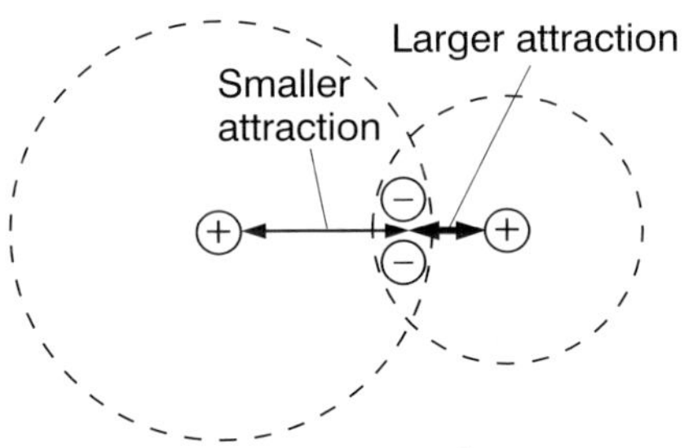

Figure 9a Shared electrons are attracted more strongly by the core of the smaller atom, which is closer

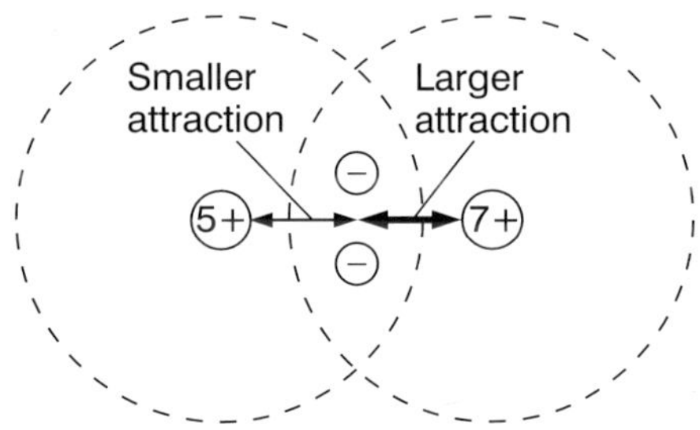

Figure 9b Shared electrons are attracted more strongly by the atom with the greater core charge

In general, different atoms attract bonding electrons unequally. One atom gets a slight negative charge because it has a greater share of the bonding electrons. The other atom becomes slightly positively charged because it has lost some of its share in the bonding electrons. Bonds like this are called **polar bonds**. An example is shown in Figure 10. The small amounts of electrical charge are shown by δ– and δ+ where δ means 'a small amount of'. Some more examples of **bond polarity** are shown in Table 1.

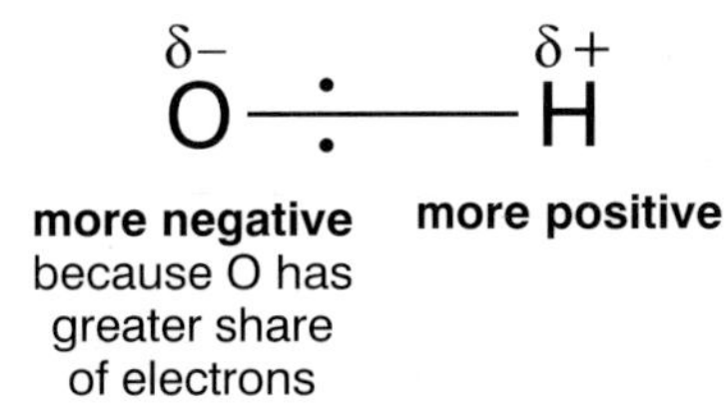

Figure 10 The O–H bond is polar

Bonded atoms	Sizes of atoms	Core charges	Bond polarity
F and Br	F smaller	same	$\overset{\delta-}{F}—\overset{\delta+}{Br}$
O and N	similar	O greater	$\overset{\delta-}{O}—\overset{\delta+}{N}$
F and P	F smaller	F greater	$\overset{\delta-}{F}—\overset{\delta+}{P}$

Table 1 Some examples of polar bonds

So bonds are like bears – some are polar and some are not. The O—H bond is a particularly important example of a polar covalent bond. It has many consequences for the chemistry of water molecules (Figure 11) (see **Section 16.4**).

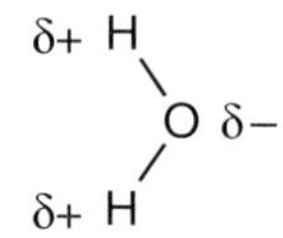

Figure 11 Polar bonding in the water molecule

Electronegativity

To decide the polarity of a covalent bond, we need a measure of each atom's attraction for bonding electrons – its 'electron pulling power'. This is called **electronegativity**. Atoms with strong electron pulling power in covalent bonds are said to be highly electronegative.

There are several ways of estimating electronegativity values. Each method gives different numbers but places the elements in a similar order – they lead to the same *relative* electronegativity values. The values in Figure 12 are derived from a method suggested by Linus Pauling, a US chemist and winner of two Nobel Prizes.

Period \ Group	1	2		3	4	5	6	7
1			H 2.1					
2	Li 1.0	Be 1.5		B 2.0	C 2.5	N 3.0	O 3.5	F 4.0
3	Na 0.9	Mg 1.2		Al 1.5	Si 1.8	P 2.1	S 2.5	Cl 3.0
4	K 0.8	Ca 1.0		Ga 1.6	Ge 1.8	As 2.0	Se 2.4	Br 2.8
5	Rb 0.8	Sr 1.0		In 1.7	Sn 1.8	Sb 1.9	Te 2.1	I 2.5
6	Cs 0.7	Ba 0.9		Tl 1.8	Pb 1.8	Bi 1.9	Po 2.0	At 2.2

Key: Li — symbol; 1.0 — electronegativity

Figure 12 Pauling electronegativity values for some main group elements in the Periodic Table

If you exclude the noble gases, the elements are more electronegative at the top of a group and at the right-hand side of the Periodic Table. The highest electronegativites correspond to reactive non-metals with small atoms; the lowest electronegativities correspond to reactive metals with large atoms.

We can use differences in electronegativity values to predict how polar a particular covalent bond will be. For example, in the C—F bond, F has a higher electronegativity value (4.0) than C (2.5) so it attracts the shared electrons more strongly, and the polarity of the bond is

$$\overset{\delta+}{C}—\overset{\delta-}{F}$$

Notice that C and H have similar electronegativities, so for practical purposes we can think of the C—H bond as non-polar. There is more about electronegativity and bond polarity later in this unit in **Section 16.1**.

Polar bonds are like covalent bonds with a bit of *ionic character* in them. The bigger the difference in electronegativity between the atoms, the more polar the bond and the greater the ionic character. The ionic and covalent models are extreme forms of bonding: polar bonds are somewhere between the two (Figure 13).

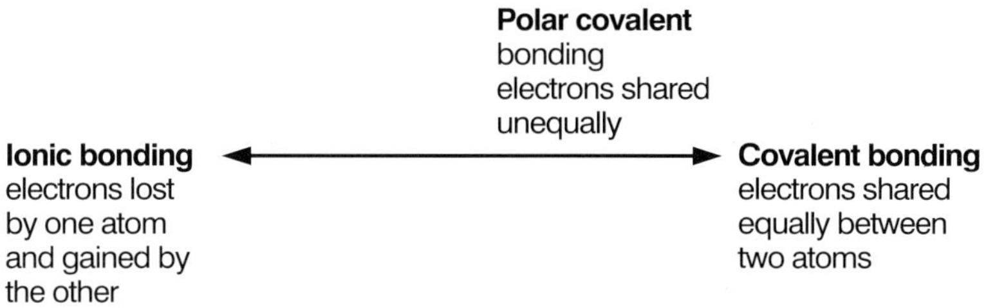

Figure 13 Bonds can vary from ionic to covalent, with polar covalent bonds in between

A cautionary note

Theories of bonding are fundamental to an understanding of chemistry. The dot-cross model is an over-simplified approach to bonding, but it is nevertheless useful. It needs to be extended before it can do more than explain simple situations. So don't be surprised if dot-cross diagrams sometimes don't work, or if some molecules seem to break the rules.

PROBLEMS FOR 13.2

1 Draw electron dot-cross diagrams for the covalent compounds shown below.

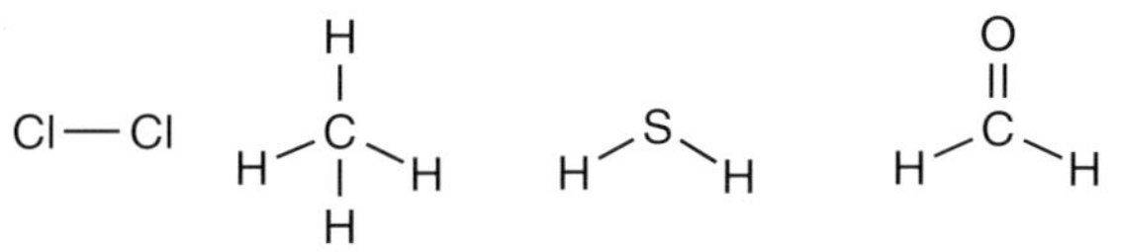

chlorine *methane* *hydrogen sulphide* *methanal*

2 **a** Describe how electronegativity values vary
 i across a period
 ii down a group
b Suggest why the noble gases are not assigned electronegativity values and so are not included in Figure 12.

3 Predict the polarity of the bonds in the following molecules:
a NO
b HF
c N_2
d HCl
e ClF
f CO

13.3 *Metallic bonding*

Metal atoms cannot achieve the stable electron arrangement of a noble gas by sharing electrons with each other or by transferring electrons from one to the other; they don't have enough outer shell electrons to allow them to do this. So what holds the metal atoms together in a sample of, say, copper or iron?

The metal atoms lose their outer electrons to form a lattice of regularly spaced positive ions each with a noble gas electron arrangement. The outer electrons from each atom contribute to a common 'pool' of electrons which move randomly throughout the lattice of positive ions. Since the electrons in the pool do not belong to any particular metal ion, they are described as **delocalised** or spread out over the lattice. Each positively charged ion in the metal is attracted to the negatively charged delocalised electrons and vice versa. It is these electrostatic attractions that are the **metallic bonds** (Figure 14).

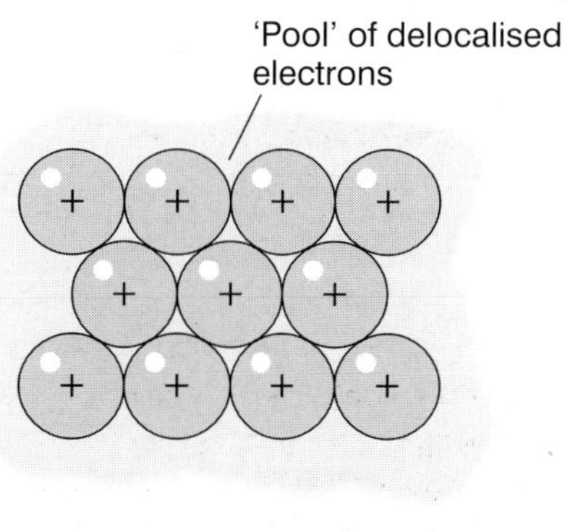

Figure 14 A model of metallic bonding

13.4 *To summarise*

The type of bonding in a *compound* is related to the position of its constituent elements in the Periodic Table:

- when metals on the left-hand side of the Periodic Table combine with non-metals on the right, ionic compounds are formed
- when non-metals on the right-hand side of the Periodic Table combine with each other, covalent compounds are formed.

In *non-metallic elements* (for example, graphite and chlorine gas) the atoms are bonded together by covalent bonds.

For *metallic elements*, a different type of bonding, called metallic bonding, gives metals their characteristic properties.

PROBLEMS FOR 13.1, 13.2 AND 13.3

1 In aluminium, how many electrons will each atom contribute to the 'pool' of delocalised electrons?

2 Suggest why metals are good conductors of electricity.

3 Predict the type of bonding present in each of the following substances:

a iron
b chlorine
c silicon tetrafluoride
d sodium bromide
e magnesium oxide
f methane

Which of the above substances would you expect to contain polar covalent bonds?

14 ENERGY CHANGES

14.1 *Energy out, energy in*

Energy changes are a characteristic feature of chemical reactions. Many chemical reactions give out energy and a few take energy in. A reaction that gives out energy and heats the surroundings is described as **exothermic**. A reaction that takes in energy and cools the surroundings is **endothermic**.

During an exothermic reaction the chemical reactants are losing energy. This energy is used to heat the surroundings – the air, the test tube, the laboratory, the car engine. The products end up with less energy than the reactants had – but the surroundings end up with more, and get hotter. We measure the energy transferred to and from the surroundings as **enthalpy change**. Enthalpy changes can be shown on a **potential energy diagram** (Figure 1).

Enthalpy (*H*)

We cannot measure the enthalpy, *H*, of a substance. What we can do is measure the *change* in enthalpy when a reaction occurs

$\Delta H = H_{products} - H_{reactants}$

The enthalpy change in a chemical reaction gives the quantity of energy transferred to or from the surroundings, when the reaction is carried out in an open container.

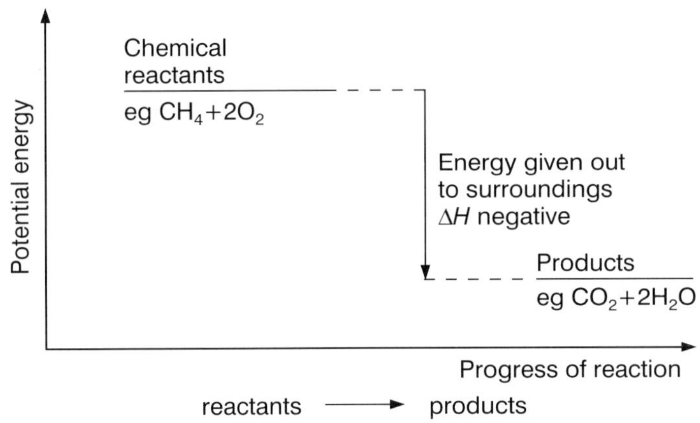

Figure 1 Potential energy diagram for an exothermic reaction, eg burning methane, $CH_4 + 2O_2 \rightarrow CO_2 + 2H_2O$

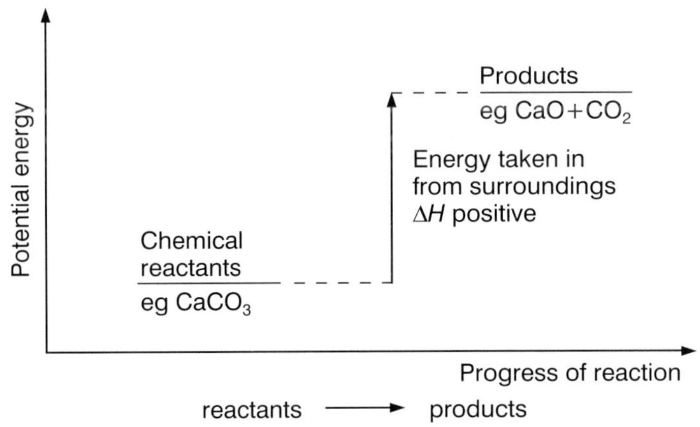

Figure 2 Potential energy diagram for an endothermic reaction, eg decomposing calcium carbonate, $CaCO_3 \rightarrow CaO + CO_2$

In an endothermic reaction, the reactants take in energy from the surroundings. This is shown in Figure 2.

How much?

As your study of chemistry becomes more advanced, you will find you need to put *numbers* to the features of chemical reactions – to make them quantitative. We measure the energy changes associated with chemical reactions by measuring enthalpy changes.

For an exothermic reaction ΔH is *negative*. This is because, from the point of view of the chemical reactants, energy has been *lost* to the surroundings. Conversely, for an endothermic reaction ΔH is *positive* – energy has been *gained* from the surroundings.

In an **exothermic** reaction, the enthalpy of the reacting system decreases.

ΔH is negative

In an **endothermic** reaction, the enthalpy of the reacting system increases.

ΔH is positive

The units of ΔH

Enthalpy changes are measured in kilojoules per mole. For example, for the reaction of methane with oxygen, we write

$$CH_4(g) + 2O_2(g) \rightarrow CO_2(g) + 2H_2O(l); \quad \Delta H = -891\,kJ\,mol^{-1}$$

This means that for every mole of methane that reacts in this way, 891 kJ of energy are transferred to heat the surroundings. If we used 2 moles of methane, we should get $2 \times 891 = 1782$ kJ of energy transferred. This assumes that all the methane is converted into products, and that none is left unburned.

When calcium carbonate is heated, it decomposes. Energy is taken in – it is an endothermic reaction. We write

$$CaCO_3(s) \rightarrow CaO(s) + CO_2(g); \quad \Delta H = +572\,kJ\,mol^{-1}$$

For every mole of $CaCO_3$ that is decomposed, 572 kJ of energy is taken in. If we decomposed 0.1 mole of $CaCO_3$, 57.2 kJ would be taken in.

Where does the energy come from?

All chemical reactions involve breaking and making chemical bonds. Bonds break in the reactants and new bonds form in the products. The energy changes in chemical reactions come from the energy changes when bonds are broken and made.

You can remind yourself of the basic ideas of chemical bonding by reading **Chapter 13**.

A chemical bond is basically an electrical attraction between atoms or ions. When you break a bond, you have to do work in order to overcome these attractive forces. To break the bond completely, you need (theoretically) to separate the atoms or ions so they are an infinite distance apart. Figure 3 illustrates this for the H—H bond in a molecule of hydrogen, H_2. So, *breaking* a bond is an endothermic process – it needs energy. When you *make* a new bond, you get energy given out.

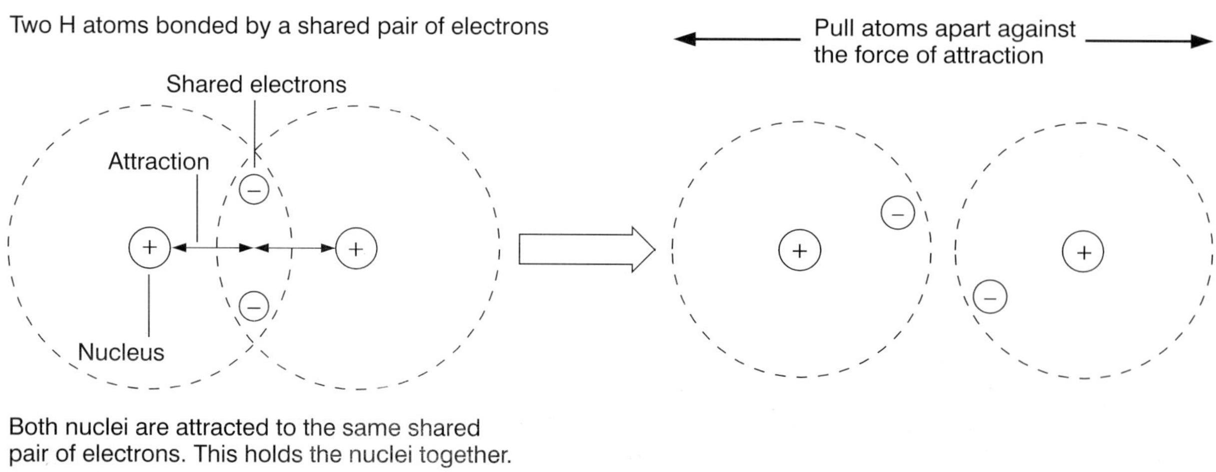

Figure 3 Breaking a bond involves using energy to overcome the force of attraction

Let's look again at the reaction that occurs when methane burns:

$$CH_4(g) + 2O_2(g) \rightarrow CO_2(g) + 2H_2O(l); \quad \Delta H = -891\,kJ\,mol^{-1}$$

The reaction involves both breaking bonds and making new bonds. First you have to break four bonds between C and H in a methane molecule, and bonds between O and O in two O_2 molecules. This bond breaking requires energy. But once the bonds have been broken, the atoms can join together again to form new bonds: two C═O bonds in a CO_2 molecule and four O—H bonds in two H_2O molecules.

Figure 4 illustrates this – but note that you do not have to break *all* the old bonds before you can make new ones. New bonds start forming as soon as the first of the old bonds have broken. In this reaction, the energy taken in during the bond-breaking steps is less than the energy given out during the bond-making steps, so the overall reaction is exothermic. (If the reverse is true, the reaction is endothermic.)

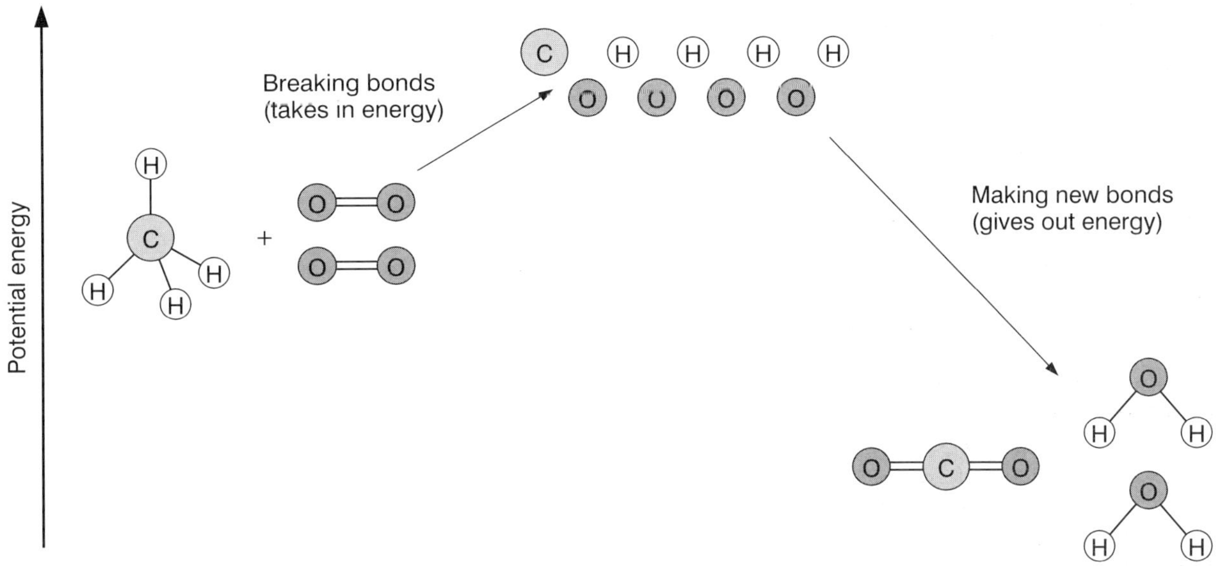

Figure 4 Breaking and making bonds in the reaction between methane and oxygen

Because you need to break bonds before product molecules can begin to form, many reactions need energy (usually in the form of heat, although sometimes light will do) to get them started. All reactions need energy to stretch and break bonds and start them off: this is called the **activation energy** (E_a). Some reactions need only a little energy, and there is enough energy available in the surroundings at room temperature to get them started. The reaction of acids with alkalis is one example. Other reactions have a higher activation energy, and they need heating to get them started.

The burning of fuels is an example. When you use a match to ignite methane, or a spark to ignite petrol vapour in a car engine, you are supplying the activation energy that is needed to stretch and break bonds so new bonds can begin to form.

It isn't necessary for *all* the bonds to break before the reaction gets going. If it was, you would have to heat things to very high temperatures to make them react. Once one or two bonds have broken, new bonds can start to form and this usually gives out enough energy to keep the reaction going, as in burning fuels. Many other reactions need continuous heating, for example reactions which are only slightly exothermic.

Energy pathways

Chemists often use values of the enthalpy change (ΔH) and the activation energy (E_a), which have been measured experimentally, to draw an **energy pathway** for the reaction.

Energy pathways are a useful way of picturing the energy changes that take place during a chemical reaction. They can give you quite a lot of information about a reaction. Look at Figure 5, which shows the potential energy diagram for a simple, one-step reaction:

A + B—C → A—B + C

The curved line is the energy pathway for the reaction. You can tell it is an exothermic reaction because the energy of the products is less than the energy of the reactants. Also, the activation energy is relatively small. An example is the reaction of chlorine atoms with ozone (O_3):

$Cl + O_3 \rightarrow ClO + O_2$

This reaction takes place high up in the atmosphere and is thought to lead to the destruction of the ozone layer. (You can find out more about this in **Unit 2, Chapter 6**.)

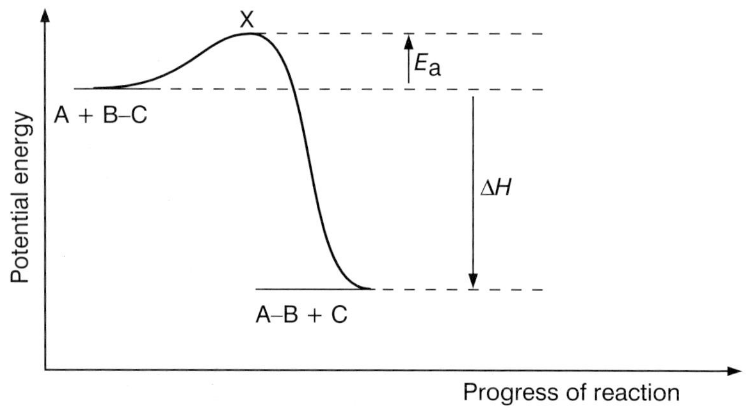

Figure 5 Potential energy diagram for the reaction: A + B—C → A—B + C, showing the energy pathway

In going from reactants to products, an unstable, intermediate arrangement of atoms is reached where old bonds are stretched and new bonds are starting to form. This arrangement is called an **activated complex**. Its formation corresponds to the highest point on the energy pathway, shown as point X in Figure 5.

A + B—C → A------B------C → A—B + C
activated complex

For the reaction of chlorine atoms and ozone, the activated complex is an arrangement in which one of the oxygen–oxygen bonds in ozone is breaking and a new bond from chlorine to oxygen is forming.

$Cl + O_3 \rightarrow$ Cl------O------O—O $\rightarrow ClO + O_2$
activated complex

Activated complexes are very unstable and exist only for a short time. Often we do not know what the activated complex for a particular reaction looks like. Many reactions, such as the combustion of methane, actually take place as a series of steps and the equation simply represents the overall reaction. In cases like this, it is meaningless to draw a single activated complex for the reaction pathway, because there will be many – one for each step in the reaction!

You will need to use your knowledge of enthalpy changes and ideas about where the energy comes from when a fuel burns when you read **Section 2.2**.

PROBLEMS FOR 14.1

1 The diagram belows shows part of a potential energy diagram for a simple, one-step chemical reaction. (The potential energy of the reactants is arbitrarily taken as zero.)

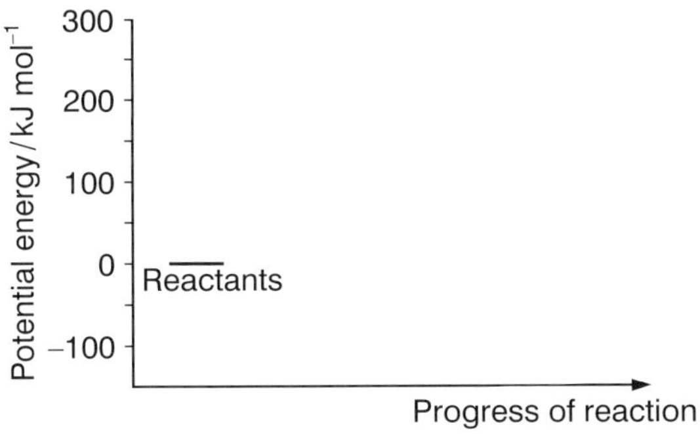

a Complete the diagram by drawing an energy pathway to show the following:
  i the energy of activation (E_a) = +200 kJ mol^{-1}
  ii the energy of the products when the enthalpy change for the reaction (ΔH) is –50 kJ mol^{-1}
  iii the position on the energy pathway where the activated complex (AC) for the reaction would be formed.

b i Explain what is meant by an 'activated complex'.
  ii Is the reaction exothermic or endothermic? Explain how you arrived at your answer.

2 The following questions refer to the potential energy diagram below. (The potential energy of the reactants is arbitrarily taken as zero.)

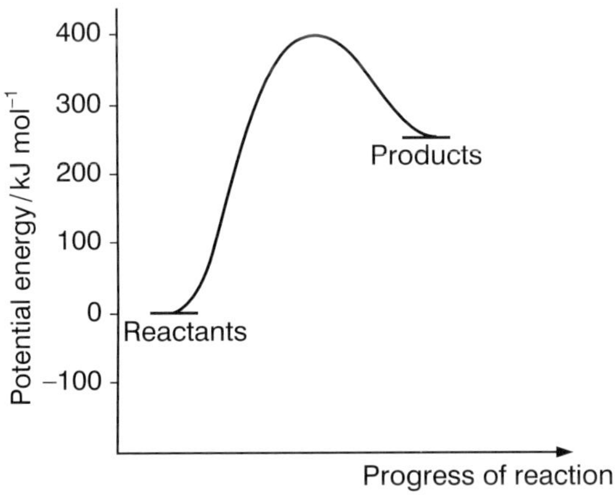

a Is the reaction exothermic or endothermic? Explain how you arrived at your answer.

b Work out E_a for the reaction.

c Work out ΔH for the reaction.

3 In the reaction below, reactant and product molecules are represented by drawings of models, where black and white spheres represent different atoms. The activated complex formed during the reaction is thought to be square-shaped.

Copy and complete the diagram by drawing a possible structure for the activated complex.

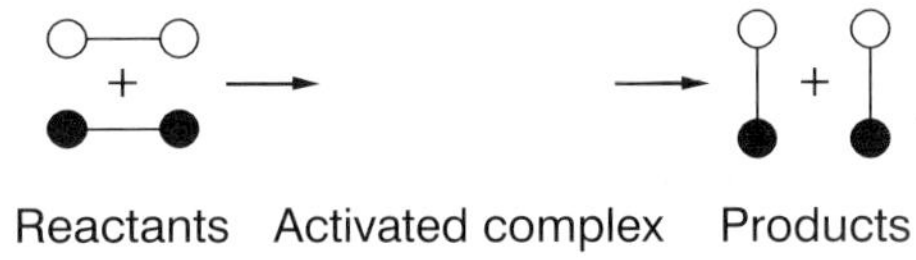

4 The potential energy diagrams A–D represent four different reactions. All the diagrams are drawn to the same scale.

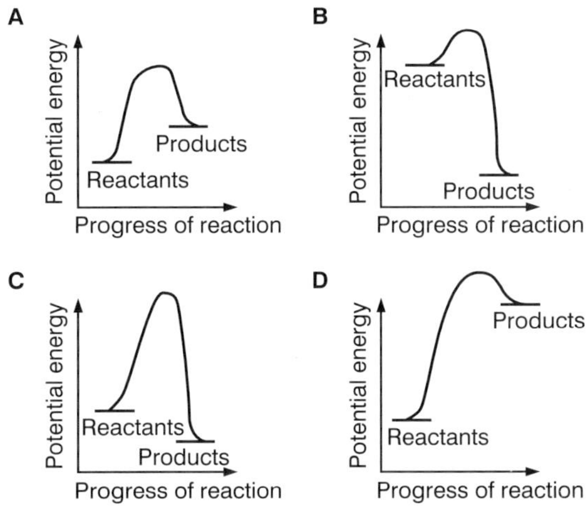

Which of the potential energy diagrams A–D:

a represent exothermic reactions?

b represent endothermic reactions?

c shows the largest activation energy?

d shows the smallest activation energy?

e represents the most exothermic reaction?

f represents the most endothermic reaction?

14.2 *Measuring enthalpy changes*

Many enthalpy changes can be measured quite simply in the laboratory. We usually do it by arranging for the energy involved in the reaction to be transferred to or from water. If it is an exothermic reaction, the water gets hotter; if it is endothermic, the water gets cooler. If we measure the temperature change of the water, and if we know its mass and specific heat capacity, we can work out how much energy was transferred to or from the water during the chemical reaction.

To do this, we need to use the relationship:

Energy transferred = $cm\Delta T$

where c is the specific heat capacity of water (4.18 kJ kg^{-1} °C^{-1})
m is the mass of water, in kg
ΔT is the change in temperature of the water

Figure 6 shows apparatus that can be used to get a more accurate value for ΔH than the crude methods you will use in the laboratory.

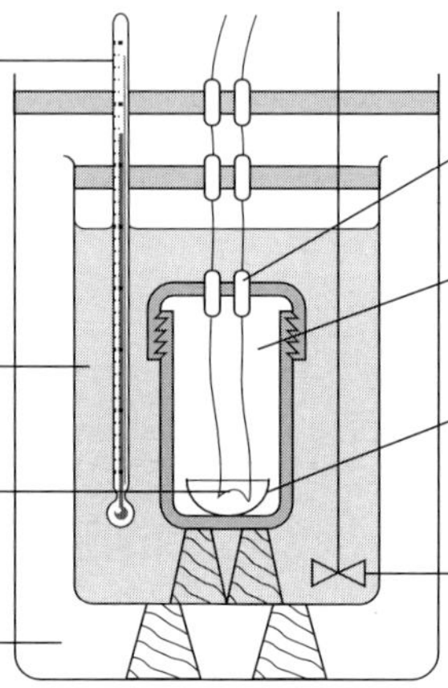

Figure 6 A bomb calorimeter for making accurate measurements of energy changes. The fuel is ignited electrically and burns in the oxygen inside the pressurised vessel. Energy is transferred to the surrounding water, whose temperature rise is measured

> **System or surroundings?**
>
> When chemists talk about enthalpy changes they often refer to the **system**. This means the reactants and the products of the reaction they are interested in.
>
> The system may lose or gain enthalpy as a result of the reaction. The **surroundings** means the rest of the world: the test tube, the air ... and so on.

> **A word about temperature**
>
> There are two scales of temperature used in science and you should be familiar with both.
>
> You will usually *measure* temperature using a thermometer marked in degrees Celsius. However, the kelvin (K) is the unit of **absolute temperature** and 0 K is called absolute zero. The kelvin is the SI unit of temperature.
>
> You can convert temperatures from the Celsius scale to the absolute scale by adding 273. Thus, the boiling point of water = 100 + 273 = 373 K. Note that the kelvin does not have a degree symbol, °.
>
> Note, too, that a *change* in temperature has the same numerical value on both scales.

Some kinds of enthalpy change are particularly important and are given special names. The ones you are likely to come across in the laboratory are:

- enthalpy of combustion
- enthalpy of solution
- enthalpy of neutralisation.

Enthalpy of combustion, ΔH_c, is the enthalpy change that occurs when *1 mole* of a fuel is burned completely in oxygen.

For example, the enthalpy change of combustion of octane, C_8H_{18}, one of the alkanes found in petrol, is –5470 kJ mol^{-1}. This is much bigger than the ΔH_c for methane, CH_4, (–891 kJ mol^{-1}) because burning octane involves breaking and making more bonds than burning methane.

> All combustion reactions are exothermic
>
> ∴ ΔH_c is always negative.

The equations for these combustions are:

$$CH_4(g) + 2O_2(g) \rightarrow CO_2(g) + 2H_2O(l)$$
$$C_8H_{18}(l) + 12\tfrac{1}{2}O_2(g) \rightarrow 8CO_2(g) + 9H_2O(l)$$

When you write an equation to represent an enthalpy of combustion, the equation must always balance and show *1 mole* of the fuel reacting, even if this means having half an oxygen molecule in the equation. You should include state symbols.

Figure 7 shows a simple arrangement that you may use to find the value for the enthalpy of combustion of a liquid fuel, such as ethanol. You will need to weigh the spirit burner before and after the experiment to find the mass of ethanol that has burned. You will also need to know the mass of water in the copper can (or beaker) and measure its change in temperature.

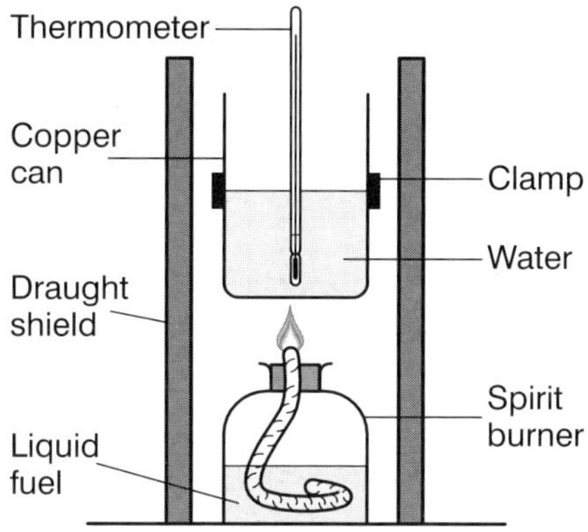

Figure 7 Simple apparatus for measuring the enthalpy of combustion of ethanol

Sample results

Mass of ethanol burned = 0.23 g
Temperature change (ΔT) = +10 °C
Mass of water absorbing heat (m) = 150 g (0.150 kg)

(Remember 1 cm^3 of water has a mass of 0.001 kg)

Energy transferred to the water = $cm\Delta T$
= $4.18\ kJ\ kg^{-1}\ °C^{-1} \times 0.150\ kg \times 10\ °C$
= 6.27 kJ

This is the energy released by burning 0.23 g of ethanol. We now need to work out the energy that would be released by burning *1 mole* of ethanol.

Ethanol (C_2H_5OH): relative formula mass = 46
So, 1 mole of ethanol has a mass of 46 g

0.23 g of ethanol releases 6.27 kJ on combustion

46 g of ethanol releases $\frac{6.27\ kJ \times 46\ g}{0.23\ g} = 1254\ kJ$

$C_2H_5OH(l) + 3O_2(g) \rightarrow 2CO_2(g) + 3H_2O(l)$

Enthalpy of combustion, $\Delta H_c = -1250\ kJ\ mol^{-1}$

Note that we have rounded off the final value to 3 significant figures in keeping with the accuracy of the experimental measurements. Note too that the sign for the enthalpy change is negative because the reaction is exothermic. All enthalpies of combustion are exothermic.

How accurate is your result? If you look in a data book, you will find that the enthalpy of combustion of ethanol is given as $-1371\ kJ\ mol^{-1}$. This will certainly be higher than the value you obtained from your experiment, because not all the heat given out from the burner was absorbed by the water. Much was lost to the surrounding air and absorbed by the container and the thermometer. This means that the temperature change you recorded was lower than it should have been.

Enthalpy of solution, $\Delta H_{solution}$, is the enthalpy change which occurs when *1 mole* of a substance dissolves completely in water. For example, the enthalpy of solution of ammonium nitrate can be represented by the equation

$NH_4NO_3(s) + aq \rightarrow NH_4^+(aq) + NO_3^-(aq)$

This is an endothermic process and energy is absorbed directly from the water being used to make the solution. The change in temperature of the water on adding the ammonium nitrate is measured as shown in Figure 8.

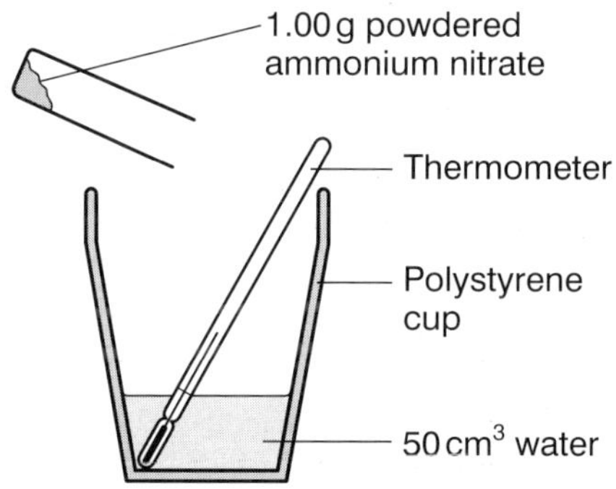

Figure 8 Measuring the enthalpy of solution of ammonium nitrate

Sample results

1.00 g of solid ammonium nitrate were dissolved in 50 cm^3 (0.050 kg) of water. There was a temperature *drop* of 1.6 °C.

Energy absorbed from water = $cm\Delta T$
= $4.18\ kJ\ kg^{-1}\ °C^{-1} \times 0.050\ kg \times 1.6\ °C$
= 0.334 kJ

(We have assumed that, since the solution is dilute, its specific heat capacity and density are close to those of water.)

This is the energy absorbed when 1.00 g of ammonium nitrate dissolves. We now need to work out the energy that would be taken in when *1 mole* of ammonium nitrate dissolves in water.

Ammonium nitrate (NH_4NO_3): relative formula mass = 80
So, 1 mole of ammonium nitrate has a mass of 80 g

1.00 g ammonium nitrate takes in 0.334 kJ when it dissolves in water

80 g ammonium nitrate takes in $\frac{0.334\text{ kJ} \times 80\text{ g}}{1.00\text{ g}} = 26.7\text{ kJ}$

$NH_4NO_3(s) + aq \rightarrow NH_4^+(aq) + NO_3^-(aq)$
Enthalpy of solution, $\Delta H_{solution} = +26.7\text{ kJ mol}^{-1}$

Note that the sign for the enthalpy change is positive because the reaction is endothermic. Enthalpies of solution can be exothermic or endothermic.

You can remind yourself about how chemists measure concentrations of solutions by reading **Section 10.4**.

Enthalpy of neutralisation, $\Delta H_{neutralisation}$, is the enthalpy change which occurs when an acid is neutralised to form *1 mole of water*. In neutralisation reactions, hydrogen ions from the acid combine with hydroxide ions from the alkali to form water molecules. The other ions in the reaction mixture do not react and are called *spectator ions*.

hydrochloric acid + sodium hydroxide → sodium chloride + water
$\mathbf{H^+(aq)} + Cl^-(aq) + Na^+(aq) + \mathbf{OH^-(aq)} \rightarrow Na^+(aq) + Cl^-(aq) + \mathbf{H_2O(l)}$

Overall change: $H^+(aq) + OH^-(aq) \rightarrow H_2O(l)$

Figure 9 shows how you could measure the enthalpy of neutralisation of hydrochloric acid by sodium hydroxide solution. When the two solutions are mixed, the energy released is transferred directly to the water present in the solution and the temperature rise of the mixture is measured.

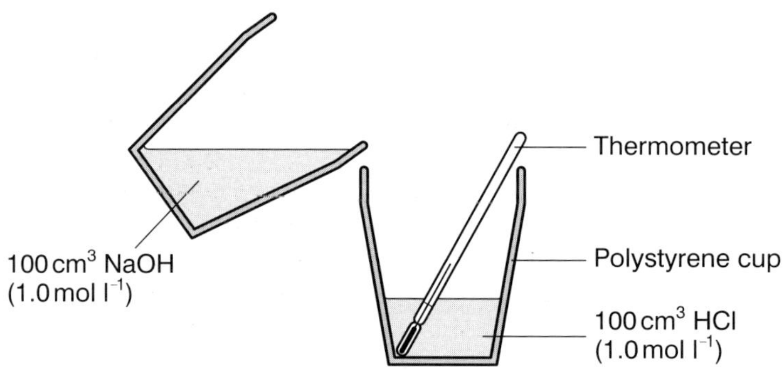

Figure 9 Measuring the enthalpy of neutralisation of hydrochloric acid and sodium hydroxide solution

Sample results

100 cm^3 of 1.0 mol l^{-1} HCl(aq) are mixed with 100 cm^3 1.0 mol l^{-1} NaOH(aq). There is a temperature *rise* of 6.5 °C.

After mixing, the total volume of the solution is 200 cm^3. We assume this has a mass of 0.200 kg. This is m. (We also assume that, since the solution is dilute, its specific heat capacity is the same as that of water.)

Energy absorbed by the water $= cm\Delta T$
$= 4.18\text{ kJ kg}^{-1}\text{ °C}^{-1} \times 0.200\text{ kg} \times 6.5\text{ °C}$
$= 5.43\text{ kJ}$

This is the energy released when 0.1 mol of HCl(aq) and 0.1 mol of NaOH(aq) react to form 0.1 mol of water. You can work this out as follows:
1000 cm^3 of 1.0 mol l^{-1} HCl(aq) contain 1 mol HCl

100 cm^3 of 1.0 mol l^{-1} HCl(aq) contain 0.1 mol HCl

Similarly, 100 cm^3 of 1.0 mol l^{-1} NaOH(aq) contain 0.1 mol NaOH
From the equation,

$HCl + NaOH \rightarrow NaCl + H_2O$

0.1 mol 0.1 mol 0.1 mol 0.1 mol

We now need to work out the energy that would be released when *1 mole* of water is formed. In this case, it means multiplying the energy released in the experiment by 10.

$HCl(aq) + NaOH(aq) \rightarrow NaCl(aq) + H_2O(l)$
Enthalpy of neutralisation, $\Delta H_{neutralisation} = -54.3\ kJ\ mol^{-1}$

Note that the sign for the enthalpy change is negative because the reaction is exothermic. Enthalpies of neutralisation are always exothermic.

The value given in a data book is $-57.3\ kJ\ mol^{-1}$. The experimental value is lower because of heat loss to the surroundings. If you look in a data book, you will notice that the enthalpy of neutralisation of any strong acid and strong alkali has the same value. This is not surprising because the same reaction is taking place in each case, regardless of the particular acids or alkalis being used.

$H^+(aq) + OH^-(aq) \rightarrow H_2O(l)$

PROBLEMS FOR 14.2

1 Copy and complete the following sentences.
 a The enthalpy of combustion of an alcohol is the enthalpy change when ...
 b The enthalpy of solution of a substance is the enthalpy change when ...
 c The enthalpy of neutralisation of an acid is the enthalpy change when ...

2 Write balanced equations to represent the following.
 a The enthalpy of combustion of methanol (CH_3OH).
 b The enthalpy of neutralisation when sulphuric acid is neutralised by potassium hydroxide.

3 Explain why the enthalpy changes for many reactions may have a negative or positive sign, but enthalpies of combustion are always negative.

4 A substance is dissolved in 0.1 kg of water causing a temperature rise of 4 °C. Calculate the quantity of energy released.

5 A certain mass of alcohol is burned in a spirit burner and used to heat up 100 g of water. The temperature of the water rises by 5 °C. Calculate the quantity of energy released by burning the alcohol.

6 When 0.02 mol of a solid is dissolved in 0.1 kg of water, the temperature of the water falls from 21 °C to 17 °C. Calculate the enthalpy of solution of the solid.

7 An alcohol burner is used to heat 200 cm^3 of water in a copper can. The temperature of the water increases from 18 °C to 22 °C and 0.002 mol of alcohol are burned. Calculate the enthalpy of combustion of the alcohol.

8 **a** In an experiment, 0.16 g of methanol (CH_3OH) is used up when a spirit burner heats 100 g of water from 17 °C to 25 °C. Calculate the enthalpy of combustion of methanol.
 b Explain why the value obtained in part **a** is much less than the value given in a data book.

9 When 14.9 g of potassium chloride (KCl) are dissolved in 200 g of water, the temperature changes from 17 °C to 11 °C. Calculate the enthalpy of solution of potassium chloride.

10 **a** When 100 cm^3 of 0.5 mol l^{-1} nitric acid (HNO_3) were mixed with 100 cm^3 of 0.5 mol l^{-1} potassium hydroxide (KOH), there was a temperature rise of 3.5 °C. Calculate the enthalpy of neutralisation.
 b The reaction in part **a** was carried out in a polystyrene container. Explain why.

11 Suppose you were asked to measure the enthalpy of combustion of methane, using a gas cooker and a saucepan.
 a What other equipment would you need?
 b What measurements would you make?
 c What would be the main sources of error?

12 The enthalpy of combustion of heptane, C_7H_{16}, is $-4817\ kJ\ mol^{-1}$.
 a What is the relative formula mass of heptane?
 b Calculate the energy transferred when the following quantities of heptane are burned:
 i 10 g **ii** 10 kg
 What assumptions have you made in this calculation?
 c What further information would you need in order to calculate the energy transferred when 1 *litre* of heptane is burned?

15 RATES OF REACTIONS

15.1 *Factors affecting reaction rates*

Different chemical reactions go at different rates. Some reactions, such as burning fuel in a cylinder of a car engine or precipitating silver chloride from solution, go very fast. Others, such as the souring of milk or the rusting of iron, are much slower. The study of **reaction rates** is an important branch of chemistry. Studying reaction rates helps chemists to find ways of speeding up or slowing down chemical processes in industry. It also helps chemists to make predictions about important reactions, such as those that occur between gases in the atmosphere, and to understand the mechanisms of chemical reactions.

What affects the rate of a reaction?

The rate of a chemical reaction may be affected by

- the **concentration** of the reactants. For example ozone (O_3) in photochemical smog is produced by the reaction of oxygen atoms with oxygen molecules.

  $O_2 + O \rightarrow O_3$

  The rate of the reaction increases as the concentration of oxygen atoms increases. In the case of solutions, concentration is measured in mol l^{-1}; in the case of gases, the concentration is proportional to the **pressure**.
- the **temperature**. Nearly all reactions go faster at higher temperatures.
- the **intensity of radiation**, if the reaction involves radiation. For example, the oxygen atoms needed for the production of ozone in photochemical smog are formed by the action of sunlight on NO_2 molecules:

  $NO_2 \xrightarrow{\text{sunlight}} NO + O$

  The NO_2 molecules are a constituent of polluted air. The reaction goes faster when the intensity of the sunlight increases.
- the **particle size** of a solid. A solid like magnesium reacts much faster when it is finely powdered than when it is in a large lump.
- the presence of a **catalyst**.

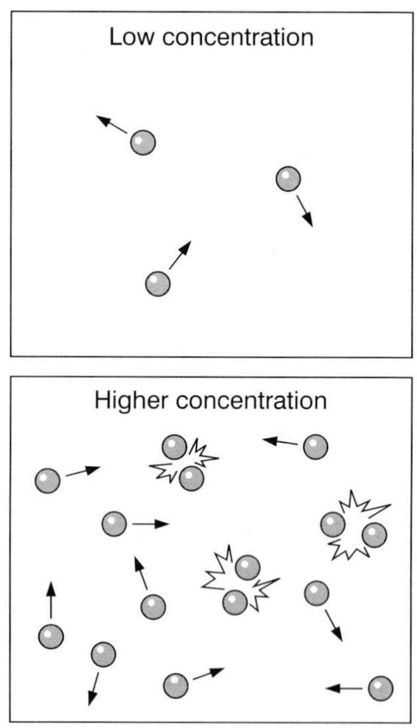

Figure 1 When particles are closer together they have a greater chance of reacting

The collision theory of reactions

We can explain the effect of these factors using a simple collision theory. The basic idea is that *reactions occur when the particles of reactants collide, provided they collide with a certain minimum kinetic energy.*

For example, imagine two particles, say an ozone molecule and an oxygen atom, moving around in the atmosphere. For a reaction to occur, the two particles must first collide so that they come into contact with each other. This will happen more often if there are more particles in a given volume, so it is easy to see why increasing the concentration of the reactants speeds up the reaction (see Figure 1). Indeed, any factor which increases the number of collisions will increase the rate of the reaction.

But, for most reactions, simply colliding isn't enough. Not every collision causes a reaction. As the particles approach and collide, kinetic energy is converted into potential energy and the potential energy of the reactants rises (Figure 2). Existing bonds start to stretch and break and new bonds start to form. The top of the **energy barrier** corresponds to the formation of an unstable **activated complex**, which may then revert back to the reactants or go on to form products. Only those pairs with enough combined kinetic energy on collision to overcome the energy barrier or **activation energy** for the reaction will go on to produce products.

> **Activation energy**
>
> Activation energy is the minimum kinetic energy required by colliding particles before reaction will occur.
> It is the energy required by colliding particles to form an activated complex.

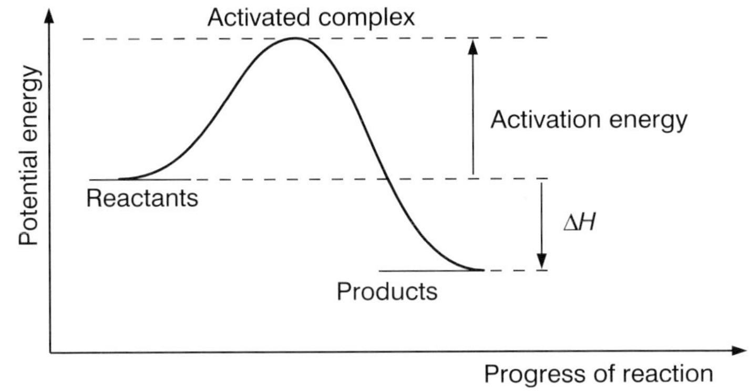

Figure 2 Energy pathway for an exothermic reaction

The proportion of colliding pairs with sufficient kinetic energy to overcome the energy barrier depends very much on the temperature of the gas. At higher temperatures, a much larger proportion of colliding pairs has enough kinetic energy to react. Temperature often has a dramatic effect on reaction rates; for example, methane and oxygen do not react at room temperature, but can react explosively when heated.

You can remind yourself about the use of potential energy diagrams and energy pathways for reactions by reading **Section 14.1**.

PROBLEMS FOR 15.1

1 For each of the following reactions, say how, if at all, you would expect its rate to be affected by the following factors.

A temperature
B total pressure of gas
C concentration of solution
D particle size of solid

a The reaction of magnesium with hydrochloric acid
$Mg(s) + 2HCl(aq) \rightarrow MgCl_2(aq) + H_2(g)$

b The reaction of nitrogen with hydrogen in the presence of an iron catalyst
$N_2(g) + 3H_2(g) \xrightarrow{Fe} 2NH_3(g)$

c The decomposition of aqueous hydrogen peroxide
$2H_2O_2(aq) \rightarrow 2H_2O(l) + O_2(g)$

2 The rate of hydrolysis of proteins is increased in the presence of an acid or an enzyme. Suggest the possible function of the acid or enzyme.

3 The reaction of chlorine atoms with ozone in the upper atmosphere (stratosphere) is thought to be responsible for the destruction of the ozone layer.

$Cl + O_3 \rightarrow ClO + O_2$

The chlorine atoms are produced by the action of high energy solar radiation on CFCs.

a Use collision theory to explain how an increase in the concentration of chlorine atoms increases the rate of reaction with ozone.

b Explain the connection between the activated complex for a reaction and the activation energy.

c Suggest a structure for the activated complex in the reaction of chlorine atoms with ozone.

d The activation energy for the reaction of chlorine atoms with ozone is relatively small. What effect would you expect a change of temperature would have on the rate of the reaction? Explain your reasoning.

15.2 *The effect of concentration on rate*

In this section we are looking at how the rates of reactions depend on concentration. Normally this means the concentration of a *solution*, though we can also apply the idea to gases. For a gas we normally measure its concentration in terms of its *pressure*.

It is useful to know how concentration affects rates, for all sorts of reasons. You might want to use changes of concentration to alter the rate of a reaction in an industrial process. Knowing how concentration affects rate can also tell us a great deal about the way that reactions occur – their **mechanisms**.

Before looking closely at the way concentration affects rate, we need to know a little about how reaction rate is measured.

The meaning of 'rate of reaction'

When we talk about the rate of something we mean the rate at which some quantity changes. Speed is rate of change of distance; the inflation rate is the rate of change of prices. Whenever you are measuring a rate you need to be clear about the units you are using. Speed is often measured in metres per second; inflation rate is often measured as percentage change in prices per year.

When we talk about the rate of a reaction we are talking about the rate at which reactants are converted into products. In the decomposition of hydrogen peroxide to form water and oxygen,

$$2H_2O_2(aq) \quad \rightarrow \quad 2H_2O(l) + O_2(g)$$

hydrogen peroxide

the rate of the reaction means the rate at which $H_2O(l)$ and $O_2(g)$ are formed, which is the same as the rate at which $H_2O_2(aq)$ is used up. We would measure the rate of this reaction in moles of product (water or oxygen) formed per second, or moles of hydrogen peroxide used up per second. Suppose it turns out that 0.0001 mol of oxygen are being formed per second (Figure 3). The rate of the reaction is

$$0.0001\,\text{mol}\,(O_2)\,s^{-1} \quad \text{or} \quad 0.0002\,\text{mol}\,(H_2O)\,s^{-1} \quad \text{or} \quad -0.0002\,\text{mol}\,(H_2O_2)\,s^{-1}$$

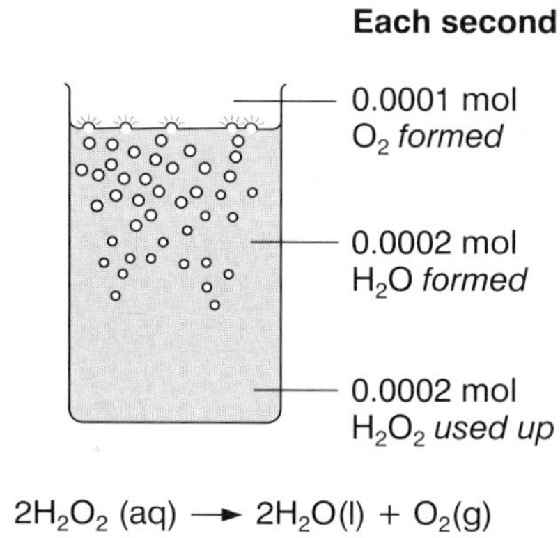

Figure 3 The rate of reaction for the decomposition of hydrogen peroxide: an example

Notice that the rate in terms of moles of H_2O is twice the rate in terms of moles of O_2: this is because two moles of H_2O are formed for each mole of O_2. Notice also that the rate in terms of H_2O_2 has a *minus* sign: this is because H_2O_2 is getting used up instead of being produced.

The units for rate of reaction are usually mol s^{-1}, though they can also be mol min^{-1} or even mol h^{-1}.

Measuring rate of reaction

When you measure the speed of a car you can do so directly, using a speedometer. Unfortunately we can't measure reaction rates directly – there is no such thing as a reaction rate meter. What we have to do is to *measure the change in amount of a reactant or product in a certain time*. In practice this means measuring a property that is related to the amount of substance. If a gas is given off, you could collect the gas and measure its *volume* at different times during the reaction, or you could carry out the reaction in a conical flask on a balance and measure the *total mass* of the reaction mixture and flask at different times. Alternatively, you might be able to follow the *concentration* of one reactant or product as the reaction proceeds. Sometimes you can do this by measuring a property related to concentration, such as colour intensity if one of the substances involved is coloured, or pH if an acid or alkali is involved.

So the procedure for measuring reaction rate is:

Step 1 Decide on a property of a reactant or product which you can measure, such as volume of gas produced or total mass of the reaction mixture.

Step 2 Measure the change in the property in a certain time.

Step 3 Find the rate in terms of $\dfrac{\text{change of property}}{\text{time}}$

Notice that the units of this will not be mol s^{-1}, but, say, (cm^3 of gas) s^{-1} if you are following the course of the reaction by measuring volume of gas produced, g s^{-1} if you are measuring mass, or mol l^{-1} s^{-1} if you are measuring concentration. However, you can convert these to mol s^{-1}.

Investigating how rate depends on concentration

Now let's look more closely at the decomposition of hydrogen peroxide in solution. This reaction proceeds slowly under normal conditions, but it is greatly speeded up by catalysts. A particularly effective catalyst is the enzyme *catalase*.

You can do an experiment to investigate the enzyme-catalysed decomposition of hydrogen peroxide solution, using the apparatus shown in Figure 4. The volume of oxygen is measured in the inverted burette. We will look at the kind of results you might get when you investigate how the rate of oxygen formation depends on the concentration of hydrogen peroxide solution.

We measure the total volume of oxygen given off at different times from the start of the experiment and plot a graph of this against time. This allows us to work out the rate of the reaction in (cm^3 of O_2) s^{-1}.

Figure 5 shows a graph of the results that were obtained by starting with a hydrogen peroxide concentration of 0.4 mol l^{-1}.

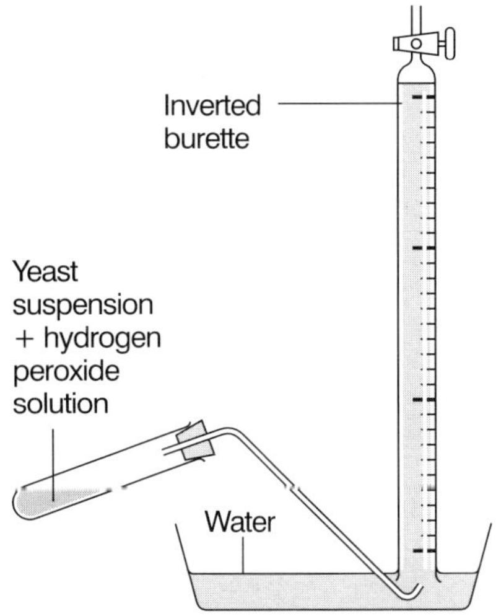

Figure 4 Apparatus for investigating the rate of decomposition of hydrogen peroxide. The yeast provides the enzyme catalase

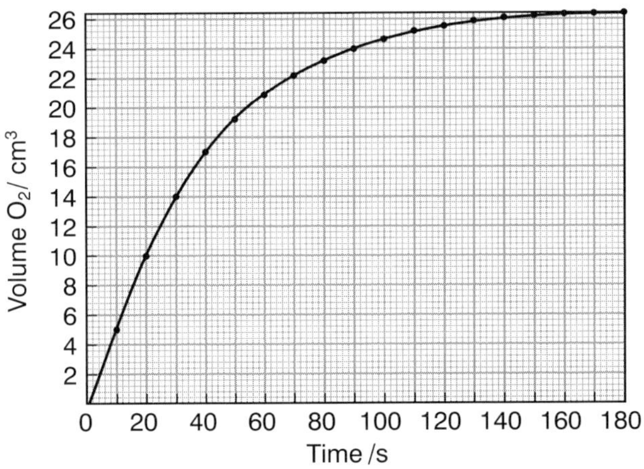

Figure 5 The decomposition of hydrogen peroxide, using a solution of concentration 0.4 mol l^{-1}. The rate is measured in terms of the volume of oxygen given off

Notice these points about the graph in Figure 5:

- *The graph is steep at first*. The gradient of the graph gives us the rate of the reaction – the steeper the gradient, the faster the reaction. The reaction is at its fastest at the start, when the concentration of hydrogen peroxide in solution is high, before any has been used up.
- *The graph gradually flattens out*. This is because, as the hydrogen peroxide is used up, its concentration falls. The lower the concentration, the slower the reaction. Eventually, the graph is horizontal: the gradient is zero, and the reaction has come to a stop.

The rate of the reaction at the start is called the **initial rate**. We can estimate the initial rate in this case by calculating the **average rate** of the reaction over the first 20 s.

$$\text{Average rate over first 20 s} = \frac{\text{change in volume of } O_2}{\text{change in time}}$$

$$= \frac{10\,cm^3 - 0\,cm^3}{20\,s - 0\,s}$$

$$= \frac{10\,cm^3}{20\,s} = 0.5\,cm^3\,s^{-1}$$

Figure 6 shows some results that were obtained when the same experiment was done using hydrogen peroxide solutions of different starting concentrations. In each case, the concentration of the enzyme catalase was kept constant, as were all other conditions such as temperature. As you would expect, the graphs start off with differing gradients, depending on the concentration of hydrogen peroxide you started with. Table 1 shows the initial rates of the experiments in Figure 6.

Concentration of hydrogen peroxide at start/mol l^{-1}	Initial rate/ (cm^3 of $O_2(g)$) s^{-1}
0.40	0.50
0.32	0.41
0.24	0.32
0.16	0.21
0.08	0.10

Table 1 Initial rates of decomposition of hydrogen peroxide

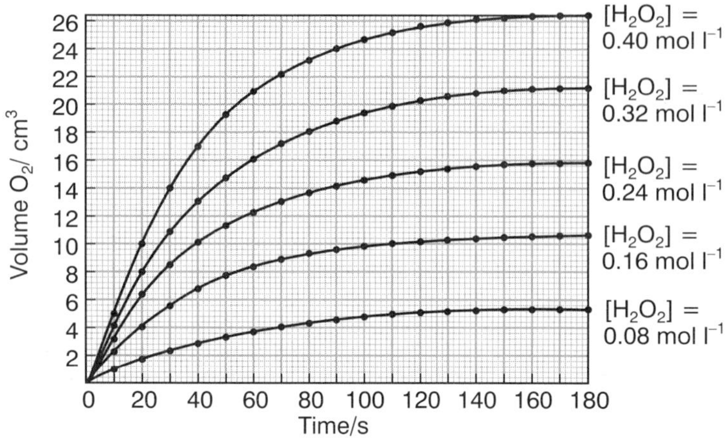

Figure 6 The decomposition of hydrogen peroxide solutions of differing concentrations

Now we are in a position to answer the question 'How does the rate of the reaction depend on the concentration of hydrogen peroxide?' Figure 7 shows the initial rates plotted against concentration of hydrogen peroxide, and you can see that it is a straight line. This means that *the rate is directly proportional to the concentration of hydrogen peroxide*. In other words

rate $\propto$ $[H_2O_2(aq)]$ (the square brackets mean 'concentration of')

Hydrogen peroxide is not the only substance whose concentration affects the rate of this reaction. It is also affected by the concentration of the enzyme catalase, but in the series of experiments shown in Figure 6 we kept the concentration of catalase constant. However, we could do another set of experiments to find the effect of changing the concentration of catalase: this time we would need to keep the concentration of hydrogen peroxide constant.

When we vary the concentration of catalase in this way, we find that the rate of the reaction is also proportional to the concentration of catalase. In other words

rate $\propto$ [catalase]

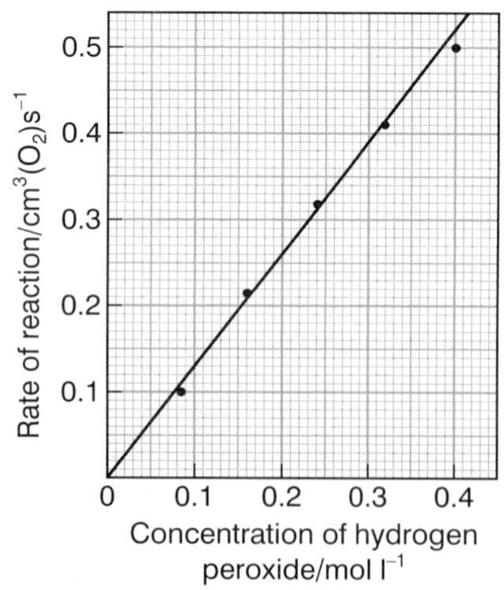

Figure 7 The initial rate of decomposition of hydrogen peroxide plotted against concentration of hydrogen peroxide

Using the reciprocal of the reaction time as a measure of rate

The rates of a particular reaction under different conditions can be compared by measuring the times taken for the *same amount* of reaction to take place. The time taken is called the **reaction time**. If the rate is high (the reaction proceeds quickly), the reaction time will be small. If the rate is low (the reaction proceeds slowly), the reaction time will be large.

A good example of this method is the reaction between sodium thiosulphate solution and hydrochloric acid.

$Na_2S_2O_3(aq) + 2HCl(aq) \rightarrow 2NaCl(aq) + SO_2(g) + S(s) + H_2O(l)$

sodium thiosulphate (under $Na_2S_2O_3$); *sulphur* (under S)

As the reaction proceeds, solid sulphur forms as a colloidal suspension of fine particles and the mixture becomes cloudy. The reaction flask is placed over a cross on a piece of white paper and viewed from above as shown in Figure 8. The cross is no longer visible when a certain amount of sulphur has formed. The reaction time to reach this point can be measured using different starting concentrations of sodium thiosulphate solution. The volume of each solution and the concentration of the hydrochloric acid must be kept constant in each experiment.

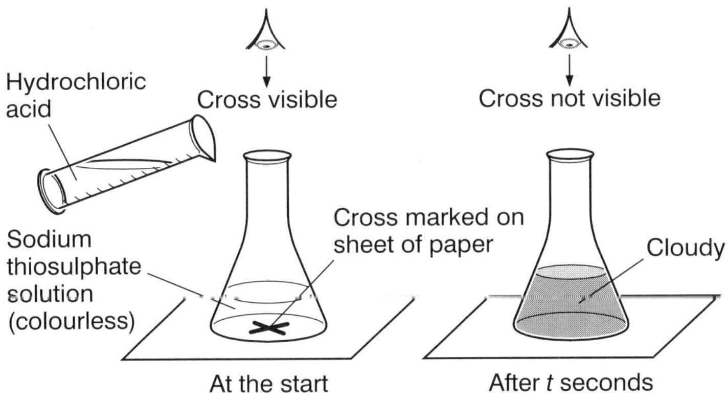

Figure 8 Following the reaction of sodium thiosulphate solution and hydrochloric acid

For each starting concentration,

average rate for this stage of the reaction =

$$\frac{\text{amount of sulphur needed to obscure the cross}}{\text{reaction time for sulphur to form}}$$

The amount of sulphur needed to obscure the cross will be the same for each experiment, so the rate of reaction is proportional to the reciprocal of the reaction time for the cross to be obscured, $1/t$.

$$\text{average rate} \propto \frac{1}{t}$$

The shorter the time taken, the faster the reaction; the longer the time taken, the slower the reaction.

If a graph of $1/t$ is plotted against the concentrations of the thiosulphate solutions used, a straight line plot is obtained, showing that the rate of the reaction is proportional to the concentration of sodium thiosulphate.

PROBLEMS FOR 15.2

1 When yeast and a solution of glucose are mixed, enzymes in the yeast convert the glucose into ethanol (alcohol) and carbon dioxide.

With the help of a labelled diagram, describe how you would carry out an experiment to follow the course of this reaction and investigate how the rate of the reaction changes as the reaction proceeds.

You may use any laboratory equipment you need.

2 You are given the following equipment and chemicals:

balance, stop clock, conical flask, measuring cylinder, marble pieces (calcium carbonate), hydrochloric acid.

Describe how you would carry out an experiment to follow the course of the reaction:

$CaCO_3(s) + 2HCl(aq) \rightarrow CaCl_2(aq) + CO_2(g) + H_2O(l)$

Explain how you could use your results to work out the average rate of the reaction over the first 25 seconds.

In questions 3, 4 and 5, you are asked to calculate the *average rate of reaction* using information from the graphs. Think carefully in each case about the correct units for your answer.

3 The graph below shows the mass of hydrogen given off plotted against time in a chemical reaction.

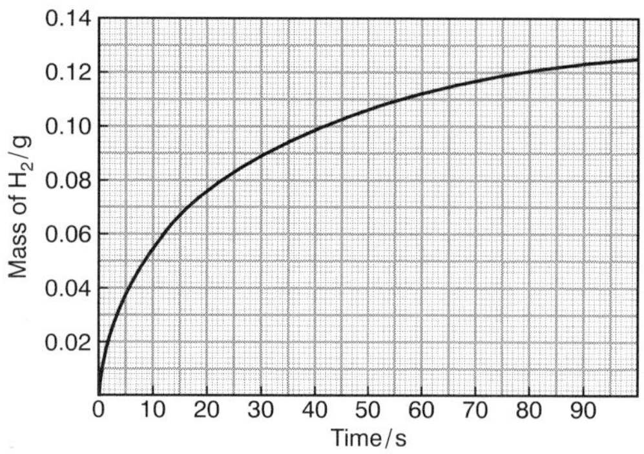

a Calculate the average rate of reaction during the first 10 s of the reaction.

b Calculate the average rate of reaction between 20 s and 40 s.

4 The graph below shows the change in concentration of an acid plotted against time during a chemical reaction.

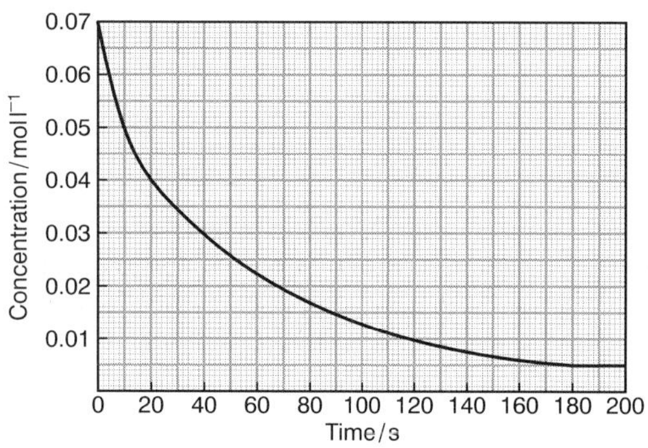

a Calculate the average rate of reaction during the first 20 s of the reaction.

b Calculate the average rate of reaction between 30 s and 70 s.

5 The graph below shows how the volume of carbon dioxide produced during a chemical reaction varied with time.

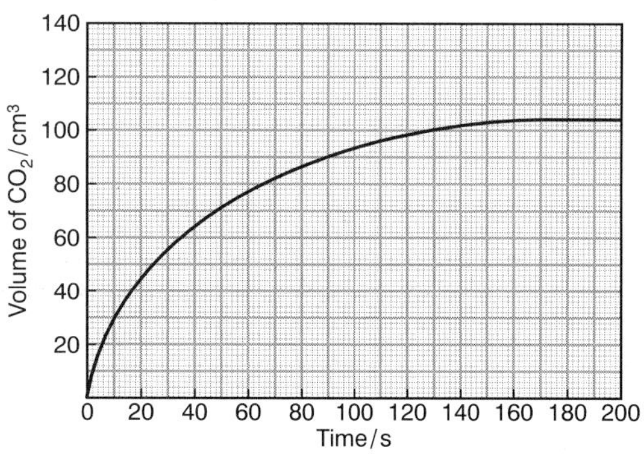

a Calculate the average rate of reaction during the first 30 s of the reaction.

b Calculate the average rate of reaction between 160 s and 180 s.

6 Excess magnesium was added to 1.0 mol l^{-1} sulphuric acid and the volume of hydrogen produced was plotted against time, as shown below.

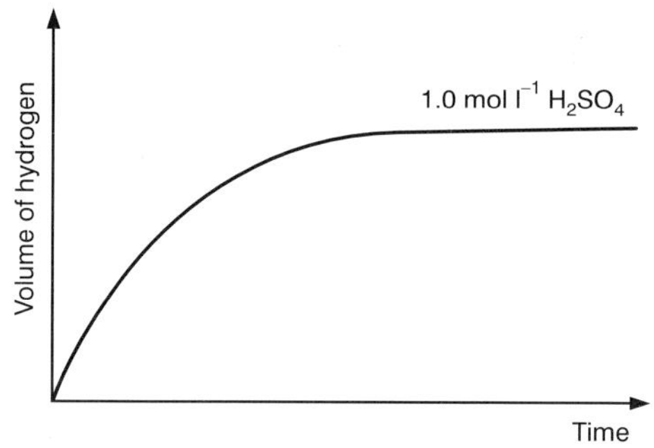

a The graph is steep at first. Explain why this is the case.

b Why does the gradient decrease as the reaction proceeds?

c Copy the graph and draw on it the curves that would be obtained

i if the experiment were repeated using the *same volume* of 0.5 mol l^{-1} sulphuric acid

ii if the experiment were repeated using *double the volume* of 0.5 mol l^{-1} sulphuric acid.

7 The graph below shows the rate of formation of a fixed amount of solid sulphur when sodium thiosulphate solution reacts with hydrochloric acid. The mixture becomes cloudy as the reaction proceeds and the time is measured for a cross on a piece of paper beneath the reaction flask to be obscured. This is done for different starting concentrations of sodium thiosulphate solution. The rate is expressed as the reciprocal of the reaction time, $1/t$.

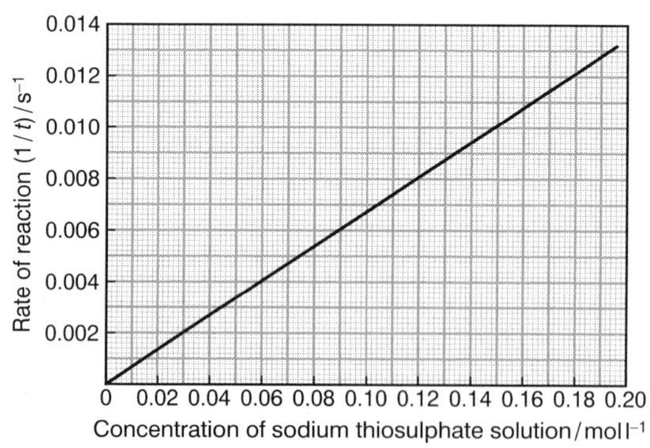

a Explain why the reciprocal of the reaction time can be taken as a measure of the rate of the reaction.

b How long did the cloudiness take to appear when the concentration of the sodium thiosulphate solution was 0.10 mol l^{-1}?

c For what concentration of sodium thiosulphate solution would the cloudiness take 80 s to appear?

15.3 *The effect of particle size on rate*

For reactions that involve a solid, the rate of reaction depends on whether the solid is in large or small lumps, or whether it is ground to a powder. Take, for example, the reaction of marble (calcium carbonate) with an acid. This reaction is responsible for much of the damage done to stonework by acid rain. The equation for the reaction is

$$CaCO_3(s) + 2HCl(aq) \rightarrow CaCl_2(aq) + CO_2(g) + H_2O(l)$$

The reaction can be followed by collecting the carbon dioxide gas given off and measuring the volume of gas as the reaction proceeds. Figure 9 shows the results for two experiments, one using small lumps of marble and one using powdered marble. The same mass of marble was used in each case and all other conditions were identical.

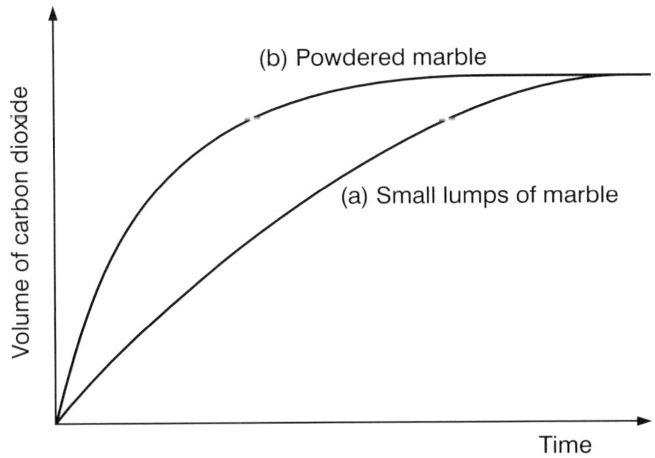

Figure 9 The reaction of marble with acid using (a) small lumps and (b) powdered marble

You can see that when powdered marble was used, the gradient of the graph is much steeper than when lumps were used, showing that the reaction took place faster. Note that the *total* volume of gas given off was the same in each experiment, because the same amount of each reactant was used.

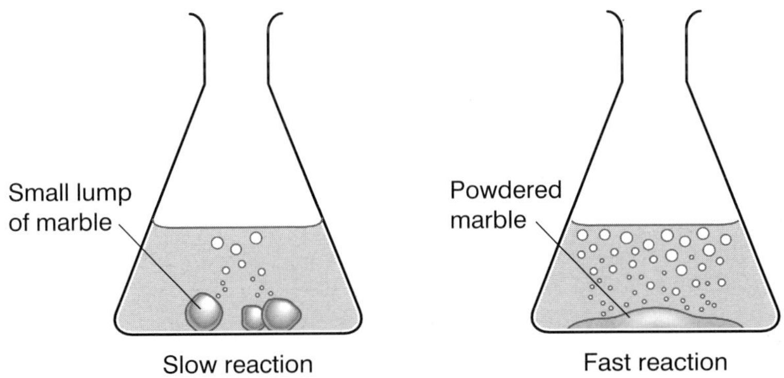

Figure 10 The smaller the particle size of a solid reactant, the faster the reaction

How can we explain this effect using collision theory? In powdered marble there is a much larger *surface area* of solid exposed for the reaction with the acid to take place on. The more collisions there are, the faster the reaction.

So, once the surface of marble stonework starts to crumble, the damage from acid rain accelerates. Fine powders can be a safety hazard in industry. The dust in flour mills is so fine that there is a risk of explosion (a very fast reaction with oxygen in the air) if the dust is not removed.

PROBLEMS FOR 15.3

1 The graph below shows how the volume of hydrogen released increases with time when 2 g of zinc lumps react with excess hydrochloric acid.

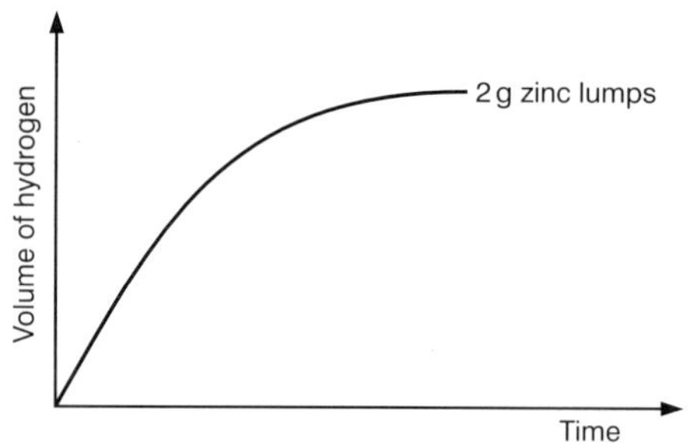

Copy the graph and add labelled curves for the reactions between the acid and

a 2 g of powdered zinc

b 1 g of zinc lumps

2 The diagrams below show one way in which the reaction between excess calcium carbonate ($CaCO_3$) and hydrochloric acid can be studied.

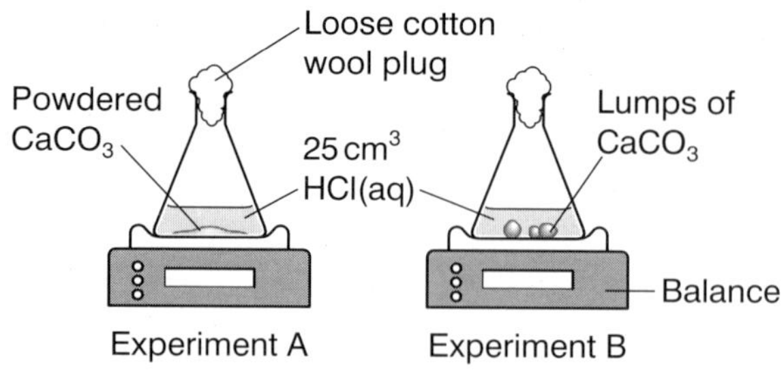

The results from Experiment A are plotted on the following graph.

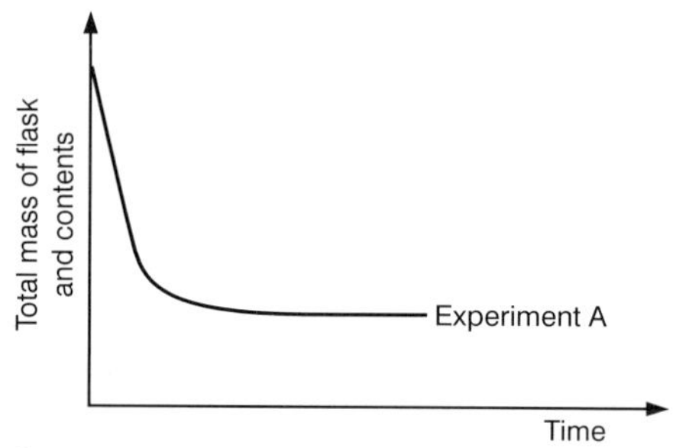

a Copy the graph and add the curve which would be obtained from Experiment B.

b Explain, using the ideas of collision theory, why powdered calcium carbonate reacts faster than lumps of calcium carbonate when added to acid.

15.4 *The effect of temperature on rate*

How can I dissolve this sugar more quickly?
How can I get this cake to cook more quickly?
How can I get this glue to set more quickly?

The answer to all these questions is likely to involve raising the temperature. But why do processes like these go faster at higher temperatures? Whatever the reason, it is a very important effect, and the chemical industry depends heavily on it. Without it, there would be no Haber process for making ammonia, for example.

If you've measured the rate of reactions at different temperatures, you'll know that temperature has a large effect. In fact, for many reactions the rate is roughly *doubled* by a temperature rise of just 10 °C.

The collision theory of reactions

You met the basic ideas of the collision theory in **Section 15.1**. In this section we will take these ideas a little further.

Think of the reaction between nitrogen and hydrogen to make ammonia:

$$N_2(g) + 3H_2(g) \rightleftharpoons 2NH_3(g)$$

The collision theory says that the reaction can only occur when N_2 and H_2 collide. The more frequently they collide, the faster the reaction. If you increase the pressure, for example, the N_2 and H_2 molecules become closer together, so they collide and react more often. That's one of the reasons the Haber process uses high pressures to make ammonia more quickly.

What about the effect of temperature? Think about what happens to the molecules when you raise the temperature. They move faster, so they collide more frequently. We can work out *how much* more frequently they collide, because the average speed of molecules is proportional to the square root of the absolute temperature. So if we increase the temperature from, say 30 °C to 40 °C (approximately 300 K to 310 K), we would expect the average speed of the molecules to increase by a factor of $(310/300)^{1/2}$, which is about 1.016, ie an increase of 1.6%. Yet we know that the rate actually increases by much more than this, sometimes by 200%–300%.

Clearly there is more to the effect of temperature than simply making the particles collide more frequently. What matters is not just *how frequently* they collide, but also *with how much energy*. Unless the molecules collide with a certain minimum kinetic energy, they just bounce off one another and stay unreacted. In fact, this is what happens to most of the molecules most of the time. In the reaction between N_2 and H_2 at 30 °C, only 1 in 10^{11} collisions results in a reaction! Even at 500 °C (a temperature used in the Haber process) only 1 in 10^4 collisions results in a reaction.

So the collision theory says that *reactions occur when molecules collide* ***with a certain minimum kinetic energy****. The more frequent these collisions, the faster the reaction.* This energy is needed to overcome the energy barrier (the **activation energy**) for the reaction.

The distribution of energies

At any temperature, the speeds – and therefore the kinetic energies – of the molecules in a substance are spread over a wide range. It is like the walking speeds of people in a street: at any moment, some are moving slowly, some fast, and the majority are moving at moderate speeds. Molecules are similar: some have high kinetic energies, many have medium energies and some have low energies.

This distribution of kinetic energies in a gas (say a mixture of N_2 and H_2 molecules) at a given temperature is shown in Figure 11. It is called the Maxwell-Boltzmann distribution.

Temperature

The temperature of a substance is related to the **kinetic energy** of its particles.

A substance feels hotter when its particles are moving more energetically. This is because movement energy can easily be passed on to the particles in you, the thermometer or anything else.

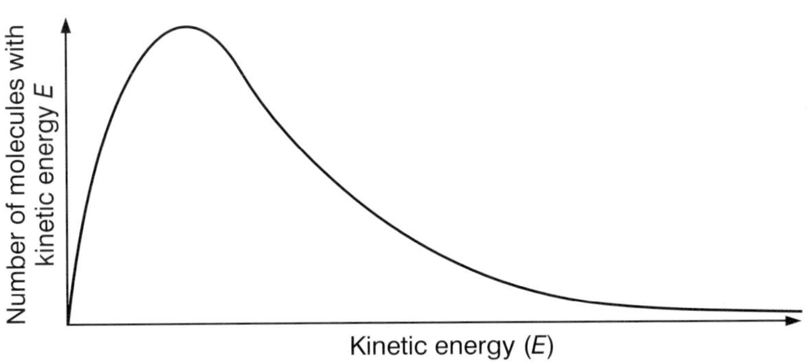

Figure 11 Distribution curve for molecular kinetic energies in a gas at a particular temperature

As the temperature increases, more molecules move at higher speeds and have higher kinetic energies. Figure 12 shows how the distribution of energies changes when you increase the temperature by 10 °C, from 30 °C to 40 °C. You can see that there is still a spread of energies, but now a greater proportion of molecules have higher energies.

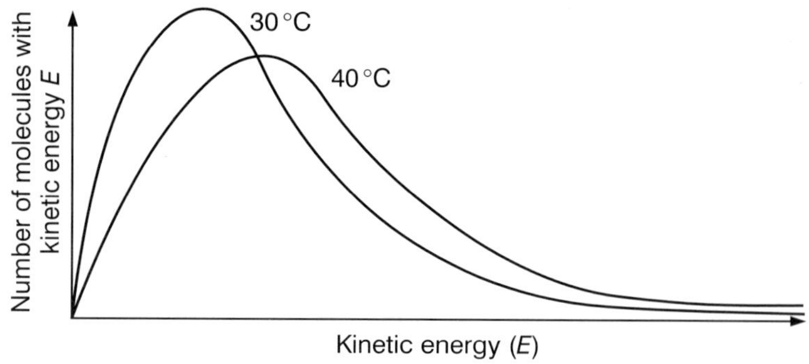

Figure 12 Distribution curves for molecular kinetic energies at 30 °C and 40 °C

Now let's look at the significance of this for reaction rates. We shall take as our example a reaction whose activation energy, E_a, is +50 kJ mol^{-1}, a value typical of many reactions. We need to think about how many *collisions* have energy greater than 50 kJ mol^{-1}, because these are the collisions that can lead to a reaction. Of course, each colliding pair of molecules would have energy far less than 50 kJ: this is the energy possessed by 6×10^{23} (1 mole) of them. The shape of the energy distribution curve for collisions is the same as for individual molecules.

Figure 13 shows the number of collisions with energy greater than 50 kJ mol^{-1}, for the reaction at 30 °C. It's given by the shaded area underneath the curve. Only those collisions with energies in the shaded area can lead to a reaction.

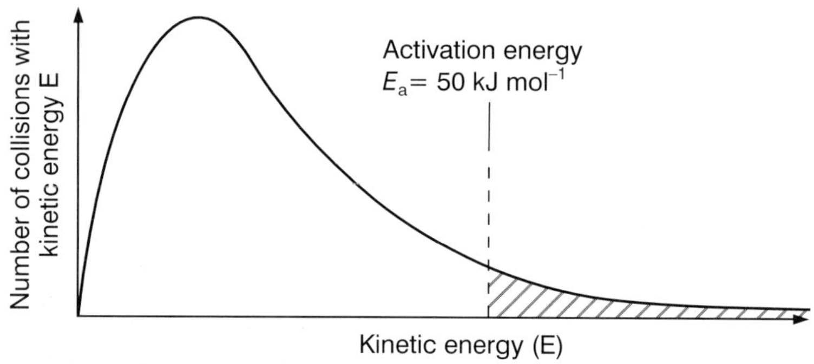

Figure 13 Distribution curve showing collisions with energy 50 kJ mol^{-1} and above

Now look at the graph in Figure 14, which shows the curves for both 30 °C and 40 °C. You can see that at the higher temperature, a significantly higher proportion of collisions have energies above 50 kJ mol^{-1} – about twice as many, in fact. This means that twice as many collisions have enough energy for reaction to occur – so the reaction goes twice as fast.

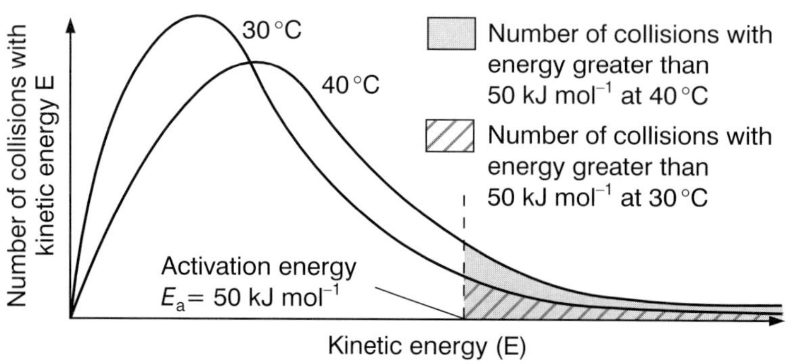

Figure 14 Distribution curves showing the effect on the proportion of collisions with energy 50 kJ mol^{-1} and above when the temperature is changed from 30 °C to 40 °C

We can summarise all this by saying that ***reactions go faster** at higher temperatures because a larger proportion of the colliding molecules have the minimum activation energy needed to react.* Increasing temperature may not make much difference to the energy of each individual molecule, but it makes a big difference to the proportion of molecules which collide with enough energy to react. In the example above, just a 10 °C increase is enough to place twice as many collisions above the minimum activation energy of $+50\,kJ\,mol^{-1}$.

For many reactions, the rate is roughly doubled by a 10 °C rise in temperature – even though everything else stays the same. It is only a rough rule, however. In fact, it only works exactly for reactions with an activation energy of $+50\,kJ\,mol^{-1}$, and for a temperature rise from 30 °C to 40 °C. Many reactions have an activation energy of about this value, and a lot of chemistry is done at about room temperature, so the rule is a reasonable rough guide. In the case of the reaction of nitrogen with hydrogen in the Haber process, the activation energy is rather greater than $+50\,kJ\,mol^{-1}$, so the rate is more than doubled for a 10 °C rise.

Photochemical reactions

With some chemical reactions, light energy can be used to speed up the reaction. Such reactions are called **photochemical reactions**.

You have already met the action of sunlight on NO_2 molecules in **Section 15.1**.

This is a key reaction in the formation of photochemical smog – see **Section 2.3**.

$$NO_2 \xrightarrow{\text{sunlight}} NO + O$$

There are many other examples of simple photochemical reactions which take place in the atmosphere. For example, ozone in the stratosphere (the upper atmosphere) absorbs high-energy ultra-violet light from the Sun and breaks down into an oxygen molecule and an oxygen atom.

$$O_3 \xrightarrow{\text{ultra-violet light}} O_2 + O$$

This is an important reaction for us. By absorbing the high-energy ultra-violet light, the ozone molecules in the stratosphere screen us from harmful solar radiation.

You will meet more examples of photochemical reactions and learn more about the chemical reactions taking place in the ozone layer in **Unit 2, Chapter 6**.

A spectacular example of a photochemical reaction in the laboratory is the reaction between chlorine and hydrogen. The overall reaction can be represented by:

$$Cl_2 + H_2 \xrightarrow{\text{sunlight}} 2HCl$$

No reaction appears to take place when the gases are mixed in the dark, but when exposed to sunlight (or a flashlight) the mixture reacts explosively.

In photosynthesis, light energy is used by green plants to convert carbon dioxide and water into glucose and oxygen.

$$6CO_2 + 6H_2O \rightarrow C_6H_{12}O_6 + 6O_2$$

The equation summarises a complex series of steps, only some of which are photochemical reactions.

PROBLEMS FOR 15.4

1 Excess zinc was added to 1 mol l^{-1} sulphuric acid at room temperature and the reaction followed by plotting the volume of hydrogen given off against time, as shown below.

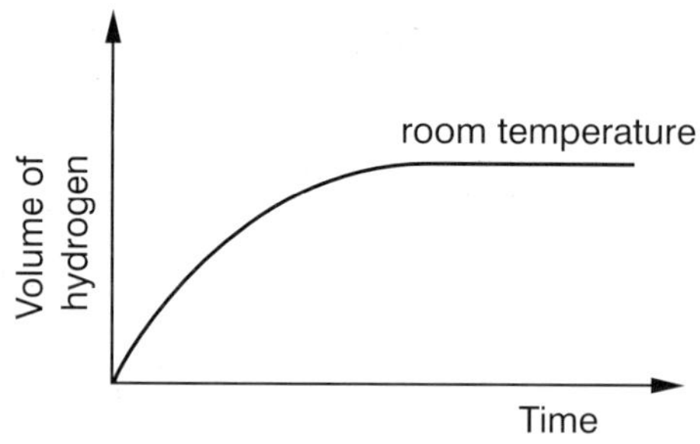

Copy the graph and add the curve that would be obtained if the reaction were carried out at a higher temperature.

2 The graph below shows how the rate of a reaction changes with temperature.

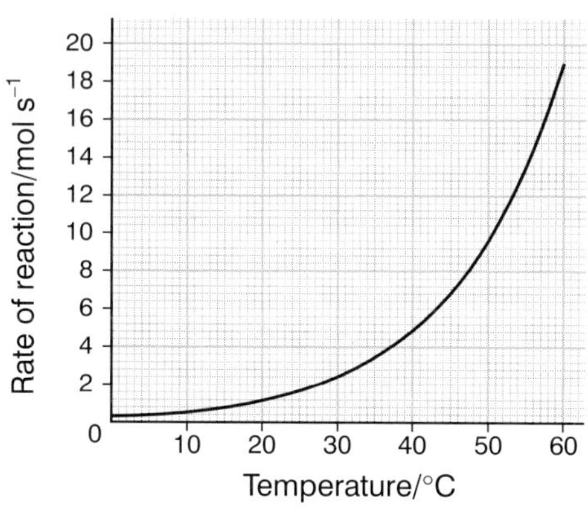

a Find the temperature rise required to double the rate of the reaction.

b Explain why a relatively small rise in temperature causes such a large increase in rate of reaction.

3 Select from the bank of words below to fill in the gaps in the following sentences.

activation temperature average kinetic enthalpy

__________ is a measure of the __________ kinetic energy of the particles of a substance.
The minimum __________ energy required by colliding particles before reaction will occur is called the __________ energy.

4 The diagram below shows the energy distribution for collisions between N_2 and H_2 molecules at temperature T_1 °C. E_a is the activation energy for the reaction of N_2 and H_2 molecules.

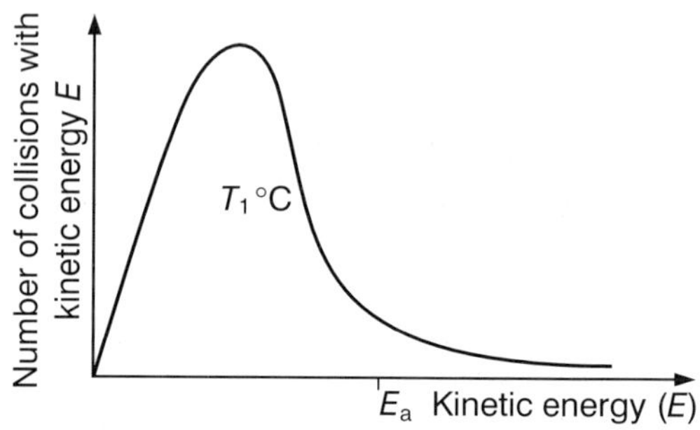

a Copy the diagram and shade the area of the graph representing the number of collisions with sufficient energy to lead to a reaction.

b On your graph, show the energy distribution for collisions between N_2 and H_2 molecules when the temperature of the reactants is increased by 10 °C to T_2 °C. Shade (using a different colour) the area under the graph representing the number of collisions with sufficient energy to lead to a reaction at T_2 °C.

15.5 *Catalysis*

A **catalyst** is a substance which alters the rate of a reaction without itself undergoing any permanent change.

Catalysts are not used up in chemical reactions and you can always get them back at the end. They are not changed chemically, though sometimes they may be changed *physically*. For example, the surface of a solid catalyst may crumble or become roughened. This suggests that the catalyst is taking some part in the reaction, but is being regenerated.

A catalyst does not appear in the overall equation for a reaction.

Only small amounts of a catalyst are usually needed. The catalyst does not affect the *amount* of product formed, only the *rate* at which it is formed.

Another interesting feature of catalysts is that they are often specific for one particular reaction. This is particularly true of biological catalysts, called **enzymes**. These are complex protein molecules which catalyse the chemical reactions that take place in the living cells of plants and animals. They affect one particular biochemical reaction strongly, but leave a similar reaction almost unaffected.

You met enzymes earlier in this unit in **Section 1.5**, as examples of 'molecules of life'. You can find out more about how their shape enables them to be so specific in **Unit 2, Section 5.10**.

Types of catalysts

If the reactants and catalyst are in the same physical state (for example, both are in aqueous solution), the reaction is said to involve **homogeneous catalysis**. Enzyme-catalysed reactions in cells take place in aqueous solution and are examples of this type of catalysis.

In contrast, many important industrial processes involve **heterogeneous catalysis**, where the reactants and the catalyst are in different physical states. This usually involves a mixture of gases or liquids reacting in the presence of a solid catalyst.

How do catalysts work?

In a chemical reaction, existing bonds in the reactants must first stretch and break. Then new bonds can form as the reactants are converted to products.

Bond breaking is an endothermic process. A pair of reacting molecules must have enough energy between them to pass over the activation energy barrier before reaction can occur. If the energy barrier is very high, relatively few pairs of molecules will have enough energy to overcome it and react to form the products – so the reaction is slow.

Catalysts speed up reactions by providing an alternative reaction pathway for the breaking and remaking of bonds that has a lower activation energy. In this case, the energy barrier is lower and more pairs of molecules can pass over to form products. This means the reaction proceeds more quickly. Note that the enthalpy change, ΔH, is the same for the catalysed and uncatalysed reactions. Figure 15 shows the energy pathways for an uncatalysed and a catalysed reaction.

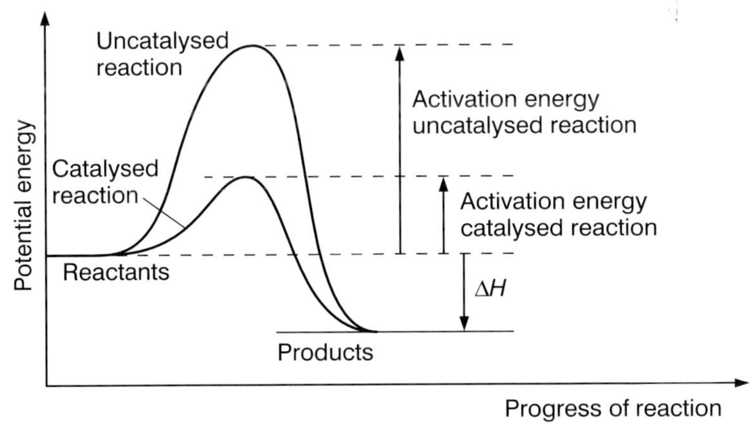

Figure 15 The effect of a catalyst on the energy pathway for a reaction

Heterogeneous catalysts

When a solid catalyst is used to increase the rate of a reaction between gases or liquids, the reaction occurs on the surface of the solid. Figure 16 shows an example.

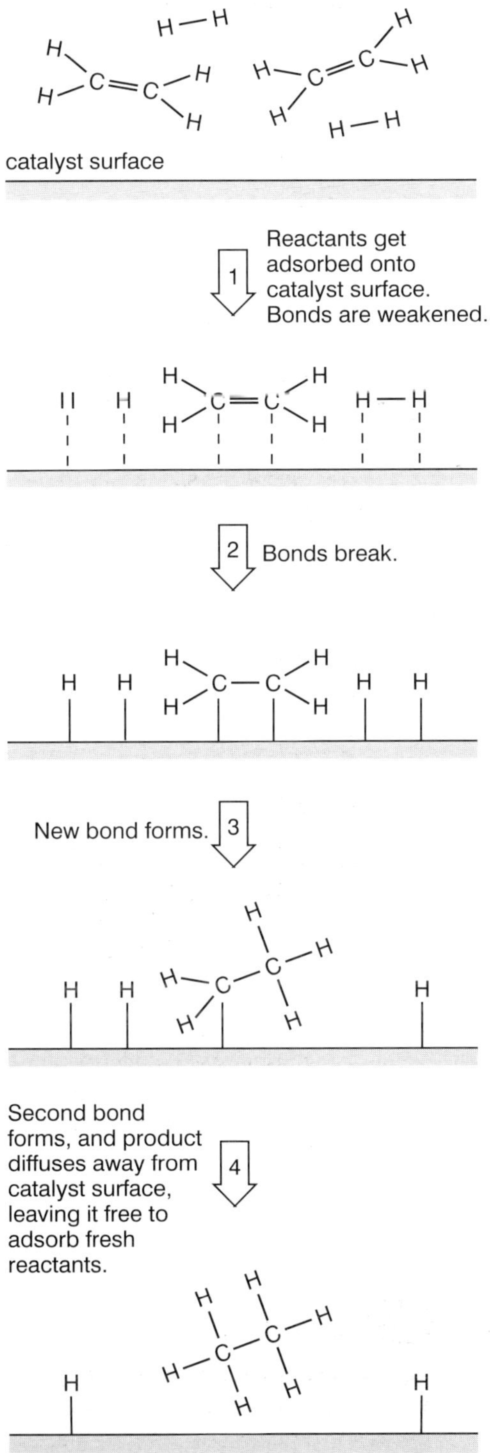

Figure 16 An example of heterogeneous catalysis. The diagrams show a possible mechanism for nickel catalysing the reaction between ethene and hydrogen to form ethane

It is important that the catalyst has a large surface area for contact with reactants. For this reason, solid catalysts are used in a finely divided form or as a fine wire mesh. Sometimes the catalyst is supported on a porous material to increase its surface area and prevent it from crumbling. This happens in the catalytic converters fitted to car exhaust systems (see **Section 2.4**).

Zeolites (crystalline aluminosilicate materials, see **Unit 2**, **Section 4.3**) are widely used in industry as heterogeneous catalysts, for example, in the cracking of petroleum fractions.

Many of the heterogeneous catalysts used in industrial processes are transition metals (metals in the central block of the Periodic Table) or transition metal compounds. You will meet a number of examples at different stages in the course. Table 2 lists three examples and tells you where to find them in the course.

Industrial process	Catalyst	Where to find out more
Reforming of petroleum fractions to produce high octane petrol components eg hexane → cyclohexane	platinum/rhenium	**Unit 2, Section 4.3**
Hydrogenation of unsaturated oils to give saturated fats eg vegetable oil + H_2 → margarine	nickel	**Unit 2, Sections 17.3 and 18.5**
Manufacture of ammonia from nitrogen and hydrogen in the Haber process $N_2 + 3H_2 \rightarrow 2NH_3$	iron	**Unit 3, Sections 9.1 and 20.2**

Table 2 Examples of the use of heterogenous catalysts in industrial processes

The use of catalysts in industry is summarised in **Unit 3, Section 23.3.**

Catalyst poisoning

Catalysts can be **poisoned** so that they no longer function properly. In fact, many substances which are poisonous to humans operate by blocking an enzyme-catalysed reaction.

In heterogeneous catalysis, the 'poison' molecules are adsorbed more strongly to the catalyst surface than the reactant molecules and the catalyst becomes inactive. This is the reason why leaded petrol cannot be used in cars fitted with a catalytic converter; lead is strongly adsorbed to the surface of the catalyst.

Catalyst poisoning is also the reason why it is not possible to replace the very costly metals, platinum, palladium and rhodium, in catalytic converters by cheaper metals, such as copper and nickel. These metals are vulnerable to poisoning by trace amounts of sulphur dioxide always present in car exhaust gases. Once the catalyst in a converter becomes inactive it cannot be regenerated. A new converter has to be fitted.

Catalyst poisoning can be a problem in industrial processes. In the UK, nearly all the hydrogen for the Haber process is prepared by steam reforming of methane. Methane reacts with steam in the presence of a nickel catalyst:

$$CH_4(g) + H_2O(g) \rightleftharpoons CO(g) + 3H_2(g)$$

If the feedstock for the process contains sulphur compounds, these must be removed first to prevent severe catalyst poisoning.

Sometimes it is possible to clean or **regenerate** the surface of a catalyst. In the catalytic cracking of long-chain hydrocarbons, for example, carbon is produced and the surface of the zeolite catalyst becomes coated in a layer of soot. This blocks the adsorption of reactant molecules and the activity of the catalyst is reduced. The catalyst is constantly recycled through a separate container where hot air is blown through the zeolite powder. The oxygen in the air converts the carbon to carbon dioxide and cleans the catalyst surface. You can read more about catalyst regeneration in **Unit 3, Section 23.3**.

Homogeneous catalysts

Homogeneous catalysts normally work by forming an intermediate compound with the reactants. This then breaks down to give the product and reform the catalyst.

An example of homogeneous catalysis in the gas state is the destruction of ozone in the stratosphere by CFCs. Ozone molecules are broken down to form oxygen molecules. The reaction is catalysed by chlorine atoms which form an intermediate, ClO. This then reacts with an oxygen atom to produce oxygen molecules and regenerate the Cl atom.

You can read more about the destruction of the ozone layer in **Unit 2, Chapter 6**.

$$\begin{aligned} & Cl + O_3 \rightarrow O_2 + ClO \\ \text{Then} \quad & ClO + O \rightarrow Cl + O_2 \\ \hline \text{Overall change:} \quad & O_3 + O \rightarrow 2O_2 \end{aligned}$$

One chlorine atom can take many molecules through this **catalytic cycle**.

In industry, most processes involve heterogeneous rather than homogeneous catalysis. Homogeneous reactions, however, can be more specific and controllable, so that fewer unwanted products are formed. A good example of a homogeneous catalyst helping to produce a single product is the manufacture of ethanoic acid from methanol. A soluble compound of rhodium is used as the catalyst, leading to a 99% conversion of methanol to ethanoic acid.

Enzymes as catalysts

Enzymes play a vital role in biological systems, enabling complex reactions to occur quickly in dilute aqueous solution at moderate temperatures and pH – the conditions in a living cell. Enzymes are so *specific* because they have a precise three-dimensional structure which exactly matches the structure of the **substrate** – the molecule which is reacting. Scientists often think of the 'lock and key' analogy, in which the enzyme is the 'lock' and the substrate is the 'key'. There may be lots of related keys, and several may fit into the lock, but only one will work.

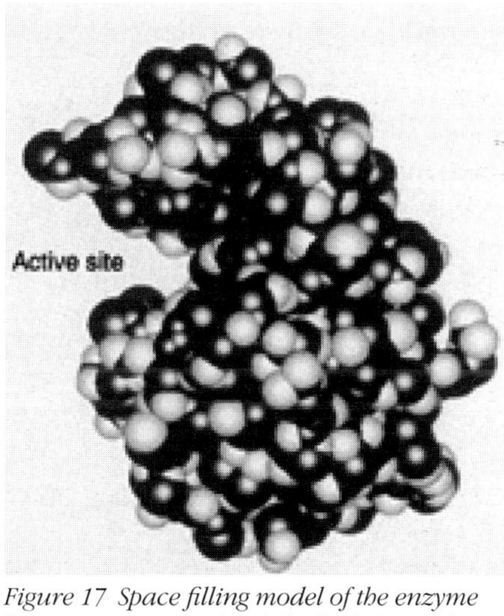

Figure 17 Space filling model of the enzyme lysozyme, showing the cleft which forms the active site

The analogy is a good one. Figure 17 shows a space filling model of the enzyme *lysozyme* which catalyses the breakdown of the cell walls of bacteria and helps protect us from infection. There is a cleft in the enzyme surface formed by the way the protein chain folds. The shape of the cleft is tailored for the substrate molecules to fit into. Within the cleft are chemical groups which bind the substrate and possibly react with it. This region of the enzyme is called its **active site**. It is where the reaction takes place.

After reaction, the product leaves the enzyme, which is then free to start again with another molecule of substrate. The whole process is summarised in Figure 18.

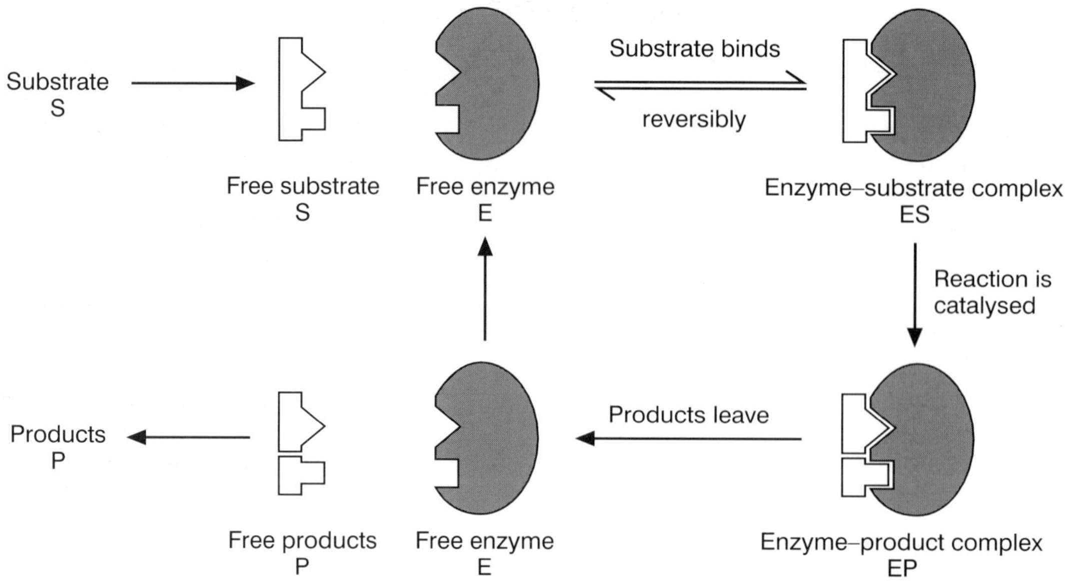

Figure 18 Illustration of the 'lock and key' model of enzyme catalysis

This is the simplest model of enzyme catalysis:

$E + S \rightleftharpoons ES \rightarrow EP \rightarrow E + P$

What happens, of course, is more complex. In many cases, scientists believe that the substrate is not quite a perfect fit, and must alter its shape to fit into the active site. This means that both the substrate and the active site will be in strained arrangements, and this helps the reaction to occur.

Enzymes at work

Large amounts of enzymes are used in the food industry and in the manufacture of washing powders. Almost all of these are *hydrolases* – enzymes which hydrolyse fats, proteins or carbohydrates.

In the food processing industry, two major uses of enzymes are

- producing glucose syrup (used as a sweetener in food products) by breaking down starch with enzymes like α-amylase
- making cheese using rennet enzymes: these break down the milk protein casein and cause the separation of the curds (solid) from the whey (liquid).

Other uses include baking, brewing and fruit processing.

Many biological washing powders contain one or more enzymes to assist in the removal of stains. The enzyme is usually a *protease* to hydrolyse proteins in blood and food, but a *lipase* may also be added to break down fats. More recently, cellulases have been added too. These break down the tiny surface fibres that give older clothes a fluffy, dull look.

Enzymes are finding applications in wider areas of waste treatment. For example, an enzyme is being used to destroy cyanide ions which are left over after gold extraction or after the production of some polymers. Enzymes can also be used to help break up oil spillages.

Figure 19 Proteases are added to 'biological' detergents to remove protein stains such as grass, blood, egg and sweat

Figure 20 Enzymes are important in the production of bread, cheese and wine

PROBLEMS FOR 15.5

1 The collision theory assumes that the rate of a reaction depends on

A the rate at which reactant molecules collide with one another

B the proportion of reactant molecules that have enough energy to react once they have collided.

Which, out of A and B, explains each of the following observations?

a Reactions in solution go faster at higher concentration.

b Solids react faster when their surface area is higher.

c Catalysts increase the rate of reactions.

d Increasing the temperature increases the rate of a reaction.

2 **Catalysts** can be classified as either **heterogeneous** or **homogeneous**.

Explain the meaning of each of the terms in bold.

3 **a** Name *two* industrial catalysts and briefly describe a process each is involved in.

b State whether each of the catalysts you chose in part **a** is heterogeneous or homogeneous.

4 The activation energy for the decomposition of hydrogen peroxide to oxygen and water is $+36.4\,\text{kJ mol}^{-1}$ in the presence of an enzyme catalyst and $+49.0\,\text{kJ mol}^{-1}$ in the presence of a very fine colloidal suspension of platinum.

a Draw a labelled potential energy diagram to show the relative effect of the two catalysts on the activation energy for the reaction.

b How will the rate of this decomposition differ for the two catalysts at room temperature? Explain your answer.

5 When carbon monoxide and nitrogen monoxide in car exhaust gases pass through a catalytic converter, carbon dioxide and nitrogen are formed.

a Write a balanced chemical equation for this reaction.

b Explain why it is important to reduce the quantities of carbon monoxide and nitrogen monoxide released into the atmosphere.

c Figure 16 shows a possible mechanism for the nickel-catalysed reaction between ethene and hydrogen. The reactants are adsorbed onto the surface of the nickel catalyst, where the reaction takes place.

i Explain the meaning of the term *adsorbed*.

ii Using (N)—(O) to represent nitrogen monoxide and (C)—(O) to represent carbon monoxide, draw out a possible mechanism for the formation of carbon dioxide, (O)—(C)—(O).

d Suggest why catalytic converters do not work effectively until the car engine has warmed up.

e If leaded petrol is used in a car fitted with a catalytic converter, the converter no longer works.

i Explain why this happens.

ii How can this be avoided?

6 Long-chain hydrocarbons can be cracked into a mixture of smaller molecules by passing them over a heated zeolite catalyst. After a while the catalyst becomes coated with carbon and the cracking reaction slows down and eventually stops. The catalyst has to be regenerated in order for it to function again.

a Why does the coating of carbon cause the reaction to stop?

b Suggest where the carbon on the surface of the catalyst came from.

c Air is blown over the hot catalyst to clean it. Suggest how this regenerates the catalyst.

7 Enzymes are used in the food industry and in the manufacture of biological washing powders.

a Explain the function of enzymes in living things.

b Describe *two* major uses of enzymes in the food industry.

c Explain why the enzyme, protease, is added to washing powders.

16
STRUCTURE AND PROPERTIES

16.1 Forces between molecules: temporary and permanent dipoles

Solid, liquid or gas?

Earlier in this unit, in **Section 13.2**, you learned how covalent bonds hold the atoms together within a molecule. At room temperature, some substances made up of covalent molecules are gases, others are liquids and the remainder are solids. What is it that determines whether the substance is a gas, a liquid or a solid?

If you look at Table 1 in **Chapter 3**, you will see that most of the gases in the Earth's atmosphere exist as *small* covalent molecules (eg N_2, O_2, and CO_2). The noble gases (eg Ar, Ne and He) exist as isolated atoms.

In a liquid or a solid there must be forces *between the molecules* causing molecules to be attracted one to another, otherwise they would move apart from each other and become a gas. These forces are called **intermolecular forces**.

If the temperature is lowered far enough, every substance, no matter how low its boiling point and melting point, will eventually solidify. When the solid melts and then boils (Figure 1), it is the intermolecular forces that are broken. The covalent bonds *within the molecules* remain intact. So, where a substance has a structure consisting of covalent molecules, the boiling point gives a good indication of how strong the attractive forces are between the molecules.

Element	Boiling point/°C
helium	–269
neon	–246
argon	–186
krypton	–152
xenon	–107

Table 1 Boiling points for the noble gases

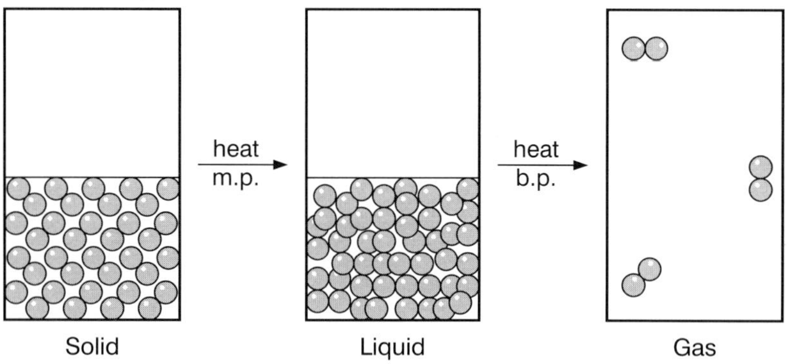

Figure 1 On heating, the molecular substance (represented by ⚭) changes from a solid to a liquid and then to a gas. Energy must be supplied to overcome the intermolecular forces. Note that the covalent bonds within the molecules remain intact

The boiling points of noble gases and alkanes

Look at the boiling points given in Table 1 for the noble gases.

When the noble gases change from liquid to gas, energy is only required to separate the atoms. There are no covalent bonds in the noble gases but there must be some forces holding the atoms together.

The low boiling points of the noble gases show that the attractive forces between the isolated atoms in the liquid state are very weak. However, the boiling points get higher as the atoms get bigger. This suggests that the larger the atoms, the stronger the attractive forces between them.

Now look at the boiling points of the alkanes in Table 2.

Alkane	Structural formula	Boiling point/°C
methane	CH_4	–162
ethane	CH_3CH_3	–89
propane	$CH_3CH_2CH_3$	–42
butane	$CH_3CH_2CH_2CH_3$	0
pentane	$CH_3CH_2CH_2CH_2CH_3$	36

Table 2 Boiling points of some alkanes

There is a pattern to these boiling points also; the longer the alkane chain the higher the boiling point. There must be stronger forces between the long molecules than between the short molecules. More energy and higher temperatures are needed to overcome these stronger forces and set the alkane molecules free. To understand how the intermolecular forces between molecules such as those of alkanes arise, and what affects their relative sizes, you need to know about the charges on molecules, called dipoles.

Dipoles

You have already come across polar bonds in **Section 13.2**: a **dipole** is simply a molecule (or part of a molecule) with a positive end and a negative end. Hydrogen chloride has a dipole; we show the molecule as

$$\overset{\delta+}{H}—\overset{\delta-}{Cl}$$

When a molecule has a dipole we say it is **polarised**. There are several ways a molecule can become polarised.

Permanent dipoles

Permanent dipoles occur when a molecule has two atoms bonded together which have substantially different electronegativity, so that one atom attracts the shared electrons much more than the other. Hydrogen chloride has a permanent dipole, because chlorine is much more electronegative than hydrogen, and so attracts the shared electrons more. Electronegativity is discussed in **Section 13.2**.

Molecules with a permanent dipole are said to be **polar molecules**.

Temporary dipoles

Some molecules do not possess a permanent dipole, because the atoms that are bonded together have the same, or very similar, electronegativity, so that the electrons are evenly shared. The chlorine molecule, Cl_2, is an example.

However, even though the molecule does not have a permanent dipole, a **temporary**, or **instantaneous dipole** can arise. We can think of the electrons in the chlorine molecule as forming a negatively charged cloud. The electrons in this cloud are in constant motion, and at a particular instant they may not be evenly distributed over the two atoms. This means that one end of the chlorine molecule has a greater negative charge than the other end; instantaneously, the molecule has developed a dipole (Figure 2).

Electron cloud evenly distributed; no dipole.

At some instant, more of the electron cloud happens to be at one end of the molecule than the other; the molecule has a temporary dipole.

Figure 2 How a dipole forms in a chlorine molecule

Left on its own, this dipole only lasts for an instant before the swirling electron cloud changes its position, cancelling out or even reversing the dipole. However, if there are other molecules nearby, the temporary dipole may affect them and produce **induced dipoles**.

Induced dipoles

If an unpolarised molecule finds itself next to a dipole, the unpolarised molecule may get a dipole induced in it. The dipole attracts or repels electrons in the charge cloud of the unpolarised molecule, **inducing** a dipole in it. Figure 3 illustrates this by showing what happens if an unpolarised chlorine molecule finds itself next to an HCl dipole.

In Figure 3, the dipole has been induced by the effect of a *permanent* dipole. A dipole can also be induced by the effect of a *temporary* dipole. This makes it possible for a whole series of dipoles to be set up in a substance that contains no permanent dipoles. How this occurs in a noble gas such as xenon is explained later in this section.

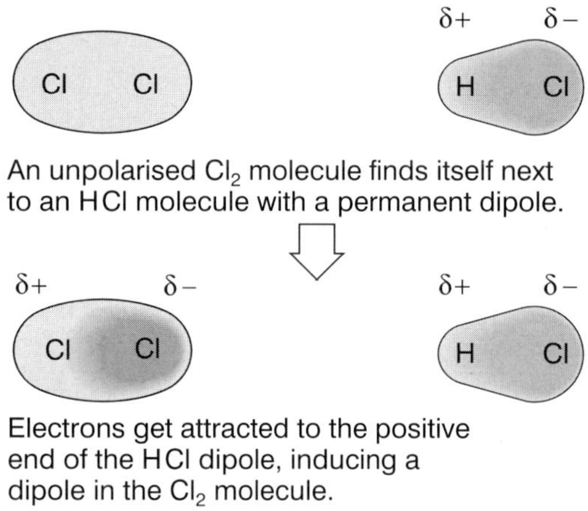

Figure 3 How a dipole can be induced in a chlorine molecule.

Dipoles and intermolecular forces

If a molecular substance contains dipoles, they can attract each other (Figure 4).

$$
\cdots \overset{\delta+}{H}-\overset{\delta-}{Cl} \cdots \overset{\delta+}{H}-\overset{\delta-}{Cl} \cdots \overset{\delta+}{H}-\overset{\delta-}{Cl} \cdots \overset{\delta+}{H}-\overset{\delta-}{Cl} \cdots
$$

Figure 4 Intermolecular forces arise from the attractions between dipoles. (The dotted lines represent the attractive forces between *HCl molecules; the solid lines represent the stronger covalent bonds* inside *each molecule.)*

All intermolecular forces arise from the attractive forces between dipoles. There are three kinds of attraction:

- **Permanent dipole–permanent dipole**, where two or more permanent dipoles attract one another. This kind of attraction occurs between the permanent dipoles in HCl.
- **Permanent dipole–induced dipole**, in which a permanent dipole induces a dipole in another molecule, then the two attract one another. This kind of attraction occurs between HCl and Cl_2 molecules.
- **Temporary dipole–induced dipole**, in which a temporary dipole induces a dipole in another molecule, then attracts it. This kind of attraction occurs between all molecules.

In the next sections we will look in more detail at two of these: temporary dipole–induced dipole forces and the forces between permanent dipoles.

Temporary dipole–induced dipole attractions

These forces are also known as **van der Waals forces**. They are the weakest type of intermolecular attraction. They act between *all* molecules, because temporary dipoles can arise in molecules which already have a permanent dipole. However, you notice them most easily in substances such as the noble gases and the alkanes because there are no other intermolecular forces present. Figure 5 shows how a temporary dipole in a xenon atom induces polarisation in a neighbouring atom.

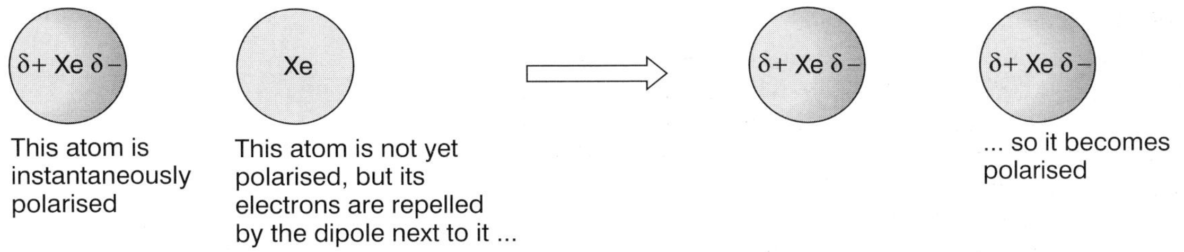

Figure 5 How an induced dipole is formed in an Xe atom

The polarisation illustrated here is not permanent. Electrons are always moving around and the original polarity will be lost and replaced by another which will be different – possibly the other way round. In the gaseous or liquid state, the atoms are always moving around. As they move past each other the dipoles constantly change position, although they are always lined up so that they attract rather than repel one another.

The forces between atoms in xenon are stronger than those between helium atoms. The xenon atom is bigger, and has more electrons. The outer electrons are less strongly held by the positive nucleus and more easily pushed around from outside. We say that xenon atoms are more **polarisable** than helium atoms. This leads to bigger induced dipoles which in turn lead to bigger forces of attraction, which is why xenon has a higher boiling point than helium.

So, van der Waals forces arise because electron movements in molecules cause instantaneous dipoles which then induce polarity in neighbouring molecules. The attractive forces between molecules are the result of electrostatic interactions between the dipoles. These interactions are normally strongest for large atoms, which have more electrons and are more polarisable than small atoms. For molecules, van der Waals forces become stronger as the number of atoms in the molecules increases. This is because larger molecules have more electrons and this increases the possibility of greater numbers of temporary dipoles being in existence at any given time.

Although van der Waals forces are relatively weak, the *overall* attractive forces between large molecules (for example, between the chains of a polymer) can be quite strong. Thus, poly(ethene) is a solid at room temperature even though van der Waals forces are the only type of intermolecular attractions present.

Remember that these forces are present in *all* substances.

Permanent dipole–permanent dipole attractions

In this section, we will look in more detail at the additional attractive forces which can arise when molecules have **permanent dipoles.**

Bond polarity and polar molecules

Section 13.2 covers bond polarity, and describes how you can use electronegativity values to decide how polar a particular bond will be.

Let's look at an important example of a polar molecule: H_2O. Oxygen is much more electronegative than hydrogen, so the O—H bond is polar:

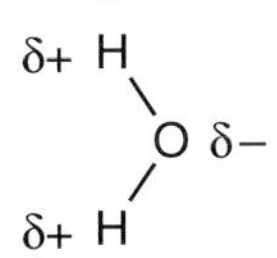

One side of the molecule is positive, the other is negative. The water molecule has a dipole with the positive charge centred between the two hydrogen atoms. We could also write it like this:

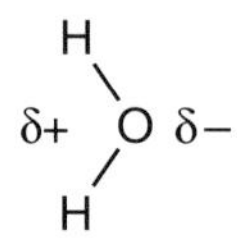

An example of a more complicated dipolar molecule is 1,1,1-trichloroethane. The three chlorine atoms make one end of the molecule negatively charged; the three hydrogen atoms at the other end are positively charged. The dipole can be seen more clearly if we draw the structure as

$$\overset{\delta+}{H_3C}-\overset{\delta-}{CCl_3}$$

When the atoms which form a molecule have no electronegativity difference, or when the electronegativity differences are small (in other words, when the bonds are non-polar), any dipole will be very small. It is also possible for a molecule to have no overall dipole even though the bonds are polar. CCl_4 is an example of such a molecule. Each Cl atom in CCl_4 carries a small negative charge and the central carbon is positive.

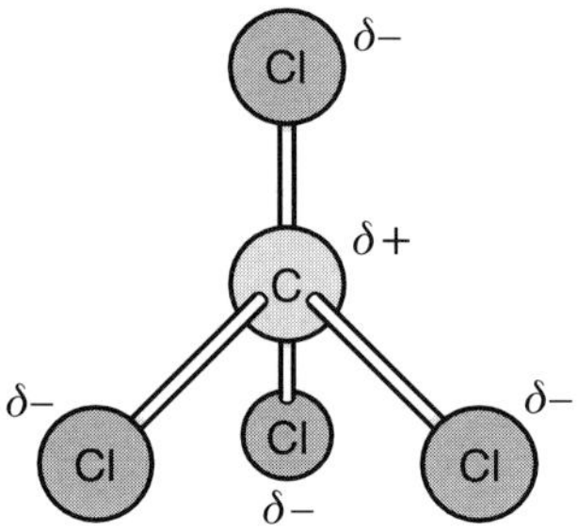

The Cl atoms are distributed symmetrically around the carbon atom, and CCl_4 has a tetrahedral shape, like that of methane.

Because the Cl atoms are distributed tetrahedrally around the carbon, the centre of negative charge is midway between all the chlorines. It is at the centre of the molecule and is superimposed on the positive charge on the carbon. The molecule has no overall dipole. We say it is **non-polar**.

So, *bond* polarity depends on electronegativity differences. A *molecular* dipole depends on electronegativity differences *and* the shape of the molecule. You may need to build a model to help you decide if a molecule has a dipole.

Dipole–dipole interactions

Let's look at the forces between molecular dipoles like those in 1,1,1–trichloroethane. This substance is a liquid at room temperature, and in the liquid phase the molecules are constantly moving and tumbling around. Sometimes the negative end of one molecule will be lined up with the positive end of another, making them attract, but often there will be two negative ends or two positive ends together, in which case they will repel (Figure 6). Overall, there will be more attraction than repulsion between the permanent dipoles, which is one reason why the molecules stay together as a liquid.

In addition to the attractive forces between the permanent dipoles, there will also be some attraction between the molecules due to van der Waals forces. But for a molecule of any given size, they will be much weaker than the dipole–dipole interactions.

In the liquid state, the molecular dipoles are constantly tumbling around.

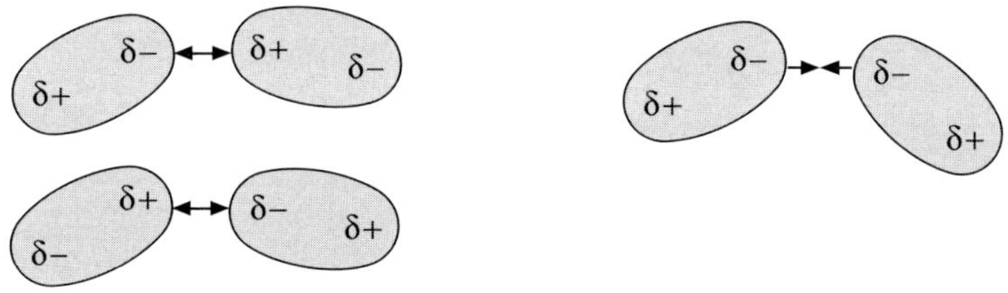

Sometimes opposite charges are next to one another, causing attraction ...

... sometimes like charges are next to one another, causing repulsion.

Figure 6 Molecular dipoles in the liquid state

In the solid state, things are rather different. In a crystalline solid the molecules are in fixed positions rather than tumbling around, and it is possible for the permanent dipoles to line up with opposite charges next to one another. The attractive forces between these aligned dipoles help to hold the molecules together in the solid state.

Remember, dipole–dipole attractions between permanent dipoles are stronger than van der Waals forces. This means that, if we have two substances, one polar and one non-polar, whose molecules have the same mass (and so approximately the same number of electrons), the polar substance will have a higher boiling point than the non-polar substance.

To sum up ...

van der Waals forces

- are forces of attraction which operate between all molecules and between the isolated atoms in noble gases
- are much weaker than other types of intermolecular forces
- are the result of electrostatic attraction between temporary dipoles and induced dipoles caused by movement of electrons.

The strength of van der Waals forces is related to the number of electrons present and hence to the size of the molecule (or isolated atom).

Permanent dipole–permanent dipole attractions

- are additional electrostatic forces of attraction between polar molecules
- are stronger than van der Waals forces for molecules of equivalent size.

A polar molecule is one that has a permanent dipole. The spatial arrangement of polar covalent bonds is important in determining whether or not a molecule as a whole is polar.

PROBLEMS FOR 16.1

1 For each pair of chemicals given below, arrange the formulae in the order in which the strength of the van der Waals forces increases.

   a Xe and Kr

   b C_8H_{18} and C_6H_{14}

   c CCl_4 and CH_4

2 Explain why noble gases have very low boiling points.

3 In a molecular substance, measuring the boiling point can give an indication of the strength of the attraction between molecules. When comparing the strength of intermolecular forces in polar and non-polar substances, explain why it is important to compare substances of similar relative molecular mass.

4 The boiling points of some molecular substances are shown in the table below.

Substance	Relative molecular mass	Boiling point/ °C
A	74	117
B	98	156
C	30	−21
D	78	80
E	110	240
F	56	8
G	32	65

a Which pairs of molecules could be used to compare the strength of the intermolecular forces present?

b For each pair, state the substance that has the stronger attractive forces between its molecules.

5 Which of the covalent bonds in the following list will be significantly polar. In cases where there is a polar bond, show which atom is positive and which is negative, using the δ+ and δ– convention.

a C—F
b C—H
c C—S
d H—Cl
e H—N
f S—Br
g C—O

6 Which of the molecules listed below possesses a permanent dipole? You will need to think about the shape of each molecule. Carbon dioxide is a linear molecule.

a CO_2
b NH_3
c SiF_4
d $CHCl_3$

7 Two compounds of silicon and sulphur have the following formulae and boiling points:
SiH_4 −112 °C; H_2S −61 °C

a i How many electrons are present in each molecule?
ii How will the strengths of van der Waals forces compare in the two compounds?
iii Does either molecule possess a permanent dipole? Explain your answer.

b Use your answers in part **a** to explain the relative magnitude of the boiling points of these compounds.

16.2 *Molecules and networks*

Carbon and silicon oxides

We breathe CO_2 and walk on SiO_2. Both are essential for life on Earth: one is used in photosynthesis, and the other produces the soil which supports growing plants.

You met both carbon dioxide, CO_2, and quartz, SiO_2, in **Chapter 3**.

Carbon and silicon are both in Group 4 of the Periodic Table and you might expect their compounds to show similar properties. Yet you could hardly imagine two more physically different substances. CO_2 is a gas at room temperature; it turns directly from a solid to a gas (sublimes) at −78 °C. SiO_2 is a hard solid with a high melting point. (Sand, for example, is impure quartz.)

The reason for this dramatic difference in physical properties is the difference in bonding between carbon and oxygen on the one hand, and silicon and oxygen on the other. Both are covalent compounds, but the small size of the carbon atom makes it possible for carbon to form double bonds with oxygen so that carbon dioxide is composed of individual molecules:

$$O{=}C{=}O$$

Intermolecular forces between the carbon dioxide molecules are weak. Little energy is needed to separate individual molecules in the solid and liquid phases to form a gas, so CO_2 is a gas at room temperature. It freezes at −78 °C to a white solid (dry ice). It is soluble in water, giving an acidic solution.

Silicon atoms are larger than carbon atoms and they normally bond to four oxygens. Quartz, SiO_2, is an extended network of SiO_4 units in which the central silicon is covalently bonded to each of four oxygen atoms. Each Si atom has a half-share in four oxygen atoms (Figure 7).

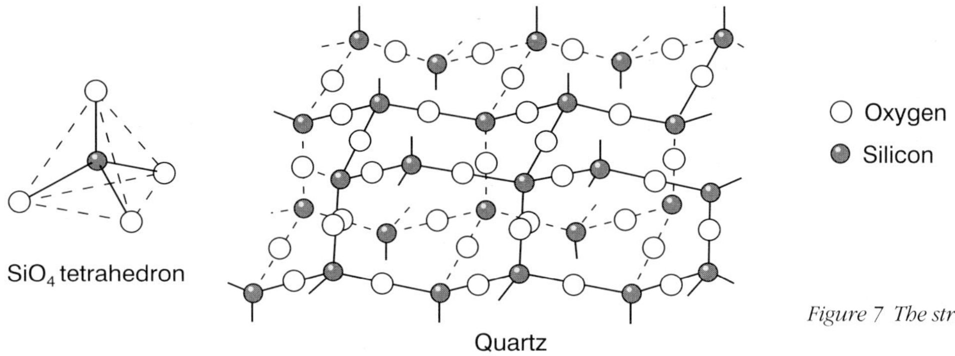

Figure 7 The structure of quartz, SiO_2

Because of its extended **network structure**, SiO_2 is insoluble in water, and has high melting and boiling points (1610 °C and 2230 °C, respectively). The reason for these properties is the strong covalent bonding which exists throughout the SiO_2 structure. Considerable energy is needed to break bonds within the structure so that a very high temperature is needed to melt quartz.

So covalent substances can exist as two types of structures depending on how the atoms are bonded. The two types of structures show very different properties.

- **Covalent molecular structures** (for example, CO_2) consist of discrete molecules. The covalent bonds within the molecules are strong. There are also weak attractive forces *between* the molecules. The strength of these intermolecular forces determines whether the substance is a gas, solid or liquid at normal temperatures (see **Section 16.1**). Covalent molecular substances often dissolve in organic solvents. Some, like carbon dioxide, dissolve in water.
- **Covalent network structures** (for example, SiO_2) consist of giant repeating lattices of covalently bonded atoms. Substances with network structures are insoluble solids with high melting and boiling points.

You can find out more about the structure of quartz and silicates in **Chapter 3**. Silicates are network structures of Si, O and Al atoms joined together in chains or sheets. Some of the oxygen atoms carry negative charges and so positive metal ions bind to the chain or to the sheet.

The examples above are covalent *compounds* containing two or more different elements. But the same types of structures are shown by non-metal *elements*; some exist as covalent molecules whereas others exist as covalent networks.

Elements with molecular structures

Some non-metal elements that exist as molecules are shown in Figure 8. Hydrogen, oxygen, nitrogen, fluorine and chlorine all exist as **diatomic molecules**, and so their chemical formulae are written as X_2. They are all gases at room temperature.

Phosphorus and sulphur are both soft, low-melting solids. Phosphorus molecules, P_4, are made up of four phosphorus atoms, whereas sulphur molecules, S_8, contain eight sulphur atoms in a ring. The structures of these molecules are shown in Figure 9. For the sake of simplicity when writing chemical equations, phosphorus and sulphur are often simply written as P and S, respectively.

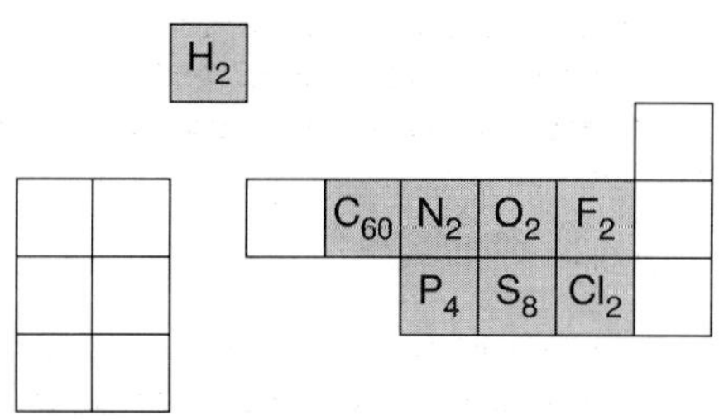

Figure 8 The position in the Periodic Table of covalent elements with molecular structures

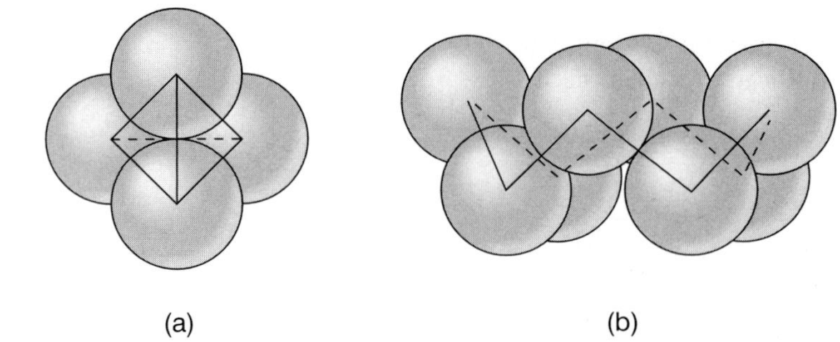

Figure 9 (a) Structure of a phosphorus molecule P_4 and (b) structure of a sulphur molecule S_8

The covalent bonds in the molecules of elements are non-polar because identical atoms are bonded to each other. In the gas state, the molecules are completely separate from one another. In the liquid and solid states, van der Waals forces cause the molecules to be attracted to each other. These are weak attractive forces and the elements have low melting points and boiling points.

Elements with network structures

Figure 10 shows some common elements that exist as covalent network structures. These elements are solids with high melting points and boiling points. The bonds between the individual atoms are strong covalent bonds (not weak van der Waals forces) and a lot of energy is needed to pull the atoms apart to form a liquid or a gas.

Figure 11a shows the structure of silicon. Each silicon is bonded to four other silicon atoms in a tetrahedral arrangement. The giant lattice repeats the structure indefinitely in all directions – even a small lump of silicon would contain billions and billions of atoms joined up in this way.

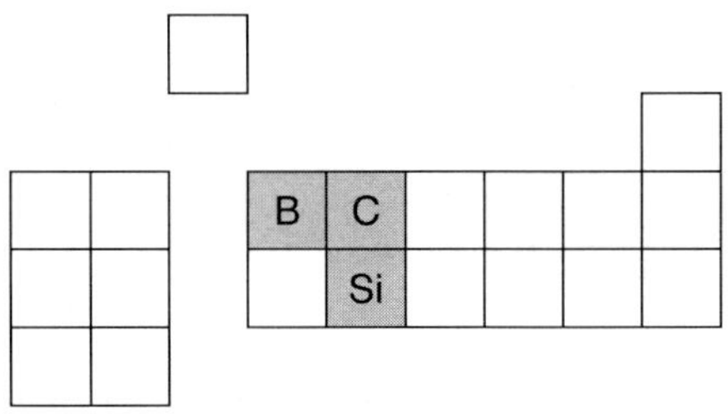

Figure 10 The position in the Periodic Table of covalent elements with network structures

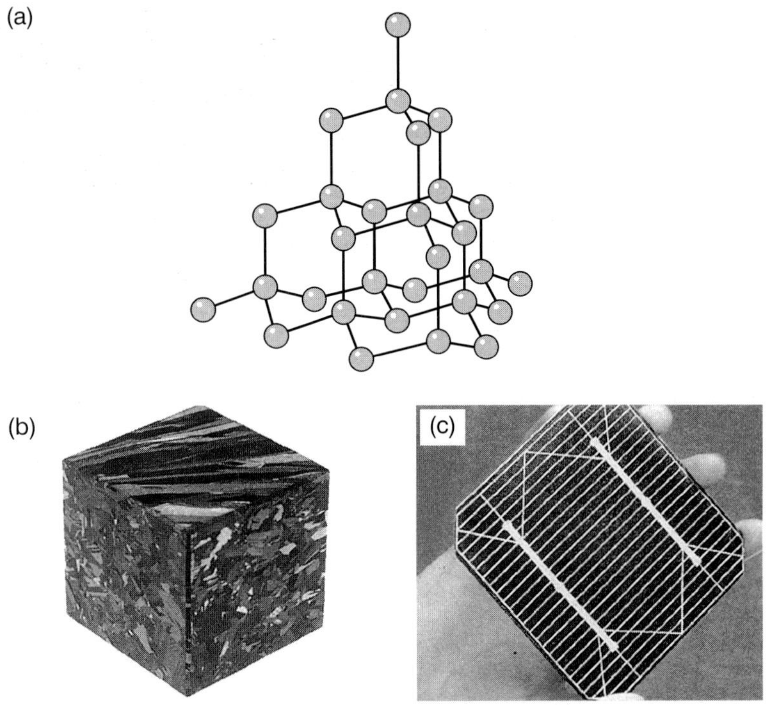

Figure 11 (a) The network structure of silicon, (b) a block of pure silicon and (c) a solar cell made from wafers of silicon about 350 μm thick

You may have noticed that carbon appears in both Figure 8 among the molecular elements and in Figure 10 among the network elements. This is not a misprint. In fact, carbon can exist in three different forms. Two of these have network structures; the third form is molecular.

Diamond and graphite

Diamond and graphite are both giant lattices made up of carbon atoms, but the atoms are arranged very differently in the two forms. Look at the structures of diamond and graphite in Figure 12.

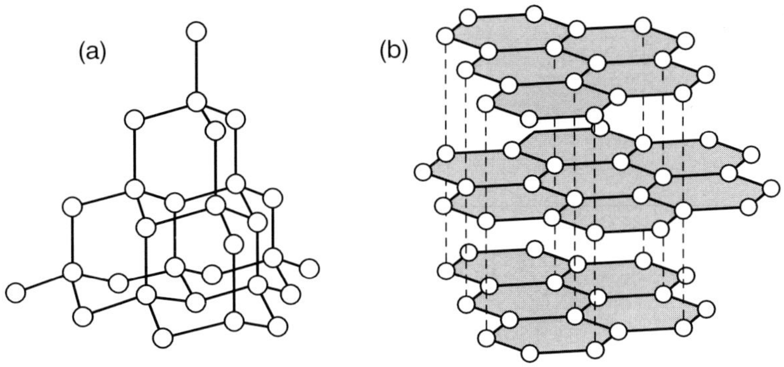

Figure 12 (a) The structure of diamond and (b) the structure of graphite

Note the similarity between the structure of diamond in Figure 12a and the structure of silicon in Figure 11a. Carbon, like silicon, is in Group 4 of the Periodic Table and can form four covalent bonds.

In **diamond**, each carbon atom is joined tetrahedrally to four other carbon atoms by strong covalent bonds forming a giant lattice. This highly symmetrical structure and the strong C—C bonds make diamond the hardest naturally occurring substance. Industrial grade diamonds are used on the tips of drills and crushed diamonds are used on high quality grinding and cutting wheels.

A less expensive alternative to diamond is **silicon carbide**, SiC, also known as carborundum. Silicon carbide is made by heating carbon and silicon together. It has the same structure as diamond but with alternating carbon and silicon atoms. It is similar in hardness and, in its powdered form, it is used to make 'wet and dry' paper, popular with DIY car repair enthusiasts.

The structure of **graphite** is quite different (Figure 12b). It is made up of giant flat layers of carbon atoms. Each layer is a two-dimensional network. In the layers, each carbon atom is joined by strong covalent bonds to three others forming rings containing six carbon atoms. This uses up only three of the four outer shell electrons in each carbon atom. The fourth electron of each atom contributes to the formation of extended delocalised electron clouds lying between the layers. This causes the layers to be held together relatively weakly.

The delocalised electrons are not attached to particular carbon atoms and are free to move over the layer. This allows graphite to be a good conductor of electricity, a property not normally associated with covalent substances. Most covalent substances do not conduct electricity because all the electrons are tightly held in covalent bonds so that they are not free to move from one molecule to the next.

Graphite is a soft, brittle solid. Because the layers are bonded weakly to each other, they can slide over one another. This is why a lump of graphite has a slippery feel to it, almost like wet soap, and also why graphite is used as a lubricant. Pencil 'leads' are made by mixing powdered graphite with clay.

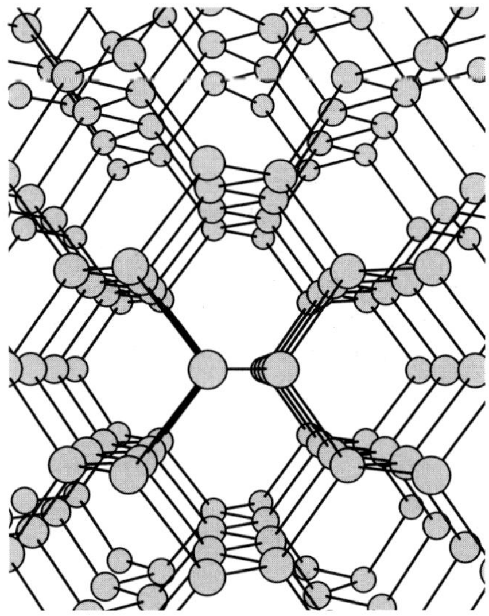

Figure 13 Imagine you are inside a diamond. The regular network structure would repeat in all directions – as far as the edge of the diamond

Fullerenes

In 1985 the existence was announced of a new form of carbon. Trace amounts of the substance were discovered by accident in the soot produced when a high power laser was used to evaporate graphite in an inert atmosphere. Five years later, larger quantities were made by passing an electric arc between graphite rods.

The new form of carbon dissolves in benzene to give a red solution, which suggests it has a molecular rather than a network structure. The molecules were shown to have the formula C_{60}.

The new substance was named **buckminsterfullerene** because the structure of the molecule was similar to a geodesic dome, a type of building designed by the architect Buckminster Fuller. Figures 14a and 14b show the structure of buckminsterfullerene, which looks like a football.

Other carbon molecules with similar structures have since been discovered and they are known collectively as **fullerenes**. Figure 14c shows a C_{70} molecule, which is shaped like a rugby ball. The fullerenes are sometimes referred to as 'buckyballs'. Sir Harry Kroto at the University of Sussex was awarded the Nobel Prize in 1996 for the discovery of the fullerenes (along with Robert Curl and Richard Smalley at the University of Texas).

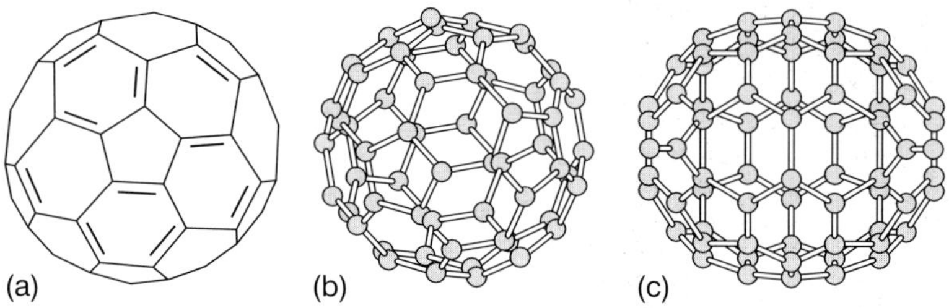

Figure 14 The fullerenes are a recently discovered molecular form of carbon. (a) shows C_{60}, named buckminsterfullerene. It is made up of a mixture of 5-membered and 6-membered rings and looks like a football (b) is another way of representing C_{60}, showing the positions of the carbon atoms (c) shows C_{70}, which is shaped like a rugby ball

Recently, elongated fullerenes looking like tiny molecular tubes have been discovered. These have been called 'buckytubes' or 'nanotubes'. If long versions of these could be manufactured the possibility exists of creating incredibly strong, lightweight fibres.

Fullerenes were discovered so recently that their properties have not been fully established. Research work is taking place to investigate possible large scale uses and there are already hundreds of fullerene patents. It has been discovered, for example, that C_{60} is an *optical limiter*. Shining light on a solution of C_{60} causes it to turn darker instantly, and the more light there is, the darker it gets, so that the intensity of the transmitted light is limited to a maximum value. This property could be put to practical use in the manufacture of protective goggles for people working with lasers.

PROBLEMS FOR 16.2

1 Explain in your own words why CO_2 is a gas at room temperature and pressure, while SiO_2 is solid.

2 Explain the following observations in terms of particles, structure and bonding.

- **a** Substances with covalent network structures are insoluble in all solvents.
- **b** Substances that are gases at room temperature always contain simple molecules or isolated atoms.

3 Silicon carbide, SiC, is a covalent compound that is almost as hard as diamond.

- **a** Draw part of the structure of silicon carbide, showing a total of about 10–15 atoms.
- **b** Suggest why silicon carbide is such a hard substance.

4 Boron nitride, BN, has a similar structure to graphite but it does not conduct electricity.

- **a** Draw part of the structure of boron nitride.
- **b** Suggest why boron nitride does not conduct electricity.

5 Buckminsterfullerene was obtained by adding benzene to the soot obtained when an electric discharge is passed between graphite electrodes. Buckminsterfullerene dissolves in benzene to give a red solution. Explain why this was a startling observation at the time and why it was taken as evidence that buckminsterfullerene has a molecular structure.

16.3 *Ions in solids and solutions*

Ionic solids

In ionic solids, ions are held together by their opposite electrical charges. Each positive ion attracts several negative ions, and vice versa. The ions build up into a giant ionic lattice, in which very large numbers of ions are arranged in fixed positions.

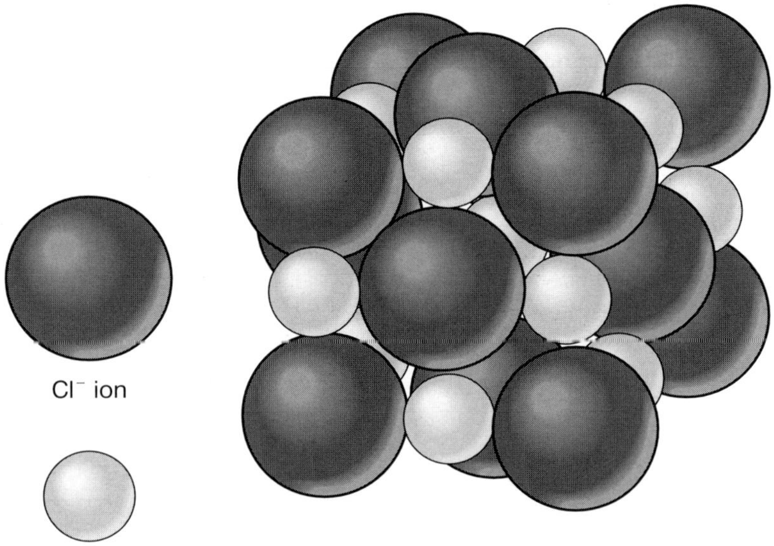

Figure 15 The sodium chloride lattice

One of the simplest examples is sodium chloride, Na^+Cl^-(s) (Figure 15). In the sodium chloride lattice, each Na^+ ion is surrounded by six Cl^- ions, and each Cl^- ion is surrounded by six Na^+ ions. Each Na^+ is attracted to the six Cl^- round it, but repelled by other Na^+ ions which are a bit further away (there are 12 of these), and attracted to the next lot of Cl^- ions (eight of them) which are further away still ... and so on. It adds up to an infinite series of attractions and repulsions, but overall the attractions are stronger than the repulsions, which is why the lattice holds together. Indeed, it holds together very strongly, which is why ionic solids are hard and have high melting and boiling points.

It is important to realise that an ionic solid like Na^+Cl^- isn't just pairs of Na^+ and Cl^- ions, but a huge lattice. You can see from Figure 15 that the lattice has a cubic shape which accounts for the fact that sodium chloride crystals are cubic.

The cubic lattice structure of sodium chloride is also found in some other compounds of Group 1 metals with halogens, such as potassium fluoride and lithium chloride. However, other ionic solids may have more complicated lattices, depending on the number and size of the different ions present.

Ionic substances in solution

Many ionic substances dissolve readily in water. When they do so, the ions become surrounded by water molecules and spread out through the water. Figure 16 illustrates this for sodium chloride. The dissolved ions, Na^+(aq) and Cl^-(aq), are no longer regularly arranged; they are scattered through the water at random. What's more, now the Na^+ and Cl^- are separated, they behave independently of each other. It is as if each has forgotten the other exists.

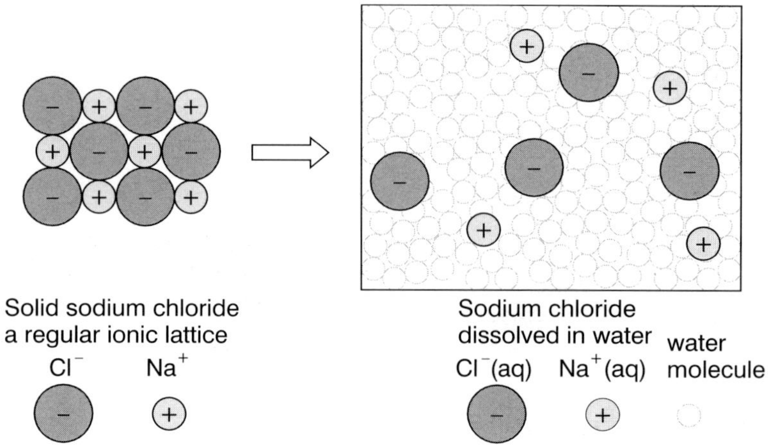

Figure 16 What happens when an ionic substance such as sodium chloride dissolves in water

This applies to all ionic substances. As soon as they are dissolved, the positive and negative ions separate and behave independently. It is best to regard sea water as a mixture of positive and negative ions dissolved in water, rather than a solution of ionic compounds.

You can read about the ions present in sea water in **Section 3.3**.

Ionic equations

Ions in solution behave independently – and this includes their chemical reactions. The reactions of an ionic substance like sodium chloride quite often involve only one of the two types of ion. The other ion does not get involved in the reaction.

For example, if you add a solution of sodium chloride to silver nitrate solution, you get a white precipitate of silver chloride. Silver ions, Ag^+, and chloride ions, Cl^-, have come together to form insoluble silver chloride, which precipitates out (Figure 17). We can write an equation for this reaction. Sodium chloride contains $Na^+(aq)$ and $Cl^-(aq)$, and silver nitrate contains $Ag^+(aq)$ and $NO_3^-(aq)$, so we write

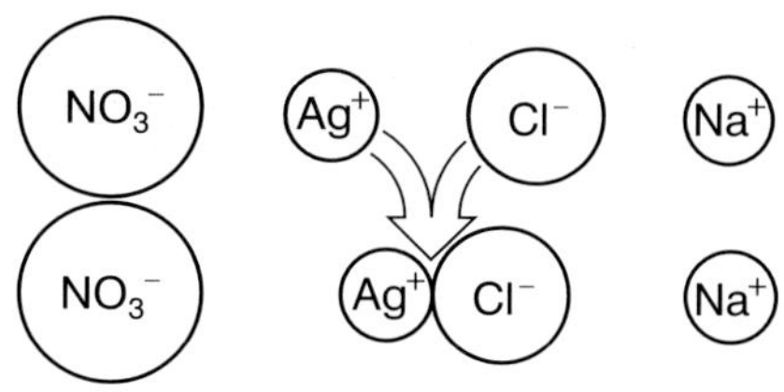

Figure 17 Making a precipitate of silver chloride by mixing silver nitrate and sodium chloride solutions. Only the $Ag^+(aq)$ and $Cl^-(aq)$ ions react

$$\underbrace{Ag^+(aq) + NO_3^-(aq)}_{\text{silver nitrate solution}} + \underbrace{Na^+(aq) + Cl^-(aq)}_{\text{sodium chloride solution}}$$

$$\rightarrow \underbrace{AgCl(s)}_{\text{silver chloride precipitate}} + Na^+(aq) + NO_3^-(aq)$$

Notice that $Na^+(aq)$ and $NO_3^-(aq)$ ions are on both sides of the equation. They do not take part in the reaction – it is as if they are on the sidelines, watching the action going on between $Ag^+(aq)$ and $Cl^-(aq)$ ions without getting involved. They are described as **spectator ions**.

Because $Na^+(aq)$ and $NO_3^-(aq)$ are not involved in the reaction, we can leave them out of the equation altogether. This simplifies the equation so that it only shows the ions that are actually taking part in the reaction:

$$Ag^+(aq) + Cl^-(aq) \rightarrow AgCl(s)$$

This type of equation, showing only the ions that take part in the reaction and excluding the spectator ions, is called an **ionic equation**.

Dissolving ionic solids

Some ionic substances dissolve in water, but others do not. What decides whether an ionic substance will dissolve? One important factor is the energy changes that are involved.

Before an ionic solid such as sodium chloride can dissolve, we need to separate the ions from the lattice so they can spread out in the solution (Figure 16). This means supplying energy to overcome the electrical attraction between the oppositely charged ions.

It takes a lot of energy to pull the ions out of the lattice, but many ionic substances *do* dissolve. So where does this energy come from?

For the moment, let us restrict our thinking to aqueous solutions where the solvent is water. Water molecules possess a dipole (Figure 18). You can find out more about dipoles in **Section 16.1**.

The tiny charges on the water molecules are attracted to the charges on ions. This happens on the surface of an ionic solid which is placed in water, so the ions are pulled into solution (Figure 19).

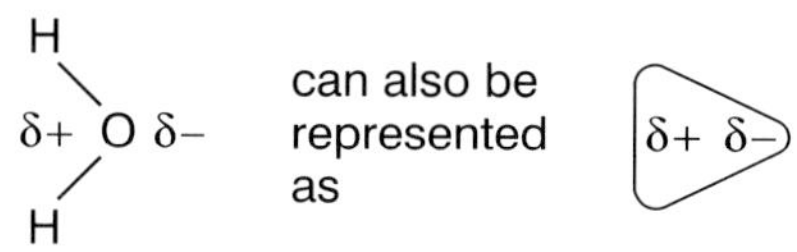

Figure 18 Water molecules possess a dipole

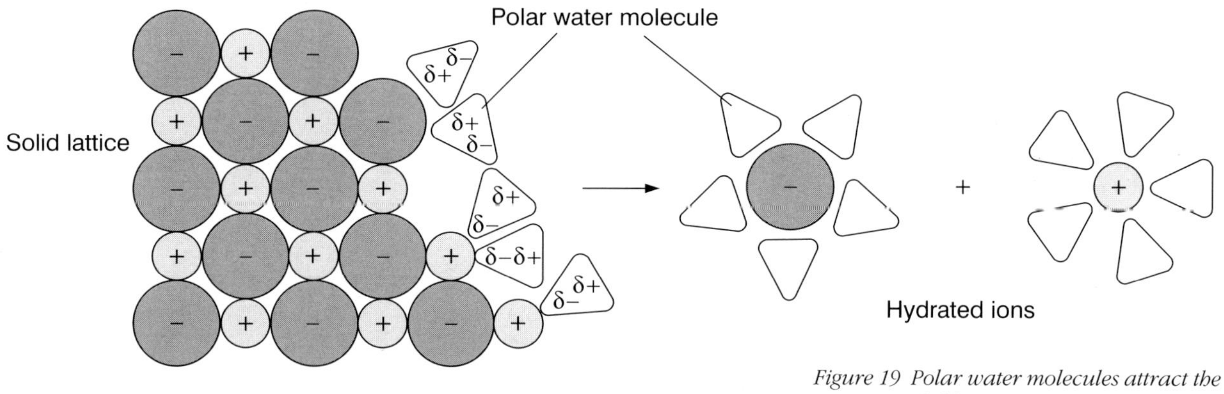

Figure 19 Polar water molecules attract the ions in a solid lattice

The ions in solution are **hydrated** – they have water molecules bound to them. Water molecules bind weakly to some ions, with the result that they are not extensively hydrated. Other ions are extensively hydrated with some water molecules bound very strongly to them.

Table 3 shows approximately how many water molecules are likely to be attached to particular positive ions in solution. You can see that the smaller and more highly charged ions attract larger numbers of water molecules. There can be several layers of water molecules around the ion – this is how Al^{3+} gets 26 water molecules fitted around it.

Ion	Average number of attached water molecules	Ion	Average number of attached water molecules
Li^+	5	Mg^{2+}	15
Na^+	5	Ca^{2+}	13
K^+	4	Al^{3+}	26

Table 3 Approximate extent of hydration for some positive ions

When bonds form between ions and water molecules, energy is released and this may supply enough energy to pull ions out of the lattice. It is not quite as simple as this, though. Water molecules are strongly attracted to each other by intermolecular attractions, called *hydrogen bonds*. (You will find out more about hydrogen bonds in **Section 16.4**.) When ions dissolve in water, some of the water molecules must be pulled apart so that they can regroup around the ions. This process requires energy too. It is the balance between a number of factors that decides whether the solid dissolves or not.

What about other solvents?

Molecules of some other solvents, such as ethanol, are also polar and can bind to ions, so some ionic solids will dissolve in ethanol.

Ionic solids, however, are insoluble in *non-polar solvents* such as hexane. The molecules in non-polar solvents have no regions of slight positive and negative charge, so they are unable to interact strongly with ions. There is not enough energy to pull the ions out of the lattice and so the ionic substance does not dissolve.

Can we predict whether a solid will dissolve in a solvent?

The answer is generally 'no' – you would need to do a test in the laboratory. Energy changes are important but they are not the only factor governing whether a solid will dissolve. What we can do is to predict the *type* of solvent the solid is *most likely* to dissolve in. A useful general rule is:

like dissolves like

Ionic compounds tend to be soluble in polar solvents such as water, but insoluble in non-polar solvents such as hexane. The same is true for polar covalent compounds with molecular structures, such as glucose.

Non-polar molecular solids, such as wax, tend to be soluble in non-polar solvents such as hexane, but insoluble in water. When a non-polar covalent solid is added to water, water molecules are only weakly attracted to the molecules of the solid. This does not provide sufficient energy to allow the water molecules to pull apart from each other and rearrange themselves around the molecules of the solid. The substance usually does not dissolve.

The reason that non-polar substances dissolve in non-polar solvents is because the energy processes involved are balanced. The molecules in the non-polar solid are held together by weak van der Waals forces so only a small amount of energy is needed to separate them. The solvent molecules, too, have van der Waals forces holding them together so only a little energy is required to reorganise them around the molecules of substance. The energy for these processes is provided by the energy released when the solvent molecules become weakly attracted to the molecules of the substance being dissolved – again these are van der Waals forces.

Solid	Solvent	Likely to dissolve?
ionic compounds and polar covalent molecules	polar non-polar	yes no
non-polar covalent molecules	polar non-polar	no yes

Table 4 Rules for predicting possible solubility

You can read more about van der Waals forces in **Section 16.1**.

PROBLEMS FOR 16.3

1 Many of the compounds of Group 2 elements are insoluble in water. Write ionic equations (with state symbols) for the precipitates formed when the following solutions are mixed.

Aqueous solutions mixed	Precipitate (colour)
a barium chloride, $BaCl_2$, and sodium sulphate, Na_2SO_4	barium sulphate (white)
b magnesium sulphate, $MgSO_4$, and sodium hydroxide, NaOH	magnesium hydroxide (white)
c calcium chloride, $CaCl_2$, and sodium carbonate, Na_2CO_3	calcium carbonate (white)
d barium nitrate, $Ba(NO_3)_2$, and potassium chromate, K_2CrO_4	barium chromate (yellow)

2 The lattice structure of caesium chloride is shown below. A caesium ion lies at the centre of a cube of chloride ions, and a chloride ion lies at the centre of a cube of caesium ions.

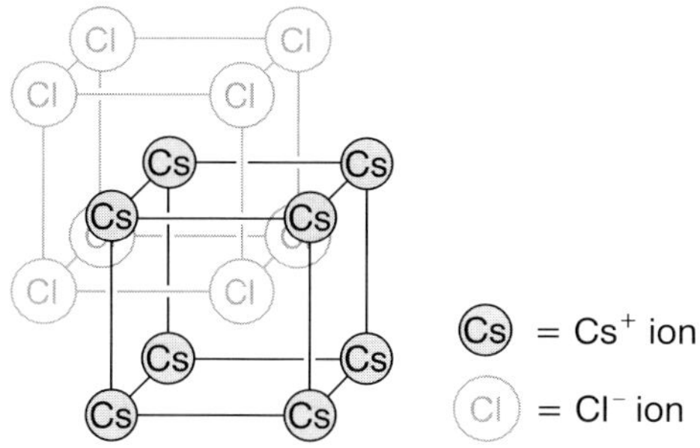

a i Predict the shape of caesium chloride crystals.
ii How many chloride ions closely surround one caesium ion?
iii How many caesium ions closely surround one chloride ion?

b Explain why caesium chloride is hard and has a high melting point (645°C).

3 Suggest why
a potassium bromide is soluble in water but insoluble in hexane
b calcium carbonate is insoluble in both water and hexane
c candle wax (made up of long-chain hydrocarbon molecules) is insoluble in water but soluble in hexane.

4 Write ionic equations (with state symbols) for the following reactions:
a hydrochloric acid with sodium hydroxide solution
b zinc metal with sulphuric acid
c copper oxide powder with dilute nitric acid
d marble chips with hydrochloric acid.

16.4 *Forces between molecules: hydrogen bonding*

Unusual properties of water

Many of the properties of water are unusual when we compare them with those of other similar compounds of low molecular mass. In particular water shows

- unusually high values for its boiling point and enthalpy change of vaporisation
- a greater specific heat capacity than almost any other liquid
- a *decrease* in density when it freezes.

The graph in Figure 20 shows how the boiling points of hydrides change as you go down a group in the Periodic Table.

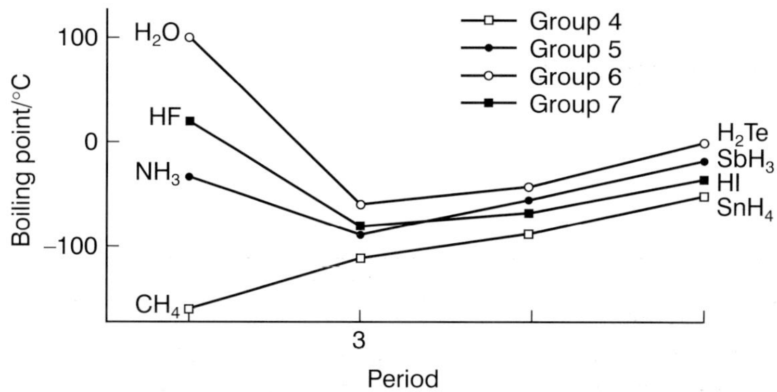

Figure 20 Variation in the boiling points of the hydrides of some Group 4, 5, 6 and 7 elements

The data for Group 4 (CH_4 to SnH_4) show how we might *expect* the boiling points to behave. The pattern corresponds to what we know about van der Waals forces and the polarisability of atoms. (These were discussed in **Section 16.1**.) The points for H_2O, NH_3 and HF are clearly out of line.

The unusual behaviour of water, ammonia and hydrogen fluoride can be explained in terms of the unusually strong attractive forces between their molecules. These attractive forces are known as **hydrogen bonding.** Hydrogen bonding holds molecules together more strongly than other types of intermolecular forces.

What is special about hydrogen bonding?

Hydrogen bonding is a particularly strong intermolecular force that involves three features:

- a *large dipole* between an H atom and a highly electronegative atom such as O, N or F
- the *small H atom* which can get very close to other atoms
- a *lone pair of electrons* on another O, N or F atom, with which the positively charged H atom can line up.

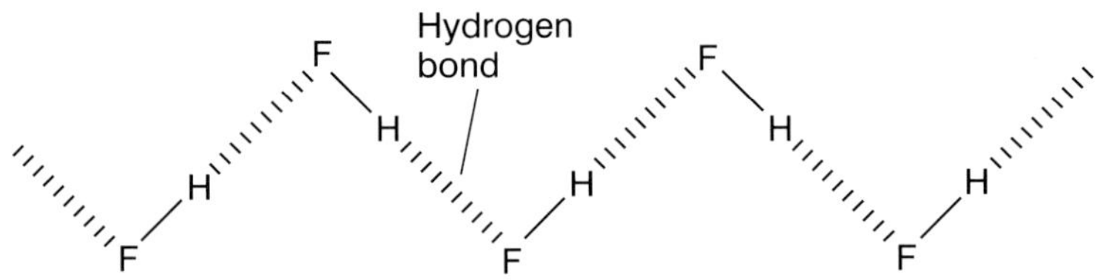

Figure 21 Hydrogen bonding in liquid hydrogen fluoride

Figure 21 illustrates hydrogen bonding in liquid hydrogen fluoride, HF. The H atoms have a strong positive charge because of the highly electronegative F atom to which they are bonded. These positive charges line up with a lone pair on another F atom, which provides a region of concentrated negative charge (Figure 22). The H and F atoms can get very close, and therefore attract very strongly, because the H atom is so small. The lining up of HF molecules is important because it means that positive and negative charges are always lined up with one another, so the forces are always attractive. As a result, the HF molecules are held together in the molecular chains you see in Figure 21.

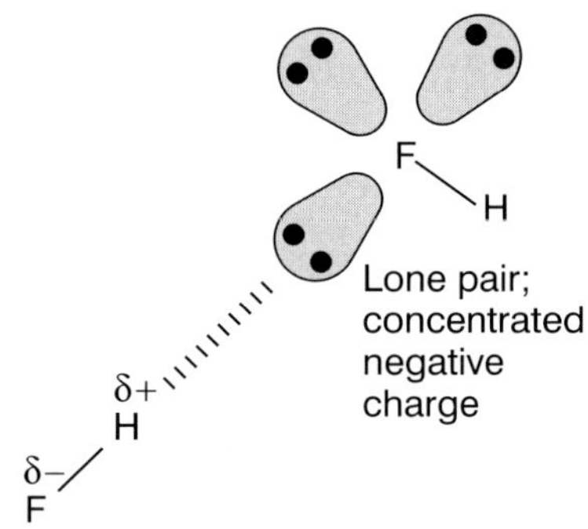

Figure 22 The positively charged H atom lines up with the lone pair on an F atom

Hydrogen bonds are much stronger than the other kinds of intermolecular forces that we have met (van der Waals and permanent dipole–permanent dipole forces).

Dipole–dipole forces are present in all the hydrogen halides (HF, HCl, HBr and HI), but the greater strength of the hydrogen bonds in HF can be seen from a comparison of the boiling points. Look at Table 5.

HF has the highest boiling point despite having the lowest relative formula mass. You can see these values plotted on a graph in Figure 20.

Intermolecular attractive forces affect the melting point and boiling point of a covalent substance because energy is needed to overcome these forces and to separate molecules when the substance melts and then boils. If the attractive forces *between molecules* are strong, then the melting point and boiling point of the substance tend to be high – whereas if forces are weak, the melting point and boiling point will tend to be low. In the case of hydrogen fluoride, the strong hydrogen bonds give it an exceptionally high boiling point compared with the other hydrogen halides.

Compound	Boiling point/°C
HF	20
HCl	–85
HBr	–67
HI	–35

Table 5 Boiling points of the hydrogen halides

The hydrogen bonding in water is shown in Figure 23. Notice that water molecules can form twice as many hydrogen bonds as hydrogen fluoride. The oxygen atom possesses two lone pairs of electrons with which positive hydrogen atoms can interact. There are twice as many hydrogens as oxygens – just the right number to maximise the bonding to the lone pairs (Figure 24).

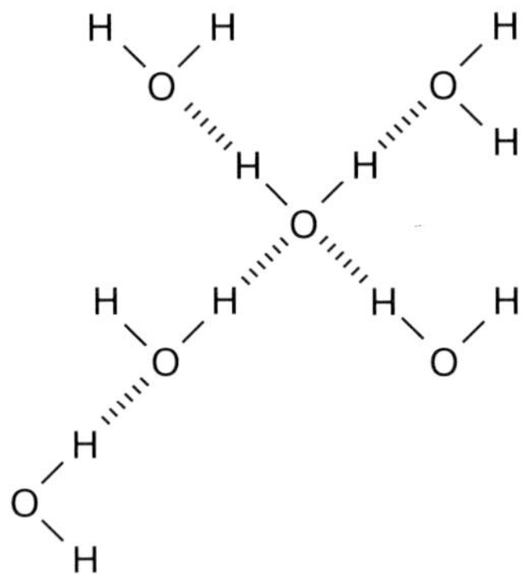

Figure 23 Hydrogen bonding between water molecules

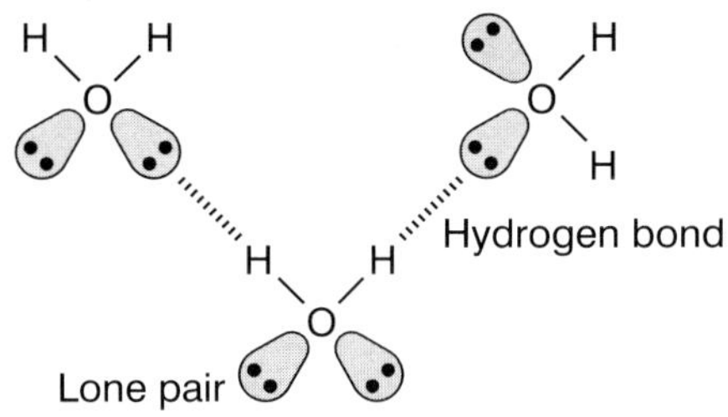

Figure 24 The positively charged H atoms line up with the lone pairs on the O atoms

Water is unique in this respect. In HF the fluorine has three lone pairs but there are only as many H atoms as F atoms. So only one-third of the available lone pairs can be used. In NH_3, another hydrogen-bonded substance, there is only one lone pair on the N so, on average, only one of the three hydrogen atoms can form hydrogen bonds.

To summarise

For hydrogen bonding to occur there needs to be a hydrogen atom made positively charged by being attached to a highly electronegative atom or group of atoms. There also needs to be a small, highly electronegative atom with at least one lone pair of electrons for the H atom to interact with. N, O and F atoms satisfy these conditions (Figure 25).

X = F, O or N
Y = F, O or N

Figure 25 Hydrogen bonds only form between particular kinds of atoms

Hydrogen bonds are stronger than other intermolecular forces, but much weaker than covalent bonds. A typical covalent bond is about 15 times stronger than a hydrogen bond.

The effects of hydrogen bonding

Many covalent substances have N—H and O—H bonds in their molecules. For example, alcohols and organic acids have O—H bonds, while amines and amides have N—H bonds. Amino acids and proteins have both types of bonds. All these substances form hydrogen bonds and hydrogen bonding plays a major role in determining their properties.

Look at Table 6. The two substances with hydrogen bonding have higher boiling points than the substances with similar relative formula masses that do not. Propane ($CH_3CH_2CH_3$) is a non-polar molecule and there are only van der Waals forces between its molecules. Ethanol (CH_3CH_2OH) is a polar molecule but, in addition to van der Waals forces and permanent dipole–permanent dipole forces, it has hydrogen bonding between its molecules. More energy is needed to overcome the intermolecular forces between ethanol molecules than between propane molecules and so the boiling point of ethanol is much higher.

Compound	Relative formula mass	Hydrogen bonding?	Boiling point/°C
propane	44	no	−42
ethanol	46	yes	78
octane	114	no	126
glycerol	92	yes	290

Table 6 The effect of hydrogen bonding on boiling point

Liquids which have hydrogen bonding have higher than expected *viscosity*. The viscosity of a liquid is a measure of how easily it flows. For a liquid to flow, the molecules must be able to move past each other, which requires the constant breaking and forming of intermolecular attractions. The stronger the attractions between the molecules, the more difficult this becomes. Glycerol is an example of a viscous liquid containing a lot of hydrogen bonding.

Substances with hydrogen bonding, such as glycerol and ethanol, are often soluble in water. Hydrogen bonds can form between water molecules and molecules of the substance and this helps the dissolving process. Ethanol and glycerol are both **alcohols** and contain —OH groups. You can learn more about alcohols in **Unit 2**, **Section 18.1**.

Hydrogen bonds are responsible for the strong intermolecular forces between the polymer chains of many fibres. In nylon, for example, there are interactions between the H atoms of the N—H groups and the O atoms of the C=O groups.

The importance of hydrogen bonding in nylon and Kevlar is described in **Unit 2**, **Chapter 5**.

Hydrogen bonding also helps fibres to absorb water. The protein chains in wool contain lots of O and N atoms that can hydrogen bond to water, which means that wool can absorb sweat and doesn't feel sticky to wear. Polythene, on the other hand, cannot form hydrogen bonds at all, so polythene clothes would feel very damp and sticky.

Water: hydrogen bonding in action

The hydrogen bonding in water is particularly strong because there are two lone pairs of electrons *and* two positively charged hydrogen atoms per oxygen atom, so intermolecular interactions are maximised (see Figure 24). Hydrogen bonding affects the properties of water in many ways which are vital to our environment and to life on Earth.

You can read about the unique role of the oceans in **Chapter 3**.

Specific heat capacity

Water has a high specific heat capacity; it takes a lot of energy to raise its temperature. Its value, together with those for some other substances, is given in Table 7.

Substance		Specific heat capacity/kJ kg^{-1} °C^{-1}
water,	$H_2O(l)$	4.18
ethanol,	$C_2H_5OH(l)$	2.43
heptane,	C_7H_{16}	2.05
copper,	Cu(s)	0.39
mercury,	Hg(l)	0.14

Table 7 Some specific heat capacities

To raise the temperature of 1 kg of copper by 1 °C requires 0.39 kJ. To do the same for water requires over ten times this quantity of energy. Water is an excellent substance for absorbing and storing energy, which is just as well for life in and around the oceans – and for the manufacturers of hot water bottles!

Water's unusually high specific heat capacity is another consequence of its hydrogen bonding. A lot of energy is used in overcoming hydrogen bonding within clusters of water molecules. This energy is not available for increasing the molecules' kinetic energies; it is not available for raising the temperature. For liquids where there is no hydrogen bonding between molecules, a greater proportion of the energy can be used for increasing the molecules' kinetic energies, and so raising the temperature; thus such liquids have lower heat capacities.

The density of water and ice

If we take a tube of water at room temperature and cool it to 4 °C, the water contracts. Most substances contract on cooling, so this is not unexpected. What is unusual is what happens next; the water *expands* as its temperature approaches 0 °C, and there is a further expansion as water freezes.

Figure 26a shows how the *density* of water changes with temperature around the freezing point. A similar graph for heptane is shown in Figure 26b. The changes in the density of heptane are typical of most solids and liquids. The changes for water are very unusual, and are due to the effects of hydrogen bonding.

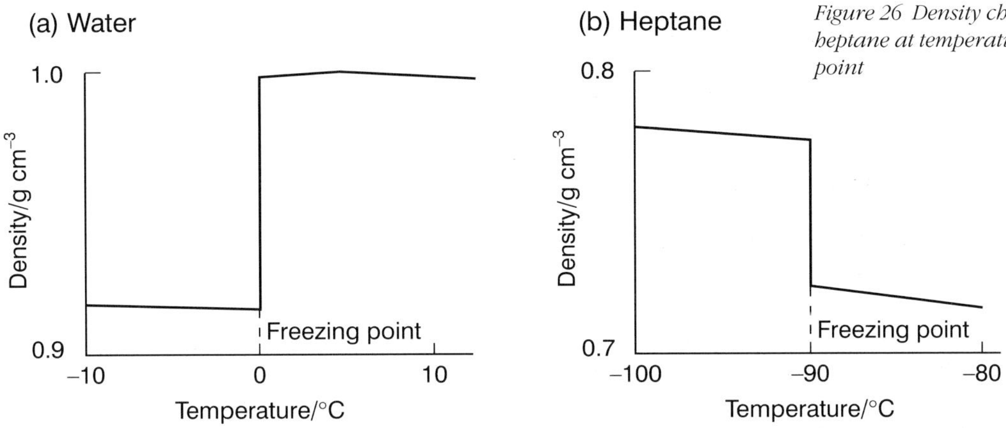

Figure 26 Density changes in water and heptane at temperatures around the freezing point

The structure of ice is shown in Figure 27. Note that the four groups around each O atom are arranged tetrahedrally in three dimensions. The arrangement of water molecules maximises the hydrogen bonding between

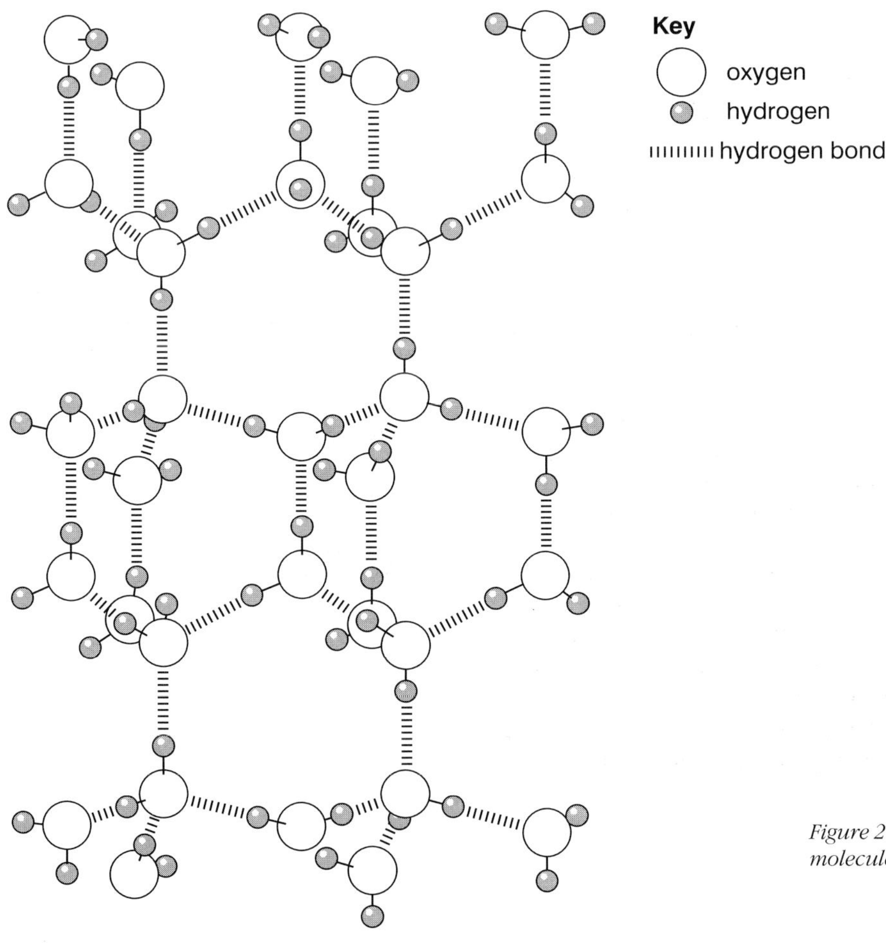

Figure 27 The arrangement of water molecules in ice

them, but leads to a very 'open' structure with large spaces in it. Therefore, the density of ice at 0 °C is less than the density of water at the same temperature. When ice melts, the open structure collapses and water molecules fall into some of the open spaces, thus giving water a greater density than the ice it came from.

When water is cooled, it seems that regions in the liquid begin to adopt the open structure of ice before the freezing point at 0 °C is reached. This explains why the fall in density begins as the liquid is cooled below 4 °C.

The lower density of ice is bad news for people who fail to insulate exposed water pipes in cold weather or who forget to add anti-freeze to the coolant system of their car. But it is good news for fish. Icebergs float and ponds freeze from the top down. The layer of surface ice insulates the water underneath. If ice sank there would be no insulation; ponds and lakes would be more likely to freeze solid, killing fish and other creatures. An occasional burst pipe is a small price to pay for the environmental advantages of the expansion of freezing water.

PROBLEMS FOR 16.4

1 a Why does the density of most solids or liquids decrease with increasing temperature?

b Explain the following properties of water in terms of hydrogen bonding:

i the greater density of liquid water compared with ice;

ii its relatively high boiling point;

iii its relatively high specific heat capacity;

iv the fact that a needle can float on the surface of water. (You can try this for yourself. Float a piece of tissue on the surface of some water in a bowl. Place a needle carefully on top of the tissue. After a few minutes the tissue should sink, leaving the needle floating on the surface until it is disturbed.)

2 a The graph below shows the boiling points for the hydrides of Group 6 elements (H_2O, H_2S, H_2Se, H_2Te). Explain why the boiling point of water appears to be abnormally high.

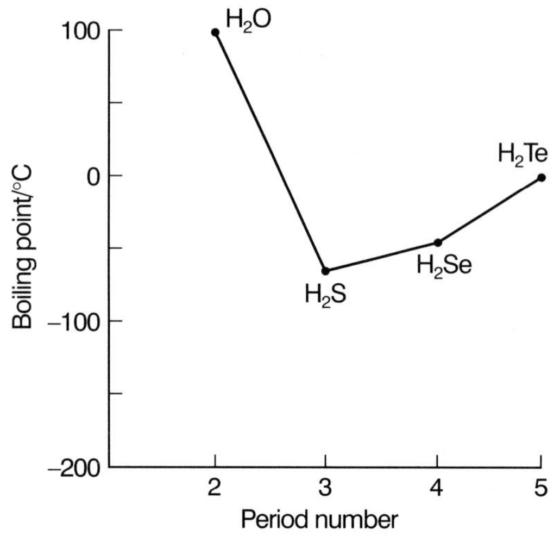

b Why does the shape of the graph for the Group 6 elements suggest that H_2O has a different type of intermolecular bonding from H_2S, H_2Se and H_2Te?

c The boiling points of the hydrides of Group 4 elements are

CH_4 −161 °C; SiH_4 −112 °C; GeH_4 −88 °C; SnH_4 −52 °C

Plot these on a graph similar to that for the Group 6 hydrides. Explain the trend in boiling points for these compounds.

3 Hydrogen bonding can occur between *different* molecules *in mixtures*. Draw diagrams to show where the hydrogen bonds form in the following mixtures.

a NH_3 and H_2O

b CH_3CH_2OH and H_2O

4 Look carefully at the structural formulae of the substances given below. (Some of them will be unfamiliar to you.) Suggest which of these substances will be able to form hydrogen bonds.

A 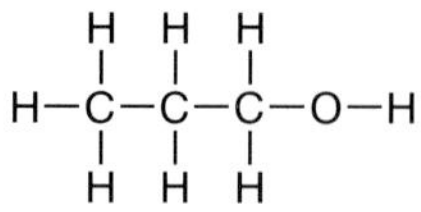

B 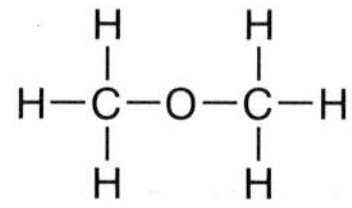

C 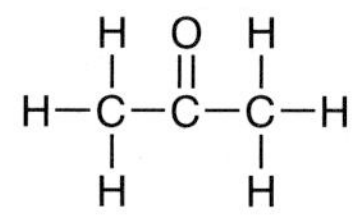

D 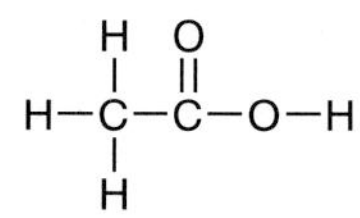

E 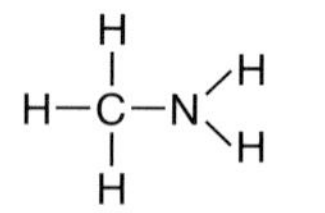

F 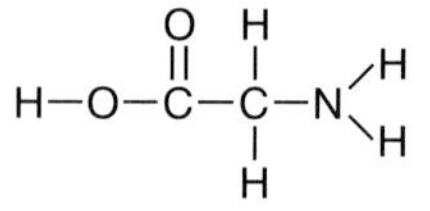

G 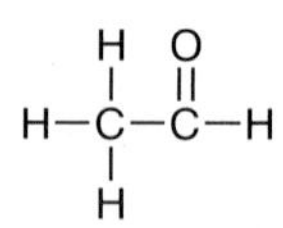

5 In this question you will need to use your knowledge of all the types of intermolecular forces you have met in **Sections 16.1** and **16.4**. List the types of intermolecular forces that will exist in each of the following compounds. The types of forces to choose from are
- van der Waals forces
- permanent dipole–permanent dipole forces
- hydrogen bonding.

(Remember, more than one type of intermolecular force may operate for each molecule.)

a NH_3

b Br_2

c

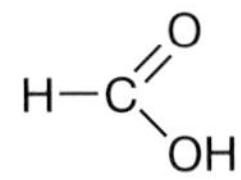

d O=C=O

e

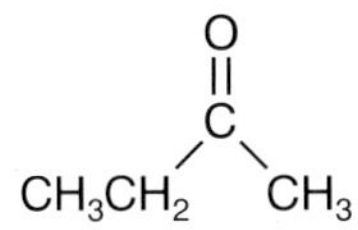

16.5 *Bonding, structure and properties: a summary*

The properties of substances are decided by their **bonding** and **structure**. Bonding means the way atoms are held together. You met ionic, covalent and metallic bonding in **Chapter 13**. You have learned about weak intermolecular bonding in this chapter (**Sections 16.1** and **16.4**).

Structure means the way the atoms are arranged relative to one another. You have already met the major types of structure in different parts of this unit. They are summarised in Figure 28.

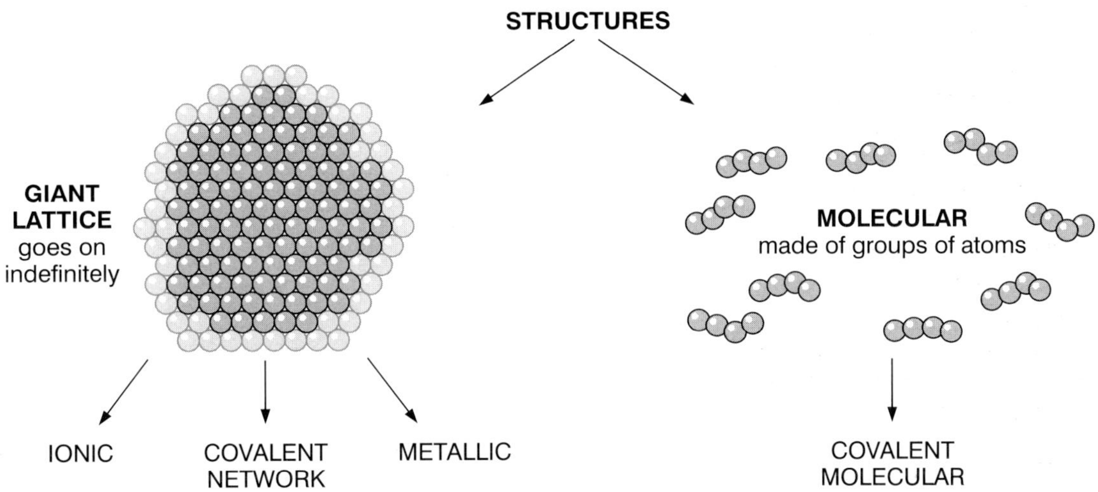

Figure 28 A summary of the structures of substances

How do bonding and structure decide properties?

The properties of a solid substance are decided by three main factors:

- *the type of particles it contains*; these may be atoms, ions or molecules.

  For example, if the substance contains ions (as in an ionic substance such as sodium chloride), it will conduct electricity when molten or dissolved. If a substance contains ions or polar molecules, it may dissolve in water.

- *the way the particles are bonded together*; this may be ionic, covalent, metallic or weak intermolecular bonds. The stronger the bonds, the higher the melting point and boiling point of the substance, and the greater its hardness.

  For example, quartz, SiO_2, has strong covalent bonds linking every atom to several others in a network structure that goes on indefinitely. This makes the atoms difficult to separate; quartz is hard, and very difficult to melt. On the other hand, although carbon dioxide, CO_2, has strong covalent bonds between C and O atoms – so it doesn't decompose into carbon and oxygen – it has only weak intermolecular forces between each CO_2 molecule. This makes the molecules easy to separate, so CO_2 has a low melting and boiling point and is a gas at room temperature.
- *the way the particles are arranged relative to one another* (the structure); the particles may be arranged in one-dimensional chains (such as poly(ethene)), two-dimensional sheets (such as clays) and many different kinds of three-dimensional arrangements.

  For example, clays with their two-dimensional sheets are flaky and slippery, while quartz with its three-dimensional network structure is very hard and gritty.

Table 8 summarises the different types of structures and their main properties. It also tells you where you can find out more details about each type.

Structure and the Periodic Table

As you go across a period of the Periodic Table, there is a trend in the structures of the elements.

Figure 29 shows the first 20 elements as they are arranged in the Periodic Table and the type of structure shown by each element. Remember that carbon exists in three different forms. Two of these (diamond and graphite) are covalent network structures, whereas the fullerenes are molecular. Note the trend across each period from metallic structures on the left, to covalent network structures in the centre, to covalent molecular structures and finally to the monatomic noble gases.

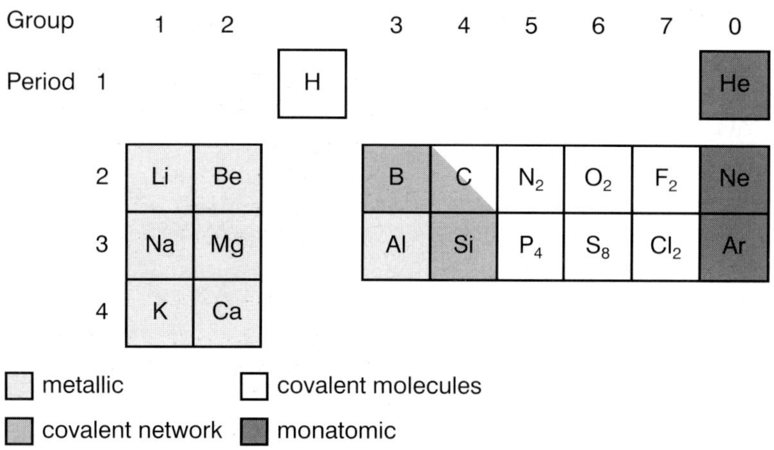

Figure 29 Trends in the structures of the first 20 elements in the Periodic Table

	GIANT LATTICE			MOLECULAR
	Ionic	**Covalent**	**Metallic**	**Covalent**
What substances have this type of structure?	compounds of metals with non-metals	some elements in Groups 3 and 4 and some of their compounds	metals	some non-metal elements and some non-metal/non-metal compounds
Examples	sodium chloride, NaCl; calcium oxide, CaO	quartz, SiO_2; diamond, C; graphite, C	sodium, Na; aluminium, Al	carbon dioxide, CO_2; chlorine, Cl_2; water, H_2O
What type of particles does it contain?	ions	atoms	positive ions surrounded by delocalised electrons	small molecules
How are the particles bonded together?	strong ionic bonds; attraction between oppositely charged ions	strong covalent bonds; attraction of atoms' nuclei for shared electrons	strong metallic bonds; attraction of atoms' nuclei for delocalised electrons	weak intermolecular bonds between molecules; strong covalent bonds between the atoms within each molecule
Melting point and boiling point	high	very high	generally high	low
Hardness	hard but brittle	very hard (if three-dimensional)	hard but malleable	soft
Electrical conductivity	conduct when molten or dissolved in water; electrolytes	do not normally conduct	conduct when solid or liquid	do not conduct
Solubility in water	often soluble	insoluble	insoluble (but some react)	usually insoluble, unless molecules can hydrogen bond with water
Solubility in non-polar solvents (eg hexane)	insoluble	insoluble	insoluble	usually soluble
Where can you find out more?	**Sections 13.1** and **16.3**	**Sections 13.2** and **16.2**	**Section 13.3**	**Sections 13.2**, **16.1**, **16.2** and **16.4**

Table 8 A summary of types of structure and their typical properties

PROBLEMS FOR 16.5

1 Which type of structure would you expect each of substances A to D to have?

- **A** A grey solid which conducts electricity and melts at 850 °C
- **B** A white solid which melts at 770 °C and conducts electricity when molten but not in the solid state.
- **C** A grey solid which melts at 1410 °C and which does not conduct electricity in either solid or molten state.
- **D** A white solid which melts at –190 °C.

2 Explain the following in terms of particles, structure and bonding.

- **a** Ionic substances do not conduct electricity when they are solid, but metallic substances do.
- **b** Substances with covalent network structures generally do not conduct electricity, except for graphite which is a good conductor.

3 The structure of ice is very similar to diamond. Explain why a diamond is harder than an ice cube.

4 What type of structures would you expect each of the following substances to have in the solid state?

- **a** Argon
- **b** Silk
- **c** Silicon carbide
- **d** An alloy of magnesium and aluminium
- **e** Rubidium bromide
- **f** Sulphur hexafluoride

5 **a** Explain why the type of bonding in a compound is related to the positions of its constituent elements in the Periodic Table.

b Name the type of bonding and the type of structure you would expect for the following compounds.

- **i** $CaCl_2$
- **ii** PCl_3
- **iii** SiC
- **iv** MgO
- **v** C_2H_6

17

ORGANIC CHEMISTRY: HYDROCARBONS

17.1 *Alkanes*

Many carbon compounds are found in living organisms, which is why their study got the name **organic chemistry**. Today, organic chemistry includes all carbon compounds whatever their origin, except CO, CO_2 and the carbonates, which traditionally are included in inorganic chemistry.

Only carbon can form the diverse range of compounds necessary to produce the individuality of living things.

> **Carbon is unique**
>
> About seven million compounds containing carbon and hydrogen are known to chemists. This is far more than the number of compounds from all the other elements put together.

Why carbon?

Carbon's electron structure is shown in Figure 1. This electron structure makes it the first member of Group 4 in the centre of the Periodic Table, and is responsible for its special properties.

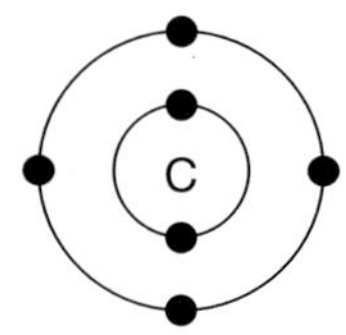

Figure 1 The electron structure of carbon

A carbon atom has 4 electrons in its outer shell. It could achieve a stable noble gas electron arrangement by losing or gaining 4 electrons; but this is too many electrons to lose or gain. The resulting carbon ions would have charges of +4 or –4, respectively, and would be too highly charged. So when carbon forms compounds, the bonds are *covalent* rather than ionic.

In methane (CH_4), for example, the carbon atom achieves a stable octet of 8 electrons by *sharing* its outer electrons with four hydrogen atoms (Figure 2). Four C—H covalent bonds are formed.

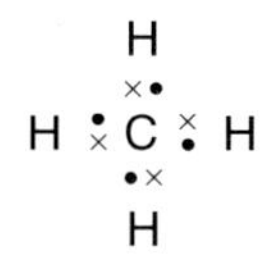

Figure 2 Covalent bonds in methane

Carbon forms strong covalent bonds with itself to give *chains* and *rings* of its atoms joined by C—C covalent bonds. This property leads to the limitless variety of organic compounds possible.

Each carbon atom can form 4 covalent bonds, so the chains may be straight or branched and can have other atoms or groups substituted in them.

Homologous series and functional groups

Chemists cope with the vast number of organic compounds by dividing them into families of related compounds, such as the alkanes and alkenes which you have already come across in your earlier studies. These families are known as **homologous series**. All the members of a series have the same general molecular formula and each member differs from the next by a —CH_2— unit. All the compounds in a series have similar chemical properties, so chemists can study the properties of the group rather than those of individual compounds. However, physical properties such as melting point, boiling point and density change gradually in the series as the number of carbon atoms in the molecules increases.

Hydrocarbons are relatively unreactive – this is particularly true of alkanes and aromatic hydrocarbons, such as benzene. They form the unreactive framework of organic compounds. But when you attach other groups, such as OH, to the hydrocarbon framework, its properties are modified. So we can think of organic compounds as having hydrocarbon frameworks with **functional groups** attached.

A functional group is a group of atoms within a molecule which gives the molecule its characteristic chemical properties. In alkenes, for example, the functional group is the C=C double bond. You'll come across more functional groups as you work your way through this chapter and the next.

Alkanes

Alkanes are **saturated** hydrocarbons. 'Saturated' means that they contain the maximum amount of hydrogen possible, with no double or triple bonds between carbon atoms.

The general molecular formula of the alkanes is C_nH_{2n+2} where $n = 1, 2, 3 \ldots$. Table 1 shows the names and formulae of some simple alkanes.

Look at the names in Table 1. The first part indicates the number of carbon atoms present. *Alk-* is the general term for *meth-*, *eth-*, *prop-* and so on. The second part *-ane* shows that the molecules are saturated.

n	Molecular formula	Name
1	CH_4	methane
2	C_2H_6	ethane
3	C_3H_8	propane
4	C_4H_{10}	butane
5	C_5H_{12}	pentane
6	C_6H_{14}	hexane
7	C_7H_{16}	heptane
8	C_8H_{18}	octane

Table 1 Some simple alkanes

The names of all alkanes end in **-ane.**

Structure of alkanes

Figure 2 shows a dot-cross formula for methane. It shows all the *outer* electrons in each atom and how electrons are shared to form the covalent bonds. For larger molecules, dot-cross formulae are rather cumbersome and the shared electron pairs can be replaced by lines representing the covalent bonds (Figure 3).

This is called a **full structural formula** and it shows all the atoms and all the bonds in the molecule.

We can also write an abbreviated version known as a **shortened structural formula**. Take heptane, for example. Its full structural formula is

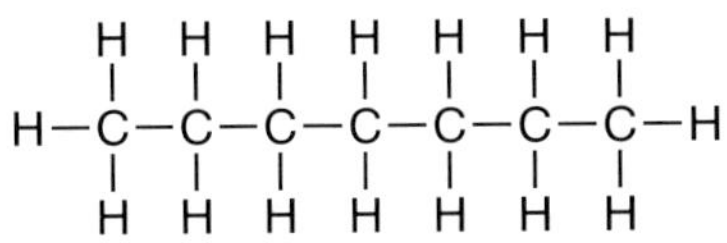

while its shortened structural formula is

$$CH_3—CH_2—CH_2—CH_2—CH_2—CH_2—CH_3$$

or

$$CH_3CH_2CH_2CH_2CH_2CH_2CH_3$$

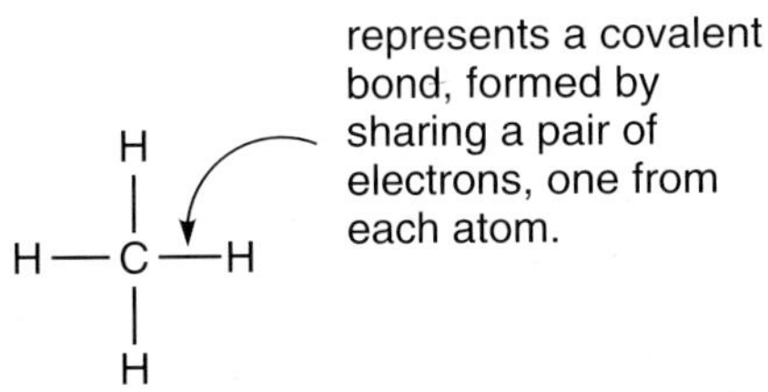

Figure 3 Using lines to represent covalent bonds

Look at Table 2. It gives the full structural formulae and also shortened formulae, for some alkanes. Note that each carbon atom is bonded to four other atoms. Each hydrogen atom is bonded to only one other atom.

Name	Molecular formula	Full structural formula	Shortened structural formula
methane	CH_4	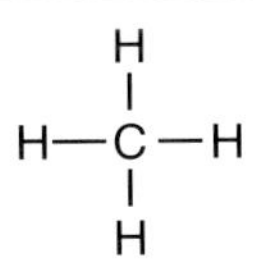	CH_4
ethane	C_2H_6	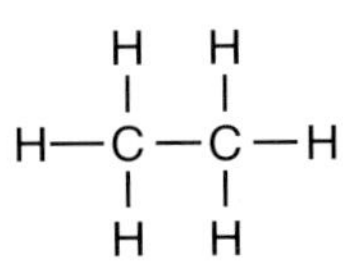	CH_3-CH_3 further shortened to CH_3CH_3
propane	C_3H_8	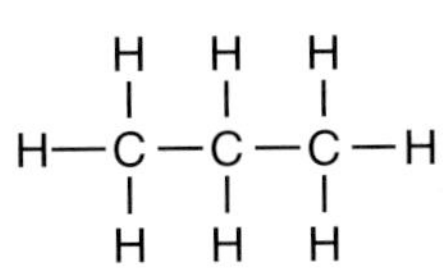	$CH_3-CH_2-CH_3$ further shortened to $CH_3CH_2CH_3$

Table 2 Structural formulae of alkanes

Branched alkanes

Alkanes may have straight chains or branched chains. So it is often possible to draw more than one structural formula for a given molecular formula. There is often a straight-chain compound and one or more branched-chain compounds with the same molecular formula.

For C_8H_{18} there are 17 possible structural formulae. Here are just two of them.

A

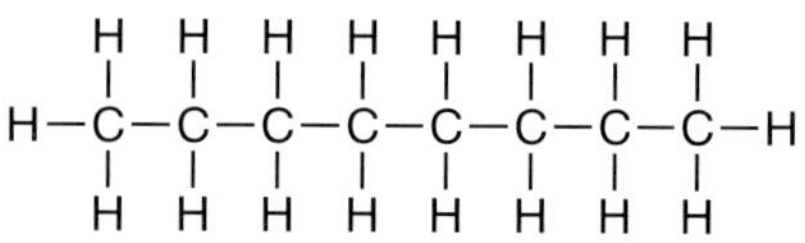

B

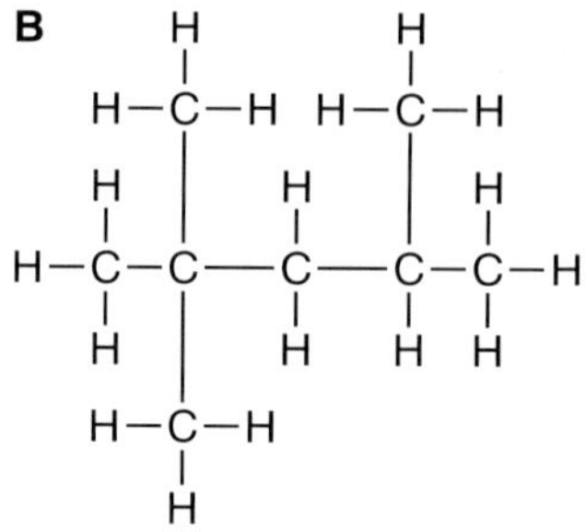

These two compounds are **isomers** because they have the same molecular formulae but different structural formulae. There is more about isomerism in **Section 17.2**.

The straight-chain isomer, **A**, is called *octane*; but how do we name the branched-chain isomer, **B**? The common name for **B** is *iso-octane*, but unless we knew the structure beforehand we could not write down its structural formula just from this name. It could be structure **B**, but it could equally well be any one of the other 15 isomeric structures. This is why it's important that every organic compound has a **systematic** name derived from an internationally agreed set of rules. So, given the systematic name of a compound and given a knowledge of the rules we are able to draw its structural formula, and vice versa.

Look again at isomer **B**. We can regard it as formed from a straight-chain alkane of five C atoms with three —CH_3 groups or branches attached.

The –CH_3 group is just methane with a hydrogen atom removed so that it can join to another atom. It is called a **methyl group**.

Groups of this kind are called **alkyl groups**. They have the general formula C_nH_{2n+1} and are often represented by the symbol **R** (see Table 3).

Alkyl group (R)	Formula
methyl	CH_3—
ethyl	CH_3CH_2—
propyl	$CH_3CH_2CH_2$—
butyl	$CH_3CH_2CH_2CH_2$—
pentyl	$CH_3CH_2CH_2CH_2CH_2$—

Table 3 Some common alkyl groups

> An alkyl group R has the general formula C_nH_{2n+1}.

The rules for naming branched-chain alkanes are as follows:

1 Pick out the longest carbon chain to get the parent name of the compound.

For B, the parent name is pentane since the longest chain contains 5 C atoms.

2 Number the C atoms in this chain starting at the end nearer a branch.

For **B**, we hit a problem because we have a branch on the second C atom in from the left and also one on the second C atom in from the right. In such cases, we ignore these two branches and apply rule 2 again. So for **B**, numbering starts at the left-hand end.

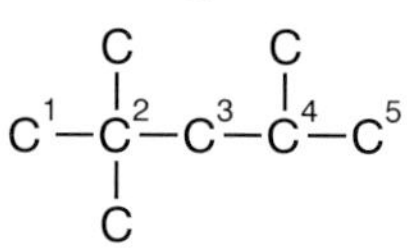

3 Identify the branches and arrange them in alphabetical order. For example, 'ethyl' before 'methyl' before 'propyl'. If there are two or more of the same branch then indicate this by using 'di', 'tri', 'tetra' and so on. Identify the numbers of the C atoms in the chain to which the branches are attached.

For **B**,
there is only one type of branch...methyl
but there are three of them..trimethyl
attached to C-2, C-2 and C-4..................................2,2,4-trimethyl

Finally, we write '2,2,4-trimethyl' in front of the parent name and arrive at

2,2,4-trimethylpentane

as the systematic name for **B**.

Now look at structure **C**, which is isomeric with **A** and **B**, and see if you can work out its systematic name.

C

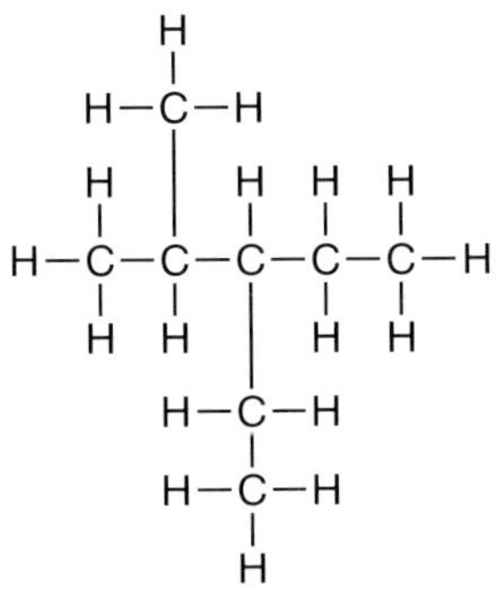

If you applied the rules correctly then you would find that it is called 3-ethyl-2-methylpentane. Notice that 'ethyl' precedes 'methyl' because the alkyl groups have to be arranged in alphabetical order according to rule 3.

Cycloalkanes

As well as open-chain alkanes, it is also possible for alkane molecules to form rings.

These molecules are called **cycloalkanes** and have the general formula C_nH_{2n}. They have two less hydrogen atoms than the corresponding alkane, because there are no CH_3— groups at the ends of the chain. Table 4 shows the full and shortened structural formulae of some cycloalkanes.

Cycloalkane	Full structural formula	Shortened structural formula
cyclopropane, C_3H_6	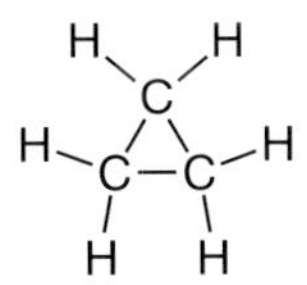	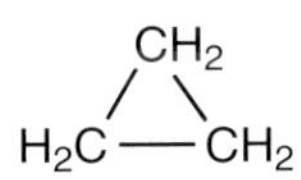
cyclobutane, C_4H_8	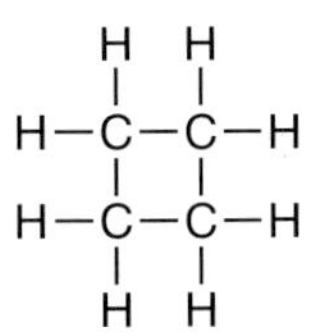	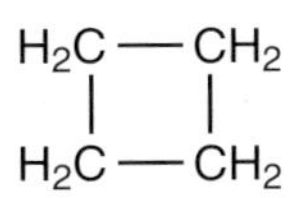
cyclohexane, C_6H_{12}	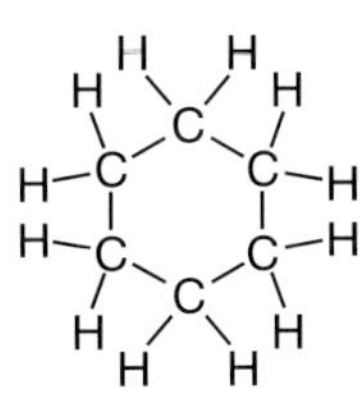	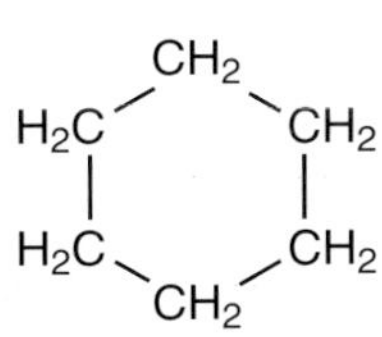

Table 4 Cycloalkanes

Shapes of alkanes

Representing structures in a two-dimensional way on paper can give a misleading picture of what the molecule looks like.

The pairs of electrons in the covalent bonds repel one another and so arrange themselves round the carbon atom as far apart as possible. Thus, the C—H bonds in methane are directed so they point towards the corners of a regular tetrahedron (Figure 4). The carbon atom is at the centre of the tetrahedron, and the hydrogen atoms are at the corners of the tetrahedron (Figure 5).

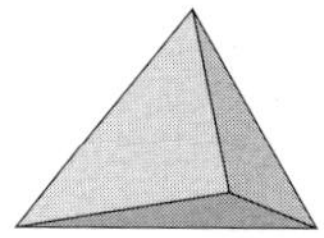

Figure 4 A regular tetrahedron

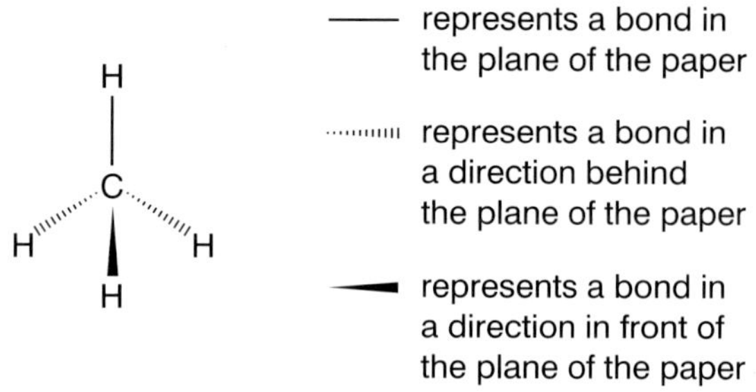

Figure 5 The three-dimensional shape of methane

The structure of ethane in three dimensions is shown in Figure 6. Each carbon atom is at the centre of a tetrahedral arrangement.

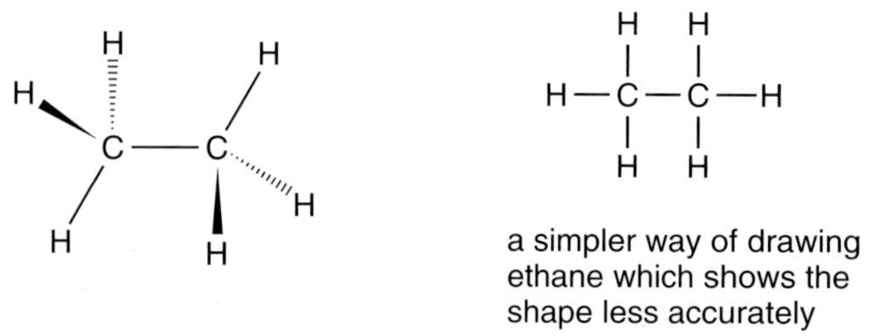

Figure 6 The three-dimensional shape of ethane

Figure 7 shows the three-dimensional structure of butane. You can see that hydrocarbon chains are not really 'straight', but a zigzag of carbon atoms.

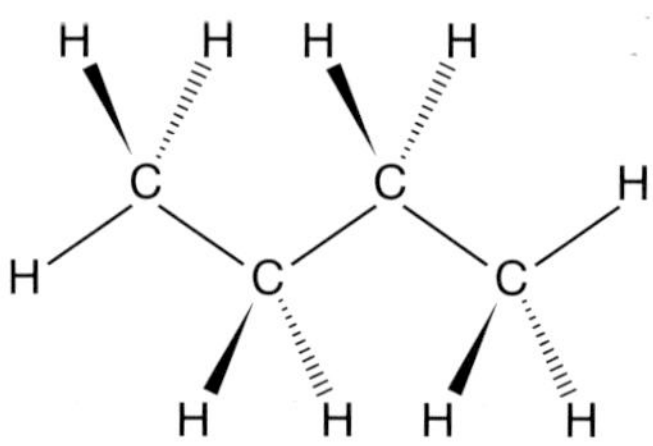

Figure 7 The three-dimensional shape of butane

It is a good idea to build models of some of these compounds.

What are alkanes like?

Whether an alkane is solid, liquid or gas at room temperature depends on the size of its molecules (see **Unit 1**, **Section 16.1**). The first members of the series (C_1–C_4) are colourless gases. Higher members (C_5–C_{16}) are colourless liquids and the remainder are white waxy solids.

Alkanes mix well with each other, but do not mix with water. The alkanes and water form two separate layers. This is because alkanes contain non-polar molecules, whereas water contains polar molecules which attract each other more strongly and prevent the alkane molecules moving in between them and mixing with them. (There is more about polarity of molecules and hydrogen bonding between water molecules in **Unit 1**, **Section 16.4**.)

Reactions of alkanes

Alkanes are very unreactive towards many laboratory reagents, for example acids, alkalis and oxidising agents such as potassium permanganate. When they do react, it is usually in the gaseous state and energy must be supplied to get the reaction started.

Oxidation of alkanes

Alkanes do not react with air at room temperature, but, if heated, they burn readily to give carbon dioxide and water. The reaction has a high activation energy and heat must be supplied to overcome this. Some of the bonds in the alkane and in O_2 must first be broken before the new bonds in CO_2 and H_2O can form.

Alkanes must be vaporised before they will burn and so the less volatile alkanes ignite less readily. This is why petrol ignites so much more easily than oil.

Once started, the combustion is very exothermic, which is why alkanes are such good fuels.

$$CH_4(g) + 2O_2(g) \rightarrow CO_2(g) + 2H_2O(l); \quad \Delta H = -891\,kJ\,mol^{-1}$$
$$C_7H_{16}(g) + 11O_2(g) \rightarrow 7CO_2(g) + 8H_2O(l); \quad \Delta H = -4817\,kJ\,mol^{-1}$$

If the supply of air is limited, the combustion may be incomplete and produce CO and C (soot) along with partially oxidised hydrocarbons.

Cracking alkanes

When alkanes are heated to high temperatures in the absence of air, they break up to form smaller molecules: hydrogen, alkanes and alkenes. Such reactions are known as **cracking**.

Two types of cracking are carried out in the petrochemical industry. They are **catalytic cracking** and **steam cracking**. Catalytic cracking normally involves breaking up the large molecules in the heavy oil fractions from crude oil, by the action of heat and in the presence of a catalyst. When the molecule splits, one of the products is an alkene and contains a C=C double bond. For example:

Catalytic cracking is an important process in making fuels such as petrol – see **Section 4.3**.

$$C_{10}H_{22} \rightarrow C_8H_{18} + CH_2{=}CH_2$$

decane *octane* *ethene*

There are many ways an alkane can break down and a complex mixture of products is obtained. The alkanes often rearrange to give branched rather than straight-chain products.

Steam cracking, on the other hand, is used to convert smaller alkanes into alkenes which then go on to be used as feedstocks in the polymer industry. No catalyst is used and the steam is there as a dilutant. It takes no part in the reaction but helps prevent the alkene molecules from recombining to form larger molecules.

The use of alkenes as feedstocks in the polymer industry is discussed in **Chapter 5**.

Ethene and **propene** are important starting materials in making polymers. In Scotland they are produced at Grangemouth and at Mossmorran in Fife by steam cracking ethane and propane.

You can read more about the processes at Grangemouth in **Unit 3, Chapter 8**.

When ethane is cracked the main products are ethene and hydrogen:

$$CH_3CH_3 \rightarrow CH_2{=}CH_2 + H_2$$

Propane, being a larger molecule, forms more products on cracking. For example:

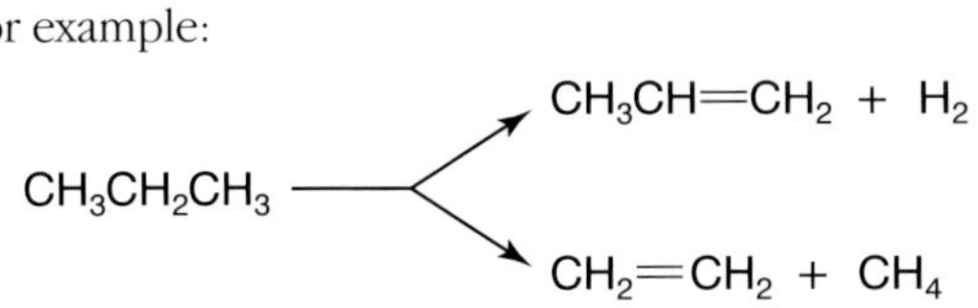

The source of the ethane and propane is the gas fraction from crude oil. It contains a mixture of methane, ethane, propane and butane, but the ethane and propane can be separated from it by fractional distillation and then steam cracked. In some plants, the gas fraction is cracked directly without first separating the ethane and propane. In other plants ethene and propene are made by steam cracking the naphtha fraction from crude oil.

PROBLEMS FOR 17.1

1 Which of the following formulae represent alkanes?
- a CH_2
- b C_2H_6
- c C_5H_{10}
- d C_7H_{16}
- e C_9H_{20}
- f $C_{13}H_{26}$

2 Write down the molecular formula of
- a an alkane which contains 12 C atoms?
- b an alkane which contains 50 H atoms?
- c cyclooctane?

3 Draw out the full structural formula of
- a $CH_3CH_2CH_2CH_2CH_3$
- b $CH_3(CH_2)_8CH_3$
- c

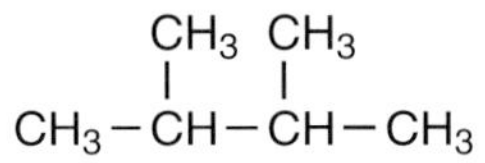

[sometimes written as $CH_3CH(CH_3)CH(CH_3)CH_3$]

4 There are two different compounds with the molecular formula C_4H_{10}. Draw out their structures.

5 Write the shortened structural formula of

a

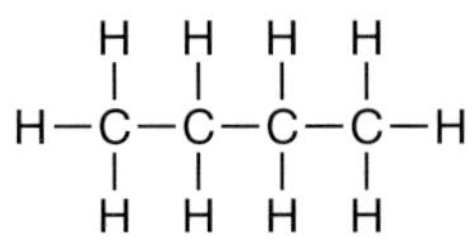

b

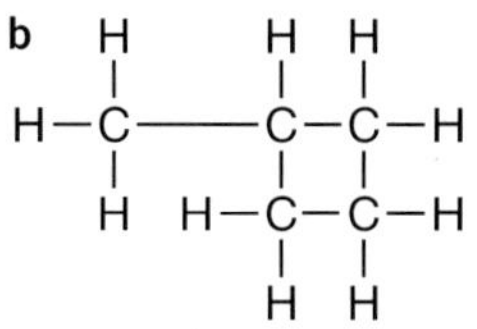

c

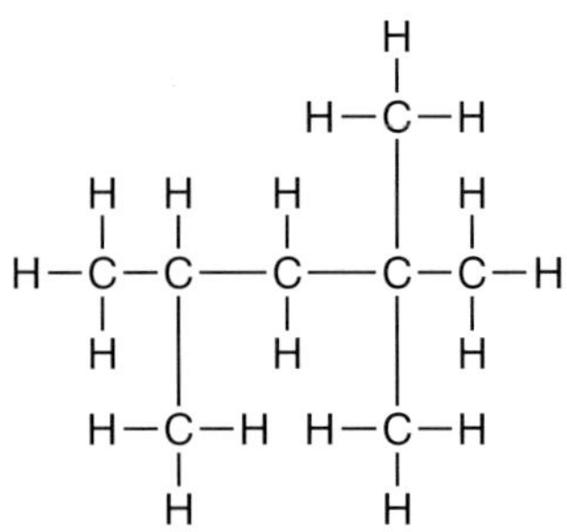

6 Write the systematic name of

a

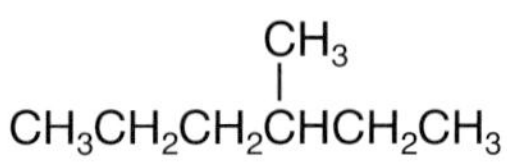

b

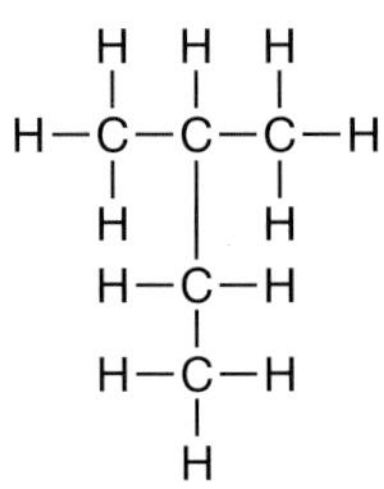

c

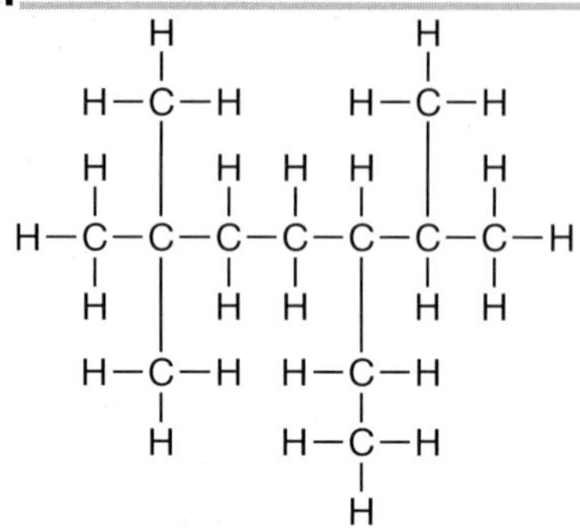

7 Draw the full structural formula of
- a pentane
- b cycloheptane
- c 4-ethyl-2,3-dimethylhexane.

8 Look at the following hydrocarbons:

A $CH_3CH_2CH_3$

B

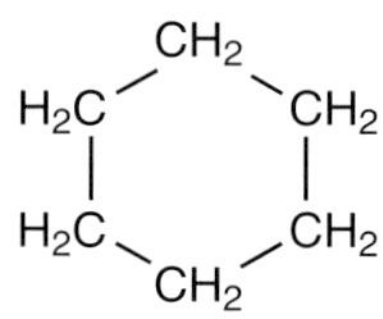

C

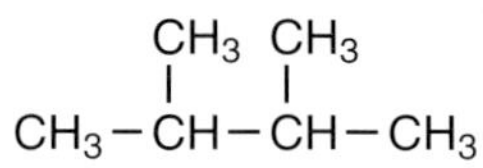

D $CH_3(CH_2)_{20}CH_3$

E $CH_3(CH_2)_4CH_3$

- a Which hydrocarbon(s) would you expect to be solid at 25 °C?
- b Which hydrocarbon would you expect to be the most volatile?
- c Which hydrocarbon does *not* fit the general formula C_nH_{2n+2}?
- d Which two hydrocarbons are isomers?
- e For which hydrocarbon is there no isomeric compound?

9 Write balanced equations for each of the following reactions.
- a Pentane, C_5H_{12}, burns in a plentiful supply of air to form carbon dioxide and water.
- b Pentane burns in a limited supply of air to form carbon monoxide and water.

10 When an alkane is cracked, each molecule forms at least two new molecules. Write the structural formulae of two pairs of compounds that could be formed by cracking butane, C_4H_{10}.

17.2 *Structural isomerism*

Two molecules which have the same molecular formula but differ in the way their atoms are arranged are called **isomers**. Isomers are distinct compounds with different physical properties, and often different chemical properties too.

The occurrence of isomers (**isomerism**) is commonest in carbon compounds because of the great variety of ways in which carbon can form chains and rings.

There are two main types of isomerism, known as structural isomerism and stereoisomerism. At this stage we'll deal only with **structural isomerism**.

Structural isomers have the same molecular formula but have their atoms bonded together in a *different order*. They have *different structural formulae*.

There are various ways in which structural isomerism can arise (Figure 8).

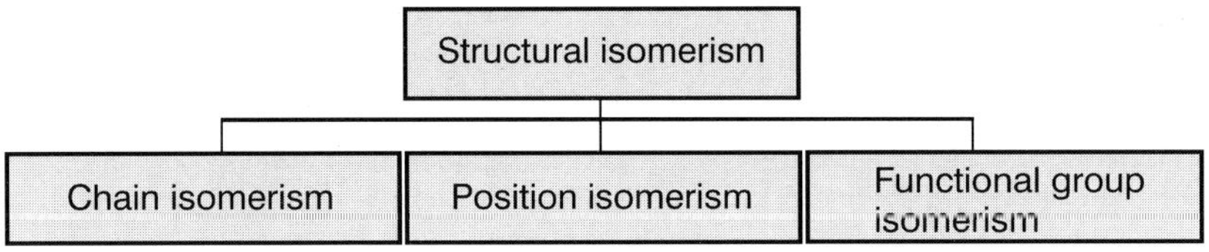

Figure 8 Types of structural isomerism

Chain isomerism

There is only one alkane corresponding to each of the molecular formulae CH_4, C_2H_6 and C_3H_8. With four or more C atoms in a chain, different arrangements are possible. For example,

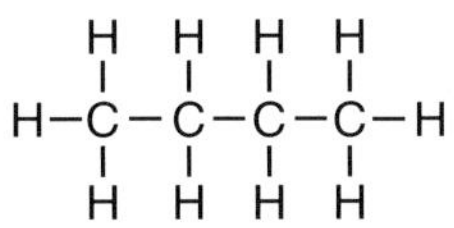

butane
boiling point 0 °C

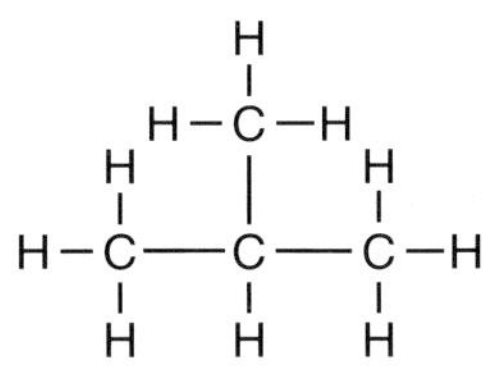

methylpropane
boiling point –12 °C

Both these compounds have the same molecular formula, C_4H_{10}. Their different structures lead to different properties. For example, the boiling point of methylpropane is 12 °C lower than that of butane.

As the number of carbon atoms in an alkane increases, the number of possible isomers increases. There are over four thousand million isomers with the molecular formula $C_{30}H_{62}$!

Position isomerism

This occurs when a functional group is situated in different positions in the molecule. For example,

a ***Isomers of C_4H_8***

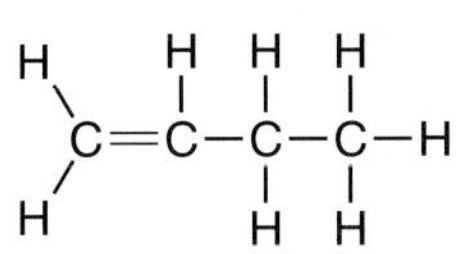

but-1-ene

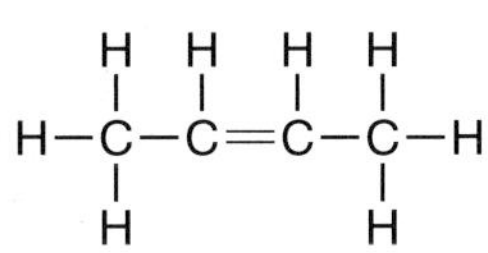

but-2-ene

Here the C=C functional group is situated at two different places in the hydrocarbon chain. (The C=C functional group is discussed in more detail in **Section 17.3**.)

b ***Isomers of $C_6H_4Cl_2$***

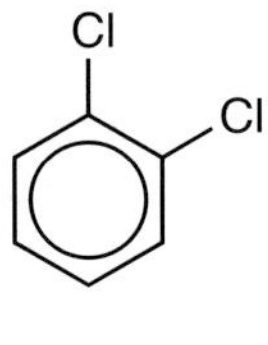

1,2-dichlorobenzene

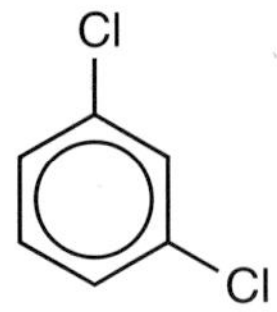

1,3-dichlorobenzene

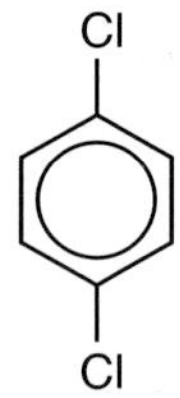

1,4-dichlorobenzene

In this example, the —Cl functional groups are situated at different positions on the benzene ring. See **Section 17.5** for more details about benzene and compounds containing benzene rings.

Functional group isomerism

It is sometimes possible for compounds with the same molecular formula to have quite different functional groups and because they have different functional groups they will belong to different homologous series.
For example,

a ***Molecular formula C_3H_8O***

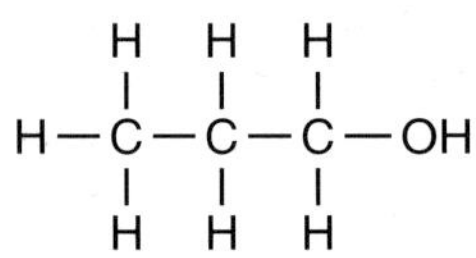

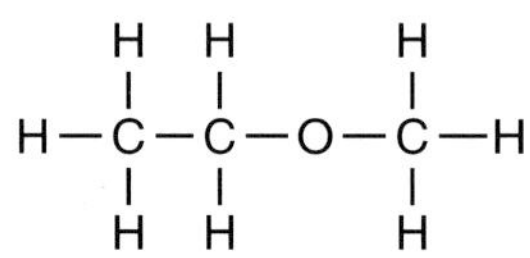

The isomer on the left is called propan-1-ol. Its functional group is —OH and it is a member of the homologous series known as the *alcohols*. The isomer on the right is methoxyethane. Its —O— functional group is quite different from that of propan-1-ol and it belongs to another family of organic compounds, known as the *ethers*. For more information about alcohols and ethers see **Section 18.1**.

As well as having different physical properties, propan-1-ol and methoxyethane will have very different chemical properties as a result of their differing functional groups.

b ***Molecular formula C_4H_8***

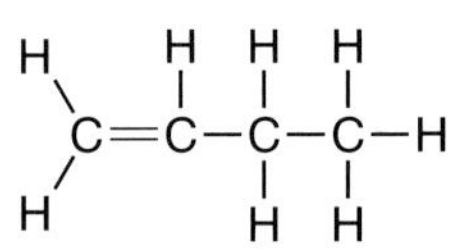

but-1-ene

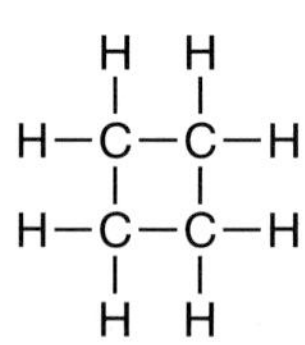

cyclobutane

But-1-ene has a C=C functional group and so belongs to the alkene family. Although cyclobutane has no functional group, we can tell from the ring of carbon atoms that it is a member of the cycloalkane series. Since they are not in the same homologous series, but-1-ene and cyclobutane will differ significantly in their chemical properties.

PROBLEMS FOR 17.2

1 Write out structural formulae for isomers of C_5H_{12}. Name each of your isomers.

2 How many isomers are there with molecular formula C_4H_9Br? Draw out their structures.

3 Which of these pairs of compounds are isomers?

a $CH_3CH_2CH_2CH_2CH_2CH_3$ and 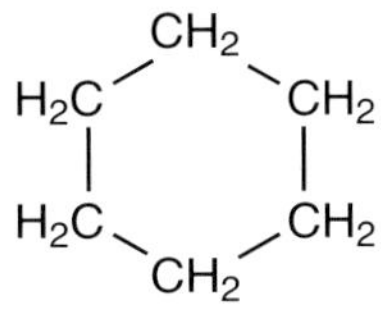

b $CH_3CH_2CH_2CH_2Cl$ and 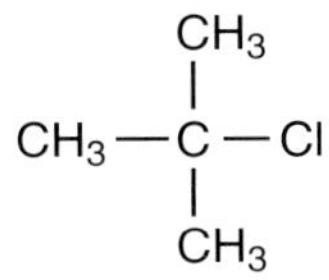

c $CH_3CH_2CH_2OH$ and CH_3COCH_3

d  and 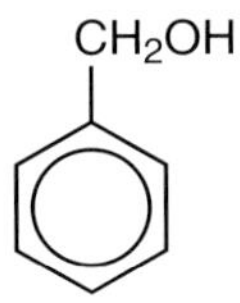

e 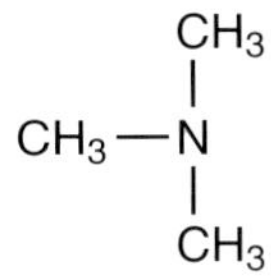 and 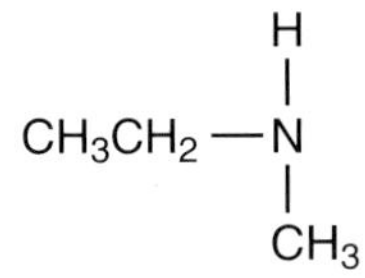

f 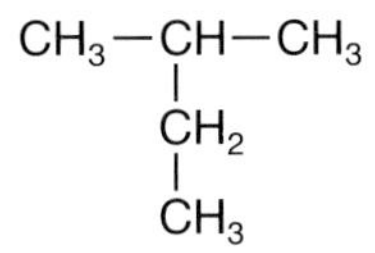 and 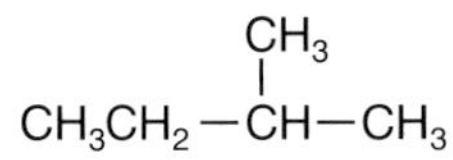

4 An organic compound containing 52.1% by mass of carbon and 13.0% by mass of hydrogen burns to produce carbon dioxide and water and no other products.

a What other element must be present in the compound?

b Calculate the empirical formula of the compound.

c The compound has a relative formula mass of 46. Write down its molecular formula.

d Chemical tests showed the molecule is an alcohol, containing an —OH functional group. Draw the full structural formula of the compound.

e Draw the structural formula of a compound that is isomeric with this one.

17.3 *Alkenes*

What are alkenes?

Ethene is the simplest member of a homologous series of hydrocarbons called **alkenes**. It has the structure

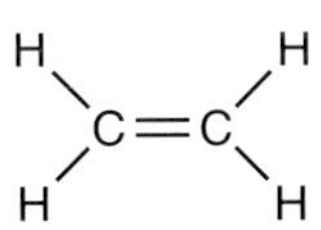

Alkenes are distinguished from other hydrocarbons by the presence of the C=C double bond. The double bond implies that they are **unsaturated hydrocarbons**.

Examples of other members of the alkene family are

propene	$CH_3CH{=}CH_2$
but-1-ene	$CH_3CH_2CH{=}CH_2$
but-2-ene	$CH_3CH{=}CHCH_3$
pent-1-ene	$CH_3CH_2CH_2CH{=}CH_2$
hex-2-enc	$CH_3CH_2CH_2CH{=}CHCH_3$
hept-3-ene	$CH_3CH_2CH_2CH{=}CHCH_2CH_3$
oct-4-ene	$CH_3CH_2CH_2CH{=}CHCH_2CH_2CH_3$

Alkenes have the general formula, C_nH_{2n}. This is the same as the general formula for cycloalkanes but the C=C functional group makes alkenes react very differently from cycloalkanes.

As with the alkanes, the boiling points of alkenes increase as the number of carbon atoms increases. Ethene, propene and the butenes are gases. After that they are liquids and eventually solids.

Notice how the alkenes are named. Take hept-3-ene as an example. The *hept-* part tells you it has the same hydrocarbon chain as heptane. The *-ene* suffix (the end part of the name) tells you it is an alkene and has a C=C double bond. The number tells you which carbon atom in the chain is the first to be involved in the double bond, assuming that you start counting from the end of the chain which is closer to the double bond.

The names of all alkenes end in **–ene**.

As well as straight-chain alkenes, such as those above, there are also branched-chain alkenes, such as the one below.

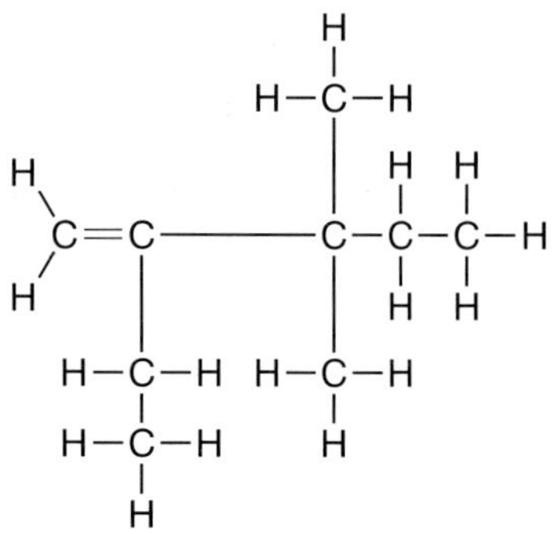

The rules for naming it are as follows.

1 Pick out the longest carbon chain containing the double bond to get the parent name.

In our example the parent name is pentene. Notice that the longest chain in the molecule contains six carbon atoms but it doesn't contain the double bond.

2 Number the C atoms in this chain starting at the end nearer the double bond. Identify the number of the C atom where the double bond starts and insert this in the parent name.

For the example above, numbering will start from the left-hand end of the molecule.

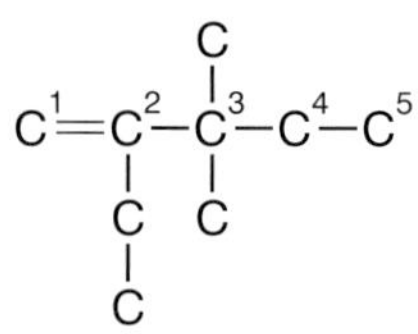

The double bond starts at C-1 and so the parent name is pent-1-ene. (If the double bond had been in the middle of the chain then numbering would start at the end nearer a branch.)

3 Identify the branches and the number of the C atoms to which they are attached.

In the example above, an ethyl group is attached at C-2 and two methyl groups are attached at C-3.

So our alkene is called 2-ethyl-3,3-dimethylpent-1-ene.

Shapes of alkenes

Ethene is a flat molecule with the shape shown in Figure 9.

The bonds around the C=C group are arranged the same way in all alkenes.

Take propene for example. Its shape is shown in Figure 10. Although there is a tetrahedral arrangement of bonds around C-3, the bonds around C-1 and C-2 are planar.

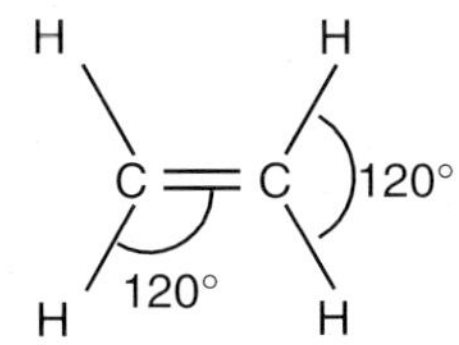

Figure 9 The ethene molecule is flat

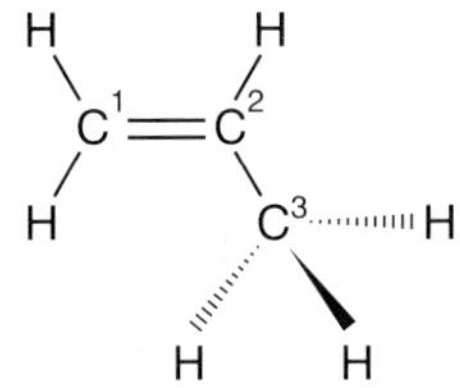

Figure 10 The three-dimensional shape of propene

Making alkenes

Alkenes can be made in the laboratory from alcohols. The vapour of the alcohol is passed over hot aluminium oxide, which acts as a catalyst. Propene, for example, can be prepared from the alcohol, propan-1-ol.

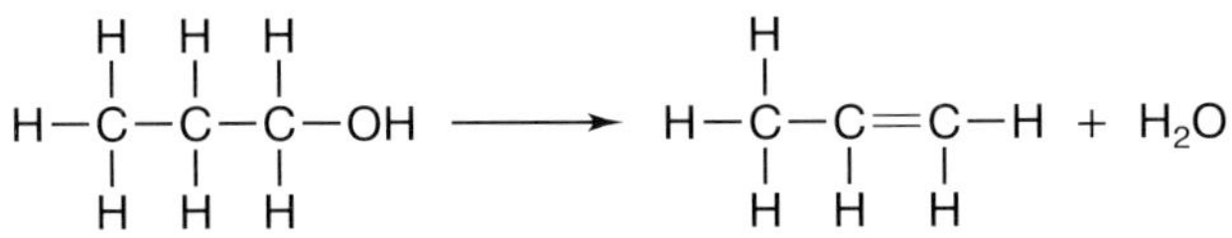

propan-1-ol *propene*

The reaction is described as **dehydration** since it involves the removal of a *water* molecule from a molecule of the reactant.

Loss of water from butan-1-ol gives a single product, but-1-ene. However, loss of water from butan-2-ol leads to the formation of *two* products. Can you explain why?

In industry, alkenes are obtained by cracking alkanes from crude oil (see **Section 17.1**).

Reactions of alkenes

Alkenes are more reactive than alkanes. This is because they contain a C=C functional group which allows them to undergo **addition reactions** with small molecules such as hydrogen, bromine, water and so on. The small molecule, say AB, adds across the double bond.

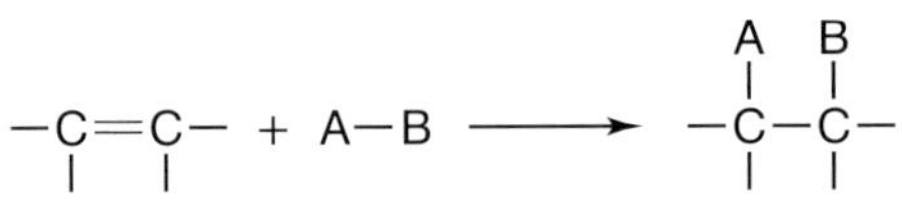

Notice that the **unsaturated** reactant is converted into a **saturated** product.

Addition of halogens

Chlorine and bromine readily add across the double bond of an alkene.

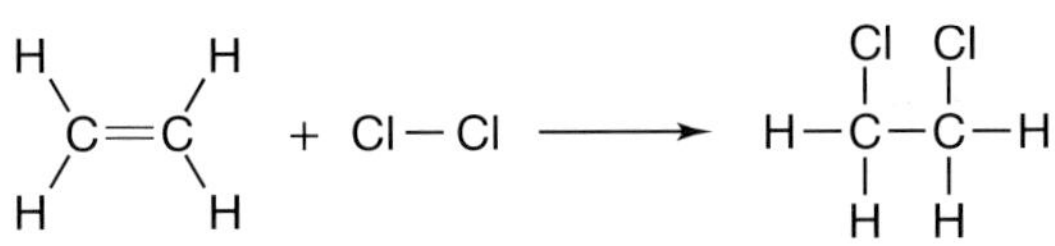

ethene *1,2-dichloroethane*

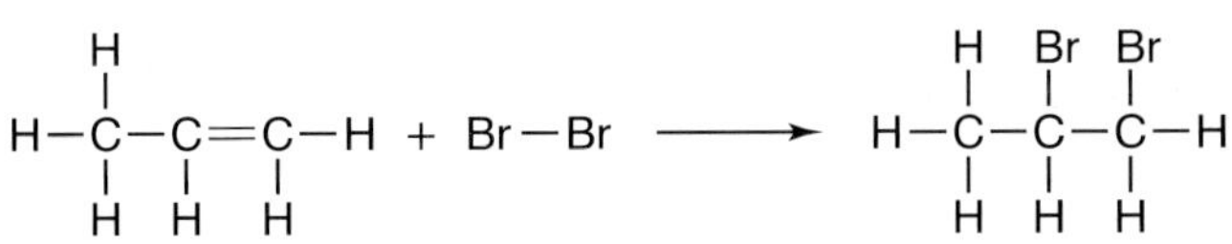

propene *1,2-dibromopropane*

The products of these reactions belong to a series of compounds called **halogenoalkanes** in which the functional groups are halogen atoms. They are basically alkanes in which one or more of the hydrogen atoms have been substituted by halogen atoms.

You will find out more about halogenoalkanes in **Section 6.3**.

The reaction between bromine and an alkene results in the decolorisation of the bromine. This can be used as a **test for unsaturation** in an organic compound. It is not specifically a test for C=C because compounds containing C≡C will also decolorise bromine (see **Section 17.4**).

Addition of hydrogen halides

The hydrogen halides, HCl, HBr and HI all add across the double bond of an alkene to form a halogenoalkane. Let's consider, for example, the addition of hydrogen bromide to but-2-ene:

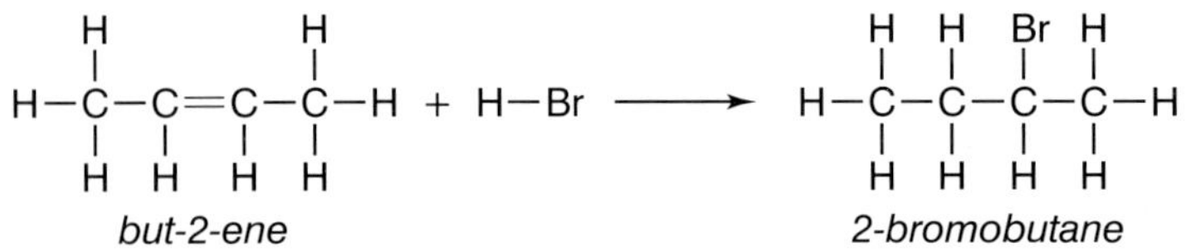

Here, only one product is formed.

Had we used but-1-ene instead then two products are possible, depending on which carbon atom of the double bond the bromine atom becomes attached to.

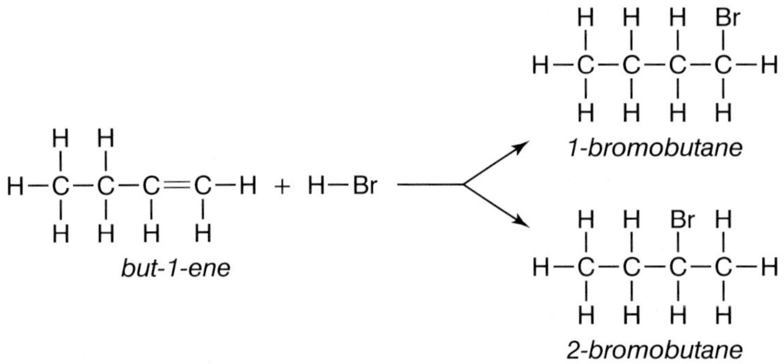

In practice, the product is made up almost entirely of 2-bromobutane.

Addition of water

In the presence of a catalyst, alkenes undergo addition reactions with water to form alcohols. For example, ethanol is formed when water adds across the double bond in ethene.

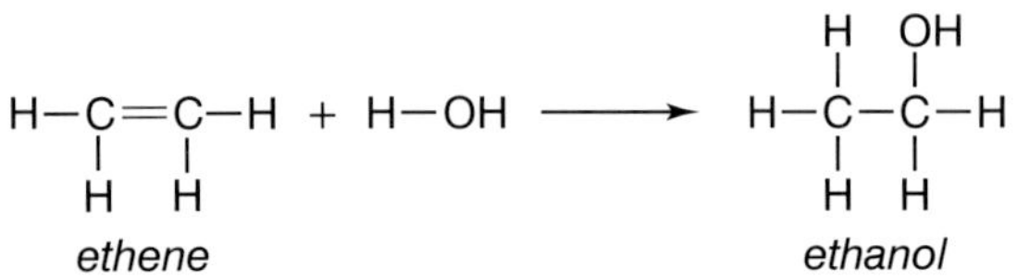

Like the addition of hydrogen halides, adding water to certain alkenes can result in the formation of two products.

The addition of water to an alkene is an example of a **hydration** reaction. (This is not to be confused with a *hydrolysis* reaction, when water splits a compound into two new compounds.)

Addition of hydrogen

Alkanes are formed when alkenes undergo catalytic addition with hydrogen. Propane, for example, is produced when hydrogen adds across the double bond in propene.

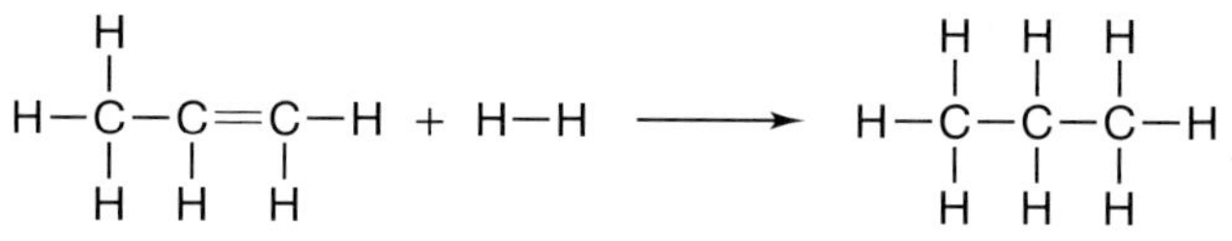

propene *propane*

The addition of hydrogen to an alkene is an example of a **hydrogenation** reaction.

Addition reactions in industry

Here are two industrial examples of addition reactions involving alkenes.

The addition of water to alkenes (hydration) is used commercially in the **manufacture of alcohols**. Ethanol, for example, is made by hydrating ethene in the presence of a catalyst (phosphoric acid). The reaction is carried out at a temperature of about 300 °C and a pressure of 60–70 atmospheres.

$$\underset{\textit{ethene}}{CH_2{=}CH_2} + H_2O \rightarrow \underset{\textit{ethanol}}{CH_3CH_2OH}$$

Ethanol used to be made by fermentation but, with the exception of that made for human consumption, the major feedstock for ethanol production is now ethene.

The addition of hydrogen to C=C double bonds is the basis of **margarine manufacture**. Hydrogen reacts with blends of plant and animal oils at 160 °C in the presence of a powdered nickel catalyst. Oil molecules are unsaturated because they contain C=C double bonds. During the conversion of oils to margarines some of the double bonds are hydrogenated. This reduces the amount of unsaturation and the soft liquid oils change into solid margarines. The hydrogenation of oils to margarines is often described as **hardening**. You can read more about the chemistry of fats and oils in **Section 18.5**.

Uses of alkenes

Alkenes are the feedstocks for a vast range of products, but their major use is in the manufacture of plastics. Ethene, for example, undergoes addition polymerisation to form poly(ethene).

You will study polymerisation in more detail later in this unit in **Chapters 5** and **19**.

$$CH_2{=}CH_2 + CH_2{=}CH_2 + CH_2{=}CH_2 \rightarrow$$
$$\text{–}CH_2\text{–}CH_2\text{–}CH_2\text{–}CH_2\text{–}CH_2\text{–}CH_2\text{–}$$

poly(ethene)

Similarly, poly(propene) is made from propene monomer units.

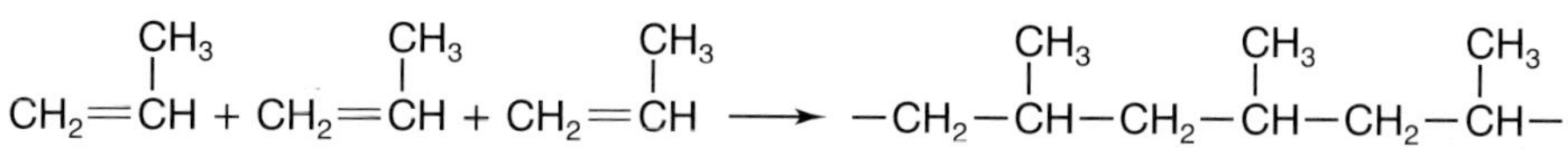

poly(propene)

Ethene can also be converted into other monomers which are subsequently polymerised. Some examples are outlined below.

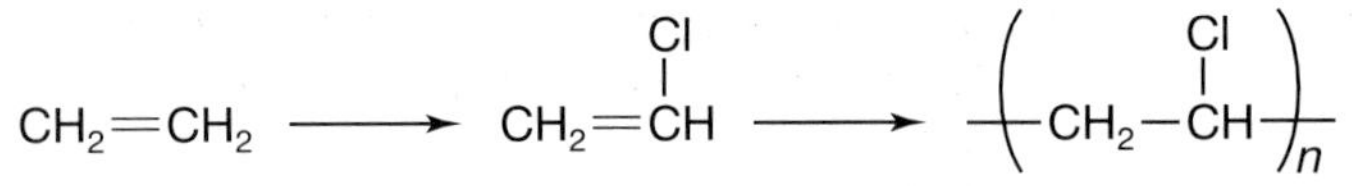

chloroethene (vinyl chloride)   *poly(chloroethene) (pvc)*

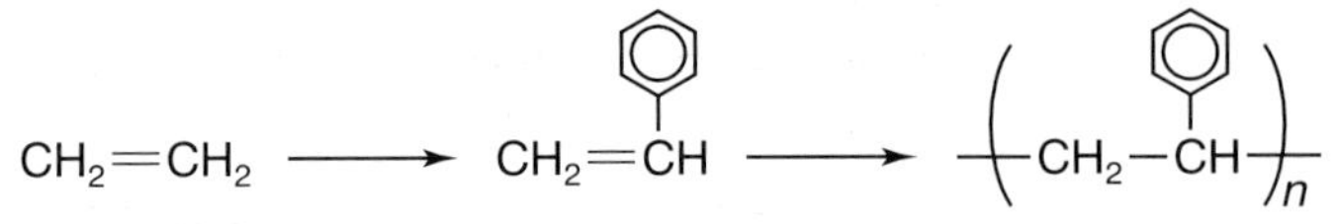

phenylethene (styrene)   *poly(phenylethene) (polystyrene)*

PROBLEMS FOR 17.3

1 Name the alkenes with the following structures:

a

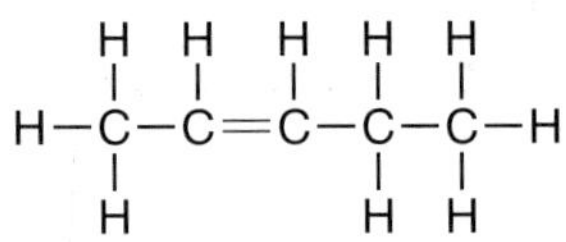

b

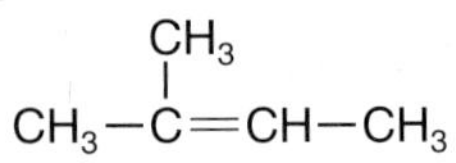

c

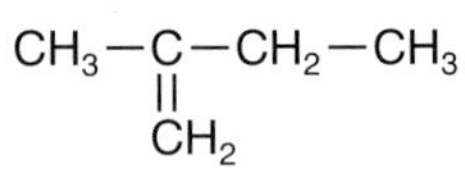

d

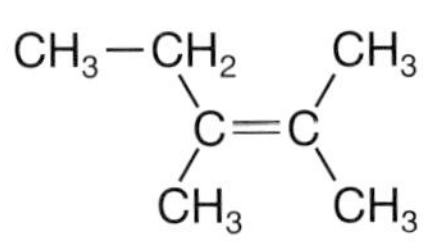

2 Draw a full structural formula for each of the following alkenes:
   - a oct-2-ene
   - b 3,3-dimethylbut-1-ene
   - c 4-ethyl-5-methylhex-1-ene.

3 But-2-ene has the structure
   $CH_3—CH=CH—CH_3$
   Draw the shortened structural formula of
   - a a straight-chain alkene which is an isomer of but-2-ene
   - b a branched-chain alkene which is an isomer of but-2-ene
   - c two cycloalkanes which are isomeric with but-2-ene.

4 The shortened structural formula for pent-2-ene is

$$\overset{1}{C}H_3-\overset{2}{C}H=\overset{3}{C}H-\overset{4}{C}H_2-\overset{5}{C}H_3$$

   - a Give the numbers of the C atoms around which the bonds are in a planar arrangement.
   - b Describe the arrangement of bonds around the other C atoms.

5 Look at the following diagram, which shows some of the reactions that propene undergoes.

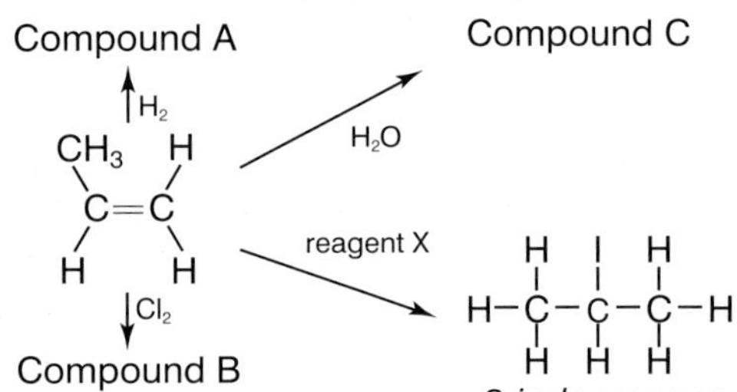

2-iodopropane

   - a Draw full structural formulae for compounds **A**, **B** and **C** (**C** is an isomer of propan-1-ol).
   - b Name reagent **X**.
   - c All the reactions in the above diagram can be described as addition reactions.
     Give another name to describe the conversion of propene to
     i compound **A**   **ii** compound **C**.

6 Draw full structural formulae for the *two* alcohols which could be dehydrated to form the alkene below.

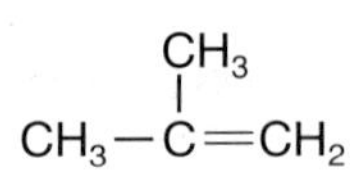

7 The addition of hydrogen chloride to but-1-ene can give two products but not in equal amounts. According to *Markovnikoff's rule* the hydrogen of the hydrogen chloride mainly attaches to the carbon atom of the double bond which already has the most hydrogens directly attached to it.
   - a Draw the full structural formula for the main product formed when hydrogen chloride reacts with but-1-ene.
   - b Why is it not necessary to consider Markovnikoff's rule when hydrogen chloride reacts with but-2-ene?

(SQA)

8 This question is about the substance *citronellol*. Citronellol belongs to a class of compounds called *terpenes*. Many closely related terpenes occur in nature where they are responsible for the fragrances of plants. We often convert them into more volatile ester derivatives for use in perfumes. Citronellol is found in geraniums and citronella grass.

Its structural formula is:

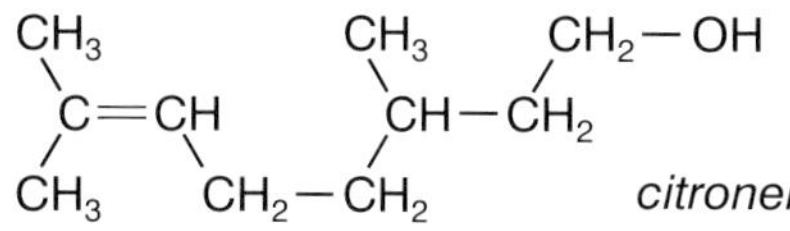

citronellol

and its skeletal formula is:

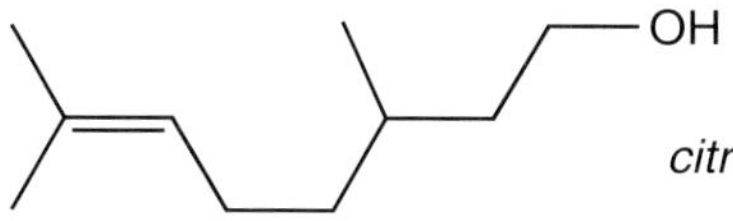

citronellol

We can think of all terpenes as being built up from the compound which is shown as **D** below.

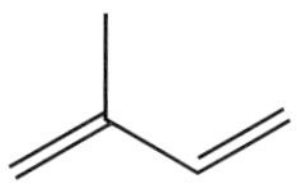

compound D

a Draw the full structural formula of compound **D**.
b Write down the molecular formula of compound **D**.
c Write down the molecular formula of citronellol.
d Name the functional groups present in citronellol.
e How many units of compound **D** combine to produce citronellol?
f Draw the full structural formula of the compound formed when compound **D** reacts with excess hydrogen.
g What would you expect to see if some citronellol was shaken with a solution of bromine dissolved in an organic solvent such as hexane?
h Draw the skeletal formula of the compound formed when citronellol reacts with bromine.

9 Poly(tetrafluoroethene) and poly(propenenitrile) are addition polymers made from monomers which contain a C=C double bond.

a Draw part of the structure of poly(tetrafluoroethene) showing at least three monomer units linked together.
b Part of the structure of poly(propenenitrile) is drawn below:

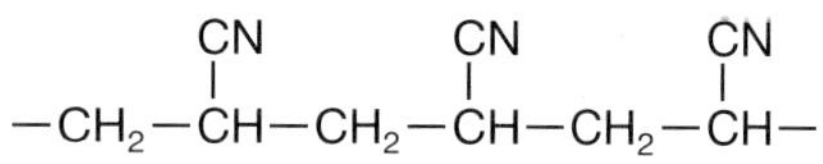

Draw the full structural formula of the monomer used to make poly(propenenitrile).

17.4 *Alkynes*

What are alkynes?

Alkynes are a homologous series of hydrocarbons in which each member has a C≡C triple bond. Ethyne is the simplest member of the series and it has the structure.

H–C≡C–H

Alkynes are unsaturated and have the general formula C_nH_{2n-2}.

Like the alkanes and alkenes, the boiling points of alkynes increase as the molecules get bigger. Alkynes are named in exactly the same way as alkenes. Consider the straight-chain alkyne below.

The names of all alkynes end in **-yne**.

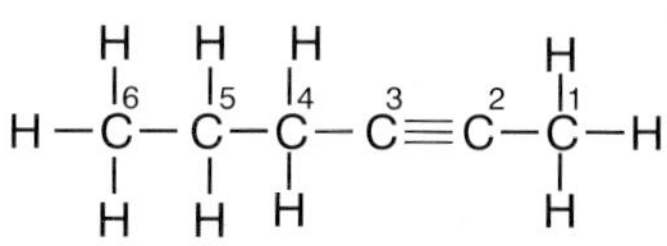

It is a hexyne since it is an unbranched chain containing six carbon atoms (*hex-*) and a C≡C triple bond (*-yne*). The carbon atoms are numbered from the right-hand end (nearer the triple bond) and the triple bond starts at C-2. So its systematic name is hex-2-yne.

Now look at the following branched-chain alkyne.

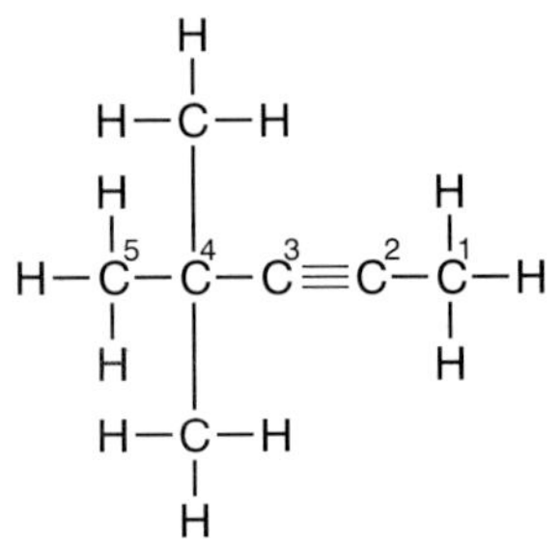

For this compound,
the longest chain containing the triple bond has five C atoms............pentyne
numbering starts from the right-hand end and the triple bond starts at C-2...pent-2-yne
there are two methyl branches.. dimethyl
both attached at C-4..4,4-dimethyl

So our branched-chain alkyne is called

4,4-dimethylpent-2-yne

Shapes of alkynes

Figure 11 shows the shape of the ethyne molecule. Notice that all four atoms in the molecule lie in a straight line. Other alkynes also contain the same linear arrangement of bonds around the C≡C group.

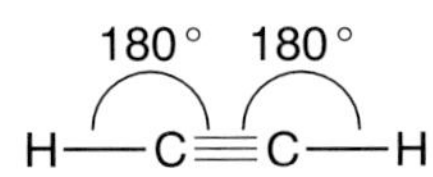

Figure 11 The shape of the ethyne molecule

Reactions of alkynes

Alkynes undergo addition reactions but, despite their greater degree of unsaturation, they are not as reactive as alkenes.

Let's see what happens when a small molecule, say AB, adds across the C≡C triple bond.

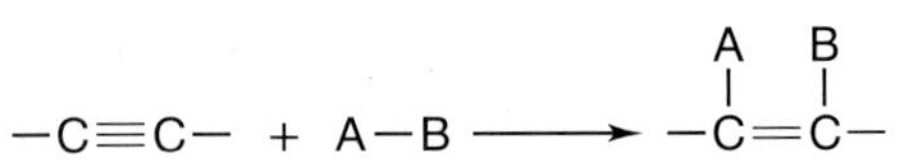

The product contains a C=C double bond and so a second addition can take place, giving a final saturated product.

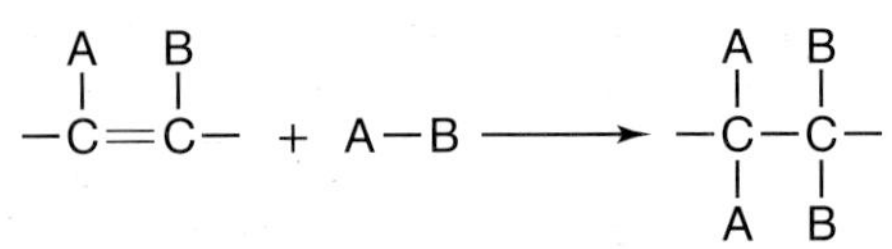

An alkyne addition reaction is therefore a two-stage process and in most cases it is possible to isolate the intermediate product before the second stage takes place.

Addition of halogens

Both chlorine and bromine add across the C≡C triple bond in an alkyne. For example:

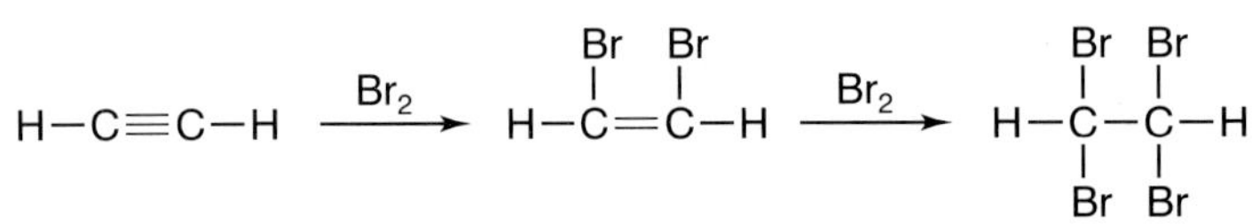

ethyne *1,2-dibromoethene* *1,1,2,2-tetrabromoethane*

When alkynes react with bromine, the bromine is decolorised. As you know, the same happens on adding bromine to an alkene. This makes the **decolorisation of bromine a good test for unsaturation** in an organic compound.

Addition of hydrogen halides

Ethyne undergoes addition reactions with the hydrogen halides HCl, HBr and HI. For example:

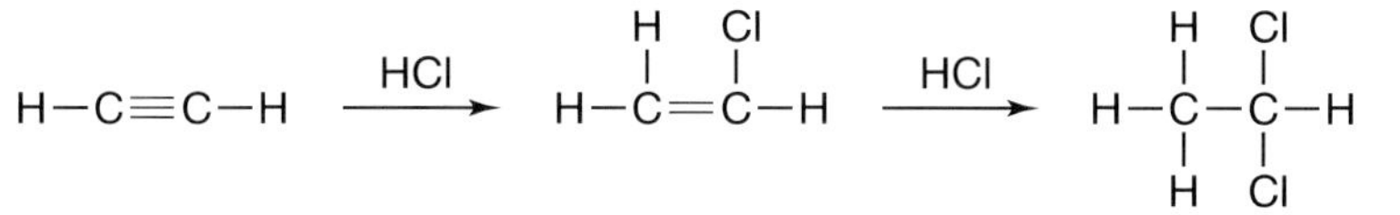

ethyne *chloroethene* *1,1-dichloroethane*

Notice at the second stage 1,2-dichloroethane could also be formed. In practice, however, 1,1-dichloroethane is the main product.

Addition of hydrogen

Hydrogen adds across the triple bond in ethyne to form ethene and then ethane. A transition metal catalyst such as platinum, palladium or nickel is needed.

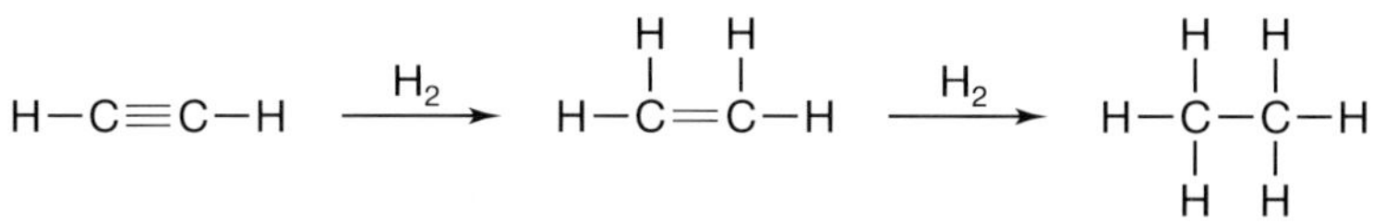

ethyne *ethene* *ethane*

PROBLEMS FOR 17.4

1 Name each of the following alkynes.

a $HC{\equiv}CCH_2CH_2CH_3$

b $CH_3{-}C{\equiv}C{-}CH(CH_3){-}CH_3$

c $CH_3CH_2CH(CH_2CH_2CH_3)C{\equiv}CH$

2 Draw the full structural formula for each of the following alkynes.

a but-2-yne

b 4,4-dimethylpent-1-yne

c 4-ethylhex-2-yne

3 The following diagram shows some of the reactions that ethyne undergoes.

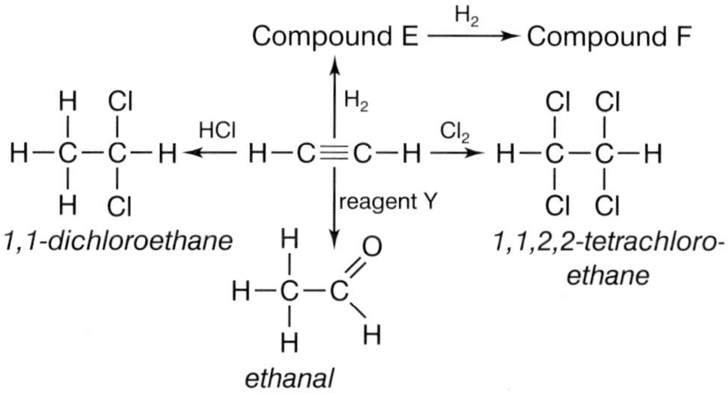

a Draw the full structural formula of the intermediate compound formed on the reaction of ethyne with
i chlorine
ii hydrogen chloride.

b Name compounds **E** and **F**.

c With reagent **Y**, ethyne undergoes addition to form an unstable intermediate which immediately rearranges to ethanal. Name reagent **Y** and draw the structural formula of the unstable intermediate.

4 The table below gives information about the addition products of different hydrocarbons with bromine.

Hydrocarbon	Molecular formula	Molecular formula of addition product
G	C_6H_{14}	no addition product
H	C_6H_{12}	$C_6H_{12}Br_2$
I	C_6H_{10}	$C_6H_{10}Br_4$
J	C_6H_{10}	$C_6H_{10}Br_2$
K	C_6H_6	no addition product

a Draw a possible structure for each of the hydrocarbons **G**, **H**, **J** and **K**.

b Draw *two* possible structures for hydrocarbon **I** with each one belonging to a different homologous series.

17.5 *Aromatic hydrocarbons*

What's special about benzene?

Benzene, the simplest aromatic hydrocarbon, is a colourless liquid – nothing special there. It has a molecular formula C_6H_6 and so it must be very unsaturated. The puzzle is that benzene resists addition reactions and does not behave like a normal unsaturated compound. It doesn't, for example, decolorise bromine solution. It is much less reactive and more stable than a typical unsaturated hydrocarbon.

The reason for this is to do with the rather special cyclic structure of benzene. Look at Figure 12 which shows the structure of the benzene ring. The benzene ring is a flat hexagon (Figure 13). The carbon–carbon bonds are equal in length and in strength and are intermediate between a carbon–carbon single bond and a carbon–carbon double bond.

Each carbon atom in the benzene ring has four outer electrons. It uses three of them to form single covalent bonds with a hydrogen atom and with two neighbouring carbon atoms. This leaves one electron on each carbon atom. Instead of overlapping in pairs to form three separate double bonds, these remaining electrons are spread out evenly and shared by all six carbon atoms in the ring. They are said to be **delocalised**. The six delocalised electrons occupy electron clouds which are situated above and below the benzene ring. The arrangement is shown in Figure 14.

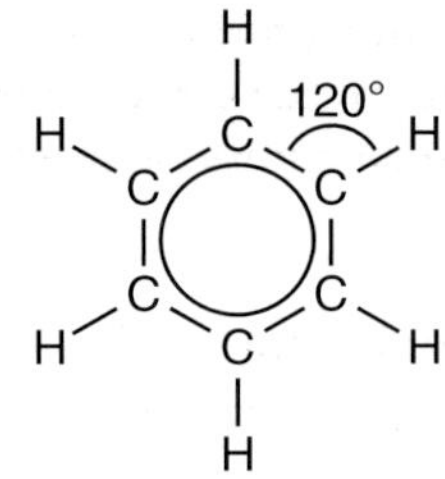

Figure 12 The structure of the benzene ring

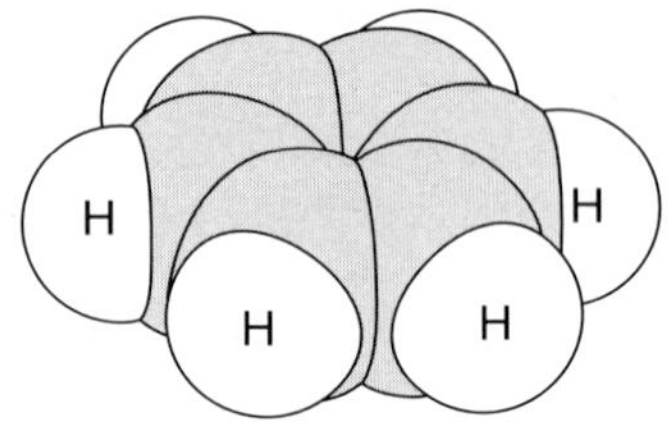

Figure 13 A diagram of a space-filling model of a benzene molecule, showing its planar shape

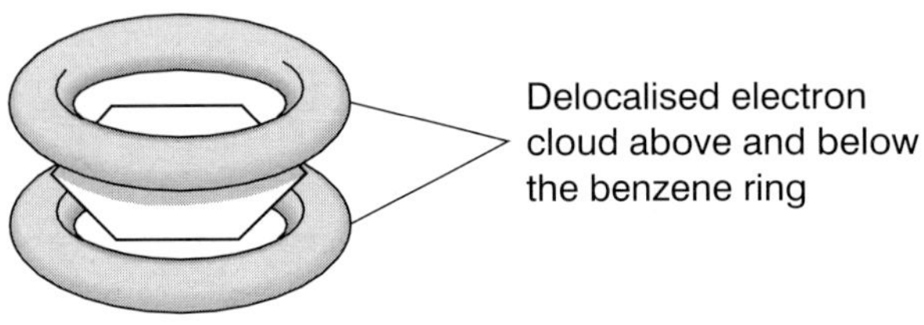

Figure 14 Delocalised electron clouds in benzene

Additional evidence for the delocalised structure comes from *electron density maps*, which can be drawn up from X-ray diffraction studies (Figure 15). Figure 15 shows a uniform electron density in all the carbon–carbon bonds.

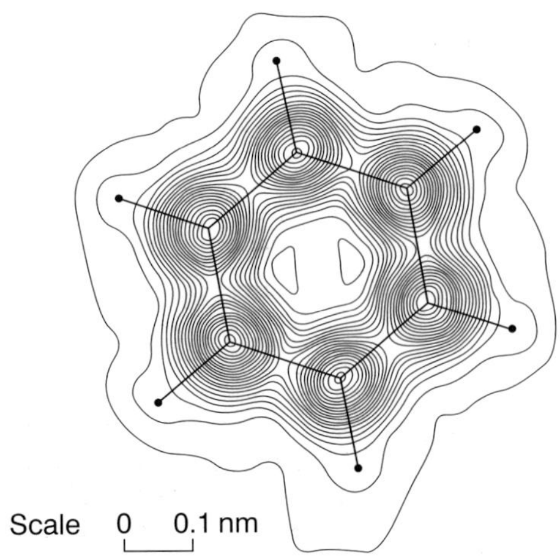

Figure 15 An electron density map for benzene at –3 °C. The lines are like contour lines on a map: they show parts of the molecule with equal electron density

In benzene's structural formula (Figure 12) the **delocalised electrons are represented by the circle drawn inside the ring**. The structural formula for benzene is quite often abbreviated to

It is the delocalised electrons which give benzene its stability and makes it reluctant to undergo addition reactions.

What are aromatic hydrocarbons?

Hydrocarbons such as benzene which contain rings stabilised by electron delocalisation are called **aromatic hydrocarbons**.

There are many aromatic hydrocarbons and some examples are drawn in Table 5. The first two are important components of high octane petrols. The third is used to make the plastic poly(phenylethene), more commonly known as polystyrene.

See **Section 4.3** for more information about octane numbers.

Table 5 Some aromatic hydrocarbons

Aromatic hydrocarbon	Molecular formula	Shortened structural formula
methylbenzene	C_7H_8	CH_3 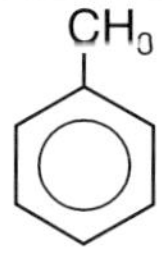
1,4-dimethylbenzene	C_8H_{10}	CH_3 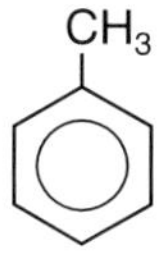
phenylethene	C_8H_8	$CH{=}CH_2$ 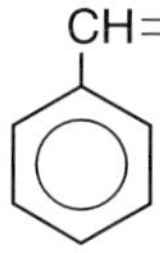

Notice the name of the third aromatic hydrocarbon. It is named from the parent compound ethene, rather than from benzene. A benzene ring in which one hydrogen atom has been substituted by another group is known as a **phenyl group**. The phenyl group has the formula C_6H_5— and the structure

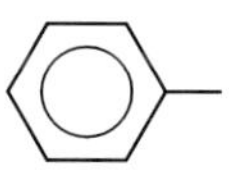

The phenyl group is related to benzene in the same way that a methyl group is related to methane.

Aromatic compounds as feedstocks

Aromatic hydrocarbons and simple aromatic compounds derived from them are important feedstocks in industry. They are used in the production of a wide variety of consumer products.

Table 6 shows just a few of these aromatic feedstocks and some examples of the consumer products manufactured from them. You are not expected to remember the structures shown in Table 6.

You will meet further examples of aromatic compounds in **Chapter 5**. Kevlar is an example of a heat-resistant polymer containing benzene rings.

Aromatic feedstock	Consumer products
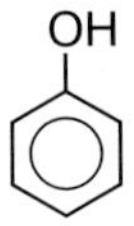 *phenol*	Germicides, eg TCP and Dettol 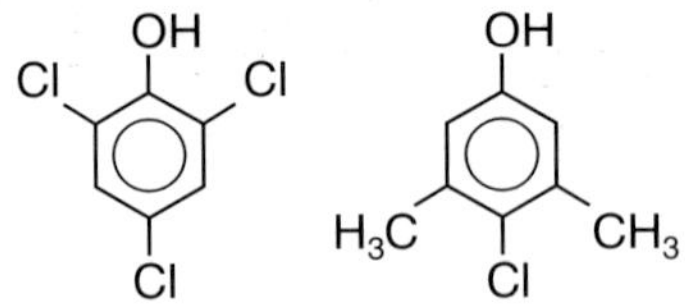 Plastics, eg phenolic resins 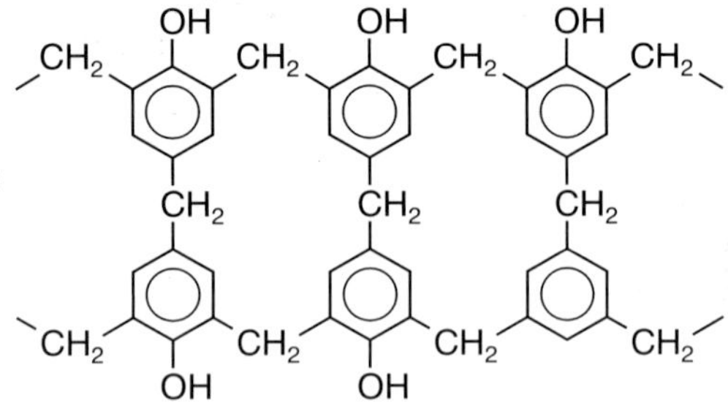
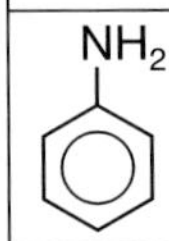 *phenylamine*	Dyes, eg aniline yellow 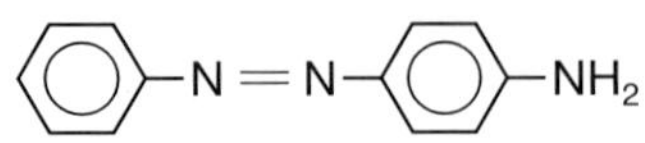
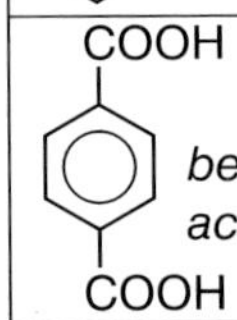 *benzene-1,4-dicarboxylic acid*	Textile fibres, eg polyesters 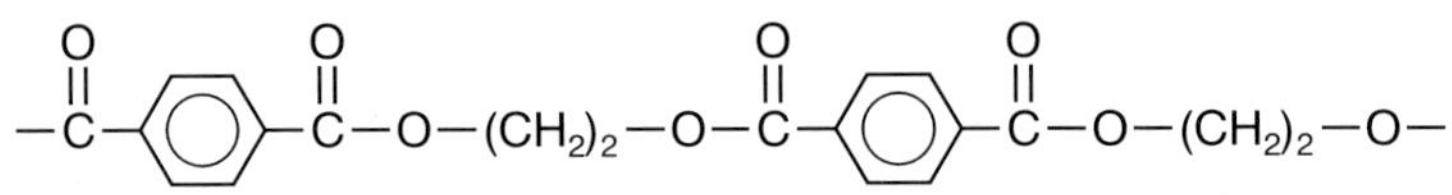
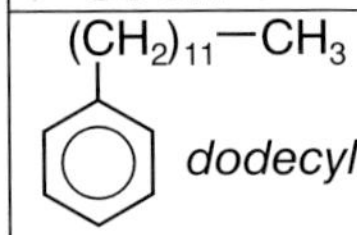 *dodecylbenzene*	Detergents, eg Fairy liquid 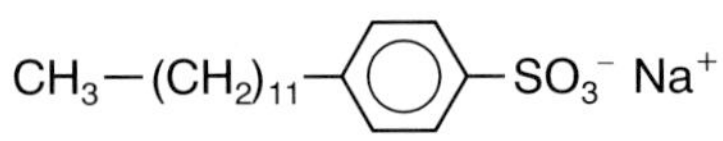
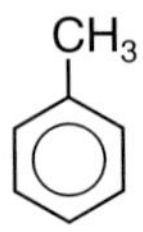 *methylbenzene*	Explosives, eg TNT 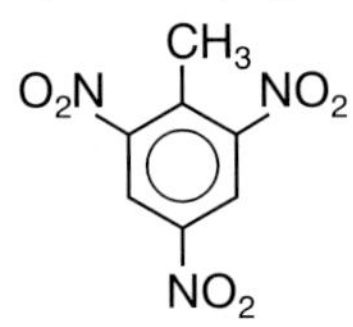

Table 6 Aromatic feedstocks and some consumer products manufactured from them

PROBLEMS FOR 17.5

1 Four aromatic hydrocarbons have the molecular formula C_8H_{10}.

- **a** Using  to represent a benzene ring, draw a shortened structural formula for each of these four aromatic hydrocarbons.
- **b** Which of your structures contains a phenyl group?

2 Aspirin has the following structural formula.

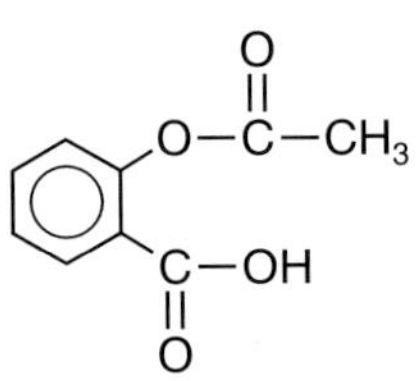

- **a** What does the circle in the benzene ring represent?
- **b** Write the molecular formula for aspirin.

3 Dewar benzene and benzvalene are isomers of benzene.

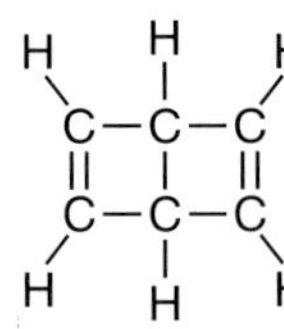
Dewar benzene

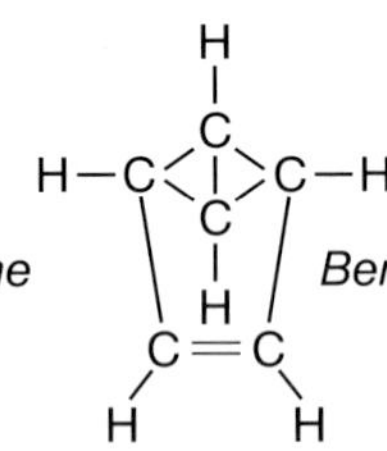
Benzvalene

- **a** Describe a chemical test which could be used to distinguish between Dewar benzene and benzene.
- **b** The skeletal structural formula for Dewar benzene is:

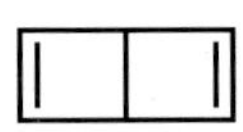

Draw the skeletal structural formula for benzvalene.

- **c** Benzene molecules, like graphite, contain delocalised electrons.
Why then does benzene *not* conduct electricity?

(*SQA*)

18

ORGANIC CHEMISTRY: OXYGEN AND NITROGEN COMPOUNDS

18.1 *Alcohols*

Ethanol is one of a group of compounds called **alcohols**. The full and shortened structural formulae of ethanol are:

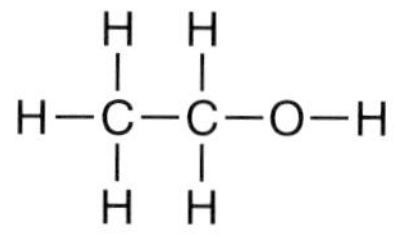

CH_3CH_2OH *ethanol*

All alcohols contain the **—OH** group of atoms, which is called the **hydroxyl group**. The atoms in this group are covalently bonded so it has no overall electrical charge.

The names of alcohols end in **-ol**.

The hydroxyl group is another example of a **functional group**. Since all alcohols have the hydroxyl functional group, they all have similar chemical properties. The name of any alcohol ends in *-ol*.

Alkanols

One group of alcohols is the homologous series called **alkanols**. These alcohols are derived from alkanes by substituting an —OH group for an —H atom. They have the general formula **ROH**, where R represents an alkyl group. They are named from the parent alkane by omitting the final *-e* and adding the ending *-ol*. For example,

CH_3OH	methanol
CH_3CH_2OH	ethanol
$CH_3CH_2CH_2OH$	propan-1-ol
$CH_3CH_2CH_2CH_2OH$	butan-1-ol
$CH_3CH_2CH_2CH_2CH_2OH$	pentan-1-ol
$CH_3CH_2CH_2CH_2CH_2CH_2OH$	hexan-1-ol
$CH_3CH_2CH_2CH_2CH_2CH_2CH_2OH$	heptan-1-ol
$CH_3CH_2CH_2CH_2CH_2CH_2CH_2CH_2OH$	octan-1-ol

For further details of naming alkanes, see **Section 17.1**.

Structure of alkanols

For alcohols in this series containing more than two carbon atoms, isomeric compounds are possible. To distinguish between these, it is necessary to label the position of the —OH group. The hydrocarbon chain is always numbered from the end which gives the lowest number for the position of the functional group. (For more information on isomerism, see **Section 17.2**.) For example, the full and shortened structural formulae of propan-1-ol are:

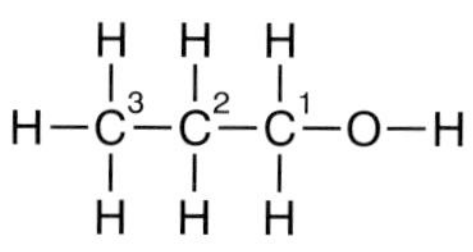

$CH_3CH_2CH_2OH$ *propan-1-ol*

The molecular formula of propan-1-ol is C_3H_8O. There is another alcohol which has the same molecular formula:

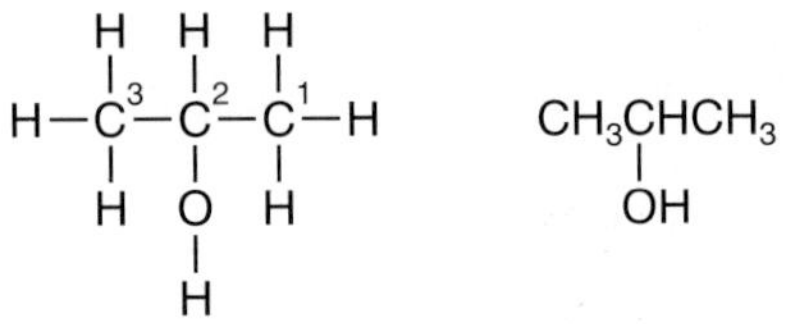

propan-2-ol

It is named propan-2-ol.

In alkanols with more than three carbon atoms, isomers can also arise due to differences in the carbon chain. Thus there are four alkanols with the molecular formula C_4H_9OH (Figure 1).

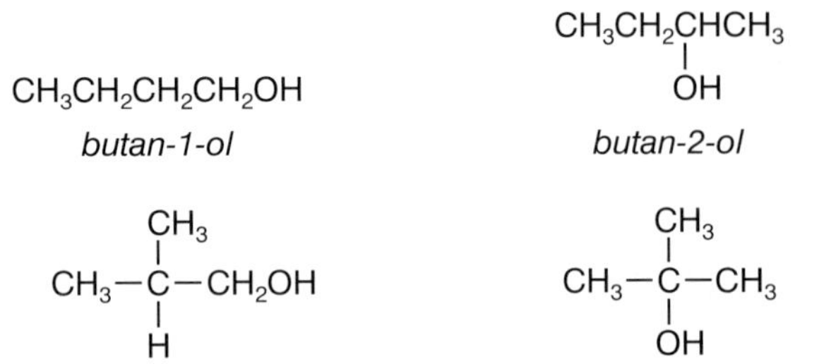

butan-1-ol *butan-2-ol*

2-methylpropan-1-ol *2-methylpropan-2-ol*

Figure 1 Structural isomers of C_4H_9OH

It can be seen that the hydroxyl group can be joined to a carbon atom with different numbers of hydrogen atoms. This means that the alcohols also differ in the number of alkyl groups attached to the hydroxyl-bearing carbon atom:

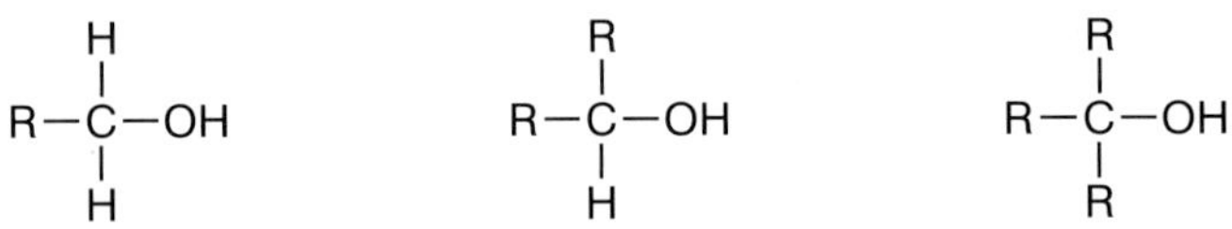

primary secondary tertiary

The three classes of alcohol have many similar chemical properties, owing to the presence of the same functional group, —OH. However, there are also differences, which are due to the different numbers of hydrogen atoms on the hydroxyl-bearing carbon atom. Propan-1-ol is a primary alcohol; propan-2-ol is a secondary alcohol. Amongst the isomers of C_4H_9OH shown above there are two primary alcohols, one secondary alcohol and one tertiary alcohol. The structures of the three classes of alcohol are summarised in Table 1.

Type of alcohol	Position of —OH group	Example
primary	at end of chain: 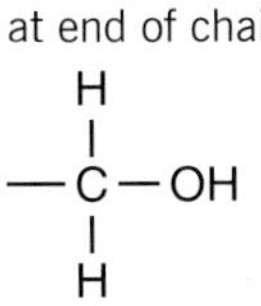	*propan-1-ol* 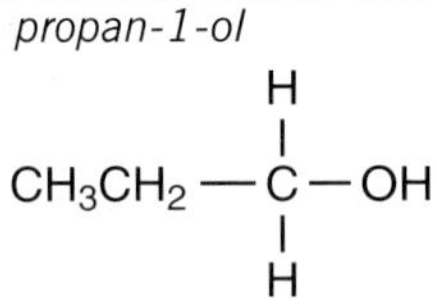
secondary	in middle of chain: 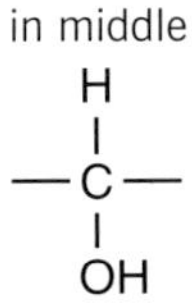	*propan-2-ol* 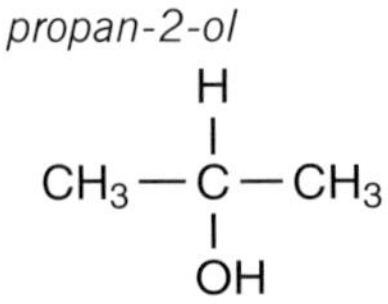
tertiary	attached to a carbon atom which carries no H atoms: 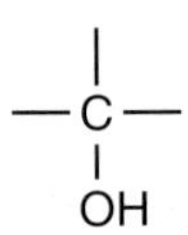	*2-methylpropan-2-ol* 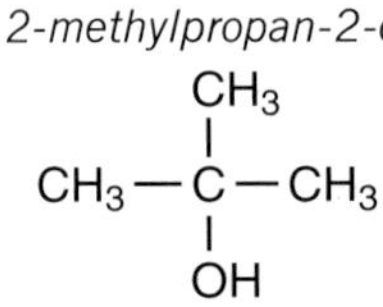

Table 1 Primary, secondary and tertiary alcohols

Some alcohols, including some which occur naturally, contain more than one —OH group in their molecules. They are known as **polyhydric** alcohols. Those with two —OH groups are known as **diols** or dihydric alcohols; for example:

$HOCH_2CH_2OH$ *ethane-1,2-diol*

Those which have three —OH groups are known as **triols** or trihydric alcohols. Propane-1,2,3-triol is an example.

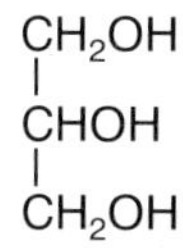

propane-1,2,3-triol (glycerol)

Physical properties of alcohols

As with the alkanes (see **Unit 1**, **Section 16.1**), the boiling points of alcohols increase fairly regularly on the addition of each $—CH_2—$ group. The increase is about 20 °C for the first members of the homologous series of alkanols (Table 2).

Alcohol	Molecular formula	Boiling point/°C
methanol	CH_3OH	64
ethanol	CH_3CH_2OH	78
propan-1-ol	$CH_3CH_2CH_2OH$	98
butan-1-ol	$CH_3CH_2CH_2CH_2OH$	118

Table 2 Boiling points of some alkanols

An increase in branching of the carbon chain is accompanied by a decrease in boiling point. This is illustrated by the four isomeric alcohols C_4H_9OH (Table 3). The structures of these alcohols are shown in Figure 1.

Alcohol	Boiling point/°C
butan-1-ol	118
2-methylpropan-1-ol	108
butan-2-ol	100
2-methylpropan-2-ol	83

Table 3 Boiling points of the alcohols C_4H_9OH

However, the boiling points of alcohols are considerably higher than those of alkanes of approximately the same relative formula mass, as shown in Table 4. This is because of the occurrence of strong attractive forces, known as **hydrogen bonds**, between the molecules of the alcohol. The bonds are electrostatic in nature; the positively charged H atom in the hydroxyl group of one molecule is attracted by a lone pair of electrons on the oxygen atom of another. This is similar to the hydrogen bonding that occurs in water (Figure 2).

Alkane	Relative formula mass	Boiling point/°C	Alcohol	Relative formula mass	Boiling point/°C
ethane	30	−89	methanol	32	64
propane	44	−42	ethanol	46	78
butane	58	−0.5	propan-1-ol	60	98

Table 4 Comparing boiling points of alkanes and alcohols

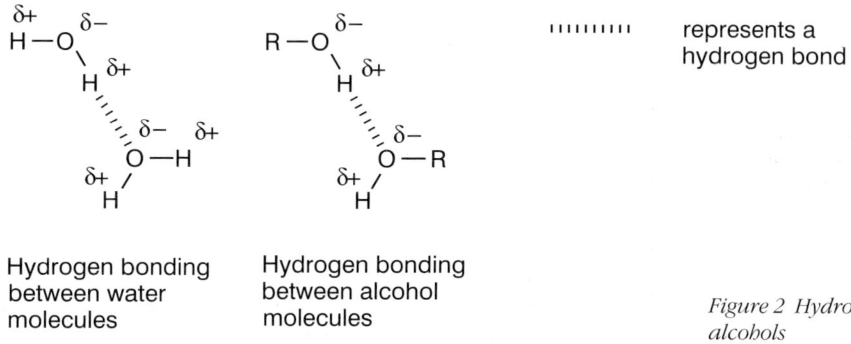

Figure 2 Hydrogen bonding in water and alcohols

It is because hydrogen bonding can occur between alcohol molecules and water molecules that the first members of the homologous series can mix with water to form a solution (Figure 3). Hydrogen bonds and other intermolecular forces are discussed in more detail in **Unit 1**, **Sections 16.1** and **16.4.**

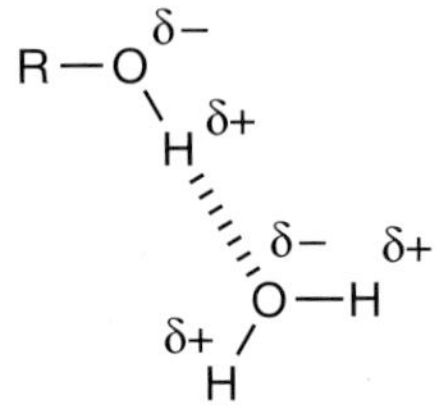

Figure 3 Hydrogen bonding between water molecules and alcohol molecules

Oxidation reactions of alcohols

The three classes of alcohols behave differently on oxidation. An oxidising agent such as an acidified solution of potassium dichromate, $K_2Cr_2O_7$, can be used. Alternatively, the alcohol vapour can be passed over heated copper. The reactions that occur under these conditions distinguish between the three types of alcohols.

Primary alcohols, such as ethanol, are oxidised to **aldehydes**, but the aldehyde is then itself oxidised to a **carboxylic acid**. The aldehyde can be extracted as an intermediate product if the reaction is done carefully. With ethanol:

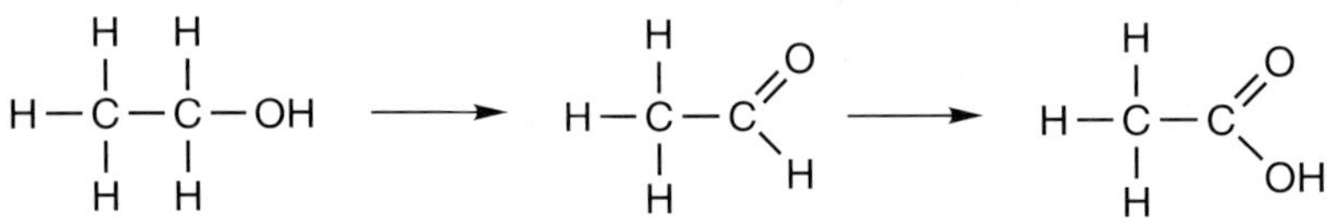

ethanol *ethanal* *ethanoic acid*

Secondary alcohols, such as propan-2-ol , are oxidised to **ketones**. For example:

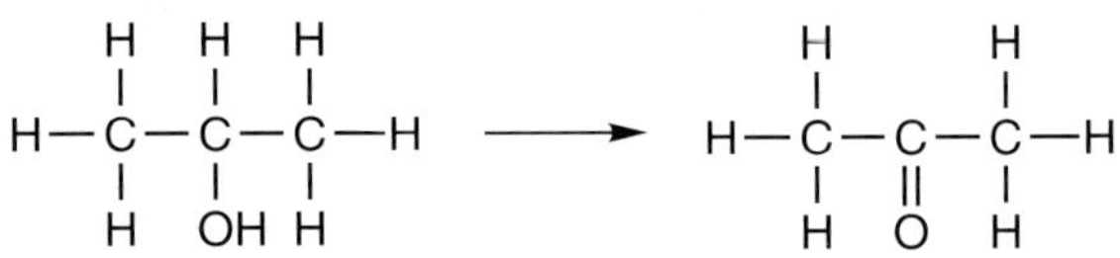

propan-2-ol *propanone*

It is difficult to oxidise the ketone further than this, because to do so would involve breaking a C—C bond.

Tertiary alcohols, such as 2-methylpropan-2-ol, are difficult to oxidise because they do not have a hydrogen atom on the carbon atom to which the —OH group is attached.

Table 5 summarises these reactions.

Type of alcohol	Product(s) on oxidation
primary	aldehyde, carboxylic acid
secondary	ketone
tertiary	does not oxidise

Table 5 Oxidation of alcohols

Uses of alcohols

Alcohols are an important group of organic compounds. It is a characteristic of organic chemistry that large complex molecules are built up from smaller, simpler ones. Because of this, chemists have found many uses for alcohols in the manufacture of pharmaceuticals, dyes, flavourings, plastics and many other types of chemicals. The smaller alkanols are also widely used as solvents.

Methanol is used as a fuel in cars, particularly where there are very strict regulations about pollutants, as in the US. Methanol can be converted into methanal, HCHO, which can be used to make Bakelite and other resins (see **Section 19.4**). About 33 million tonnes of methanol are manufactured worldwide each year.

The use of ethanol and methanol as alternative fuels is discussed in **Section 4.4**.

Ethanol is best known in alcoholic drinks, but it is also widely used (often as 'meths') for its solvent properties.

Ethane-1,2-diol, sometimes known as ethylene glycol or simply glycol, is best known as a coolant for car radiators. However, much more is used to make polyesters. Polyesters are used to make clothes ('Terylene') and containers (PET), for example for carbonated drinks.

You can read more about polyesters in **Sections 5.6** and **19.3**.

Manufacture of methanol

Methanol is manufactured from **synthesis gas**, a mixture of carbon monoxide and hydrogen gases.

$$\underset{\textit{synthesis gas}}{CO(g) + 2H_2(g)} \xrightarrow[250\,^{\circ}C]{\text{Cu catalyst}} \underset{\textit{methanol}}{CH_3OH(l)}$$

Synthesis gas is produced from natural gas (see **Section 9.1**) and other hydrocarbons, and from coal, by process known as **steam reforming**. The manufacture of methanol from synthesis gas is discussed in more detail in **Unit 3**, **Section 20.2**.

Ethers

Ethers are derived from alkanes by substituting an alkoxy group (—OR) for an H atom. For example,

$CH_3CH_2—O—CH_2CH_3$	*ethoxyethane*
$CH_3CH_2CH_2—O—CH_3$	*methoxypropane*

General formula of ethers R—O—R'

Note that the longer hydrocarbon chain is chosen as the parent alkane.

Physical properties of ethers

You can think of ethers as being derived from water by replacing *both* the H atoms by alkyl groups.

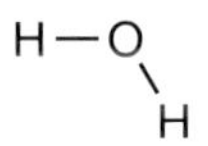

water

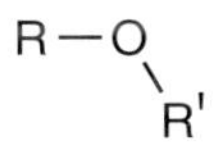

an *ether*

Ether molecules are only slightly polar and the attractive forces *between* molecules are relatively weak. There are no H atoms attached to the oxygen to form hydrogen bonds between ether molecules.

The boiling point of an ether is similar to that of the alkane with corresponding relative formula mass. Like alkanes, the lower ethers are very volatile and dangerously flammable.

Ethers are only slightly soluble in water, but mix well with other non-polar molecules such as alkanes.

See **Section 4.3** for details of the use of MTBE as an ingredient in different blends of petrol.

One ether, methyl tertiary butyl ether (MTBE) is widely used in petrol.

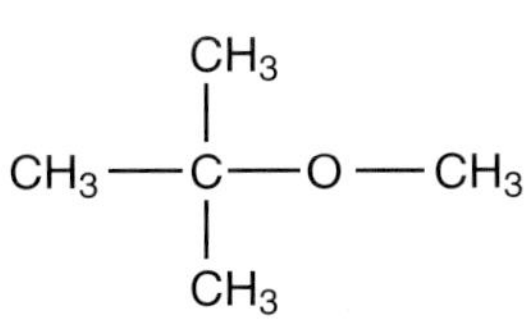

MTBE

PROBLEMS FOR 18.1

1 Draw the structural formulae of the following isomers of pentan-1-ol:
 a pentan-2-ol
 b 2-methylbutan-1-ol
 c 2-methylbutan-2-ol
 d 3-methylbutan-2-ol
 e 2,2-dimethylpropanol.
 In each case state whether the alcohol is a primary, secondary or tertiary alcohol.

2 Explain why the boiling point of propane-1,2,3-triol is much higher than that of heptan-1-ol, even though the relative formula masses of the two substances are similar.

3 Name the following alcohols:
 a 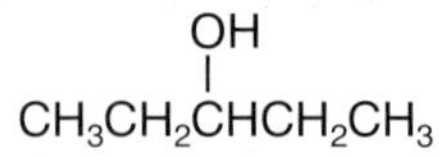
 b 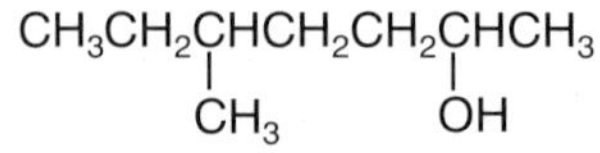
 c 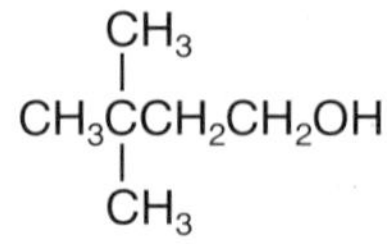
 In each case state whether the alcohol is a primary, secondary or tertiary alcohol.

18.2 *Aldehydes and ketones*

Aldehydes and ketones are two series of organic compounds which both contain the **carbonyl group** $>$C=O .

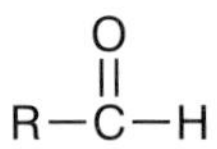

aldehyde

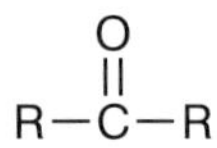

ketone

The carbonyl functional group has characteristic properties which are shown by both classes of compound, so that it is convenient for the two homologous series to be considered together. However, the attachment of a hydrogen atom to the carbonyl group in an aldehyde gives aldehydes certain properties which ketones do not possess and which enable the two classes of compound to be distinguished from one another.

Aldehydes are named using the suffix *-al*. For example:

The names of aldehydes end in **-al**.

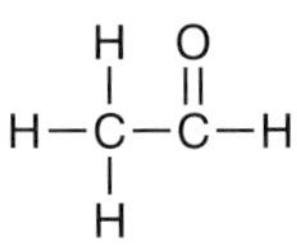 CH_3CHO *ethanal*

The homologous series of aldehydes based on alkanes are sometimes called **alkanals**. An alkanal is named in the same manner as an alkanol (see **Section 18.1**) except that the name ends in *-al*. For example:

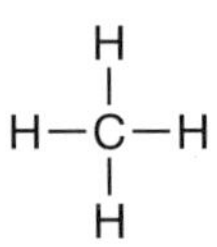

methane

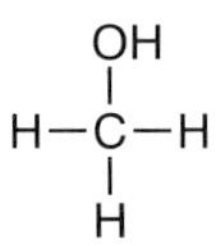

methanol

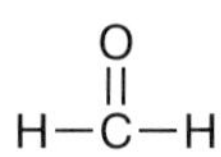

methanal

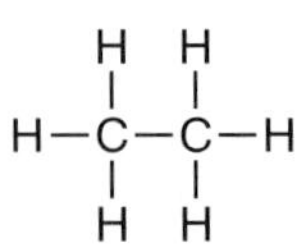

ethane

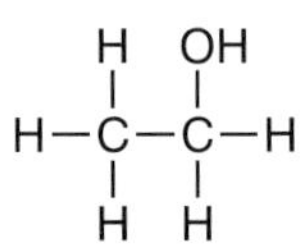

ethanol

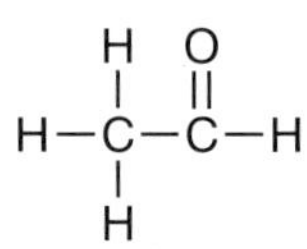

ethanal

Ketones are named using the suffix *-one*. For example:

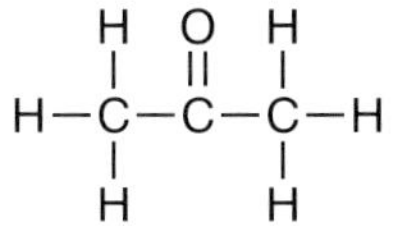

CH_3COCH_3 *propanone*

The names of ketones end in **-one**.

The homologous series of ketones based on alkanes are sometimes known as **alkanones**. The method of naming follows the usual rules and the name always ends in *-one*. For example:

$CH_3CH_2CH_3$
propane

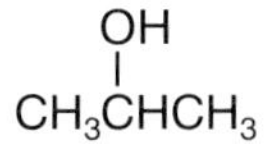

propan-2-ol

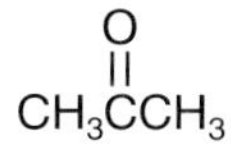

propanone

$CH_3CH_2CH_2CH_2CH_3$
pentane

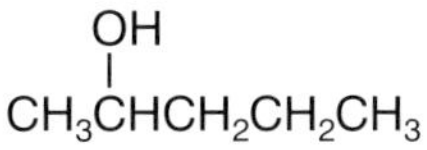

pentan-2-ol

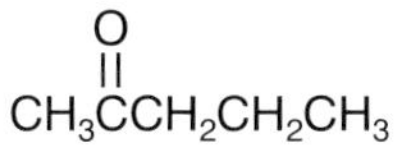

pentan-2-one

Alkanones with four or more carbon atoms have isomers. The position of the carbonyl group is specified by inserting the number of its carbon atom counting from the nearer end of the chain. For example:

When naming aldehydes and ketones, note that the carbon atom in the carbonyl group counts as one of the carbon atoms for the name.

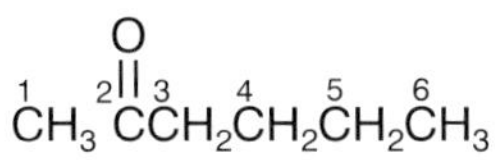

hexan-2-one

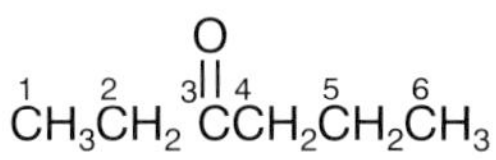

hexan-3-one

Physical properties of aldehydes and ketones

Methanal is a gas, other aldehydes and ketones of relatively low relative formula mass are liquids and the remainder are solids. Methanal dissolves readily in water; a 40 per cent solution is know as formalin. The liquid aldehydes and ketones of low relative formula mass are also readily soluble in water (for example, ethanal and propanone mix with water in all proportions), but the solubility of alkanals and alkanones decreases as the carbon chain lengthens.

Preparation of aldehydes and ketones

Aldehydes and ketones can be produced in the laboratory by gentle oxidation of alcohols (see **Section 18.1**). The —OH group can be oxidised by oxidising agents such acidified potassium dichromate solution. The orange dichromate ion, $Cr_2O_7^{2-}$, is reduced to green Cr^{3+} in solution. Alternatively, the alcohol vapour can be passed over heated copper. The basic reaction is:

the carbonyl group

Aldehydes are produced from *primary* alcohols and ketones are formed from *secondary* alcohols. For example:

ethanol $CH_3CH_2OH \longrightarrow CH_3CHO$ *ethanal*

propan-2-ol

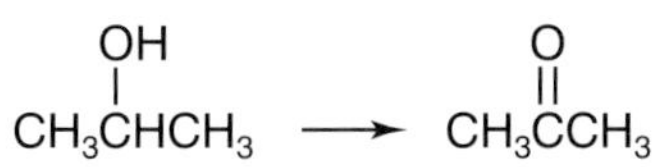

propanone

Notice how the nomenclature changes from *-ol* to either *-al* or *-one*.

Reactions of aldehydes and ketones

Many of the reactions of aldehydes and ketones are similar, because they both contain a carbonyl functional group. However, there are key differences because the aldehydes contain a hydrogen atom next to the carbonyl group.

The hydrogen atom next to the carbonyl group means that aldehydes are easily oxidised, even by weak oxidising agents such as the copper(II) ion, Cu^{2+}, in Fehling's solution. The aldehyde is oxidised to an acid and the blue copper(II) ion, in solution, is reduced to copper(I) oxide, which is precipitated as an orange brown solid. For example, ethanal is oxidised to ethanoic acid by Fehling's solution:

ethanal

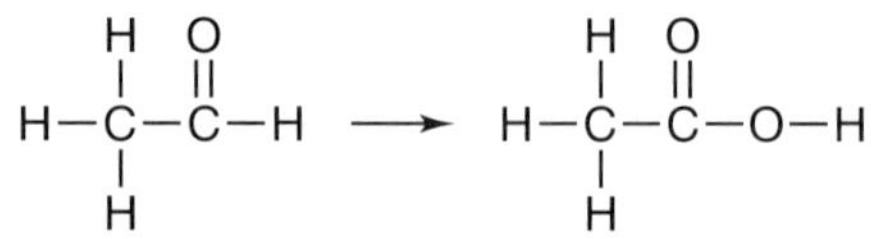

ethanoic acid

Ketones, on the other hand, are not readily oxidised by either Fehling's solution or an acidified solution of potassium dichromate.

> Reaction with Fehling's solution is used as a test to distinguish between aldehydes and ketones.

Uses of aldehydes and ketones

The most used aldehydes and ketones are methanal, HCHO, and propanone, CH_3COCH_3. **Methanal** is used to make thermosetting polymers (see **Section 19.4**). Small amounts are used in solution as formalin, a disinfectant. **Propanone** is also used to make polymers, for example Perspex.

PROBLEMS FOR 18.2

1 Name the following alkanals:
 a HCHO
 b CH_3CH_2CHO
 c 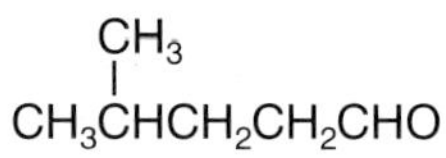

2 Name the following alkanones:
 a $CH_3COCH_2CH_3$
 b $CH_3CH_2COCH_2CH_3$

3 Draw the full structural formulae of the following compounds:
 a pentanal b pentan-2-one.

4 Which of the following alcohols can be oxidised to an alkanal or an alkanone? In each case state which type of carbonyl compound would be formed and draw its full structural formula.
 a $CH_3CH_2CH(OH)CH(CH_3)CH_3$
 b $CH_3C(CH_3)_2CH_2OH$
 c $CH_3CH_2CH(CH_3)CH(OH)CH_3$

These formulae are sometimes used as an alternative to the other shortened form.
Thus **b** is

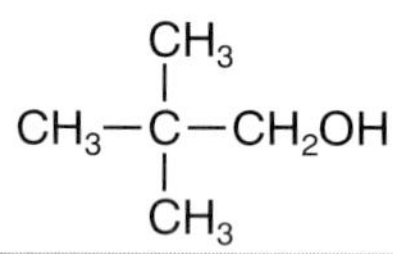

18.3 Oxidation and reduction of carbon compounds

This is a good time to review what is meant by the terms oxidation and reduction in organic chemistry.

The idea of oxidation and reduction of ions in terms of loss and gain of electrons is discussed in **Unit 3**, **Chapter 24**. In the case of carbon compounds, we are usually dealing with neutral molecules, and so we need a different way of deciding whether a substance has been oxidised or reduced.

For **oxidation** to occur, oxygen must be added or hydrogen must be removed from the substance. This means that **the ratio of oxygen to**

hydrogen is increased or that hydrogen is removed. Some examples of oxidation are:

CH_3CH_2OH *ethanol*	$\rightarrow$	CH_3CHO *ethanal*
CH_3CH_2CHO *propanal*	$\rightarrow$	$CH_3CH_2CO_2H$ *propanoic acid*
CH_3CH_3 *ethane*	$\rightarrow$	$CH_2{=}CH_2$ *ethene*

It is important to note that the converse of this statement is not true. Not all reactions in which the ratio of oxygen to hydrogen is increased can be classified as oxidation. For example, ethanol is manufactured from ethene and steam (see **Section 17.3**):

$CH_2{=}CH_2 + H_2O$ *ethene*	$\rightarrow$	CH_3CH_2OH *ethanol*

Although the proportion of oxygen to hydrogen has increased, such a process is known as **hydration** because the elements of water have been added.

Reduction is the reverse of oxidation. A substance has been reduced if **the ratio of hydrogen to oxygen has been increased or if hydrogen has been added.** Some examples of reduction are:

butanone → *butan-2-ol*

$CH_3CH{=}CH_2$ *propene*	$\longrightarrow$	$CH_3CH_2CH_3$ *propane*

PROBLEMS FOR 18.3

1 For each of the following reactions, (i) give the name of the reactant and products and (ii) state whether an oxidation or a reduction has occurred.

a $CH_3CHO \rightarrow CH_3COOH$

b $CH_3CH_2OH \rightarrow CH_3CHO$

c 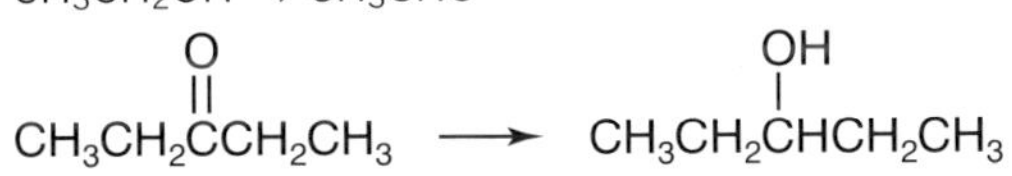

d $CH_3CHO \rightarrow CO_2$

e $CH_3C{\equiv}CH \rightarrow CH_3CH_2CH_3$

2 State the oxidising or reducing agent you would use for the following reactions.

a $CH_3OH \rightarrow HCHO$

b 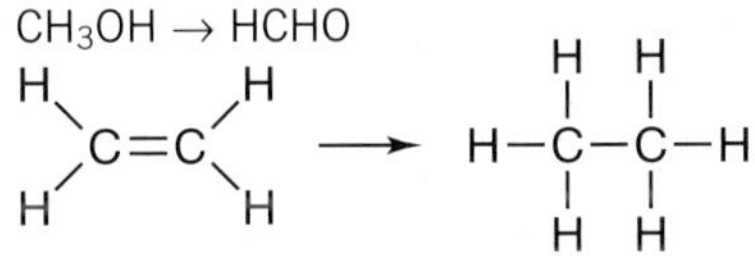

c $CH_3CHO \rightarrow CH_3COOH$

18.4 *Carboxylic acids*

This section provides an introduction to carboxylic acids. You will find more information about some of the reactions of carboxylic acids in **Sections 5.5**, **18.5**, **18.6** and **19.3**. Carboxylic acids get their name from their characteristic functional group, the **carboxyl group**.

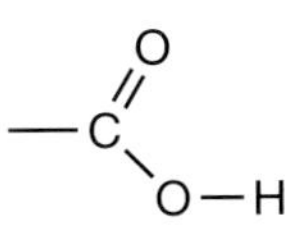

carboxyl group

This formula is often abbreviated to —COOH, although the two oxygen atoms are *not* joined together.

The structure of the rest of the molecule can vary widely and this gives rise to a large number of different carboxylic acids. When it is an alkyl group the acids are also sometimes known as **alkanoic acids** and can be represented by the general formula **RCOOH**.

Alkanoic acids are named from the parent alkane by omitting the final *-e* and adding the ending *-oic acid*. For example:

One C atom:	CH_4 *methane*	HCOOH *methanoic acid*
Two C atoms:	C_2H_6 *ethane*	CH_3COOH *ethanoic acid*
Three C atoms:	C_3H_8 *propane*	CH_3CH_2COOH *propanoic acid*

Note that the carbon atom in the —COOH group counts as one of the carbon atoms for the name.

Branched-chain carboxylic acids are also named in a similar way to alkanes.

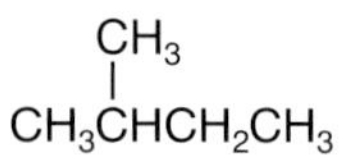
2-methylbutane

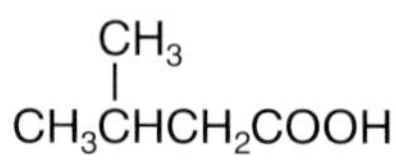
3-methylbutanoic acid

The first two members of the alkanoic acid series, methanoic acid and ethanoic acid, are often still called by their older names, formic acid and acetic acid, so it is as well to know these too.

The names of carboxylic acids end in **-oic acid**.

Manufacture and uses of carboxylic acids

The uses of some of the most important carboxylic acids are outlined in this book.

Ethanoic acid, manufactured from methanol, is used as a feedstock for ethenyl ethanoate, which is then polymerised. The polymer is widely used in paints:

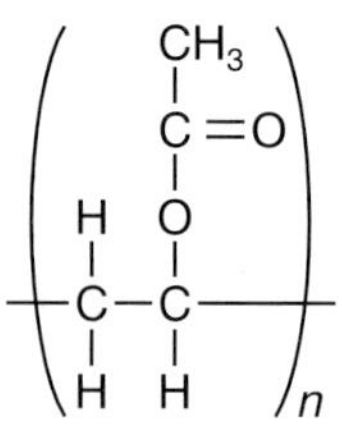
poly(ethenyl ethanoate)

Ethanoic acid, in dilute solution, is more familiar as vinegar.

Long-chain carboxylic acids are used to make soaps (**Section 18.5**). Two dicarboxylic acids are used to make very well-known polymers. Benzene-1,4-dicarboxylic acid is used in the manufacture of Terylene, a polyester, and hexanedioic acid in the manufacture of nylon-6,6.

The discovery and development of nylon is discussed in **Section 5.5**. Polyesters are introduced in **Section 5.6**. Details can also be found in **Section 19.3**.

PROBLEMS FOR 18.4

1 Name the following carboxylic acids:

a $CH_3CH_2CH_2COOH$

b 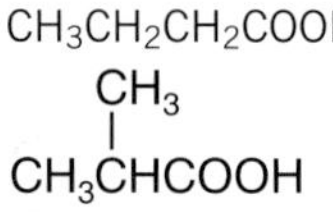

2 Draw the structural formulae of the following carboxylic acids:

a hexanoic acid

b 3-methylbutanoic acid.

18.5 *Esters, oils and fats*

What are esters and how are they named?

Esters are formed when an alcohol reacts with a carboxylic acid.

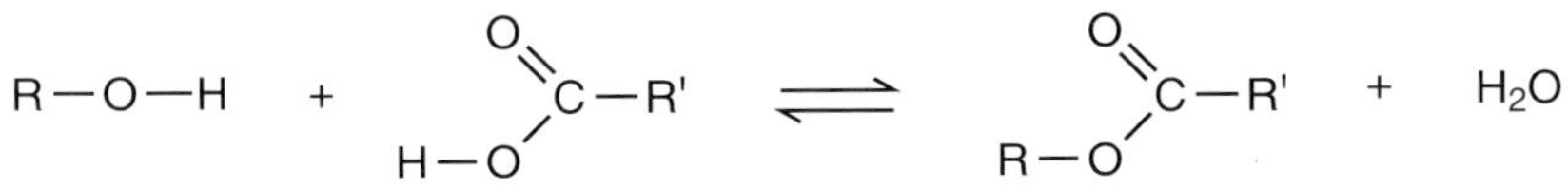

(R is an *alkyl group* such as CH_3 or C_2H_5.)

Because water is produced, just as when we breathe on a cold surface, the process was christened 'condensation' and the name has stuck. A **condensation reaction** can be thought of as two or more molecules reacting together to form a larger molecule with the elimination of small molecules such as water.

The ester link is formed by the reaction of the hydroxyl group in the alcohol and the carboxyl group in the acid. The process is known as **esterification**.

Esters are named after the alcohol and acid from which they are derived. For example, ethyl ethanoate is formed from ethanol and ethanoic acid:

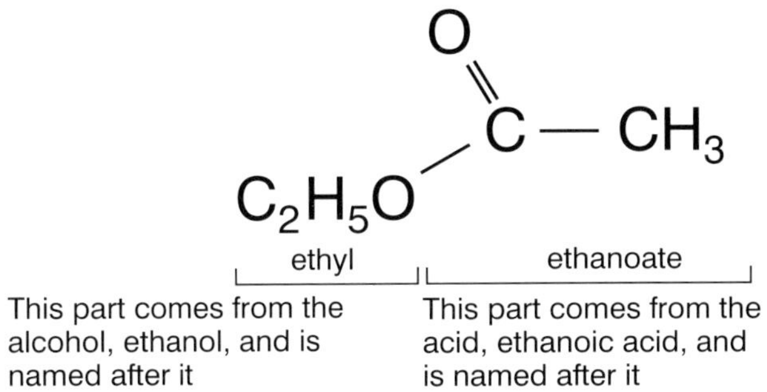

The ending *–oate* instead of the *–oic* from the parent acid shows clearly that the compound is an ester.

The names of esters end in **-oate**.

You have to be careful to get the 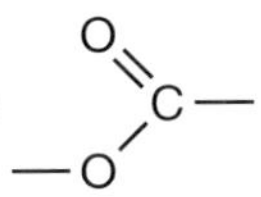 group the right way round.

For example, ethyl ethanoate 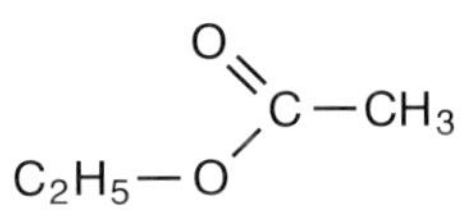

is very different from its isomer, methyl propanoate 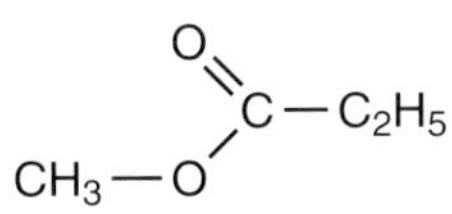

Making models of these two esters will help you understand how they are named.

The names of esters can be arrived at in several ways.

- *From the names of the parent alkanol and alkanoic acid*
  For example, meth|anol and butan|oic acid
  ↓ methyl ↓ butanoate
  ↘ methyl butanoate ↙

- *From the full structural formula*

For example,

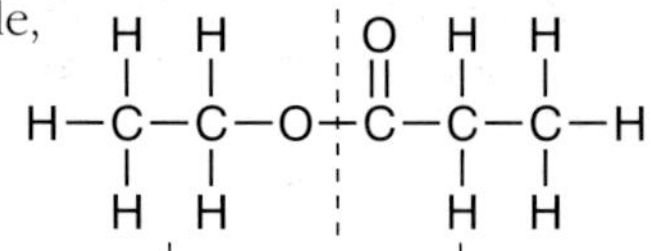

ethyl propanoate

ethyl → ethyl propanoate; propanoate → ethyl propanoate

- *From the shortened structural formula*

For example, $C_3H_7O|OCCH_3$

propyl ethanoate

propyl → propyl ethanoate; ethanoate → propyl ethanoate

You can draw shortened and full structural formulae for esters given either of the following:

- *The names of the parent alkanol and alkanoic acid*

For example, methanol + butanoic acid

CH_3OH C_3H_7COOH

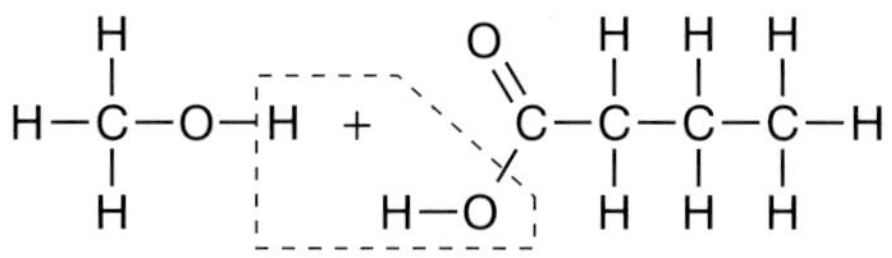

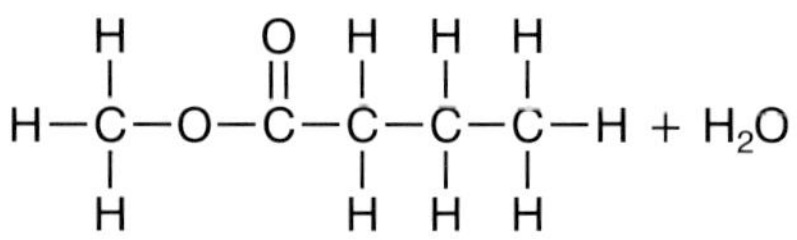 Full structural formula

$CH_3OOCCH_2CH_2CH_3$ Shortened structural formula

- *The name of the ester*

For example, ethyl methanoate

ethanol methanoic acid

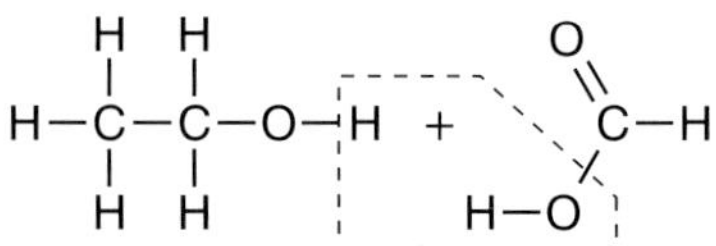

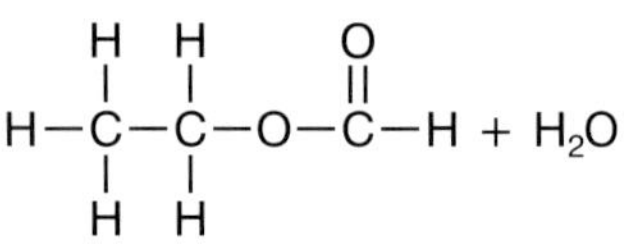 Full structural formula

CH_3CH_2OOCH Shortened structural formula

Preparation of esters

Esters are often prepared by heating the parent acid and alcohol together under reflux. For example:

$$\underset{\textit{methanol}}{CH_3OH} + \underset{\textit{ethanoic acid}}{CH_3COOH} \rightleftharpoons \underset{\textit{methyl ethanoate}}{CH_3OOCCH_3 + H_2O}$$

This is a reversible reaction and it eventually comes to equilibrium. To improve the yield from a given amount of acid, an excess of alcohol is added (see **Unit 3**, **Section 20.2**). Concentrated sulphuric acid is added as a catalyst.

Radioisotopes were used to work out the mechanism of this reaction – see **Unit 3**, **Section 9.4**.

Uses of esters

Esters have strong sweet smells which are often floral or fruity. Many naturally occurring esters are responsible for well-known fragrances. Some examples are given in Table 6. Esters can be used in food flavourings and in perfumes.

Table 6 Some ester fragrances

Ester	Fragrance
ethyl methanoate	raspberries
propyl ethanoate	pears
methyl butanoate	apples
pentyl butanoate	apricots

Many organic compounds dissolve readily in esters, so that esters are often used as solvents.

Reactions of esters

One of the most important reactions of esters is hydrolysis, a reaction that is the reverse of the way they are made. For example:

$$\underset{\textit{ethyl ethanoate}}{C_2H_5OOCCH_3} + H_2O \rightleftharpoons \underset{\textit{ethanol}}{C_2H_5OH} + \underset{\textit{ethanoic acid}}{CH_3COOH}$$

This can be shown with full structural formulae:

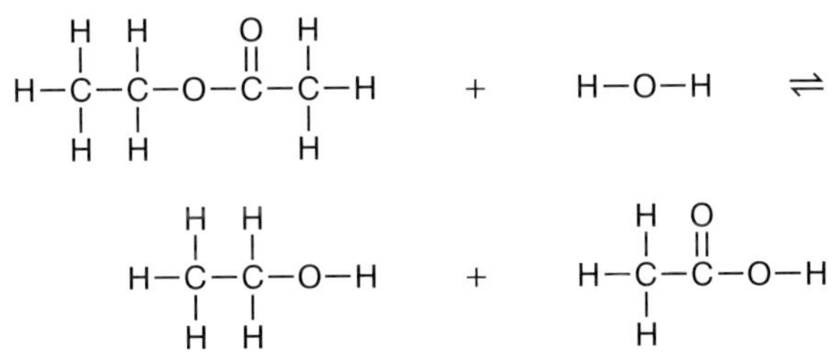

This is a **hydrolysis** reaction. However, on their own, water and an ester react very slowly, but the process can be speeded up by catalysis. For example, dilute sulphuric acid is a catalyst. Excess water displaces the equilibrium to the right and the yield of products is improved.

Another way of hydrolysing an ester is to add an alkali, such as sodium hydroxide solution:

$$\underset{\textit{ethyl ethanoate}}{C_2H_5OOCH_3} + NaOH \rightleftharpoons \underset{\textit{ethanol}}{C_2H_5OH} + \underset{\textit{sodium ethanoate}}{CH_3COO^-Na^+}$$

The salt of ethanoic acid, sodium ethanoate, is produced. This hydrolysis reaction, in which an alkali is used, is sometimes called **saponification**.

Esters from 2-hydroxybenzoic acid

A look at the structure of 2-hydroxybenzoic acid shows that there are two ways of esterifying it. Either the —OH group could be reacted with a carboxylic acid, or the —COOH group could be reacted with an alcohol. *Aspirin* is the product of esterifying the —OH group to form 2-ethanoylhydroxybenzoic acid. It is quite soluble in water, so it can be absorbed into the bloodstream through the stomach wall.

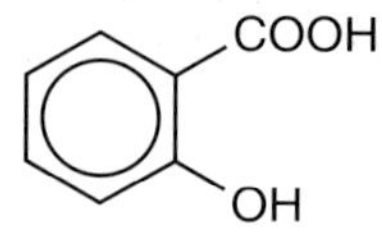
2-hydroxybenzoic acid

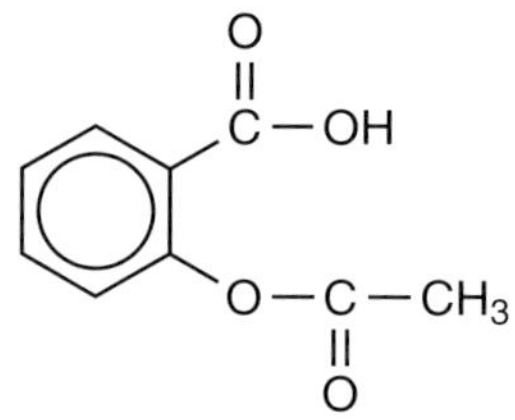
aspirin

The product of reacting the —COOH group with methanol is called methyl 2-hydroxybenzoate. This is better known as *oil of wintergreen* which is used as a liniment. It is soluble in fat rather than water so it is absorbed through the skin. Like aspirin, it reduces pain and swelling.

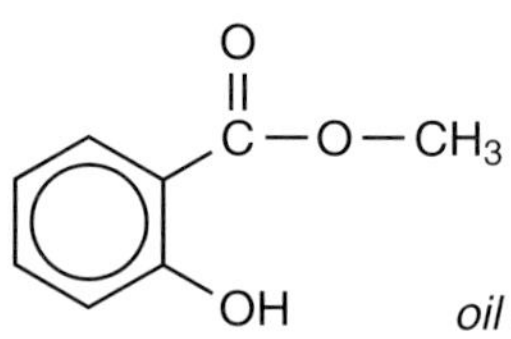
oil of wintergreen

Polyesters

Polyesters are condensation polymers. They are made by reacting a *diol* (containing two —OH groups) with a *dicarboxylic acid* (containing two —COOH groups). For example, a common polyester called Terylene is made from ethane-1,2-diol and benzene-1,4-dicarboxylic acid.

The manufacture of polyesters is discussed in **Section 5.6**. Their structure and properties are described in **Section 19.3**.

$HOCH_2CH_2OH$

ethane-1,2-diol

benzene-1,4-dicarboxylic acid

Oils and fats: examples of esters

Oils and fats provide an important way of storing chemical energy in living systems. Most of them have the same basic structure. The only difference is that oils are liquid at room temperature whereas fats are solid. A selection of natural oils and fats is listed in Table 7.

Oils and fats are a more concentrated source of energy than carbohydrates; see **Unit 1**, **Section 2.2**.

Animal fats	Marine oils	Vegetable oils
beef (dripping)	whale	sunflower
pig (lard)	herring	olive
mutton	cod liver	rapeseed

Table 7 Classification of fats and oils, with some examples

Most oils and fats are **esters** of propane-1,2,3-triol (commonly called *glycerol*) with long-chain carboxylic acids, RCOOH. In *palmitic acid*, for example, R is a $CH_3(CH_2)_{14}$— group.

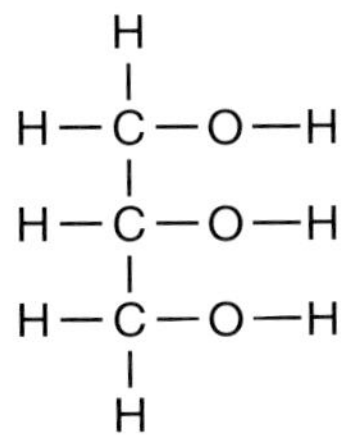

propane-1,2,3-triol (glycerol)

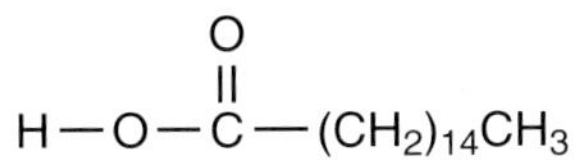

palmitic acid

Because glycerol has three alcohol groups in each molecule, *three* carboxylic acid molecules can form ester linkages with each glycerol molecule to form a **triester** (sometimes called a **triglyceride**). The triester formed from glycerol and palmitic acid has the following structure:

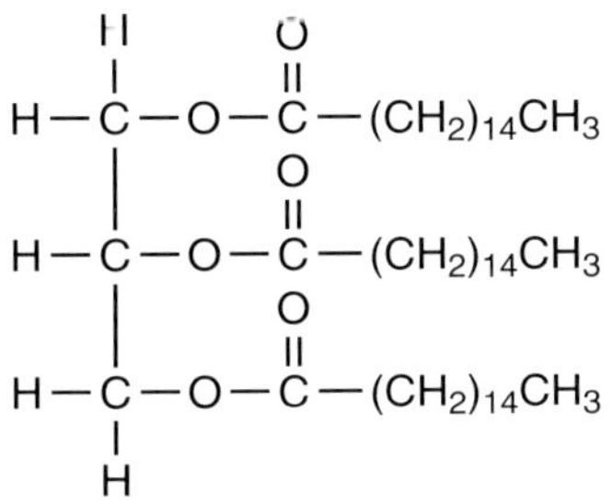

triester formed from *glycerol* and *palmitic acid*

The triesters found in natural oils and fats are often **mixed triesters** in which the three acid groups are not all the same. Figure 4 summarises the structure of triesters.

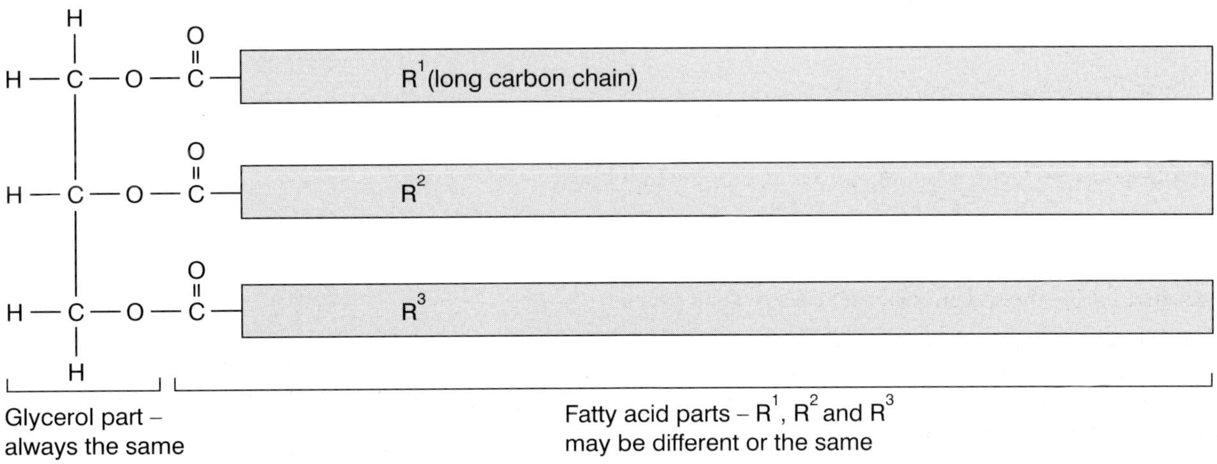

Figure 4 The general structure of the triesters found in fats and oils

Fats and fatty acids

The carboxylic acids in fats and oils usually have unbranched, straight-chain, hydrocarbon chains. They contain an even number of carbon atoms, ranging from C_4 to C_{24}, but often contain 16 or 18 carbon atoms. They are sometimes called **fatty acids** because of their origin. The alkyl groups, R^1,

R^2 and R^3, are either fully saturated or contain one or more double bonds. Table 8 shows some common fatty acids. There are many others, some of which contain three or four double bonds.

Structure	Traditional name	Origin of name
$CH_3(CH_2)_{14}COOH$	palmitic acid	palm oil
$CH_3(CH_2)_{16}COOH$	stearic acid	suet (Greek: *stear*)
$CH_3(CH_2)_7CH{=}CH(CH_2)_7COOH$	oleic acid	olive oil
$CH_3(CH_2)_4CH{=}CHCH_2CH{=}CH(CH_2)_7COOH$	linoleic acid	oil of flax (Latin: *linum*)

Table 8 Common fatty acids

Many doctors believe that the more saturated fats – the ones with few double bonds – tend to cause blockage of blood vessels and so may lead to heart disease. They believe that a diet containing *polyunsaturated* fats with many double bonds may be healthier. Figure 5 shows a representation (called a *skeletal formula*), of typical saturated and unsaturated fatty acids.

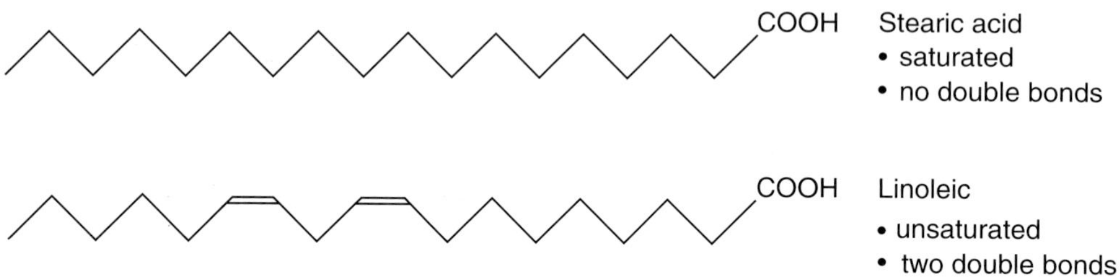

Figure 5 Typical saturated and unsaturated fatty acids

The modern systematic names for fatty acids can get quite long (for example, oleic acid is octadec-9-enoic acid), so they are still known by their old names. These names are often derived from one of the oils or fats from which they are obtained (see Table 8).

The *proportions* in which the acid groups occur are more or less constant for a particular oil or fat. Human body fat, for example, contains mainly four acids:

oleic acid	47%
palmitic acid	24%
linoleic acid	10%
stearic acid	8%

An oil or fat can be identified by breaking it down into propane-1,2,3-triol (glycerol) and its fatty acids and then measuring the amount of each acid present.

Like all esters, oils and fats can be split up by **hydrolysis**. This is usually done by heating the oil or fat with concentrated sodium hydroxide solution to give the triol and the sodium salts of the acids. For example:

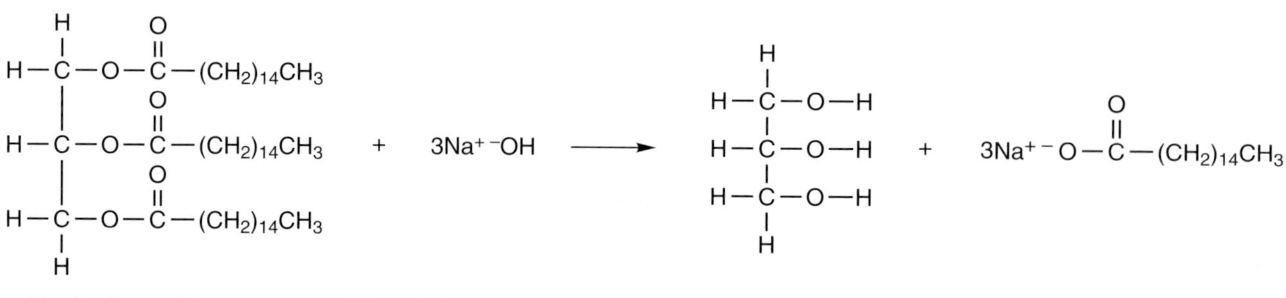

triester formed from *propane-1,2,3-triol* and *palmitic acid* — *propane-1,2,3-triol* — *sodium palmitate*

This is how soap is made. Soaps are sodium or potassium salts of fatty acids and they are made by heating oils and fats with sodium or potassium hydroxide. What oils do you think *Palmolive* soap is made from?

The free fatty acids can be released from the sodium salts by adding a dilute mineral acid such as hydrochloric acid.

Solid or liquid?

Any natural oil or fat contains a mixture of triesters. The nature of the fatty acids present affects the properties and determines whether the substance is a liquid oil or a solid fat.

Saturated and unsaturated triglyceride molecules can be represented as shown in Figure 6.

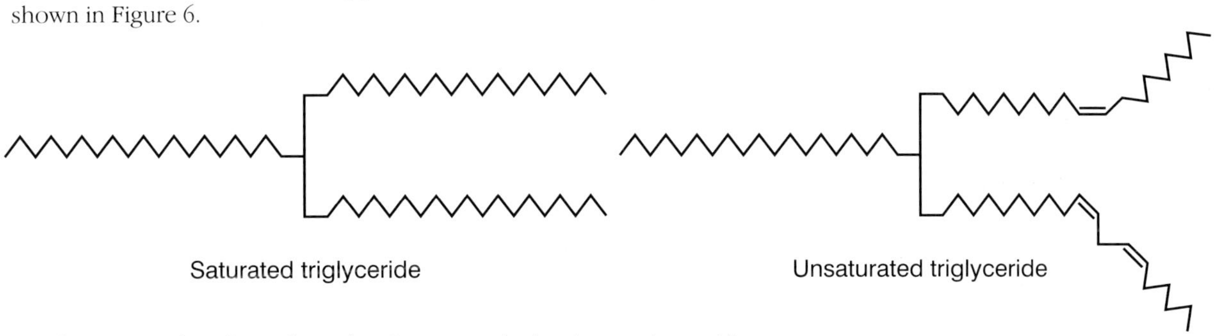

Figure 6 Representation of triglycerides

The saturated triglyceride molecules can pack closely together, with intermolecular van der Waals forces holding the molecules together. The unsaturated triglyceride molecules cannot pack so closely because the presence of double bonds causes the molecules to 'kink'! This means the van der Waals forces between molecules will be weaker than between the saturated molecules.

This means that the unsaturated molecules require less energy to separate them and so will have a lower melting point than the corresponding saturated molecules. As a result, the more unsaturated fatty acid molecules there are in a tryglyceride mixture the more likely it is to be a liquid at room temperature, ie an oil. The oils shown in Table 7 have a higher proportion of unsaturated triglycerides than the fats.

Converting oils to fats

Although many natural oils and fats can be eaten directly, most require processing in order to make them fit a number of requirements we have regarding taste, texture and so on. One of the commonest processes is the **hydrogenation** of unsaturated oils to make margarine.

The addition of hydrogen to unsaturated molecules reduces the number of double bonds, ie the molecules become less unsaturated and more saturated.

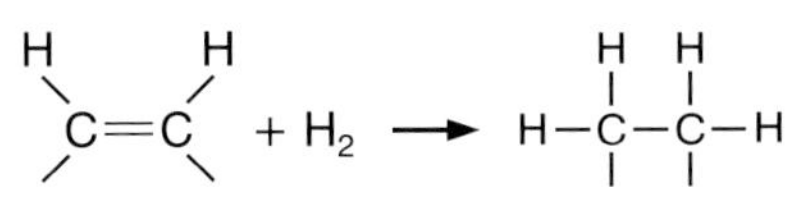

double bond in
fatty acid chain

This in turn means the molecules become solid. Too much hydrogenation causes the resulting fats to be too hard and brittle, so the hydrogenation has to be carefully controlled.

Modern margarines are made by passing hydrogen through heated oil containing a nickel catalyst. Fat-free milk is added, along with flavourings, salt and colour, and sometimes vitamins. Margarine manufacture is summarised in Figure 7.

Hydrogenation is an example of an addition reaction. You can read about the addition reactions of alkenes in **Section 17.3**.

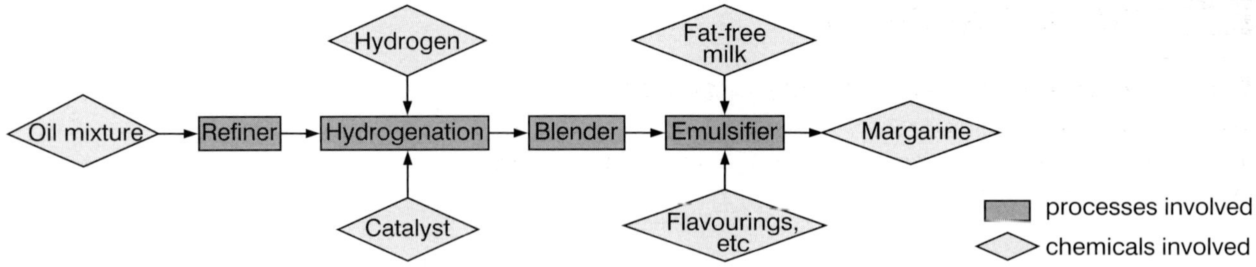

Figure 7 Flow chart showing margarine manufacture

PROBLEMS FOR 18.5

1 **a** Write down the formulae for methyl ethanoate and ethyl methanoate.

b How are they related?

2 Look at the esters in Table 6.

a Write down the names of the alcohols and acids from which they are derived.

b Write down the structures of these alcohols, acids and esters.

3 Name and draw the structures for the esters formed when the following are heated with a concentrated sulphuric acid catalyst under reflux:

a ethanol and propanoic acid

b ethanoic acid and propanol.

4 Name and draw the structures of the products when the following are heated in dilute sulphuric acid under reflux:

a ethyl ethanoate

b propyl methanoate.

5 Name and draw the structures of the products when the following are heated in dilute sodium hydroxide solution under reflux:

a ethyl ethanoate

b propyl methanoate.

6 **a** What structural feature of a molecule identifies a compound as an ester?

b Write down the full structural formula of the triester made from propane-1,2,3-triol (glycerol) and ethanoic acid. Make a model of this triester.

c Give an equation to show what happens when this ester is heated with hot concentrated sodium hydroxide solution.

7 Below is the structure of a triester present in a naturally occurring oil.

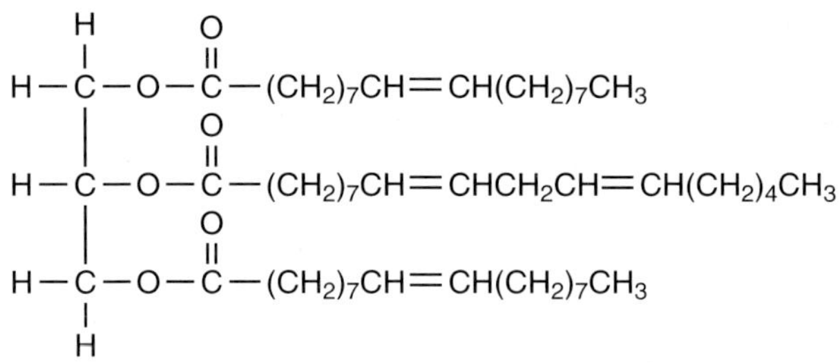

The triester was hydrolysed by heating with hot concentrated sodium hydroxide solution, and the resulting solution was neutralised by addition of hydrochloric acid.
Name the hydrolysis products obtained. How many moles of each product would be obtained from 1 mole of the triester?

8 Explain the difference between the following terms which often appear on food labels:

a saturated fat

b monounsaturated fat

c polyunsaturated fat.

9 4-Hydroxybutanoic acid has the structure $HOCH_2CH_2CH_2COOH$.
Individual molecules of 4-hydroxybutanoic acid can undergo 'internal esterification'. Draw a structure for this ester. It will help if you build a model.

18.6 *Amines and amides*

What are amines and how are they named?

Amines are the organic chemistry relatives of ammonia. Their structures resemble ammonia molecules in which alkyl groups take the place of one, two or all three hydrogen atoms.

ammonia primary amine secondary amine tertiary amine

Notice the following points about naming amines.

- Amines with *one* alkyl group are called **primary amines**; **secondary amines** have *two* alkyl groups, and **tertiary amines** have *three* alkyl groups.
- Lower primary amines are named as follows:

  CH_3NH_2 — *methylamine* $CH_3CH_2NH_2$ — *ethylamine*

- Lower secondary and tertiary amines are named as follows:

dimethylamine *trimethylamine*

- Higher homologues are often named using the prefix 'amino' and the name of the alkane from which they appear to be derived:

$CH_3CH_2CH_2CH_2CH_3$ *pentane*

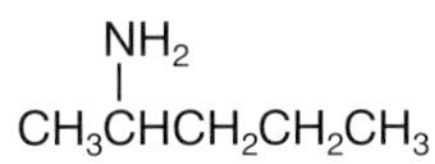

2-aminopentane

Physical properties of amines

Amines with low relative formula masses are gases or volatile liquids. The volatile amines also resemble ammonia in having strong smells. The characteristic smell of decaying fish comes from amines such as ethylamine and trimethylamine. Rotting animal flesh gives off the diamines $H_2N(CH_2)_4NH_2$ and $H_2N(CH_2)_5NH_2$, which are sometimes called by the names putrescine and cadaverine, respectively, although their proper names are 1,4-diaminobutane and 1,5-diaminopentane, respectively.

Reactions of amines

The most important properties of amines are that they are, like ammonia, bases. They react with acids to form salts. For example:

$$CH_3NH_2 + HCl \rightarrow CH_3\overset{+}{N}H_3\ Cl^-$$

methylamine → *methylammonium chloride*

They also react with compounds containing a carboxyl group, such as carboxylic acids, to form **amides**.

What are amides?

Amides contain the functional group

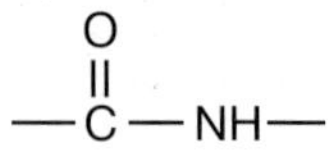

Primary amides have the formula

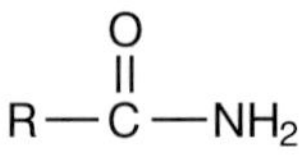

where the hydroxyl group, —OH, in a carboxylic acid has been replaced by an amino group, $—NH_2$. For example:

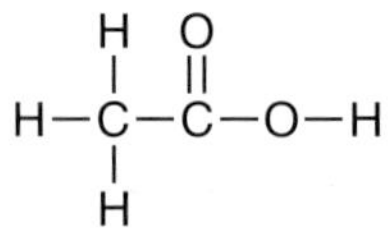

ethanoic acid

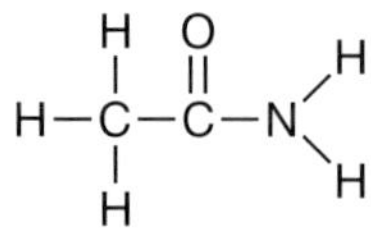

ethanamide

Secondary amides have the formula

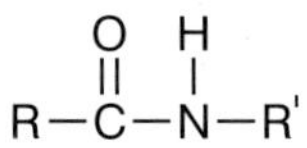

They can be considered to be formed from an amine and a carboxylic acid. For example:

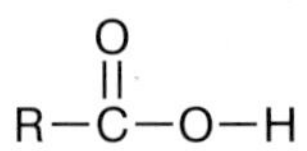

carboxylic acid

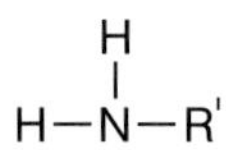

amine

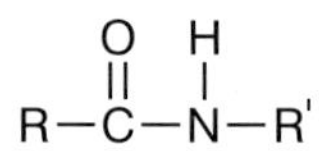

secondary amide

The 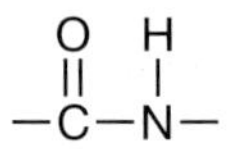 group is known as a **secondary amide** group. It is the link in condensation polymers (*polyamides* or *nylons*) and in proteins, naturally occurring condensation polymers of amino acids (see **Section 18.7**). In proteins, the secondary amide link is known as a **peptide** link.

See **Sections 5.5** and **19.3** for more information about nylon.

18.7 *Amino acids: the building blocks of proteins*

Amino acids contain at least one amino group and one carboxylic acid group. The reactions of amino acids resemble reactions of amines (see **Section 18.6**) and carboxylic acids (see **Section 18.4**).

The simplest amino acid is aminoethanoic acid, which is known as glycine.

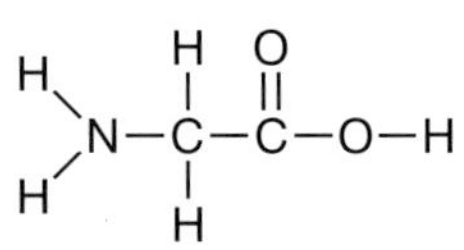

glycine

The next member of the series is aminopropanoic acid: two isomers exist.

$H_2NCH_2CH_2COOH$

β-aminopropanoic acid

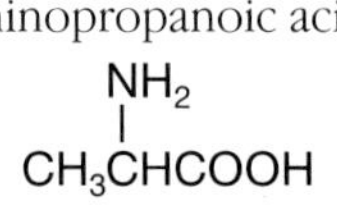

α-aminopropanoic acid

When the amino group is attached to the carbon atom next to the carboxylic acid group it is called an **α-amino acid**.

All the proteins in the world are made from just 20 α-amino acids, which have the same general structure as shown in Figure 8.

The 20 animo acids which occur in proteins each have a different side chain, labelled R. Table 9 shows the structures of these R groups and the names of the amino acids, together with their abbreviated symbols.

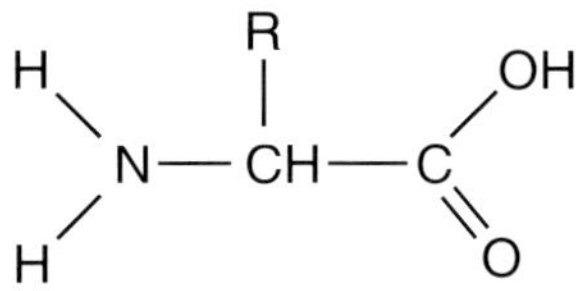

The amino group ($—NH_2$) and the carboxylic acid group ($—COOH$) are both attached to the same carbon atom: the α-carbon. That is why they are called α-amino acids.

Figure 8 General structure of α-amino acids

Amino acid	Abbreviation	R group	Amino acid	Abbreviation	R group
glycine	Gly	$—H$	cysteine	Cys	$—CH_2—SH$
alanine	Ala	$—CH_3$	methionine	Met	$—CH_2—CH_2—S—CH_3$
valine	Val	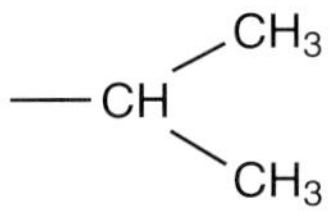	aspartic acid	Asp	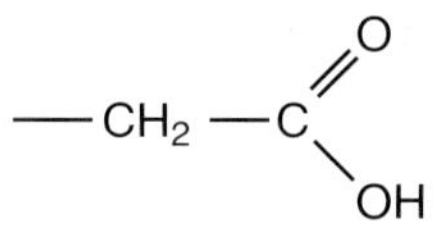
leucine	Leu	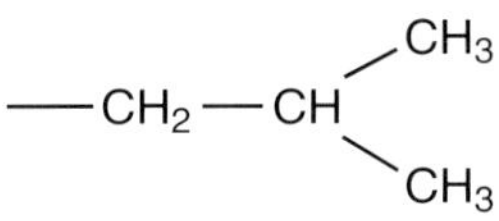	glutamic acid	Glu	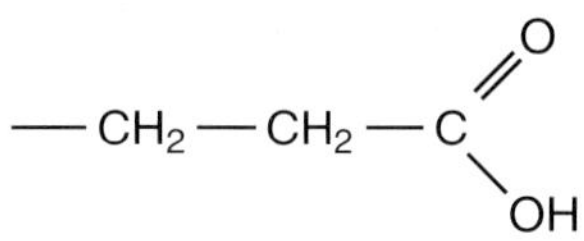
isoleucine	Ile	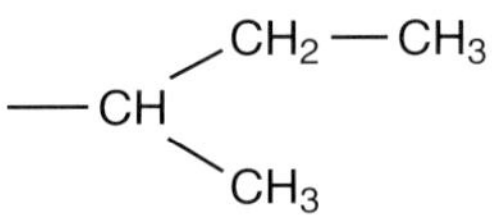	asparagine	Asn	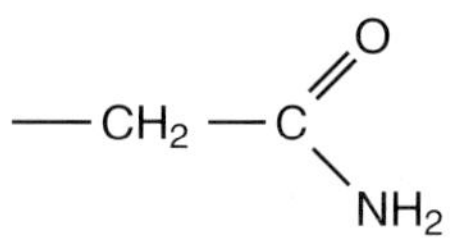
phenylalanine	Phe	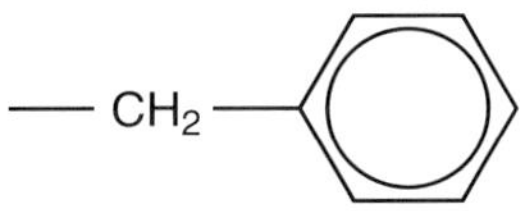	glutamine	Gln	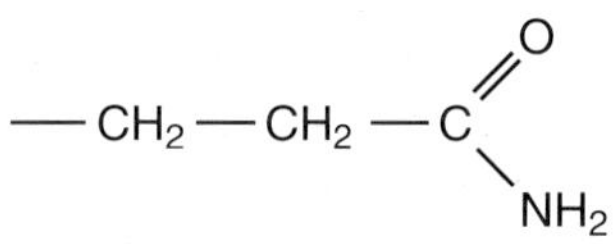
proline	Pro	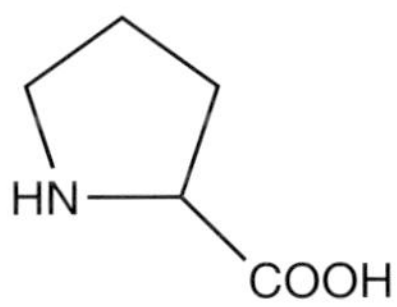	tyrosine	Tyr	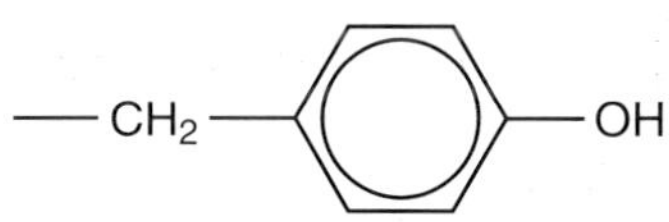
tryptophan	Trp	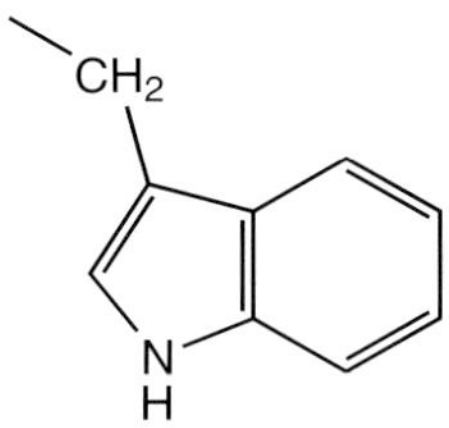	histidine	His	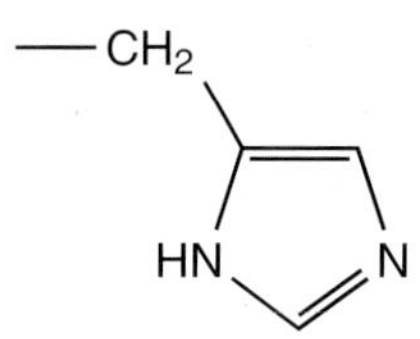
serine	Ser	$—CH_2—OH$	lysine	Lys	$—CH_2—CH_2—CH_2—CH_2—NH_2$
threonine	Thr	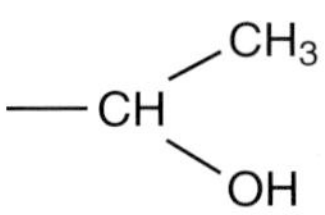	arginine	Arg	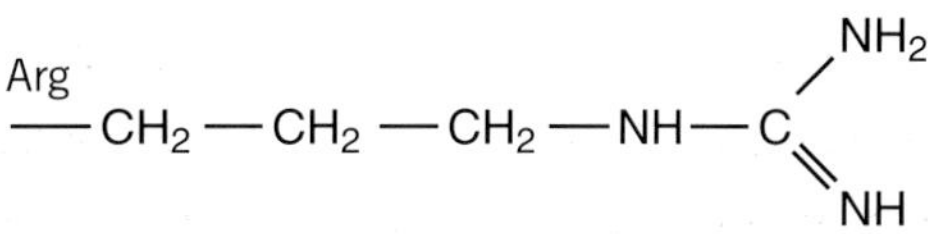

Table 9 The 20 amino acids which make up proteins (for clarity, the whole amino acid has been drawn out in the case of proline)

In all of us, even adults who have stopped growing, proteins need to be continuously replaced. This is obvious when your hair and your nails grow after they have been cut. It is also important that hormones and some enzymes (like the digestive enzymes) are made when needed, then destroyed once they have done their job, so that they do not go on producing their effects after the need for them has passed.

We replace our proteins from the food we eat, but we do not need to eat proteins which are identical to the ones being replaced. We do not need to eat human hair orf finger nails to grow our own! In fact, we don't even have to eat animal protein; we can supply our needs by eating plants.

For example, one molecule of the protein insulin is made from 54 amino acid molecules, 14 of them being different. They are joined to one another by **peptide links**. Thus, in a protein such as insulin, the carboxylic acid group on one amino acid joins on to the amino group on the next, and a molecule of water is lost from between them. This type of process is called a **condensation reaction** (see **Section 18.6**).

You can find out more about insulin in **Section 5.9**.

During digestion our bodies break proteins down by hydrolysis into their constituent 'building blocks' – amino acids. Eight of these are usually classified as **essential amino acids** – our bodies cannot synthesise them: we must take them in as food. The other 12 can be made from carbohydrate and other amino acids in the body.

The amino acids are then reassembled to make our own collection of proteins. For each of us, this collection is unique: most of your proteins will be indentical to those found in other humans, but some will be different. Human proteins are also different from the proteins of other animals. However, in some cases very similar proteins are found not just throughout the animal kingdom but in plants and micro-organisms as well: for example, the enzymes which oxidise glucose in cells' metabolism.

All these millions of proteins are built up from the same small number of amino acids. What makes each protein different is the order in which the amino acids are joined to one another. This is called the **primary structure**.

Details of the secondary, tertiary and quaternary structure of insulin can be found in **Section 5.9**.

Making peptides

To put 40 or more amino acids together in the right order is a wondrous process. Scientists are now able to break down a protein into its constituent amino acids. They can also assemble amino acids in precisely the right order. To do this, they need to be able to assemble little 'bits' of the final protein, known as **peptides**.

As discussed above, amino acids are joined together by peptide links. For example:

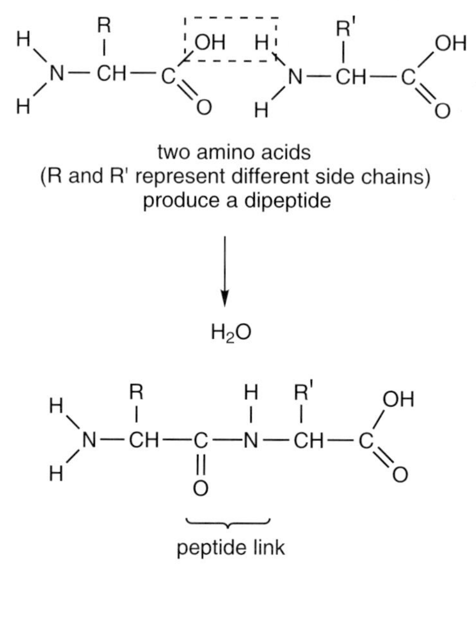

The dipeptide formed from glycine (Gly) and alanine (Ala) is

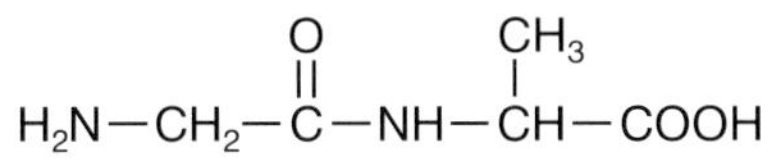

This dipeptide is abbreviated to GlyAla.

From a dipeptide, scientists can make a tripeptide by adding another α-amino acid molecule. A **polypeptide** contains up to 40 amino acid residues. A protein contains more than about 40 amino acid residues, although the distinction between a polypeptide and a protein is an arbitrary one (the term 'residue' is used for an α-amino acid which has lost the elements of water in forming a peptide or protein).

PROBLEMS FOR 18.6 AND 18.7

1 Name the amines with the following structures:
   - a $C_2H_5NH_2$
   - b $(CH_3)_2NH$
   - c 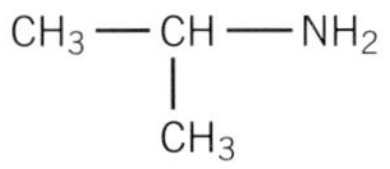
   - d $(CH_3)_2NC_2H_5$

2 Draw structures for the following amines:
   - a propylamine
   - b diethylamine
   - c diethylmethylamine.

3 Draw structures for the following compounds:
   - a α-aminobutanoic acid
   - b ethanamide
   - c phenylamine
   - d the peptide GlySer. (Use Table 9.)

4 Aspartame is used as a sugar substitute in diet drinks. It is produced from the dipeptide AspPhe. Draw out the structure of this dipeptide and the structures of the amino acids formed when it is hydrolysed in the stomach.

5 A protein chain contains the following section

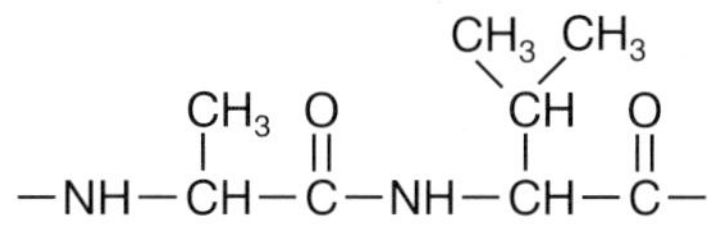

Draw out the structures of the amino acids obtained from the hydrolysis of this section of the protein chain. Use Table 9 to name the amino acids formed.

18.8 *Calculating percentage yield*

Balanced equations allow us to calculate the quantity of products which can be obtained from known amounts of reactants. In the calculations carried out so far it has been assumed that 100% conversion takes place. The maximum quantity of product that can be obtained is known as the **theoretical yield**. The **actual yield** obtained is usually less than the theoretical yield. There are a number of reasons for this:

- part of the reactants may not react
- there may be side reactions, ie more than one product
- not all of the product is separated out.

The actual yield can be expressed as a percentage:

$$\text{percentage yield} = \frac{\text{actual yield}}{\text{theoretical yield}} \times 100$$

Example

Carbon monoxide and hydrogen react to form methanol:

$$CO + 2H_2 \rightarrow CH_3OH$$

When 22.4 kg of carbon monoxide react with excess hydrogen, 19.2 kg of methanol is produced.

a Calculate the theoretical yield.

b Calculate the percentage yield.

Answer

a Theoretical yield

Step 1: Balanced equation and mole ratio

CO	+ $2H_2$	$\rightarrow CH_3OH$
1 mole	2 mole	1 mole

Step 2: Convert to mass

$$28\,g\ CO \rightarrow 32\,g\ CH_3OH$$

$$1\,g\ CO \rightarrow \frac{32}{28}\,g\ CH_3OH$$

$$1\,kg\ CO \rightarrow \frac{32}{28}\,kg\ CH_3OH$$

$$22.4\,kg\ CO \rightarrow \frac{32}{28} \times 22.4 = 25.6\,kg\ CH_3OH$$

Theoretical yield of CH_3OH is 25.6 kg.

b Percentage yield

$$\text{Percentage yield} = \frac{\text{actual yield}}{\text{theoretical yield}} \times 100$$

$$= \frac{19.2}{25.6} \times 100$$

$$= 75\%$$

PROBLEMS FOR 18.8

1 When methane reacts with nitric acid, nitromethane, a fuel used in Indy car racing, is produced.

$CH_4 + HNO_3 \rightarrow CH_3NO_2 + H_2O$

When 160 g of methane is reacted with excess nitric acid, 500 g of nitromethane is produced.

a Calculate the theoretical yield.

b Calculate the percentage yield.

2 Iron(III) oxide can be reduced to iron by carbon monoxide:

$Fe_2O_3 + 3CO \rightarrow 2Fe + 3CO_2$

When 200 kg of iron(III) oxide reacts, 120 kg of iron is produced.

a Calculate the theoretical yield.

b Calculate the percentage yield.

19

THE STRUCTURE AND PROPERTIES OF POLYMERS

19.1 *What is a polymer?*

A **polymer** molecule is a long molecule made up from lots of small molecules called **monomers**. If all the monomer molecules are the same, and they are represented by the letter A, an A–A polymer forms:

$$\ldots\ A + A + A + A\ \ldots \rightarrow \ldots\ \text{–A–A–A–A–}\ \ldots$$

Poly(ethene) and poly(chloroethene), pvc, are examples of A–A polymers.

If two different monomers are used, an A–B polymer is formed, in which A and B monomers alternate along the chain:

$$\ldots\ A + B + A + B\ \ldots \rightarrow \ldots\ \text{–A–B–A–B–}\ \ldots$$

Polyamides (nylons) and polyesters are examples of this type of A–B polymer.

More information about polymers can be found in **Chapter 5**.

19.2 *Addition polymerisation*

Many polymers are formed by the reaction known as **addition polymerisation**. The monomers usually contain C=C double bonds, for example, in alkenes. The addition polymerisation of propene is typical:

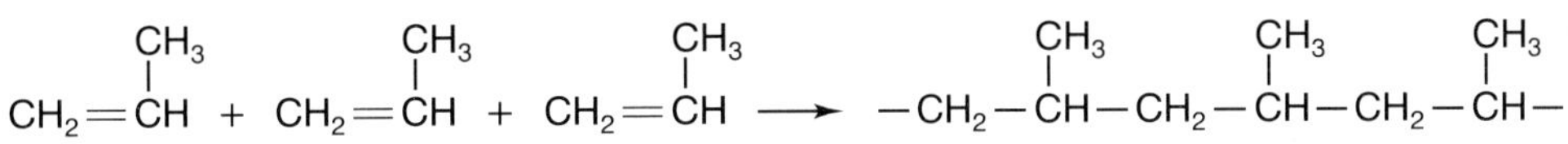

propene monomer

poly(propene) polymer

In the chain, the same basic unit is repeated over and over again, so the polymer structure can be shown simply as

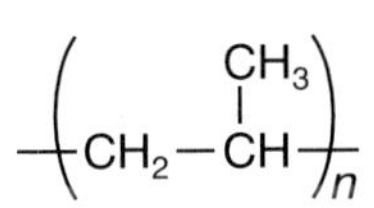

n = a very large number

the repeating unit of poly(propene)

Addition polymers have also recently been developed based on alkynes:

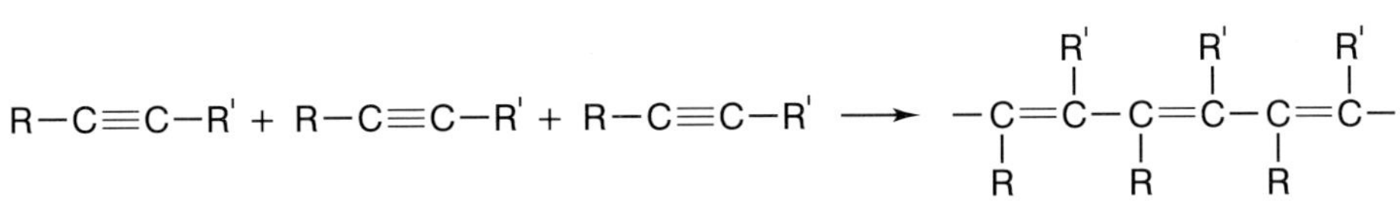

alkyne monomer

poly(alkyne) polymer

and the repeating unit is

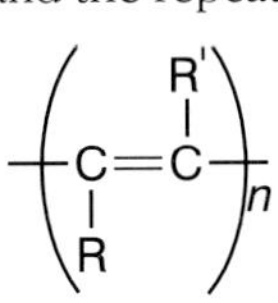

The structures of other addition polymers are shown in **Section 17.3**.

Poly(ethyne), where R and R' = H, is a conducting polymer. You can read about its discovery and properties in **Section 5.4**.

19.3 Condensation polymerisation

Polymers can also be formed by the condensation reaction that is used to make esters. The monomers need to have at least two suitable functional groups per molecule for a condensation polymer to be produced.

The fibre, Terylene, a **polyester**, is a typical condensation polymer. It is formed from two monomers, as shown below:

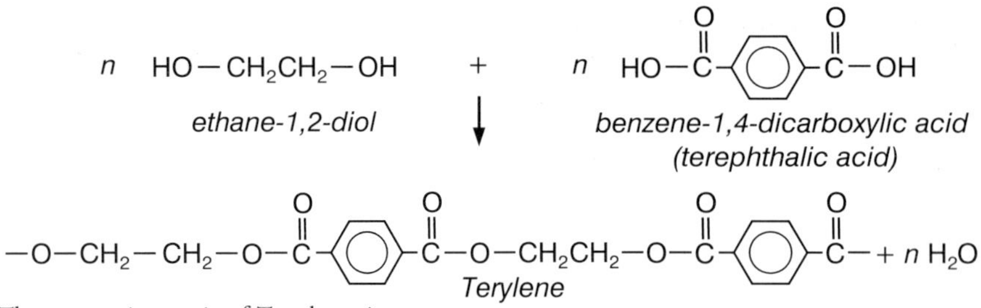

The repeating unit of Terylene is

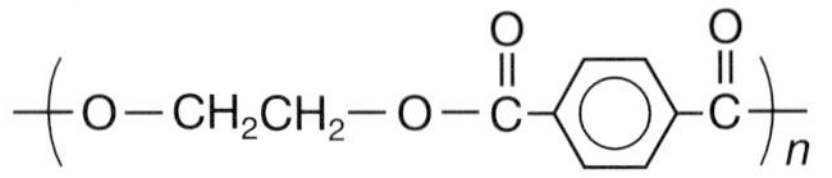

Nylon and Terylene are examples of linear polymers and their structure makes them ideal for use as **fibres**. Nylons are polyamides formed by the condensation polymerisation of a diamine with a dicarboxylic acid. For example, when 1,6-diaminohexane and hexane-1,6-dioic acid are polymerised, nylon-6-6, a polyamide, is formed and water is eliminated.

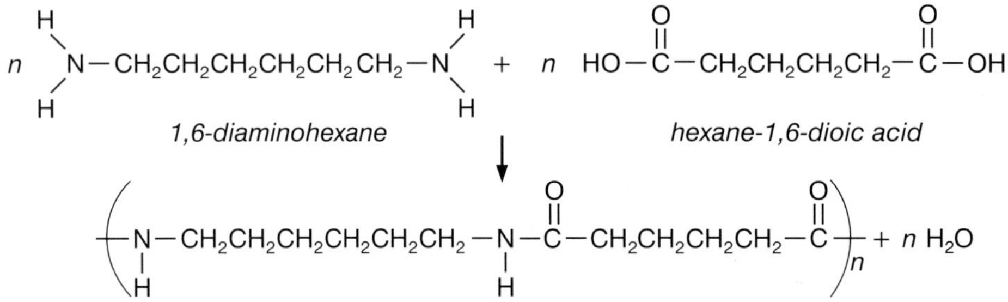

nylon-6,6

If monomers have more than two functional groups per molecule then cross-links can be formed between adjacent polymer chains and a **resin** is the result. Resins are hard and are used where a rigid polymer structure is required.

Resins are examples of **thermosetting polymers**, which are widely used because they do not melt when heated. Many thermosetting polymers are formed by condensation reactions involving **methanal** (see **Section 19.4**).

You can find out more about nylons, polyesters and other condensation polymers, such as Kevlar, are described in **Chapter 5**.

Why nylon-6,6?

A nylon is named according to the number of carbon atoms in the monomers. The first digit indicates the number of carbon atoms in the diamine and the second digit indicates the number of carbon atoms in the dicarboxylic acid.

19.4 Polymer properties

Elastomers, plastics and fibres

Polymer properties vary widely. Polymers which are soft and springy, which can be deformed and then go back to their original shape, are called **elastomers**. Rubber is an elastomer.

Poly(ethene) is not so springy and when it is deformed it tends to stay out of shape, undergoing permanent or plastic deformation. Substances like this are called **plastics**.

Stronger polymers, which do not deform easily, are just what you want for making clothing materials. Some can be made into strong, thin threads which can then be woven together. These polymers, such as nylon, are called **fibres**.

Poly(propene) is on the edge of the plastic/fibre boundary. It can be used as a plastic like poly(ethene), but it can also be made into a fibre for use in carpets.

What decides the properties of a polymer?

The physical properties of a polymer, such as its strength and flexibility, are decided by the characteristics of its molecules. The particularly important characteristics are:

- *chain length*: in general, the longer the chains the stronger the polymer
- *side groups*: polar side groups give stronger attraction between polymer chains, making the polymer stronger
- *branching*: straight, unbranched chains can pack together more closely than highly branched chains, giving polymers that are more ordered and therefore stronger
- *cross-linking*: if polymer chains are linked together extensively by covalent bonds, the polymer is harder and more difficult to melt. Thermosetting polymers have extensive cross-linking.

In poly(ethene), the attractive forces between polymer chains are weak van der Waals forces. In poly(chloroethene), there are additional permanent dipole–permanent dipole attractions due to the polar side groups. In nylons, hydrogen bonding can also occur between polymer chains. You can read about these intermolecular forces in **Unit 1**, **Sections 16.1** and **16.4**.

Thermoplastics and thermosets

We can put polymers into two groups, depending on how they behave when they are heated.

Thermoplastics are polymers without cross-links between the chains. The intermolecular forces between the chains are much weaker than the covalent cross-links in a thermoset and so the attractive forces can be broken by warming. The chains can then move relative to one another (Figure 1a).

When this happens the polymer changes its shape – it is deformed – and thermoplastics can be moulded. When the polymer is cool the intermolecular forces reform, but between different atoms.

Over 80% of all the plastics produced are thermoplastics. This includes poly(ethene), poly(propene), polyesters such as Terylene and polyamides such as nylon.

Figure 1 Thermoplastics and thermosets

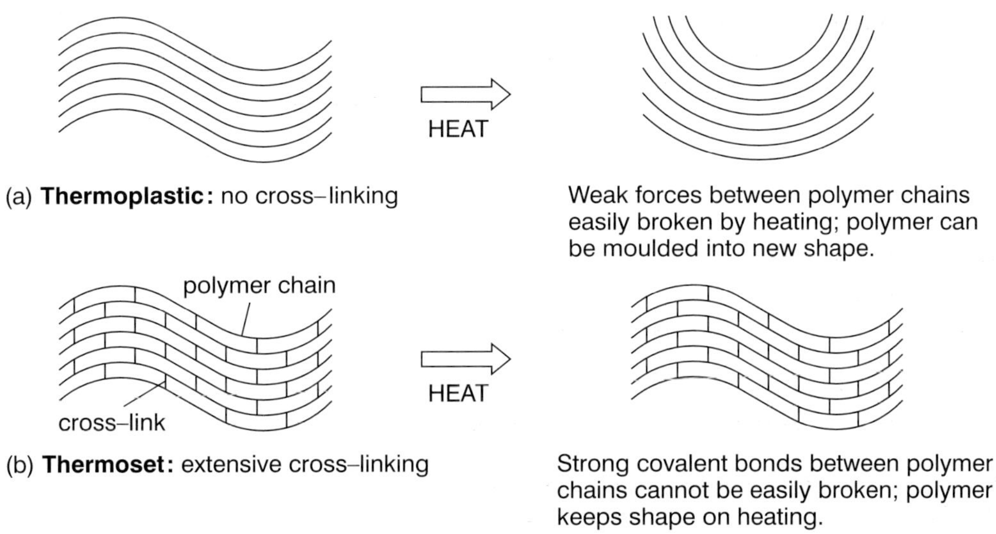

Thermosetting polymers or **thermosets** have extensive cross-linking. The cross-links prevent the chains moving very much relative to one another, so the polymer stays set in the same shape when it is heated (Figure 1b).

Bakelite is an example of a thermosetting polymer. It is made from phenol and methanal. When heated together, a three-dimensional structure is formed. This is a network with strong covalent bonds (Figure 2).

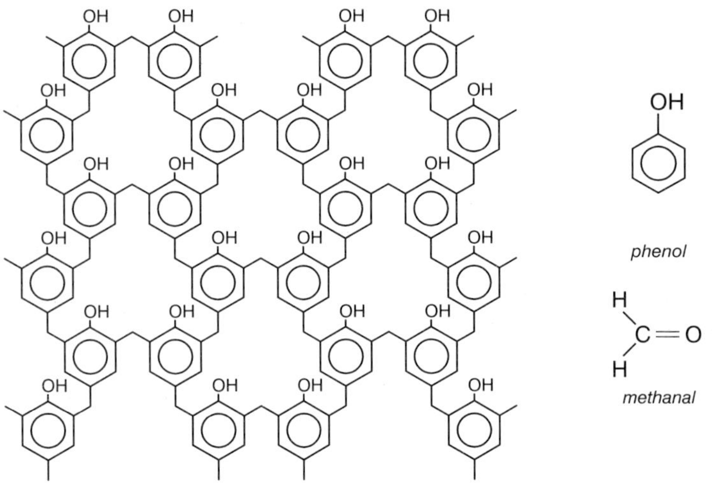

Figure 2 The structure of Bakelite, a thermosetting polymer. (This is only one possible arrangement. Note that ∧ is a shorthand way of writing $/CH_2\backslash$*)*

This huge cross-linked structure would require a massive amount of energy to break it down. Instead of melting when it is heated, Bakelite chars and decomposes.

The structure would also have to be broken down into much smaller pieces before it could dissolve. Bakelite is a very insoluble substance because the energy needed to do this is not normally available.

Bakelite was the first synthetic polymer to be made. It is still used occasionally, for example in the brown plastic used in some electrical insulators.

PROBLEMS FOR 19.1–19.4

1 Write down the structure of a short length of
  a poly(ethene)
  b nylon-6,6

2 pvc is made from a monomer $H_2C{=}CHCl$. Write down the structure of a short length of the polymer.

3 Name the two alkenes that were used to make the following polymer:

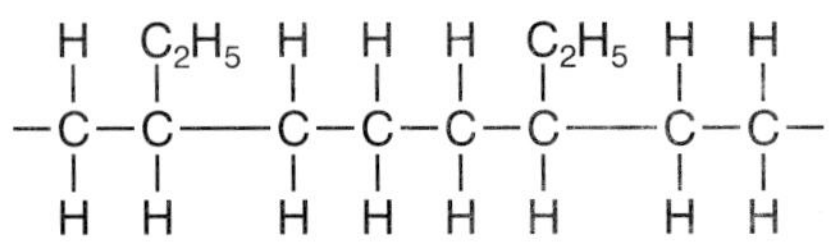

(SEB Chemistry paper)

4 Bucrylate is an ester which is used in surgical operations. It instantly polymerises when it comes into contact with solutions containing ions, for example, blood. The structure of the monomer is shown below:

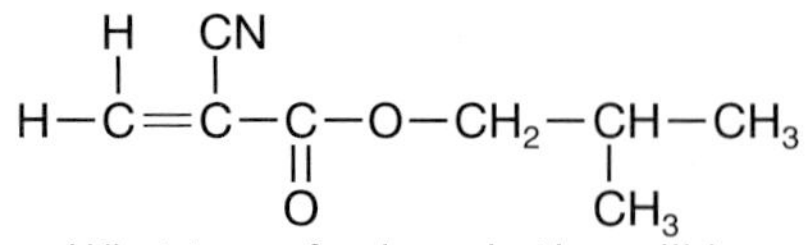

  a What type of polymerisation will bucrylate undergo?
  b Draw the structure of the repeating unit in poly(bucrylate).

(Adapted from SEB Chemistry paper)

20

EQUILIBRIUM IN CHEMISTRY

20.1 *Dynamic equilibrium*

The general meaning of the term *equilibrium* is a state of balance in which nothing changes. For example, a see-saw with two people of equal mass, sitting one on each side, is in a state of equilibrium.

In chemistry, a state of equilibrium is also a state of balance, but it has a special feature: chemical equilibrium is a **dynamic equilibrium**.

Consider the sealed bottle of soda water shown in Figure 1. The bottle and its contents make up a **closed system**. Nothing can enter or leave the bottle. In this system, carbon dioxide is present in two states: as a gas above the liquid, $CO_2(g)$, and as an aqueous solution in water, $CO_2(aq)$.

Suppose the bottle has been standing at a steady temperature for some time. If you measure the pressure of carbon dioxide gas above the water you find it is constant. If you measure the concentration of carbon dioxide dissolved in the water, you find that is constant too. The system is at equilibrium, and nothing appears to be changing, at least on the **macroscopic scale** – the scale which we humans are used to.

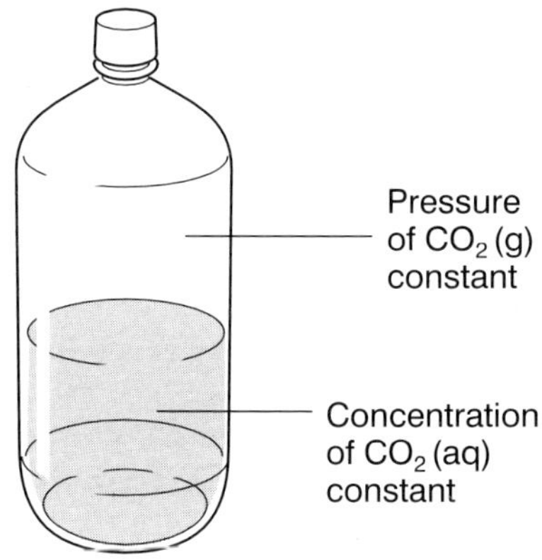

Figure 1 The carbon dioxide equilibrium on a macroscopic scale

If you were able to see how the individual molecules are behaving – that is on a **microscopic scale** – the picture would be rather different.

In any gas, the molecules are constantly moving. They move rapidly in random directions, and inevitably they collide with the molecules in the surface of the water (Figure 2). Some bounce back into the gas phase, but some dissolve and enter the aqueous phase.

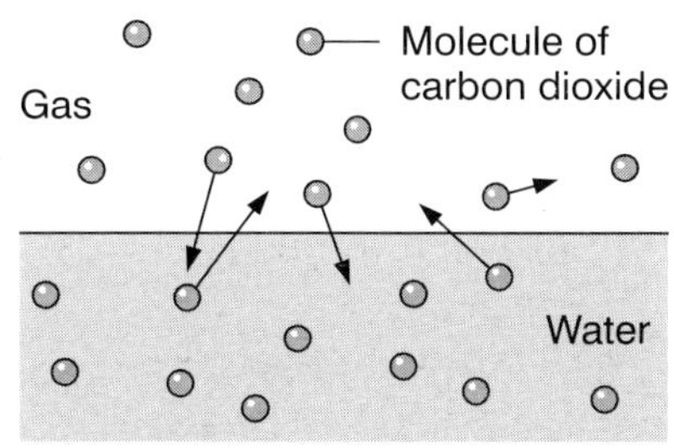

Figure 2 The carbon dioxide equilibrium on a microscopic scale

At the same time, the molecules of carbon dioxide in the water are also constantly moving around, colliding with water molecules and with each other. Near the surface of the water, some of these molecules escape into the gaseous phase (Figure 2).

So there are molecules entering the aqueous phase, and molecules leaving it. It is a **reversible change**. *When the system is at equilibrium, the molecules enter and leave at the same rate*. On the macroscopic scale it *seems* as though nothing is changing, but, on the molecular scale, molecules are constantly moving from one phase to the other. That is why the situation is described as *dynamic* equilibrium. It can be represented by the equation

$$CO_2(g) \rightleftharpoons CO_2(aq)$$

The $\rightleftharpoons$ sign represents dynamic equilibrium.

Other examples of dynamic equilibrium

Dynamic equilibrium is a basic principle of chemistry. You can find examples everywhere – and you can find two more without going any further than the bottle of soda water shown in Figure 1.

Water in equilibrium

The water itself is present in two phases: liquid and gas. Water molecules from the liquid are constantly escaping into the gaseous phase. At the same time, molecules from the gas are constantly colliding with the surface of the liquid and condensing. In a closed system (the bottle with the top on), molecules reach the situation after a while where they are escaping and condensing at the same rate. Then the system is in dynamic equilibrium, and it is represented by

$$H_2O(l) \rightleftharpoons H_2O(g)$$

Carbon dioxide reacts with water

Neither of the two examples just mentioned involve chemical reactions. They are physical processes involving the liquid/gas phase change. But when carbon dioxide dissolves in water, a small proportion of it actually *reacts* with the water instead of just dissolving. Hydrogencarbonate ions and hydrogen ions are formed:

$$CO_2(aq) + H_2O(l) \rightleftharpoons \underset{\textit{hydrogencarbonate ion}}{HCO_3^-(aq)} + H^+(aq)$$

Only a small proportion of the carbon dioxide reacts in this way. Most of it remains as $CO_2(aq)$, with only a small amount converted to $HCO_3^-(aq)$.

On the molecular scale, $CO_2(aq)$ molecules are constantly reacting with $H_2O(l)$ molecules to form $HCO_3^-(aq)$ and $H^+(aq)$ ions. At the same time, $HCO_3^-(aq)$ and $H^+(aq)$ ions are constantly reacting together to reform $CO_2(aq)$ and $H_2O(l)$. It is a **reversible reaction**.

Many chemical reactions are reversible reactions. When such reactions are represented by equations, like the one above, the reaction going from left to right is known as the **forward reaction**. The reaction going in the opposite direction is known as the **reverse reaction**. Although the reaction is reversible, it is the practice to refer to substances on the left of the equilibrium sign as the **reactants** and those to the right as the **products**.

Chemical equilibria

If a mixture of nitrogen dioxide, NO_2, and carbon monoxide, CO, is heated in a sealed container, the following reversible reaction takes place:

$$NO_2(g) + CO(g) \rightleftharpoons NO(g) + CO_2(g)$$

At the start, no molecules of NO or CO_2 are present so there will be no reverse reaction to begin with. As soon as some NO and CO_2 are produced, the reverse reaction will start to occur; slowly at first but with increasing rate as the concentrations of NO and CO_2 increase. At the same time, the concentrations of NO_2 and CO will be decreasing and so the rate of the forward reaction will start to decrease. Eventually a point is reached when the rate of the forward reaction equals the rate of the reverse reaction. At this point, reactants are turning into products as quickly as products are turning into reactants. The reversible reaction is now at equilibrium (see Figure 3). However, it is a dynamic equilibrium because both forward and reverse reactions are still occurring, although the concentrations of both products and reactants now remain constant.

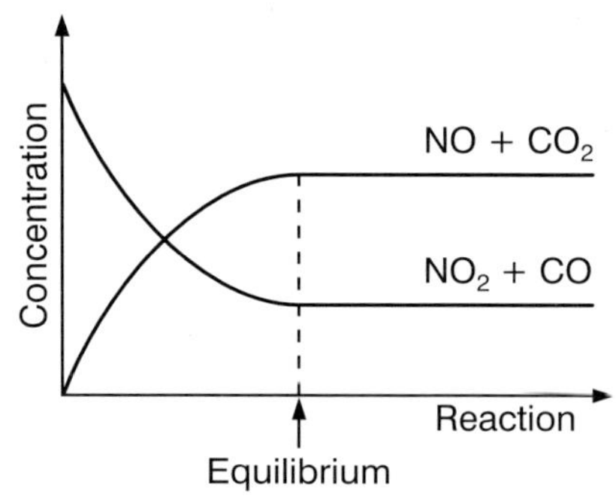

Figure 3 Reaction of NO_2 and CO

Similarly for the reaction between carbon dioxide and water, when the system in the soda water bottle has had a chance to settle down, the forward and reverse reactions go at the same rate. Nothing *seems* to change on the overall, macroscopic level. If you measure the pH of the soda water, you find it is constant; the concentration of $H^+(aq)$ ions is constant, and so are the concentrations of all the other substances involved. It is another example of dynamic equilibrium.

Position of equilibrium

There are many equilibrium mixtures possible for a given reaction system, depending on the concentrations of the substances you mix and the conditions, such as temperature and pressure. We often use the term **position of equilibrium** to describe one particular set of equilibrium concentrations for a reaction. If one of the concentrations is changed, the system is no longer in equilibrium, and the concentrations of all the substances will change until a new position of equilibrium is reached.

Although the rates of the forward and reverse reactions are the same when equilibrium is reached, this does not necessarily mean that the concentrations of the reactants and the concentrations of the products are the same at equilibrium. In fact, they rarely are. Usually, at the point of equilibrium there are more products than reactants or vice versa.

If the forward reaction is nearly complete before the rate of the reverse reaction has increased sufficiently to establish equilibrium, we say that the position of equilibrium lies to the products or right-hand side of the reaction. In other cases, little of the reactants will have changed to products when the rate of the reverse reaction becomes equal to the rate of the forward reaction and we say that the position of equilibrium lies to the reactants or left-hand side of the reaction.

Once a reaction system is at equilibrium, it is impossible to tell whether the equilibrium mixture was arrived at by starting with reactants or products. Under the same conditions, the same equilibrium position is arrived at whether you start with the reactants or the products. The reversible reaction of hydrogen and iodine to form hydrogen iodide can be used to illustrate this:

$$H_2(g) + I_2(g) \rightleftharpoons 2HI(g)$$

Suppose we heat a mixture of equal amounts of hydrogen and iodine (1 mole of each) in a sealed reaction vessel at 500 °C and measure the amounts of H_2, I_2 and HI during the course of the reaction. If we then plot these results as graphs we obtain Figure 4a. Equilibrium is reached at the point on the graphs where the lines become horizontal.

If we now repeat the experiment with hydrogen iodide (2 moles) we obtain the graphs shown in Figure 4b. Notice that we have ended up with identical amounts of HI and identical amounts of H_2 and I_2 in both graphs. This shows that we reach the same equilibrium position irrespective of whether we start with the reactants or the products.

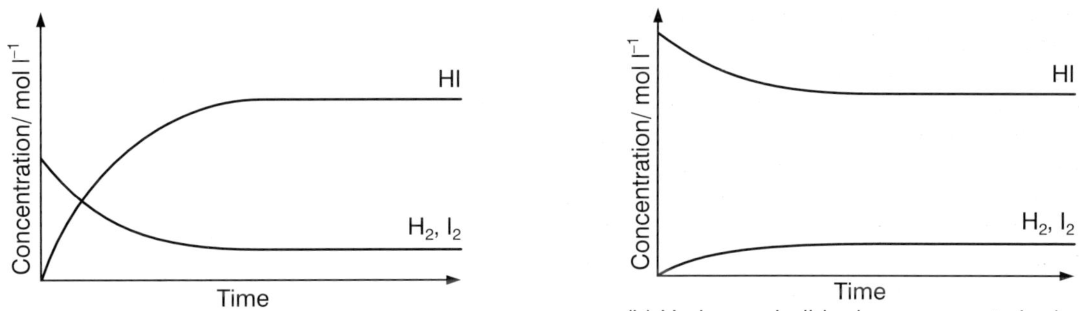

(a) A mixture of pure hydrogen and iodine reaches equilibrium.

(b) Hydrogen iodide decomposes to hydrogen and iodine at the same temperature.

Figure 4 Equilibrium can be reached from either a mixture of hydrogen and iodine, or from pure hydrogen iodide

20.2 *Shifting the equilibrium position*

The position of an equilibrium can be altered by changing:

- the concentrations of reacting substances if dealing with solutions
- the pressures of reacting gases
- the temperature.

By studying data from many reactions it has been possible to propose rules which enable you to make qualitative predictions about the effect of each of these changes on an equilibrium. It has been found that each of these rules is an example of the simple general rule that:

If a system is at equilibrium, and a change is made in any of the conditions, then the system responds to counteract the change as much as possible.

As you work through the following sections you will be able to see how each of the separate rules is an example of this general rule.

Using a catalyst does not shift the position of an equilibrium. A catalyst lowers the activation energy of a reaction. Since the reaction is reversible, the activation energy of both the forward and reverse reactions will be reduced by the same amount so both will speed up by the same proportion (see **Unit 1**, **Section 15.5**). This means that the same equilibrium position is reached. The catalyst causes the equilibrium position to be reached much more quickly than it would be if no catalyst were used.

Changing the concentration

Yellow iron(III) ions, $Fe^{3+}(aq)$, and colourless thiocyanate ions, $SCN^-(aq)$, react in solution to give the deep red ion, $[Fe(SCN)]^{2+}(aq)$. The reaction is reversible and the equilibrium set up is

$$\underset{\text{yellow}}{Fe^{3+}(aq)} + \underset{\text{colourless}}{SCN^-(aq)} \rightleftharpoons \underset{\text{deep red}}{[Fe(SCN)]^{2+}(aq)}$$

The intensity of the red colour of the solution can be used as a measure of how much $[Fe(SCN)]^{2+}(aq)$ is present. If the solution gets darker it means that the equilibrium is shifting to the right; if lighter, it is shifting to the left.

If we increase the concentration of either $Fe^{3+}(aq)$ or $SCN^-(aq)$, the solution gets darker, showing that the position of equilibrium has moved to the right.

The concentration of $Fe^{3+}(aq)$ ions in the solution can be reduced by adding ammonium chloride. (Chloride ions from the ammonium chloride react to form $[FeCl_4]^-(aq)$.) If we do this, the red colour becomes paler, indicating that the equilibrium is being moved to the left.

From these observations, and others involving a variety of reversible reactions, we can arrive at rules for deciding how concentration changes will affect the position of an equilibrium. These are:

- increasing the concentration of reactants causes the equilibrium to move to the product side
- increasing the concentration of products causes the equilibrium to move to the reactant side
- decreasing the concentration of reactants causes the equilibrium to move to the reactant side
- decreasing the concentration of products causes the equilibrium to move to the product side.

Since chemistry is all about turning reactants into products, chemists spend a lot of time working out ways of shifting equilibrium positions as far to the right as possible when dealing with reversible reactions. When a gas is one of the products of a reversible reaction the simplest way to achieve this is to keep removing the gas from the reaction chamber. This ensures that the reaction is being constantly shifted to the product side.

Changing the pressure

Many important industrial processes involve reversible reactions which take place in the gas phase. For these processes it is essential that conditions are identified which ensure that the equilibrium is shifted as far to the right as possible.

From the study of equilibria in gas-phase reactions a simple rule has emerged which is:

- increasing the pressure moves the equilibrium to the side of the equation with fewer gas molecules as this tends to reduce the pressure
- decreasing the pressure moves the equilibrium to the side of the equation with more gas molecules as this tends to increase the pressure.

The steam reforming of methane, which you have already met in **Unit 2, Section 18.1**, provides a good illustration of these rules. This process is used to make methanol. In the first stage, methane is reacted with steam to form CO and H_2.

$$\underset{\text{2 molecules}}{CH_4(g) + H_2O(g)} \rightleftharpoons \underset{\text{4 molecules}}{CO(g) + 3H_2(g)}$$

There are more molecules on the product side of the equation. So as the rules predict, a reduction in the pressure results in the position of equilibrium shifting to the right. However, the rate depends on the concentration (in this case pressure) of the reactants and for this reaction if too low a pressure were used the reaction would go too slowly. For these reasons the process is carried out at a pressure of 30 atmospheres which, from the economic point of view, gives the optimum amount of product at an optimum rate.

The second stage of methanol manufacture involves the reaction

$$\underset{\text{3 molecules}}{CO(g) + 2H_2(g)} \rightleftharpoons \underset{\text{1 molecule}}{CH_3OH(g)}$$

For this reaction, increasing the pressure will move the equilibrium to the product side; the higher the pressure, the more product will be obtained. Also, increasing the pressure will increase the rate at which equilibrium is established. Originally, this reaction was carried out at 300 atmospheres pressure but the development of better catalysts now allows it to be carried out at a pressure of 100 atmospheres, which results in reduced costs.

Changing the temperature

Heating a reaction always makes it go faster. How does heating (or cooling) affect the position of an equilibrium? For a reversible reaction, if the forward reaction is exothermic then the reverse reaction will be endothermic to the same extent, and vice versa.

Nitrogen dioxide, NO_2, is a dark brown gas which exists in equilibrium with its colourless dimer, dinitrogen tetraoxide, N_2O_4.

$$\underset{\text{brown}}{2NO_2(g)} \underset{\text{endothermic}}{\overset{\text{exothermic}}{\rightleftharpoons}} \underset{\text{colourless}}{N_2O_4(g)}$$

The reaction that forms N_2O_4, the forward reaction, gives heat out. The reverse reaction, forming NO_2, takes heat in.

If a sealed container of the brown equilibrium mixture is heated in boiling water it becomes darker. On cooling in ice, it turns almost colourless. From this evidence we can conclude that changing the temperature shifts the position of this equilibrium. It also illustrates the rule for deciding how a change of temperature affects the position of an equilibrium. This is:

- heating a reversible reaction at equilibrium shifts the reaction in the direction of the endothermic reaction
- cooling a reversible reaction at equilibrium shifts the reaction in the direction of the exothermic reaction.

When a description of a reversible reaction includes a ΔH value, it is accepted practice that the ΔH value given is always for the forward reaction.

The Haber process

See **Section 9.1** for further details about the Haber process.

The Haber process is an important industrial process that is used to manufacture ammonia, NH_3. Ammonia has many uses, one of the most important being the production of artificial fertilisers. This process provides a good illustration of how industrial chemists adjust the conditions of a reversible reaction so that the process can be operated in the most cost-effective way.

The Haber process reaction is

$$\underset{\text{4 molecules}}{N_2(g) + 3H_2(g)} \underset{\text{endothermic}}{\overset{\text{exothermic}}{\rightleftharpoons}} \underset{\text{2 molecules}}{2NH_3(g)} \qquad \Delta H = -92\,\text{kJ mol}^{-1}$$

At room temperature and atmospheric pressure the reaction is extremely slow and the equilibrium lies completely to the reactant side.

There are fewer molecules on the product side so if the pressure were increased this would favour the production of ammonia and would also help to increase the rate of the reaction. For this reason the industrial process is carried out at pressures between 25 and 150 atmospheres.

Heating would also increase the rate but, unfortunately, the forward reaction is exothermic, so the higher the temperature, the less the product side is favoured. A temperature of between 400 °C and 500 °C is used. The reaction goes faster at the higher temperature but less ammonia is produced. Figure 5 shows the percentage of ammonia produced for a range of different temperatures and pressures.

Although a catalyst does not affect the equilibrium position, it increases the rate of a reaction enormously. The Haber process uses an iron catalyst. Nitrogen and hydrogen are passed over the heated catalyst and about 15% is converted into ammonia. The mixture of ammonia and unreacted nitrogen and hydrogen is then chilled. The ammonia condenses and is separated from the hydrogen and nitrogen which are then recycled through the reaction process.

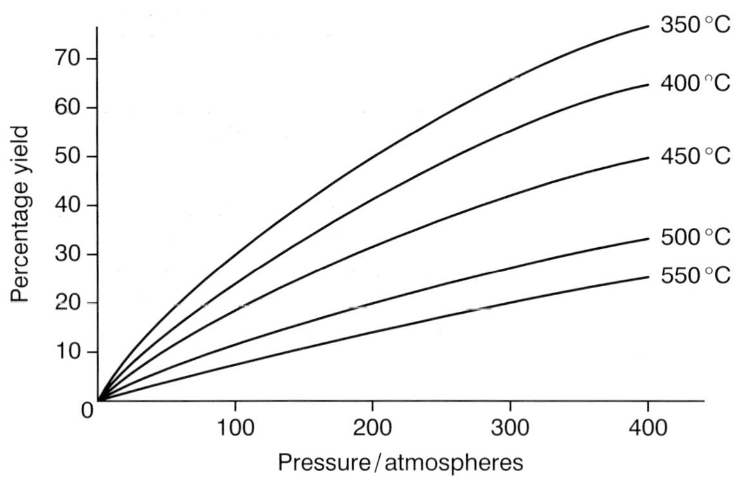

Figure 5 The Haber process – the effect of pressure on percentage yield at selected temperatures

PROBLEMS FOR 20.1 AND 20.2

1 You will be very familiar with the equilibrium
$H_2O(l) \rightleftharpoons H_2O(g)$

a Why is this described as a *dynamic* equilibrium?

b One place where this equilibrium is established is in a closed plastic bag full of wet socks. Use the idea of dynamic equilibrium to explain why the socks in the bag never get dry.

c You can dry the socks by taking them out of the bag. Once they are out of the bag the system is no longer in equilibrium. Explain why.

d In which direction does this equilibrium move
 i when dew forms in the evening?
 ii when you demist a car window?

2 In a reversible reaction, which of the following statements is true?
Equilibrium is reached when

A molecules of reactants cease to change into molecules of products

B the concentrations of reactants and products are equal

C the concentrations of reactants and products are constant

D the activation energy of the forward reaction is equal to that of the reverse reaction

(SQA (adapted))

3 State the direction in which the position of equilibrium of each system would move (if at all) if the pressure were increased by compressing the reaction mixture. Give your answer as 'left → right', 'right → left', or 'no change'.

a $2NO(g) + O_2(g) \rightleftharpoons 2NO_2(g)$

b $C_2H_6(g) \rightleftharpoons C_2H_4(g) + H_2(g)$

c $2HI(g) \rightleftharpoons H_2(g) + I_2(g)$

d $2NO_2(g) \rightleftharpoons N_2O_4(g)$

e $2CO(g) + O_2(g) \rightleftharpoons 2CO_2(g)$

4 In the equilibrium

$N_2(g) + O_2(g) \rightleftharpoons 2NO(g) \qquad \Delta H = +180\,\text{kJ mol}^{-1}$

how would the equilibrium concentration of NO be affected by

a increasing the temperature

b decreasing the pressure

c decreasing the concentration of oxygen?

(SQA (slightly adapted))

5 The hypochlorite ion, ClO^-, acts as a bleaching agent in solution. When chlorine is added to water, the following equilibrium is set up

$Cl_2(aq) + H_2O(l) \rightleftharpoons 2H^+(aq) + ClO^-(aq) + Cl^-(aq)$

Why does the addition of sodium hydroxide increase the bleaching efficiency of the solution?

(SQA (part))

6 Ethanol is produced industrially at about 70 atmospheres pressure and 300 °C by the following reaction. The reaction needs a catalyst.

$C_2H_4(g) + H_2O(g) \rightleftharpoons C_2H_5OH(g)$

$\Delta H = -46\,\text{kJ mol}^{-1}$

Which of the following would move the equilibrium to the right?

A increasing the temperature

B increasing the concentration of steam

C decreasing the pressure

D adding more catalyst.

7 In a reversible reaction the concentrations of reactants and products were measured during the reaction and the following graph was drawn.

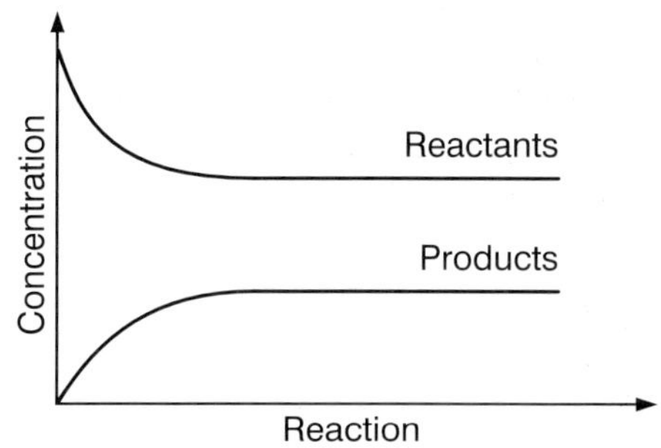

a To which side, reactants or products, does the position of equilibrium lie in this reaction? Explain how you arrived at this decision.

b At which point on the graph is equilibrium reached?

21

ACIDS AND BASES

21.1 *Acid–base reactions*

What do we mean by acid and base?

We say that hydrochloric acid is an 'acid' because of the things it does. It turns universal indicator red, reacts with carbonates to give CO_2 and is neutralised by bases – all properties that we expect of acids.

Chemists try to *explain* properties in terms of what goes on at the level of atoms, molecules and ions. They explain the characteristic properties of acids by saying that acids have *the ability to transfer H^+ ions to something else.* The substance which accepts the H^+ ions is called a **base**.

> **Properties of acidic solutions**
> - turn litmus red
> - neutralised by bases
> - pH < 7
> - liberate CO_2 from carbonates

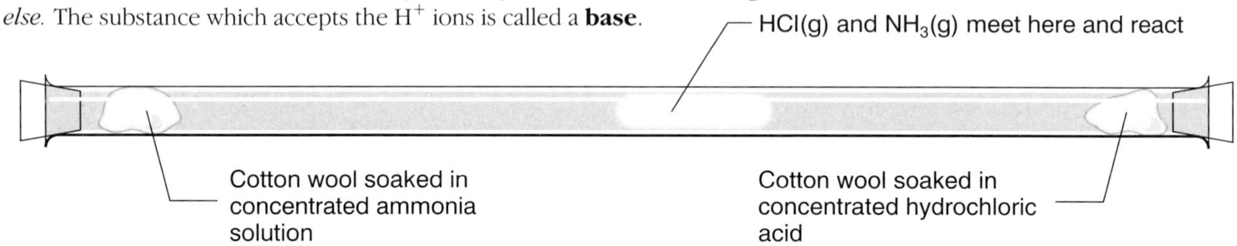

Figure 1 The reaction of hydrogen chloride with ammonia – an acid–base reaction

For example, take the reaction of hydrogen chloride with ammonia. You may have seen this reaction demonstrated in apparatus like that shown in Figure 1. The reaction forms a salt, ammonium chloride:

$$NH_3(g) + HCl(g) \rightarrow NH_4^+Cl^-(s)$$

In this reaction, the HCl transfers H^+ to NH_3 (Figure 2). The HCl is behaving as an acid, and the NH_3 is behaving as a base.

The general definition of an acid is that it is a substance which donates H^+ in a chemical reaction. The substance that accepts the H^+ is a base. The reaction in which this happens is called an **acid–base reaction**.

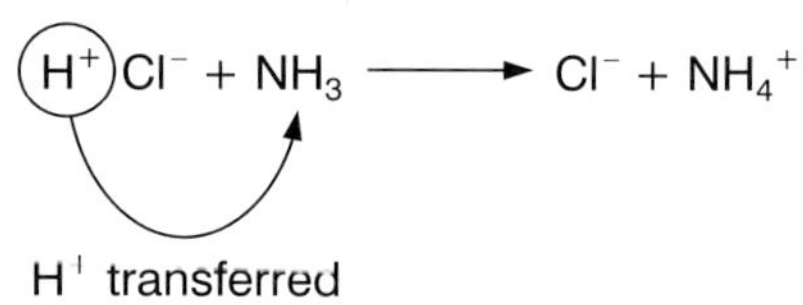

Figure 2 HCl transfers an H^+ ion to NH_3

Solutions of acids and bases

Hydrogen chloride is a gas and contains HCl molecules. Water is almost totally made up of H_2O molecules. Yet hydrochloric acid, a solution of hydrogen chloride in water, readily conducts electricity, so it must contain ions. There must be a reaction between the hydrogen chloride molecules and water molecules which produces these ions. The reaction is

$$HCl(aq) + H_2O(l) \rightarrow H_3O^+(aq) + Cl^-(aq)$$

H_3O^+ is called the **oxonium ion**, but the formula $H_3O^+(aq)$ is usually shortened to $H^+(aq)$ and the dissociation of HCl(aq) into ions represented by

$$HCl(aq) \rightarrow H^+(aq) + Cl^-(aq)$$

When an acid is dissolved in water, $H^+(aq)$ ions form in solution.

An **alkali** is a base that dissolves in water to produce hydroxide ions, $OH^-(aq)$. This is illustrated in Figure 3. Some alkalis, such as sodium hydroxide and potassium hydroxide, already contain hydroxide ions, whereas others, such as sodium carbonate and ammonia, form $OH^-(aq)$ ions when they react with water:

$$CO_3^{2-}(aq) + H_2O(l) \rightleftharpoons HCO_3^-(aq) + OH^-(aq)$$
$$NH_3(aq) + H_2O(l) \rightleftharpoons NH_4^+(aq) + OH^-(aq)$$

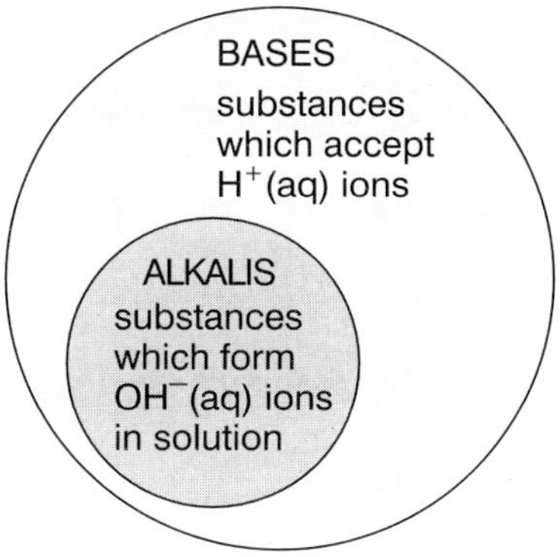

Figure 3 The relationship between alkalis and bases

The water equilibrium

We normally write the chemical formula for water as H_2O, representing covalent molecules. This implies that water should not conduct electricity. However, even ultra pure water conducts electricity very slightly. This is due to the presence of the following equilibrium:

$$H_2O(l) \rightleftharpoons H^+(aq) + OH^-(aq)$$

No matter how pure we make the water, there will always be a small amount of H^+(aq) and OH^-(aq) present. They are not impurities, but result from the dissociation of water to establish the dynamic equilibrium shown above.

In considering aqueous solutions of acids and alkalis, we need to look more closely at how they affect the concentration of H^+(aq) and OH^-(aq).

In pure water and in neutral solutions every water molecule which is dissociated will produce one hydrogen ion and one hydroxide ion, so $[H^+(aq)] = [OH^-(aq)]$. The position of the water equilibrium does change slightly with temperature, but at *normal temperature* (25 °C)

$[H^+(aq)] = [OH^-(aq)] = 1 \times 10^{-7}\,mol\,l^{-1}$.

Furthermore, for all aqueous solutions, at any given temperature, the product $[H^+(aq)] \times [OH^-(aq)]$ is a constant, known as the **ionic product** of water. At 25 °C, since both $[H^+(aq)]$ and $[OH^-(aq)]$ have a value of $1 \times 10^{-7}\,mol\,l^{-1}$, it follows that the value of the ionic product of water must be $1 \times 10^{-14}\,mol^2\,l^{-2}$. This means that if we know either $[H^+(aq)]$ or $[OH^-(aq)]$ in a solution, we can calculate the concentration of the remaining ion by making the negative powers of ten add up to 14 (assuming that the temperature is the normal 25 °C).

For example,
if $[H^+(aq)] = 1 \times 10^{-5}\,mol\,l^{-1}$ then $[OH^-(aq)] = 1 \times 10^{-9}\,mol\,l^{-1}$,
if $[H^+(aq)] = 1 \times 10^{-11}\,mol\,l^{-1}$ then $[OH^-(aq)] = 1 \times 10^{-3}\,mol\,l^{-1}$,
and so on.

> $[H^+(aq)]$ is used by chemists as shorthand for 'the concentration of H^+(aq)'. Similarly, $[OH^-(aq)]$ means 'the concentration of OH^-(aq)'.

The pH scale

Acids and alkalis are involved in many chemical processes, including the biochemical reactions that take place inside our bodies. Chemists often find it useful to know the concentration of H^+(aq) or OH^-(aq) present in a solution. Because it is not easy to work with figures such as 1×10^{-8} and 1×10^{-4} all the time, they use a measuring system called the pH scale.

The pH scale was devised at the beginning of the 20th century by a Danish chemist called Soren Sorensen. He wanted a simple way to tell how much acid or alkali was present in a solution. The 'p' in pH stands for *potens*, which is Latin for power: so pH is measuring the power of H^+(aq) in a solution, ie its concentration.

> An acidic solution is defined as one in which $[H^+(aq)] > [OH^-(aq)]$.
>
> An alkaline solution is defined as one in which $[H^+(aq)] < [OH^-(aq)]$.

Figure 4 shows how the pH of a solution is related to $[H^+(aq)]$ in a solution. The pH value is the same as the negative power of 10 that relates to $[H^+(aq)]$. When $[H^+(aq)]$ changes by a factor of 10, the pH changes by 1. This is something known as a *logarithmic scale*. Solutions with a pH of 7 are neutral. Acids have a pH of less than 7 and alkalis have a pH of more than 7.

Many solutions of acids or alkalis have a concentration that is less than 1 mol l^{-1} and so the pH values of these solutions lie between 0 and 14. An acid where $[H^+(aq)]$ was more than 1 mol l^{-1} would have a pH of less than 0, ie a negative pH. Similarly, an alkaline solution where $[OH^-(aq)]$ was greater than 1 mol l^{-1} would have a pH greater than 14.

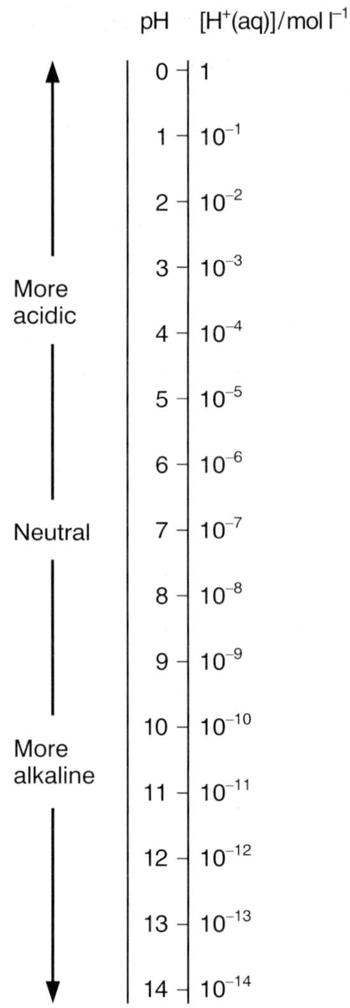

Figure 4 The pH scale

21.2 *Strong and weak acids*

A concentrated solution is one in which a relatively large amount of solute has been dissolved. A dilute solution contains relatively little solute. In everyday use we often use the words 'strong' and 'weak' when we really mean 'concentrated' and 'dilute'. We might talk about a strong cup of coffee or a cup of weak tea. However, in chemistry, the words strong and weak have particular meanings in relation to acids and bases.

A **strong acid** is one which is completely dissociated into ions when it is dissolved in water. Hydrochloric acid is an example of a strong acid. It is made by dissolving covalent hydrogen chloride, HCl, molecules in water. Once dissolved, all the HCl molecules split up into $H^+(aq)$ and $Cl^-(aq)$ ions.

$$HCl(aq) \rightarrow H^+(aq) + Cl^-(aq)$$

We say that a strong acid such as this is completely dissociated in solution.

A **weak acid**, on the other hand, is one which is only partially dissociated when dissolved in water. If a weak acid with the general formula HA is dissolved in water, only some of the HA molecules will dissociate into ions and the following equilibrium is set up:

$$HA(aq) \rightleftharpoons H^+(aq) + A^-(aq)$$

There are many more weak acids than strong acids and they vary in strength depending on the extent to which the equilibrium lies to the right-hand side. The more it lies to the right-hand side, the stronger is the acid. Most carboxylic acids (see **Unit 2**, **Section 18.4**) are weak acids. The dissociation of ethanoic acid, for example, establishes this dynamic equilibrium:

$$CH_3COOH(aq) \rightleftharpoons H^+(aq) + CH_3COO^-(aq)$$

The position of this equilibrium is well to the left so ethanoic acid is a weak acid.

Carbon dioxide and sulphur dioxide both react with water to form weakly acidic solutions. For example:

$$CO_2(aq) + H_2O(l) \rightleftharpoons H^+(aq) + HCO_3^-(aq)$$
$$HCO_3^-(aq) \rightleftharpoons H^+(aq) + CO_3^{2-}(aq)$$

This equilibrium system, where $HCO_3^-(aq)$ is the product of the first equilibrium and the reactant of the second, is sometimes simplified by combining the two equilibria:

$$H_2CO_3(aq) \rightleftharpoons 2H^+(aq) + CO_3^{2-}(aq)$$

In this equation, the solution of carbon dioxide in water is called carbonic acid and is represented by $H_2CO_3(aq)$.

Comparing strong and weak acids

When comparing strong and weak acids it is important to compare solutions whose concentrations are identical with respect to potential $[H^+(aq)]$. Strong acids may be distinguished from weak acids in a number of ways.

1 pH

A 0.1 mol l^{-1} solution of a strong acid such as hydrochloric acid has a pH value of 1. A weak acid of the equivalent concentration has a pH value that is greater than 1, due to a lower concentration of $H^+(aq)$ ions in the solution. For example, a 0.1 mol l^{-1} solution of sulphur dioxide has a pH of about 1.5, a 0.1 mol l^{-1} solution of ethanoic acid a pH of about 3 and a 0.1 mol l^{-1} solution of carbon dioxide a pH of about 4.

2 Conductivity

A strong acid has a much higher conductivity than a weak acid. This is because there are fewer ions present in the solution of the weak acid due to the position of the dissociation equilibrium being well to the left.

3 Reaction rate

Acids can react with metals, alkalis, metal oxides and metal carbonates, and all of these reactions involve the $H^+(aq)$ ion. The rates of these reactions decrease as $[H^+(aq)]$ decreases. No matter which reaction you choose, a weak acid of a certain concentration reacts more slowly than a strong acid of the same potential $H^+(aq)$ concentration. This is because the position of the equilibrium of a weak acid means that, at any given time, there are relatively few $H^+(aq)$ ions available for reaction.

For example, in the reaction

$$2HX(aq) + Mg(s) \rightarrow MgX_2(aq) + H_2(g)$$

where HX is HCl or ethanoic acid, CH_3COOH, the reaction with the strong hydrochloric acid is much faster than the reaction with the weak ethanoic acid. But if we add excess magnesium to identical volumes of the two acids and if the two acids are of identical concentration then we get identical volumes of hydrogen produced.

Why don't we get less hydrogen with the weak acid? This is because, when the reaction starts, the concentration of $H^+(aq)$ ions is reduced as they react with the magnesium. In the weak acid this means that the equilibrium is unbalanced so more weak acid molecules dissociate into ions, producing more $H^+(aq)$ ions, which then react. This process continues until eventually all of the acid molecules have dissociated and reacted. Although a weak acid reacts more slowly than a strong acid, with metals, alkalis and other substances, it will eventually react with the stoichiometric quantities indicated by the overall equation for the reaction. Both acids will neutralise equal amounts of alkali.

21.3 *Strong and weak bases*

A solution is basic if it contains more $OH^-(aq)$ ions than $H^+(aq)$ ions. A solution of this type is also described as being alkaline.

If a base B is dissolved in water then the following equilibrium is established:

$$B(aq) + H_2O(l) \rightleftharpoons BH^+(aq) + OH^-(aq)$$

If B is a strong base, then the equilibrium will lie completely to the right-hand side; there will be complete dissociation. If B is a weak base then only partial dissociation will take place. Both sides of the equilibrium will be present in solution, with the extent of dissociation depending on how strong the base is.

The most commonly encountered weak base is probably ammonia, NH_3. When dissolved in water we get the equilibrium shown below:

$$NH_3(aq) + H_2O(l) \rightleftharpoons NH_4^+(aq) + OH^-(aq)$$

The pH of a 1 mol l^{-1} solution of ammonia is over 11, which indicates that $[OH^-(aq)]$ is only about 1×10^{-3} mol l^{-1}. This means that more than 99% of the ammonia molecules remain undissociated.

Other examples of weak bases are amines (see **Unit 2, Section 18.6**), which are derived from the ammonia molecule by replacing one or more of its hydrogen atoms by, for example, alkyl groups.

Completely dissociated strong bases tend to be alkalis such as sodium hydroxide, Na^+OH^-, or calcium hydroxide, $Ca^{2+}(OH^-)_2$. Calcium hydroxide is classed as a strong base, because in solution it is completely dissociated into ions, despite the fact that a saturated calcium hydroxide solution has a pH of only 12. This is due to the fact that calcium hydroxide is not very soluble in water, which causes $[OH^-(aq)]$ to be low.

Comparing strong and weak bases

As with acids, any comparison of strong and weak base solutions must be made on the basis of the solutions having the same potential $OH^-(aq)$ concentration.

1 pH

A solution of a strong base has a higher pH than a weak base due to the higher concentration of $OH^-(aq)$ ions in the solution of the strong base.

2 Conductivity

A solution of a strong base has a higher conductivity than a weak base. This is because of the greater concentration of ions in the strong base solution.

3 Reaction rate

As with acids, the strength or weakness of a base has no effect on the extent to which it is able to react with other chemicals. For example, 1 mole of hydrochloric acid can be neutralised either by 1 mole of sodium hydroxide (a strong base) or by 1 mole of ammonia (a weak base), but the reaction rate will be less for ammonia than it is for sodium hydroxide.

21.4 *pH of salt solutions*

Salts

Salts are produced when an acid and base react together. If the base is an alkali then the general reaction is:

acid + alkali → salt + water

Since this is called a **neutralisation reaction**, we might expect that the solution of the salt (assuming it is soluble) would have a neutral pH of 7. If you look at Table 1, which gives the pH of some salt solutions, you will see that this is not always the case.

Some salt solutions are neutral but others may be either acidic or alkaline. Is there any pattern that will allow us to make sense of this and lead us to an explanation? There seems to be. The first two salts, giving neutral solutions, are produced by reacting a strong acid and a strong base. The next two, made by reacting a weak acid and a strong base, give alkaline

Salt	pH
sodium chloride	7
potassium nitrate	7
sodium carbonate	10
potassium ethanoate	9
ammonium chloride	5

Table 1 Approximate pH values of some salt solutions (1 mol l^{-1})

(basic) solutions. The last, a salt of a strong acid and a weak base, gives an acidic solution. Other salts give the same results. For example, soap (see **Unit 2**, **Section 18.5**) gives an alkaline solution since it is the salt of a strong base (sodium hydroxide) and a weak acid (a fatty acid).

The rules for predicting whether a solution of a salt is neutral, acidic or alkaline are summarised in Table 2.

In order to explain this pattern we need to consider how salts affect the water equilibrium.

Acid	Base	Salt solution
strong	strong	neutral
weak	strong	alkaline (basic)
strong	weak	acidic

Table 2 How to predict the pH of a salt solution

Salts of weak acids and strong bases

Example: sodium carbonate

When sodium carbonate dissolves in water, the solution contains $Na^+(aq)$ and $CO_3^{2-}(aq)$ ions (from sodium carbonate) and $H^+(aq)$ and $OH^-(aq)$ ions (from the dissociation of water molecules in the water equilibrium). The following equilibria are established:

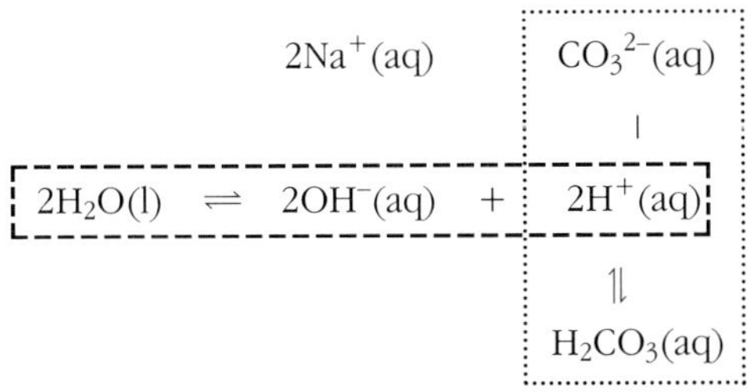

The $H^+(aq)$ and $CO_3^{2-}(aq)$ ions form one half of the weak carbonic acid equilibrium (see **Section 21.2**). This equilibrium lies largely to the undissociated $H_2CO_3(aq)$ side so $H_2CO_3(aq)$ molecules are instantly formed in the solution. This removes $H^+(aq)$ ions from the solution. The reduction in $[H^+(aq)]$ upsets the water equilibrium so more water molecules dissociate, producing more $H^+(aq)$ and $OH^-(aq)$ ions. More $H_2CO_3(aq)$ forms, removing more $H^+(aq)$ ions, and more water molecules dissociate until the two dynamic equilibria are established. The dissociation of water molecules has also been producing OH^- ions, so $[OH^-(aq)]$ is now greater than $[H^+(aq)]$ and the salt solution is alkaline.

Salts of strong acids and weak bases

Example: ammonium chloride

When ammonium chloride dissolves in water, the solution contains $NH_4^+(aq)$ and $Cl^-(aq)$ (from ammonium chloride) and $H^+(aq)$ and $OH^-(aq)$ (from water). The following equilibria are established:

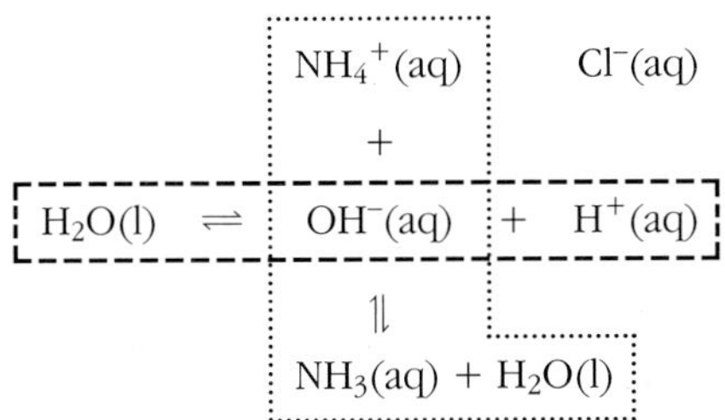

The $NH_4^+(aq)$ and $OH^-(aq)$ ions form one half of the ammonia equilibrium (see **Section 21.3**). Since ammonia is a weak base, the equilibrium lies largely to the neutral covalent $NH_3(aq)$ side, so $NH_3(aq)$ and $H_2O(l)$ molecules are instantly formed in the solution. This removes $OH^-(aq)$ ions from the solution. The reduction in $[OH^-(aq)]$ now upsets the water equilibrium so more water molecules dissociate, producing more $H^+(aq)$ and $OH^-(aq)$ ions. More $NH_3(aq)$ is then formed and more water molecules dissociate until the two equilibria are in balance. However, this time the dissociation of water molecules has also been producing $H^+(aq)$ ions, so $[H^+(aq)]$ is now greater than $[OH^-(aq)]$ and the salt solution is acidic.

Salts of strong acids and strong bases

Example: sodium chloride

When sodium chloride dissolves in water, the solution contains $Na^+(aq)$, $Cl^-(aq)$, $H^+(aq)$ and $OH^-(aq)$ ions. Thus, in solution we have:

$$Na^+(aq) \qquad Cl^-(aq)$$

$$+$$

$$H_2O(l) \rightleftharpoons OH^-(aq) + H^+(aq)$$

Neither $Na^+(aq)$ nor $Cl^-(aq)$ form any kind of equilibrium system with $OH^-(aq)$ or $H^+(aq)$ and so do not interfere with the water equilibrium. This means that $[H^+(aq)]$ remains equal to $[OH^-(aq)]$ and the solution is neutral.

PROBLEMS FOR 21.1–21.4

1 Which of the following are the same for equal volumes of equimolar solutions of sodium hydroxide and ammonia?
- a pH of the solution
- b mass of solute present
- c conductivity of the solution
- d moles of an acid needed for complete reaction.

(*SQA*)

2 Which of the following is the best description of a 0.1 mol l^{-1} solution of nitric acid?
- a dilute solution of a weak acid
- b dilute solution of a strong acid
- c concentrated solution of a weak acid
- d concentrated solution of a strong acid.

(*SQA*)

3 What is the pH of the following solutions?
- a $[H^+(aq)] = 1 \times 10^{-6}$ mol l^{-1}
- b $[H^+(aq)] = 1$ mol l^{-1}
- c $[OH^-(aq)] = 1 \times 10^{-2}$ mol l^{-1}
- d $[OH^-(aq)] = 1 \times 10^{-9}$ mol l^{-1}

4 Some solid ammonium chloride is added to a dilute solution of ammonia. Which of the following ions will decrease in concentration as a result?
- a ammonium
- b hydrogen
- c hydroxide
- d chloride.

(*SQA*)

5 Which of the following soluble salts dissolve in water to give alkaline solutions?
- a sodium nitrate
- b potassium carbonate
- c calcium ethanoate
- d ammonium nitrate.

6 The following table gives the pH of various salt solutions and information about the acids and bases used to make the salts.
- a Copy and complete the table.
- b Use ideas of equilibria to explain why a solution of ammonium sulphate is acidic and a solution of potassium ethanoate is alkaline.

Salt	pH	Acid	Base	Which is stronger?
sodium chloride	7	hydrochloric acid	sodium hydroxide	
ammonium sulphate	4		ammonia	
potassium ethanoate	9	ethanoic acid		
ammonium carbonate	8			base
sodium citrate	9	citric acid		
potassium nitrate	7			neither

22

HESS'S LAW AND ENTHALPY CYCLES

22.1 *Calculating enthalpy changes*

In **Unit 1, Chapter 14** you found out how to use experimental results to calculate enthalpy changes such as the enthalpy of combustion, ΔH_c, of ethanol, the enthalpy of solution, $\Delta H_{solution}$, of ammonium nitrate and the enthalpy of neutralisation, $\Delta H_{neutralisation}$, of hydrochloric acid and sodium hydroxide solution. In each of these examples, it is possible to record the temperature change, ΔT, of a known mass of water, m, which is caused by the energy change associated with the reaction. The energy transferred can be calculated using the formula:

$$\text{energy transferred} = cm\Delta T$$

where c is the specific heat capacity of water. The enthalpy change for the required molar quantities can then be calculated.

For example, the enthalpy change when water is made by burning hydrogen in oxygen can be measured directly.

$$H_2(g) + \tfrac{1}{2}O_2(g) \rightarrow H_2O(l); \quad \Delta H = -286\,\text{kJ mol}^{-1}$$

Notice that if you are interested in the enthalpy change associated with the formation of a compound from its elements in their standard states, it is usual to quote this for 1 mole of the compound. For this reason, only $\tfrac{1}{2}$ mole of oxygen is needed in the above equation.

However, chemists are often interested in the enthalpy changes for reactions that, unlike the examples given above, cannot be measured directly. For example, the enthalpy change for the formation of 1 mole of methane from its elements

$$C(s) + 2H_2(g) \rightarrow CH_4(g); \quad \Delta H = ????$$

cannot be measured directly, as the reaction does not occur under nomal conditions. A value for the enthalpy change is obtained *indirectly* by calculating it from other data, using an **enthalpy cycle**.

Standard state

The standard state of a substance is its most stable state under standard conditions of pressure (1 atmosphere) and temperature (25 °C).

This may be a pure solid, liquid or gas.

Enthalpy cycles

Figure 1 shows an enthalpy cycle, also known as an energy cycle. There is both a direct and an indirect way to turn graphite (C) and hydrogen (H_2) to methane (CH_4). We can't measure the enthalpy change for the *direct* route. The *indirect* route goes via CO_2 and H_2O, and involves two enthalpy changes which we *can* measure.

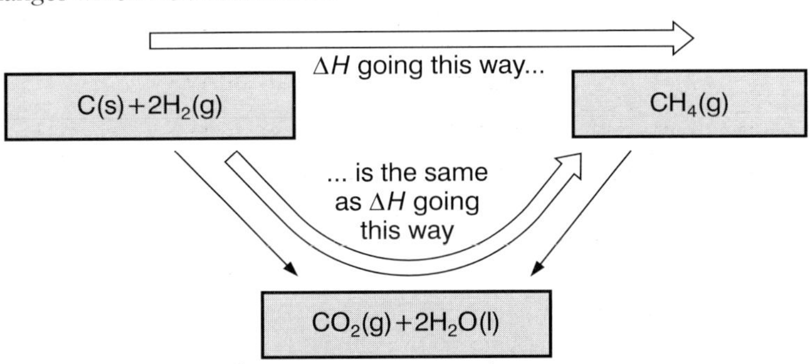

Figure 1 An enthalpy cycle for finding the enthalpy change for the formation of methane, CH_4

The key idea is that the total enthalpy change for the indirect route is the same as the enthalpy change via the direct route. This makes sense if you think about it; energy cannot be created or destroyed – this is the law of Conservation of Energy. So *as long as your starting and finishing points are the same, the enthalpy change will always be the same no matter how you get from start to finish*. This is one way of stating **Hess's law**, and an enthalpy cycle like the one in Figure 1 is called a **Hess cycle**.

If you know the enthalpy changes involved in two parts of the cycle, you can work out the enthalpy change in the third (Figure 2).

Hess's law

The enthalpy change for any chemical reaction is independent of the intermediate stages, provided the initial and final conditions are the same for each route.

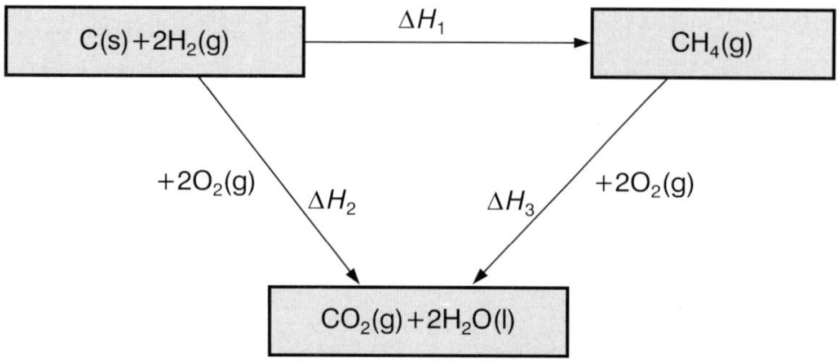

Figure 2 Using an enthalpy cycle to find ΔH_1. $\Delta H_1 = \Delta H_2 - \Delta H_3$

So if we can measure ΔH_2 and ΔH_3, we can find ΔH_1 – which is the enthalpy change that cannot be measured directly. Quite simply,

$$\Delta H_1 = \Delta H_2 - \Delta H_3$$

(It has to be minus ΔH_3 because the reaction to which ΔH_3 applies actually goes in the opposite direction to the way we want it to go in order to produce CH_4. To find the enthalpy change for the reverse reaction, we must reverse the sign.)

Now ΔH_2 is the sum of the enthalpy changes of combustion of 1 mole of carbon and 2 moles of hydrogen, and ΔH_3 is the enthalpy change of combustion of methane. ΔH_1 is the enthalpy change for the formation of methane, which is the quantity we are trying to find.

$$\Delta H_2 = \Delta H_c(C) + 2\Delta H_c(H_2)$$
$$\Delta H_3 = \Delta H_c(CH_4)$$

Putting in the values of the enthalpies of combustion, which we can measure, we get

$$\begin{aligned}\Delta H_1 &= \Delta H_c(C) + 2\Delta H_c(H_2) - \Delta H_c(CH_4)\\ &= -394\,\text{kJ mol}^{-1} + 2(-286)\,\text{kJ mol}^{-1} - (-891)\,\text{kJ mol}^{-1}\\ &= -75\,\text{kJ mol}^{-1}\end{aligned}$$

$\Delta H_c(C) = -394\,\text{kJ mol}^{-1}$
$\Delta H_c(H_2) = -286\,\text{kJ mol}^{-1}$
$\Delta H_c(CH_4) = -891\,\text{kJ mol}^{-1}$

Enthalpy changes for the formation of many compounds and also aqueous solutions of ions, from their elements in their standard states, are now available in data books. Some of these have been measured directly, others have been calculated from enthalpy changes for other reactions, as shown above for methane. These data are very useful as they can be used to calculate the enthalpy change for a reaction, rather than attempt to do an experiment to measure it. For example, if you wanted to know the enthalpy of solution of sodium chloride, you could use data from tables to calculate it. $\Delta H_{solution}$ for sodium chloride is the enthalpy change for

$$NaCl(s) + aq \rightarrow Na^+(aq) + Cl^-(aq)$$

This can be included in an enthalpy cycle (Figure 3).

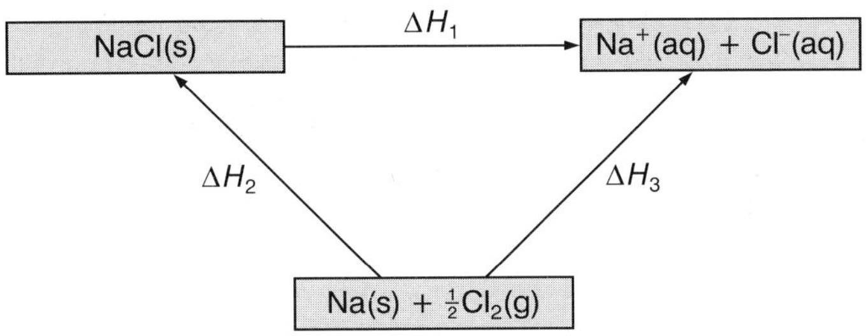

Figure 3 Using an enthalpy cycle to find $\Delta H_{solution}$ for sodium chloride, NaCl

The cycle shows two routes to form the ions in solution. Hess's law states that the enthalpy changes for both routes are the same, therefore,

$$\Delta H_2 + \Delta H_1 = \Delta H_3$$

We want to find ΔH_1, so rearranging this equation gives,

$$\Delta H_1 = \Delta H_3 - \Delta H_2$$

$\Delta H_1 = \Delta H_{solution}(NaCl)$
$\Delta H_2 = \Delta H$ for the formation of NaCl(s)
$\Delta H_3 = \Delta H$ for the formation of $Na^+(aq)$ + ΔH for the formation of $Cl^-(aq)$

From data tables we can find that the enthalpy changes for the formation of NaCl(s), $Na^+(aq)$ and $Cl^-(aq)$ are as follows:

NaCl(s) = $-411\ kJ\ mol^{-1}$
$Na^+(aq)$ = $-240\ kJ\ mol^{-1}$
$Cl^-(aq)$ = $-167\ kJ\ mol^{-1}$

Putting in the values for the enthalpy changes:

$$\Delta H_{solution}(NaCl) = -240\ kJ\ mol^{-1} + (-167)\ kJ\ mol^{-1} - (-411)\ kJ\ mol^{-1} = +4\ kJ\ mol^{-1}$$

This is a very small enthalpy change, which you would find difficult to measure experimentally.

In **Section 7.2** you can use a similar enthalpy cycle to find the enthalpy of solution of the calcium carbonate shells of marine animals. Shells are more soluble in the cold water at the bottom of the oceans.

PROBLEMS FOR 22.1

1 a Write an equation to represent the formation of propane, C_3H_8, from its elements in their standard states.

b Draw an enthalpy cycle, similar to the one in Figure 1, to show the relationship between the formation of propane from carbon and hydrogen, and the combustion of these substances to give carbon dioxide and water.

c Use your enthalpy cycle to calculate a value for the enthalpy change for the formation of propane. You may use *only* the following data on enthalpy changes of combustion.

$\Delta H_c(C) = -394\ kJ\ mol^{-1}$
$\Delta H_c(H_2) = -286\ kJ\ mol^{-1}$
$\Delta H_c(C_3H_8) = -2220\ kJ\ mol^{-1}$

2 Suppose you need to find the enthalpy change for the formation of carbon monoxide. You cannot measure this directly, because you cannot normally convert carbon to carbon monoxide. However, you *can* convert both C and CO to CO_2, and you can use the enthalpy changes for these reactions to find the enthalpy change for the formation of CO.

$C(s) + O_2(g) \rightarrow CO_2(g)$; $\Delta H_c(C) = -394\ kJ\ mol^{-1}$

$CO(g) + \frac{1}{2}O_2(g) \rightarrow CO_2(g)$; $\Delta H_c(CO) = -283\ kJ\ mol^{-1}$

Draw an appropriate enthalpy cycle and use it to calculate the enthalpy change for the formation of CO(g).

3 The formula of blue crystals of hydrated copper(II) sulphate is $CuSO_4.5H_2O(s)$. The enthalpy change for the formation of these crystals from anhydrous copper(II) sulphate cannot be measured directly.

$CuSO_4(s) + 5H_2O(l) \rightarrow CuSO_4.5H_2O(s)$ ΔH_1

However, it is easy to measure the enthalpy changes for two other changes involving these substances:

$CuSO_4.5H_2O(s) + aq \rightarrow Cu^{2+}(aq) + SO_4^{2-}(aq)$ ΔH_2

$CuSO_4(s) + aq \rightarrow Cu^{2+}(aq) + SO_4^{2-}(aq)$ ΔH_3

a Draw an enthalpy cycle incorporating these three changes.

b A student obtained the following experimental results:
On adding 6.25 g (0.025 mol) of $CuSO_4.5H_2O(s)$ to 50 cm^3 of water in a polystyrene beaker, she found that the temperature fell by 1.3 °C.
On adding 4.00 g (0.025 mol) of $CuSO_4(s)$ to another 50 cm^3 of water in a polystyrene beaker, she found that the temperature increased by 7.8 °C.
Use these experimental results to calculate the enthalpy changes ΔH_2 and ΔH_3 (in kJ mol^{-1}).
The specific heat capacity of water is 4.18 kJ kg^{-1} °C^{-1} and the density of water is 1 g cm^{-3}.

c Use the enthalpy cycle drawn in part **a** and your values for ΔH_2 and ΔH_3 to calculate ΔH_1.

23 MANAGING A CHEMICAL PROCESS

23.1 *The operation of a chemical manufacturing process*

Batch or continuous

The chemical manufacturing process is the recipe by which a product is made. It includes specific instructions relating to the use of the equipment (plant), quantities and qualities of raw materials, mixing sequences, temperatures, pressures and so on. Most chemical processes involve a sequence of events (called *unit operations*) which can be represented as in Figure 1.

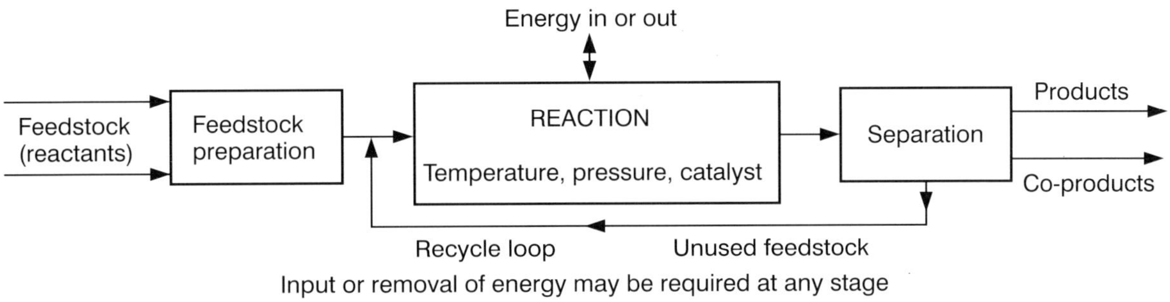

Figure 1 Sequence of unit operations in a chemical plant

For a particular process, this sequence may be organised in one of two ways: **batch** or **continuous.** Sometimes a mixture of batch and continuous unit operations may be applied to achieve a given product. Any organic preparations you have performed so far are batch processes.

In a **batch process**, the starting materials are put into a vessel and allowed to react together. The reaction is monitored and, when complete, the reaction phase is terminated. The product is separated from the reaction mixture. The process is then repeated in an identical manner, batch by batch, until the required amount of product is manufactured (Figure 2).

In a **continuous process**, the starting materials are fed in at one end of the plant and product is withdrawn at the other end of the plant in a continuous flow (Figure 2). The process may proceed through various unit operations in specifically designed parts of the plant, each dedicated to a particular step in the sequence.

Plants capable of operating batch processes are generally more versatile than continuous plants. It is usually possible to carry out a variety of chemical reactions in batch operation using the same vessel with no, or only minor, modifications. Batch processes are most cost effective when relatively small volumes of products are required, although returns on sales of small tonnages would not usually be adequate to justify building a dedicated, single product plant. Batch processes are also more applicable to slow reactions where long residence time in the vessel would make adaptation to continuous operations difficult.

For an example of batch processing, see the manufacture of organic bromine compounds, discussed in **Section 9.2**.

There are drawbacks to batch processes, relating to safety and contamination. In a batch reaction relatively large reacting masses might not easily be controllable in the event of an exothermic reaction. It may not be

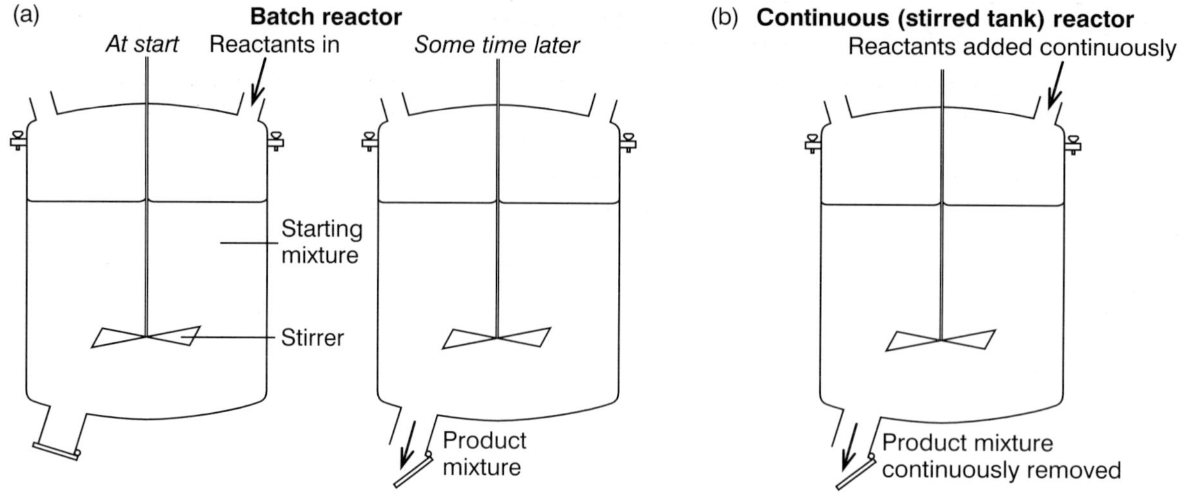

Figure 2 Comparison of (a) batch and (b) continuous tank reactors

easy to cool the contents of the reaction vessel. Continuous processes tend to operate with relatively low volumes of reactants together at any time, allowing faster removal of thermal energy and better control.

Contamination from batch to batch and product to product is also a potential problem in batch processing. Detailed clean-out procedures, which are time consuming and costly, must be followed to minimise cross-contamination.

Continuous processes are more suited to high-tonnage production of a single product. Once steady conditions have been established, the process can be run with minimal labour but relying heavily on instrumental and automatic control. To operate a continuous process, the plant is designed to produce product of consistent quality under optimum conditions, thereby minimising undesirable by-products and waste. Ethene, ammonia and many other high-tonnage chemicals are made this way (Table 1).

Table 1 Examples of chemicals manufactured by batch or continuous processes

Batch	Continuous
copper refining **(Section 9.3)**	ethene, propene **(Section 4.3)**
organic bromine compounds **(Section 9.2)**	poly(ethene) **(Section 5.2)**
iodine-123 **(Section 9.4)**	ammonia **(Section 9.1)**
pharmaceuticals, for example aspirin	bromine **(Section 9.2)**
dyes	sulphuric acid

There are often drawbacks to continuous processes, however. In particular, they require tailor-made plant which is therefore less flexible. Nevertheless, there are new ingenious ways in which some flexibility can be incorporated by changing the feedstock or the conditions of the process (see **Section 23.3**).

Table 2 summarises the advantages and disadvantages of the two types of process.

Plant construction materials

It is essential to choose construction materials which do not react with the feedstocks, catalysts, solvents and products involved in the process. The wrong choice can lead to lower efficiency, hazardous reactions, and product contamination – not to mention holes in the reaction vessels or pipes.

Glass-lined vessels, alloys or glass-reinforced plastics are widely used in place of steel components when there is a corrosion risk.

Advantages	Disadvantages
Batch	
cost effective for small quantities; capital cost of plant much lower – possible to buy plant 'off the shelf'	charging and emptying the reactor is time consuming; no more product is being formed in this **shut-down** time
slow reactions can be catered for	larger workforce needed
a range of products can be made in the same vessel	contamination more likely
a greater percentage conversion is achieved compared with the same-sized continuous reactor for the same time	exothermic reactions can be more difficult to control
Continuous	
suited to high-tonnage production	very much higher capital cost before any production can occur
greater throughput; shut-down may not occur for months or even years	not cost effective when run below full capacity
more easily automated and process is better	
controlled as fine adjustments are possible	
contamination risk is low when used for one product	contamination risk is higher when used for two or more products
consistent quality ensured	
requires minimal labour	

Table 2 Advantages and disadvantages of batch and continuous processes

PROBLEM FOR 23.1

Classify the following processes as **batch** or **continuous**:

- **a** the removal of carbon monoxide, oxides of nitrogen and unburnt hydrocarbons in a catalytic converter on a petrol-driven motor vehicle
- **b** cake making in the home
- **c** the distillation of crude oil to provide a range of fractions for different purposes
- **d** the conversion of glucose into a mixture of glucose and fructose by passing the glucose through a bed of the immobilised enzyme, glucose isomerase. (The product is called 'high fructose syrup'. This syrup is sweeter than either fructose or glucose. The food and soft drinks industry use millions of tonnes of high fructose syrup every year.)

23.2 *Feedstocks*

The feedstocks are the reactants which go into a chemical process. The raw materials usually have to be 'prepared' or treated to ensure that they are sufficiently pure and present in the correct proportion to use as feedstock. For example, by far the largest part of an ammonia plant is concerned with this 'feedstock preparation', making the nitrogen and hydrogen mixture for direct conversion to ammonia. Methane, air and water, the raw materials, are first purified by filtration and chemical treatment to remove potential catalyst poisons such as sulphur. Further chemical treatment is then carried out to produce the correct mixture of nitrogen and hydrogen.

The operation of an ammonia plant is discussed in **Section 9.1**.

An important part of feedstock preparation is getting the feedstock into a form in which it is easy to handle. Transferring gases and liquids is relatively easy because they can be transported by pipes within the chemical plant or even across the country (see Figure 3). Even so, the cost of pumping may be high and so every effort is made to keep the number of pumps and length of piping down to a minimum.

Solids are expensive to handle. Sometimes they are melted and maintained as hot liquids to reduce the transportation costs. For example, sulphur used in the manufacture of sulphuric acid is often delivered by ship to the plant as a molten liquid, and is then also used in the plant as a liquid. Another way of handling solids is to mix them with a liquid to form a **slurry**; they can then be pumped for many kilometres along pipelines.

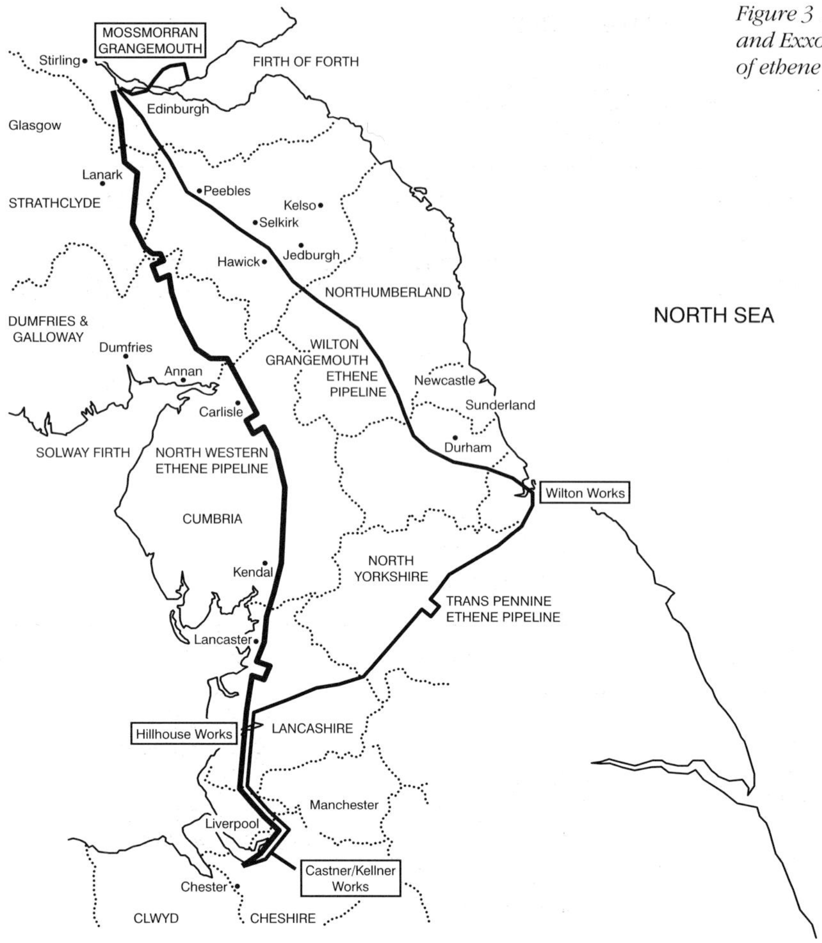

Figure 3 Pipelines from BP Grangemouth and Exxon, Mossmorran, for the distribution of ethene

Inorganic chemicals such as sulphuric acid, ammonia, nitric acid, sodium carbonate and chlorine are mainly produced from metallic ores and minerals, the gases in the atmosphere and water. There is usually little choice of feedstock for such chemicals (Table 3).

Table 3 Feedstocks for some inorganic chemicals

Inorganic chemical	Feedstock
sulphuric acid	sulphur, air
ammonia	nitrogen, hydrogen
nitric acid	ammonia, air
sodium hydroxide	sodium chloride
chlorine	sodium chloride
sodium carbonate	sodium chloride, calcium carbonate, ammonia

A vast proportion of the organic chemicals produced today are derived from oil and natural gas. Natural gas is mainly methane, but ethane, propane and butane are often present as well. These are steam cracked to produce ethene and propene (Figure 4). Steam cracking of alkanes is described in **Unit 2, Section 17.1**.

Distillation of oil produces a variety of fractions, in particular, LPG (liquified petroleum gas), naphtha and gas (or diesel) oil. These fractions are converted into a variety of building blocks from which many chemicals are made. These building blocks are often alkenes. However, branched-chain alkanes, cycloalkanes and aromatic hydrocarbons are also produced for use in unleaded petrol (Figure 5).

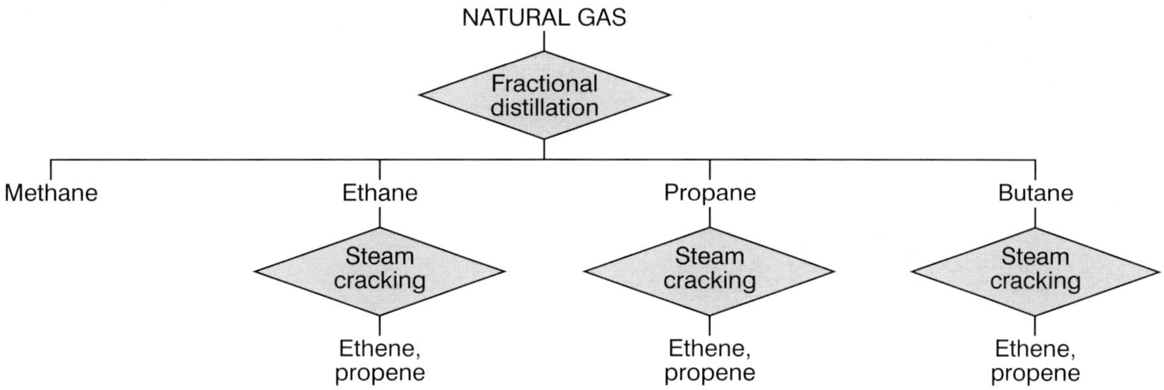

Figure 4 Feedstocks from natural gas

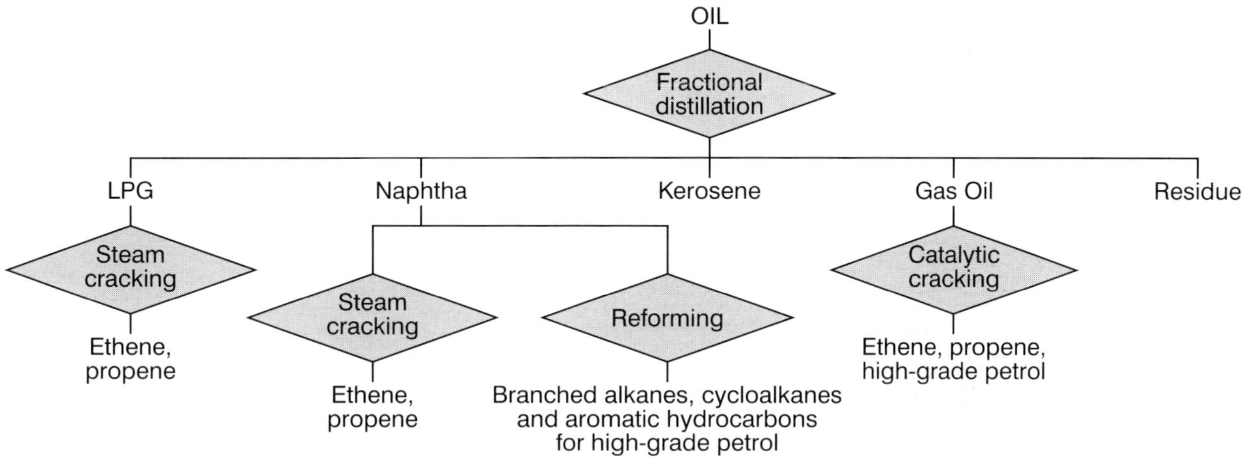

Figure 5 Feedstocks from oil

PROBLEM FOR 23.2

Until recently in the UK, naphtha was used as the feedstock for making ethanoic acid. Naphtha is one of the fractions obtained in refining petroleum. It is principally a mixture of hydrocarbons which contain between five and nine carbon atoms per molecule.

Naphtha is oxidised by air at a temperature of about 200 °C and a pressure of about 50 atmospheres. No catalyst is needed and the reaction takes place in one step. The product contains about 50% ethanoic acid together with a range of compounds including propanone, methanoic acid, propanoic acid and butane-1,4-dioic acid. Some of these co-products are useful; others are simply burned at the chemical plant to provide energy.

Naphtha oxidation has been replaced by a process in which methanol reacts with carbon monoxide in the presence of a rhodium/iodine catalyst.

$CH_3OH(l) + CO(g) \rightarrow CH_3COOH(l)\ \ \Delta H = -135\,kJ\,mol^{-1}$

This process takes place at a similar temperature to the one used in naphtha oxidation but the pressure is lower (about 30 atmospheres). The catalyst is expensive but the process is very specific. A 99% yield of ethanoic acid is obtained. Methanol is cheap and readily available from a number of sources. It can be made from petroleum or natural gas.

One way of investigating the success and efficiency of a chemical process is to look at the advantages and disadvantages associated with factors such as:

- the conditions employed
- the feedstock
- the product
- the co-products.

a Taking the four factors in turn, suggest one advantage in each case which the methanol/carbon monoxide process has over the naphtha oxidation.

b Suggest one disadvantage which the methanol/carbon monoxide process has compared to naphtha oxidation.

(Part of Salters Paper 2, slightly modified)

23.3 The best conditions for the process

The choice of temperature, pressure and catalyst are the conditions most likely to affect the yield of a chemical process. In **Section 23.4** we examine other possibilities of saving resources, for example:

- recycling reactants left behind after removing the product
- recovering and using heat in order to save energy.

Temperature

The key reaction in the Contact process, which produces sulphuric acid, is the reaction between sulphur dioxide and oxygen:

$$2SO_2(g) + O_2(g) \rightarrow 2SO_3(g); \quad \Delta H = -196\,\text{kJ mol}^{-1}$$

Because it is an exothermic reaction, too high a temperature reduces the yield. However, too low a temperature reduces the rate of reaction. A compromise has to be made to allow a reasonable yield in a reasonable time. Such a compromise regarding temperature is also required in the Haber process for making ammonia.

See **Sections 9.1** and **20.2** for details about the conditions for making ammonia in the Haber process.

When a chemical process leads to a number of products, the temperature of the process may have a profound effect on the relative yields of each product. For example, when steam cracking naphtha to produce alkenes, a change of temperature, residence time and the relative concentrations of naphtha and steam alters the relative concentrations of products (Table 4). Residence time is the average time taken by the hydrocarbon molecules to pass through the reaction.

Table 4 Effect of changing reaction conditions on the steam cracking of naphtha

Product (%)	**Low severity cracking** for example, 750 °C, 0.5 s residence time, steam:naphtha ratio 0.3:1	**High severity cracking** for example, 850 °C, 0.1 s residence time, steam:naphtha ratio 0.5:1
methane	15	18
ethene	19	32
propene	16	13
C_4 hydrocarbons	10	9
C_5–C_9 hydrocarbons	36	18

Thus, if there is a demand for ethene and propene, to make polymers, high severity conditions are used; if there is a demand for C_5–C_9 hydrocarbons, for petrol, low severity conditions are used. Another way to cope with a variation in demand is to change the feedstock.

See **Section 8.8** for information about the choice of feedstock.

Pressure

The effect of pressure on gas-phase processes may also have a profound effect. An increase in pressure will increase the rate of the reaction taking place. However, the yield of product may be increased or decreased. An increase in yield will occur on increasing the pressure for gas-phase reactions where there are fewer product molecules than reactant molecules. For example, ethanol is made industrially by the addition of steam to ethene in the presence of a solid acid catalyst. The equation for the reaction is:

$$CH_2{=}CH_2(g) + H_2O(g) \rightleftharpoons C_2H_5OH(g)$$

The yield of ethanol is increased as the pressure is raised. A pressure of 60 atmospheres is used for this process.

For the same reason, high pressures are used in the Haber process.

Catalysts

A catalyst is a substance which alters the rate of a reaction by providing an alternative reaction route with a lower activation energy. Many catalysts are made from transition metals such as platinum, palladium and rhodium. These are very costly, being even more expensive than silver and gold. Because they are not used up, it is economical to use them despite the high cost of the metal. Table 5 gives some examples of industrially important catalysts.

Table 5 Some important examples of industrial catalysis

Reaction	Catalyst used
hydrogen and nitrogen to produce ammonia	iron
hydrogenation of edible oil for margarine	nickel
carbon monoxide and hydrogen to produce methanol	copper
oxidation of ammonia to give nitrogen monoxide (NO) for making nitric acid	platinum–rhodium alloy
oxidation of sulphur dioxide to sulphur trioxide for making sulphuric acid	vanadium(V) oxide, V_2O_5
hydration of epoxyethane to give ethane-1,2-diol for polyesters	0.5% sulphuric acid or sodium hydroxide solution

If a catalyst can be found for the reaction, it may be possible to lower the temperature at which it takes place. This in turn will save much external heating and the consequential energy costs. Further, if it is an exothermic reaction, lowering the temperature will increase the yield.

Most of the examples given in Table 5 are *heterogeneous* catalysts. They are solids which bring about catalysis when their surfaces come into contact with the reactants in liquid or gas streams.

When the catalytic material is in the same phase as the reactants and can mix with them, it is called a *homogeneous* catalyst. Acids and alkalis are the most common homogeneous catalysts.

The performance of a catalyst is judged in terms of

- the effect it has on the reaction rate
- its life expectancy
- the percentage reclaimed when it is renewed.

The performance of catalysts deteriorates because of reactions with impurities in the feedstock (catalyst poisoning). Also, at high operating temperatures, a heterogeneous catalyst may melt and lose its high surface area. A catalyst may get covered in soot or other surface coating, and may need to be cleaned in a regenerator.

There is more about catalysts, including the use of enzymes as industrial catalysts, in **Unit 1**, **Section 15.5**.

PROBLEMS FOR 23.3

1 Catalysts can be held in two main types of reactors. In **fixed bed** reactors, the reactants are passed over a bed of solid catalyst. The catalyst may be in a large vessel filled with porous pellets, or as a thin mesh of metal through which the reactants are blown.

Fluidised bed reactors have catalysts in the form of a fine powder. Reactant gases are passed through the reactor at such a speed that the particles become suspended in the gas stream and behave like a fluid. In this type of reactor, the catalyst is regenerated by making the 'fluid' flow into a separate container for the regeneration process.

An example of a fluidised bed reactor is shown in Figure 6.

a Draw simple outline sketches to represent a fluidised bed reactor and a fixed bed reactor.

b Draw up a table to compare the likely effectiveness of the two types of reactor in terms of

- **i** simplicity of design
- **ii** ease of operation
- **iii** ease of contact between catalyst and reactant
- **iv** ease with which the catalyst can be regenerated
- **v** heat transfer and temperature control.

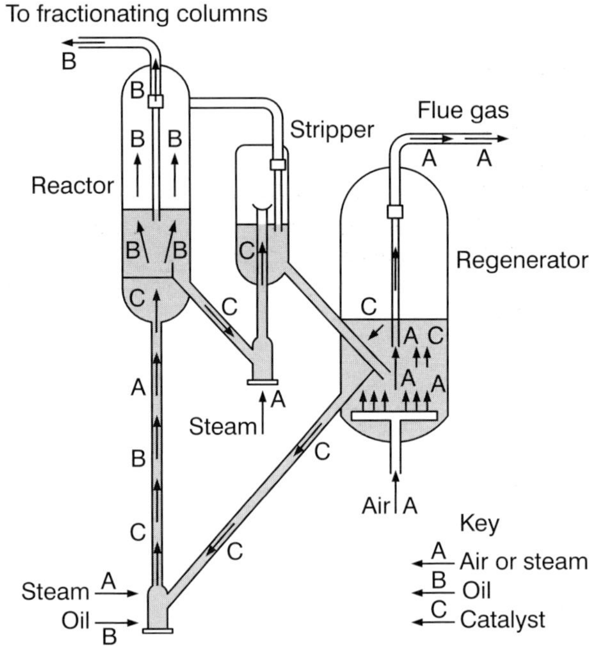

Figure 6 A 'cat-cracker' using a fluid catalyst

c Explain the benefits of
  i maximising contact between the catalyst and reactants
  ii designing a reactor so that the catalyst can be easily regenerated.

2 Before you tackle this question, you will need to have studied **Unit 1**, **Chapter 15** on reaction rates and **Chapter 20** on equilibria.

ICI's low pressure technology for the industrial production of methanol

The manufacture of methanol looks deceptively simple:

$CO(g) + 2H_2(g) \rightarrow CH_3OH(g)$

However, methanol production on an industrial scale is far from simple. Feedstock, catalysts, enthalpy changes, reaction rates, temperature and pressure all need to be considered. ICI's low pressure technology involves three key stages:

- steam reforming of methane to produce synthesis gas (a mixture of CO and H_2)
- conversion of synthesis gas to methanol
- distillation of the methanol to give a high-purity product.

We will consider the conversion of synthesis gas to methanol. It involves several reactions. These include

$CO(g) + 2H_2(g) \rightleftharpoons CH_3OH(g)$; $\Delta H = -90.7\,\text{kJ mol}^{-1}$

The conversion is carried out over a catalyst at a temperature of 260 °C–270 °C and a pressure of 100 atmospheres. The methanol converter is designed to maintain this temperature. One way in which this is done is by using an internal heat exchanger. About 10% conversion to methanol is obtained.

a Decide what happens to the percentage of methanol in the mixture as
  i pressure increases
  ii temperature increases.

b Explain the reasons for the choice of reaction conditions in terms of achieving
  i a high yield
  ii a high rate.

23.4 *Becoming even more efficient*

The efficiency of a chemical process depends on various physical factors such as temperature, pressure, rate of mixing, etc. When choosing the temperature and pressure for a particular reaction step it might seem appropriate to go for the highest possible, to maximise the rate at which the product is formed. However, it is not always as simple as this, and it is necessary to find the conditions that give the most economical conversion. For example, very high temperatures and pressures require very specialised, expensive plant to attain, and add to the difficulty in controlling the chemical reactions.

Even more important, the choice of conditions can have a significant impact on the *yield*.

Further, the costs of the process can be reduced considerably if an effective catalyst can be found. Huge savings can also be made by recycling unreacted feedstock and in saving energy that would otherwise be lost to the environment.

Recycling

The reclamation and recycling of unreacted materials is an important aspect of process development for both economic and environmental reasons.

Recycling means separating unreacted feedstock from the reaction mixture (see Figure 1), and this is not always easy. It is important that impurities do not get recycled along with the unreacted feedstock. If an impurity is recycled but does not react, its concentration will build up in the reaction step and it will interfere with the reaction. For example, in the manufacture of ammonia, argon from the air is recycled along with the unreacted nitrogen and hydrogen. As argon is unreactive, it is difficult to remove it and so it is not possible to recycle continuously. The argon-enriched nitrogen and hydrogen mixture is bled off at intervals.

The Haber process for the manufacture of ammonia is discussed in **Section 9.1**.

Saving energy

The efficient use of energy is significant in most chemical processes. During the past few years, often in response to high energy costs and environmental considerations, chemical companies have markedly reduced their energy consumption.

Many chemical reactions involve the release of thermal energy and this can be conserved by lagging pipes and by using heat exchangers (Figure 7). Energy from exothermic parts of the process can be used to supply energy to endothermic parts. The energy released in an exothermic reaction can be used to raise the temperature of the reactants.

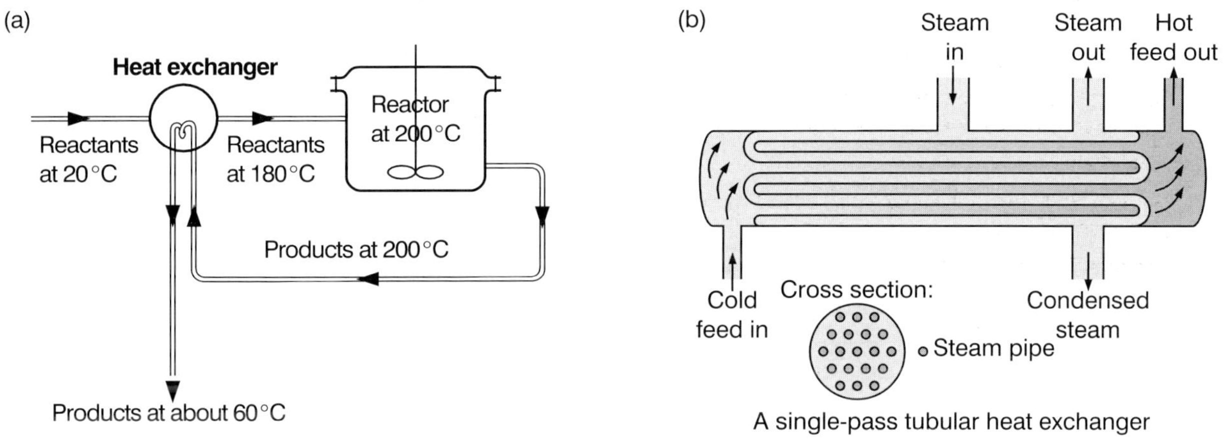

Figure 7 (a) An example of how energy can be used in a chemical process. (b) A diagram to illustrate a heat exchanger

Water or steam are the most usual agents for transferring energy in a chemical plant. Steam is a much safer alternative to oil, gas or electricity where flammable substances are involved.

In an integrated plant, energy (transferred in steam or hot liquid or gas streams) from one process can be used in a completely different part of the plant. This reduces the quantity of oil, gas and electricity that has to be purchased. By contrast, a small site would have to raise steam in a special boiler using purchased fuel oil, gas or electricity. The price of energy in the small plant would therefore be considerably greater than that in an integrated plant.

Co-products and by-products

When feedstock is passed through a reactor, a number of things might happen to it.

1. The reaction may form only one product. For example:

   nitrogen + hydrogen ⇌ ammonia

   $N_2(g) + 3H_2(g) \rightleftharpoons 2NH_3(g)$

2. The reaction may form two products. For example, phenol is manufactured from (1-methylethyl)benzene (often called cumene). The overall reaction can be represented as

   $C_6H_5CH(CH_3)_2 + O_2 \rightarrow C_6H_5OH + (CH_3)_2CO$

   cumene *phenol* *propanone*

   The propanone formed in this reaction is an example of a **co-product**. Six tonnes of propanone are produced for every 10 tonnes of phenol and the proceeds of its sale make a significant contribution to profits. The route would become uncompetitive if demand for propanone was to fall.

3. A reaction other than the one that was intended may occur. For example, epoxyethane, $(CH_2)_2O$, can be made in a one-step process in which ethene is mixed with oxygen and passed over a silver catalyst at 300 °C and a pressure of about 3 atmospheres. Not surprisingly, under these conditions there is also the possibility of the ethene being completely oxidised:

   $C_2H_4 + 3O_2 \rightarrow 2CO_2 + 2H_2O$

   The carbon dioxide and water formed in this unwanted reaction are called **by-products**.

4. Some feedstock may remain unreacted.

PROBLEM FOR 23.4

Ethanol is made by the addition of steam to ethene. The equation for this exothermic reaction is

$CH_2{=}CH_2(g) + H_2O(g) \rightleftharpoons C_2H_5OH(g)$

$\Delta H = -87.5\ \text{kJ mol}^{-1}$

In one plant the conditions for this process are 300 °C and 70 atmospheres pressure. The catalyst is a solid impregnated with phosphoric acid. When the ratio of reactants is 1 mole of ethene to 2 moles of steam, the proportion of ethene converted at equilibrium is 40%.

However, the process is operated so that a 10% conversion of ethene is achieved following one pass of the reactants through the catalyst.

a With the aid of a diagram, suggest how some of the thermal energy released in the reaction can be put to good use in the process.

b Explain why it is more desirable to achieve a 10% conversion of ethene than a figure closer to the equilibrium value at this temperature and pressure.

23.5 Costs

All of the factors considered so far contribute to the cost of the process. Major costs – such as research and development, plant design and construction, and initial production – are incurred before any product is sold into the market. Sales of the product have to generate enough return to offset these initial costs and generate a profit for the company (Figure 8).

The profit generated is the difference between the selling price and the costs of production. Production costs are made up of two elements: **fixed costs** and **variable costs.**

Fixed (or indirect) **costs** are those incurred by the company whether they produce 1 tonne or tens of thousands of tonnes of product. For example, as soon as it has been built, the production plant starts to lose value, or depreciate, regardless of how much product is made. Other fixed costs include labour costs, land purchase or rental, sales expenses, telephone bills, etc. Hence the fixed cost element in the production cost is calculated by spreading the total annual charges over the number of units produced per year. If only 1 tonne of product is produced per year then the fixed cost element of production cost will be significantly higher than if 100 tonnes are produced with the same fixed costs. As fixed costs are allocated against a product, it is best to divide the cost against the maximum output of product possible. This will only occur when the plant is operating at maximum capacity.

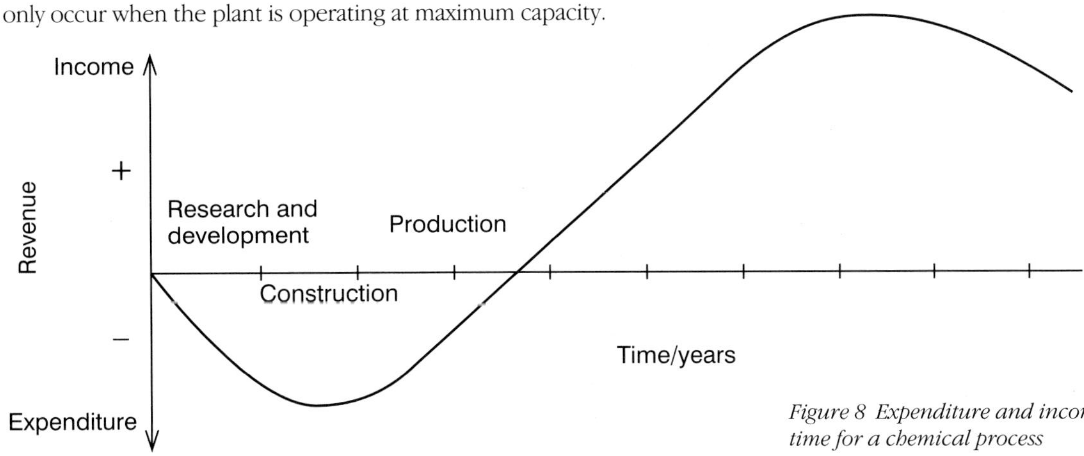

Figure 8 Expenditure and income against time for a chemical process

Variable (or direct) **costs** relate specifically to the unit of production. Raw materials are the most obvious variable cost, along with costs of effluent treatment and disposal, and the cost of distributing the product. If no production occurs then these variable costs will not be incurred, whereas fixed costs will still have to be paid.

Capital costs relate to establishing chemical plant, buildings and infrastructures around which manufacture is based. These capital costs are recovered as part of the fixed cost element (depreciation). In general, for normal accountancy purposes, the life of a plant is assumed to be 10 years, after which it may be said to be 'written off'. This depreciation happens because chemical plants corrode, fall to pieces or are made obsolete by technological advances. In practice, many plants have a perfectly satisfactory life beyond 10 years. In estimating the profitability of a plant, it will be necessary to recoup the investment in capital costs (as well as covering the fixed and variable costs) before the plant is written off.

PROBLEMS FOR 23.5

1 Explain why
   a the capital cost of a 200 tonne day^{-1} chemical plant is less than twice that of a 100 tonne day^{-1} plant
   b if you double the output of a chemical plant, you may double the variable costs, but you don't double the fixed costs.

2 Calculate the value of the sales per employee for the countries in Table 6.
   a How does the UK compare with other European countries in terms of value of sales per employee?
   b Which country has the greatest value of sales per employee?
   c Suggest what affects these figures.

	Sales/ 10^9 US$	Number of employees
Germany	112.9	501 000
France	81.0	239 000
UK	55.4	238 000
Italy	52.8	189 000
Belgium	38.4	94 000
Spain	31.0	123 000
The Netherlands	30.2	79 000
Switzerland	24.5	64 000
Ireland	14.1	21 000
Sweden	9.6	34 000

Table 6 European chemical industry sales, 1997

24 REDOX

24.1 *Oxidation and reduction*

When we set fire to a piece of magnesium ribbon in air, the magnesium is oxidised – it gains oxygen. The product, magnesium oxide, is an ionic compound.

$$Mg(s) + \frac{1}{2}O_2(g) \rightarrow Mg^{2+}O^{2-}(s)$$

If you think carefully about what is happening, you will see that the reaction is composed of two *half-reactions* which can be described by two *half-equations*:

$$Mg \rightarrow Mg^{2+} + 2e^- \quad \text{and} \quad \frac{1}{2}O_2 + 2e^- \rightarrow O^{2-}$$

(Half-equations are often called **ion-electron equations**.) The magnesium is losing electrons in one half-reaction, and the oxygen is gaining them in the other. Another way of looking at oxidation is to say that *oxidation is the loss of electrons*.

Reduction is the opposite of oxidation, so *reduction* must occur when *electrons are gained*. The oxygen is being reduced in this example.

Reduction and oxidation occur together when magnesium burns in air: it is a reduction–oxidation or **redox** reaction (Figure 1).

If you had burned magnesium in chlorine, you would have performed another redox reaction. The ion-electron equations show this:

$$Mg \rightarrow Mg^{2+} + 2e^- \quad \text{and} \quad Cl_2 + 2e^- \rightarrow 2Cl^-$$

Magnesium loses electrons and is oxidised; chlorine gains electrons and is reduced.

Redox is a very important type of chemical reaction. Many reactions can be classified as redox reactions using the ideas in the box on the right.

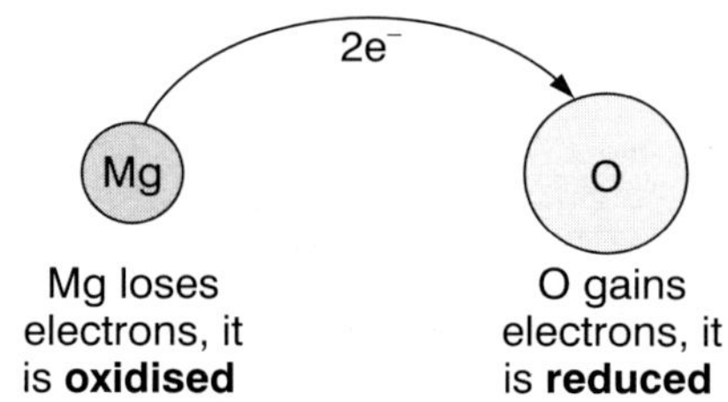

Figure 1 A redox reaction occurs when electrons are transferred from magnesium to oxygen

Oxidation is loss of electrons.	**Reduction is gain of electrons.**
An **oxidising agent** removes electrons from something else.	A **reducing agent** gives electrons to something else.

Displacement reactions

You will recall that when zinc is added to copper(II) sulphate solution a *displacement reaction* takes place. The blue colour of the solution fades and a brown solid is deposited on the zinc. We say that the zinc has displaced the copper from the solution. The overall equation for the reaction is

$$Zn(s) + CuSO_4(aq) \rightarrow Cu(s) + ZnSO_4(aq)$$

but this is not the most helpful way of showing what is happening. Look at Figure 2 which shows the reaction in diagrammatic form.

One reactant, copper(II) sulphate solution, contains a mixture of copper(II) ions and sulphate ions, which move around independently of each other. The same is true for the zinc ions and sulphate ions in the product, zinc sulphate solution. Clearly the sulphate ions have not changed and take no part in the reaction. They are **spectator ions**. On removing them from the overall equation we obtain

$$Zn(s) + Cu^{2+}(aq) \rightarrow Cu(s) + Zn^{2+}(aq)$$

Since the *zinc atoms* are changing into *zinc ions* they must lose electrons.

$$Zn(s) \rightarrow Zn^{2+}(aq) + 2e^-$$

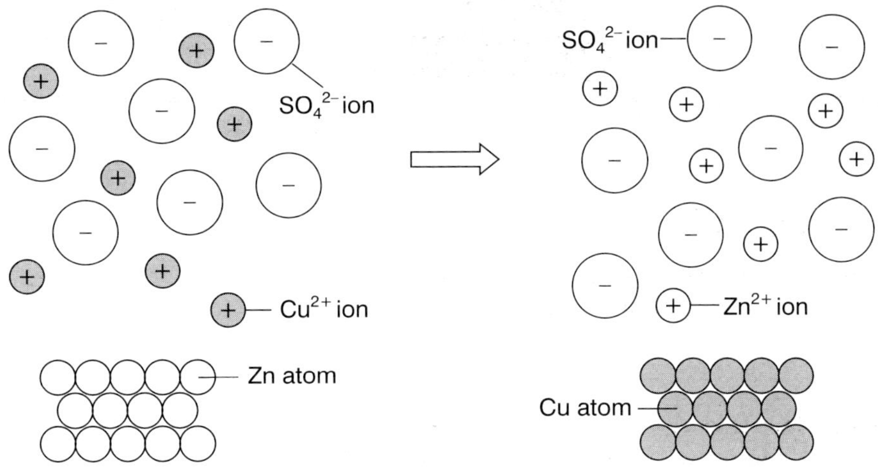

Figure 2 The reaction of zinc with copper(II) sulphate solution

The zinc atoms are being oxidised to zinc ions.

The electrons that the zinc atoms lose are gained by the copper(II) ions when they form copper atoms.

$Cu^{2+}(aq) + 2e^- \rightarrow Cu(s)$

The copper(II) ions are being reduced to copper atoms. The overall reaction involves zinc atoms *transferring electrons* to copper(II) ions.

Notice that in an oxidation ion-electron equation the electrons are on the products side. In a reduction they are on the reactants side. Notice also that in an ion-electron equation, the total charge on the reactants side equals that on the products side.

For example, in the above reduction, copper atoms are neutral and so the charge on the products side is zero. On the reactants side the 2+ charge of the copper ion cancels out the 2– charge of the electrons, giving a net charge of zero.

In the following displacement reaction the metal ions are the spectator ions.

When chlorine is bubbled through potassium iodide solution, iodine is displaced. The overall equation for the reaction is:

$Cl_2(g) + 2KI(aq) \rightarrow 2KCl(aq) + I_2(aq)$

On removing the 'spectating' $K^+(aq)$ ions we obtain the redox equation:

$Cl_2(g) + 2I^-(aq) \rightarrow 2Cl\ (aq) + I_2(aq)$ redox

This can be written as two ion-electron equations:

$2I^-(aq) \rightarrow I_2(aq) + 2e^-$ oxidation

$Cl_2(aq) + 2e^- \rightarrow 2Cl^-(aq)$ reduction

The electrons lost by the iodide ions are accepted by the chlorine molecules, which makes the chlorine molecules the **oxidising agent** in this reaction. The iodide ions must therefore be the **reducing agent** and this is confirmed by the fact that they donate the electrons needed to reduce the chlorine molecules.

In **Section 9.2** you will meet a similar redox reaction between chlorine and bromide ions, which is used to displace bromine from sea water.

Balancing ion-electron equations

When you write an ion-electron equation, you must check two things. Firstly, that the numbers of each type of atom are the same on both sides and, secondly, that the charges balance.

Example 1

When iron(II) sulphate solution reacts with acidified potassium dichromate solution, both iron(III) and chromium(III) ions are formed. The iron(II) ions in the iron(II) sulphate solution have changed to iron(III) ions:

$Fe^{2+}(aq) \rightarrow Fe^{3+}(aq)$ *(charge unbalanced)*

To balance the charges an electron has to be added to the right-hand side:

$Fe^{2+}(aq) \rightarrow Fe^{3+}(aq) + e^-$

The iron(II) ions have been oxidised and so the chromium(III) ions must have been formed by the reduction of the dichromate ions ($Cr_2O_7^{2-}$) present in the potassium dichromate solution:

$Cr_2O_7^{2-}(aq) \rightarrow Cr^{3+}(aq)$

(chromium and oxygen atoms and charge unbalanced)

Working out the balanced ion-electron equation for this reduction process is not quite as straightforward as it was for the oxidation process. First of all we must balance the chromium atoms:

$Cr_2O_7^{2-}(aq) \rightarrow 2Cr^{3+}(aq)$ *(oxygen atoms and charge unbalanced)*

The next step is to balance the oxygen atoms. We need oxygen on the right-hand side of the equation and this is introduced as water molecules. Seven H_2O molecules will be required to balance the seven oxygen atoms in the dichromate ion:

$Cr_2O_7^{2-}(aq) \rightarrow 2Cr^{3+}(aq) + 7H_2O(l)$

(hydrogen atoms and charge unbalanced)

Hydrogen is now needed on the left-hand side of the equation and this is introduced as hydrogen ions. 14 H^+ ions will be required:

$Cr_2O_7^{2-}(aq) + 14H^+(aq) \rightarrow 2Cr^{3+}(aq) + 7H_2O(l)$ *(charge unbalanced)*

The final step is to balance the charge. As it stands, the total charge on the reactants side is 12+ (14+ from the 14 H^+ ions and 2– from the $Cr_2O_7^{2-}$ ion) and 6+ on the products side (6+ from the two Cr^{3+} ions). So six electrons must be added to the left-hand side and this completes the ion-electron equation:

$Cr_2O_7^{2-}(aq) + 14H^+(aq) + 6e^- \rightarrow 2Cr^{3+}(aq) + 7H_2O(l)$

Notice that the electrons appear on the reactants side, confirming that it is a reduction half-reaction. Notice too that hydrogen ions are needed in the reduction of dichromate ions. This explains why the potassium dichromate solution initially present has to be acidified.

The two ion-electron equations for the iron(II) sulphate/acidified potassium dichromate reaction are therefore:

$Fe^{2+}(aq) \rightarrow Fe^{3+}(aq) + e^-$ oxidation

$Cr_2O_7^{2-}(aq) + 14H^+(aq) + 6e^- \rightarrow 2Cr^{3+}(aq) + 7H_2O(l)$ reduction

These two equations show that electrons are being transferred from the iron(II) ions to the dichromate ions.

Now that we have the two ion-electron equations, we can combine them to get the equation for the overall redox reaction. But we have to exercise some care because the redox equation has to be balanced with respect to the electrons. In other words, the number of electrons lost in the oxidation step must equal the number of electrons gained in the reduction step.

In this case we must multiply the oxidation ion-electron equation by 6.

$$6Fe^{2+}(aq) \rightarrow 6Fe^{3+}(aq) + 6e^-$$
$$Cr_2O_7^{2-}(aq) + 14H^+(aq) + 6e^- \rightarrow 2Cr^{3+}(aq) + 7H_2O(l)$$

Now that the electrons are balanced we simply add together these equations. The electrons cancel and we obtain the overall ionic equation for the redox reaction.

$$6Fe^{2+}(aq) + Cr_2O_7^{2-}(aq) + 14H^+(aq) \rightarrow 6Fe^{3+}(aq) + 2Cr^{3+}(aq) + 7H_2O(l)$$

Example 2

In another redox reaction involving $H^+(aq)$ and $H_2O(l)$, sodium sulphite solution is added to acidified potassium permanganate solution and the purple colour of the permanganate ions disappears as they are changed into manganese(II) ions:

$MnO_4^-(aq) \rightarrow Mn^{2+}(aq)$ *(unbalanced)*

Try applying the rules outlined above to build up the full ion-electron equation for this change. It is:

$$MnO_4^-(aq) + 8H^+(aq) + 5e^- \rightarrow Mn^{2+}(aq) + 4H_2O(l)$$

Notice that the electrons are on the left-hand side of the equation showing that the permanganate ions have been reduced.

Tests on the final reaction mixture show that sulphate ions are present. This implies that the sulphite ions have changed into sulphate ions:

$SO_3^{2-}(aq) \rightarrow SO_4^{2-}(aq)$ *(unbalanced)*

It follows that the balanced ion-electron equation for this change is:

$$SO_3^{2-}(aq) + H_2O(l) \rightarrow SO_4^{2-}(aq) + 2H^+(aq) + 2e^-$$

So the sodium sulphite/acidified potassium permanganate equation is made up of the two ion-electron equations:

$SO_3^{2-}(aq) + H_2O(l) \rightarrow SO_4^{2-}(aq) + 2H^+(aq) + 2e^-$ oxidation

$MnO_4^-(aq) + 8H^+(aq) + 5e^- \rightarrow Mn^{2+}(aq) + 4H_2O(l)$ reduction

The sulphite ions are oxidised and the electrons they lose are transferred to the permanganate ions which are, in turn, reduced.

Now that we have both ion-electron equations, we can combine them to get the equation for the overall redox reaction. We must first multiply the oxidation equation by 5 and the reduction equation by 2:

$5SO_3^{2-}(aq) + 5H_2O(l) \rightarrow 5SO_4^{2-}(aq) + 10H^+(aq) + 10e^-$ oxidation

$2MnO_4^-(aq) + 16H^+(aq) + 10e^- \rightarrow 2Mn^{2+}(aq) + 8H_2O(l)$ reduction

Now that the electrons are balanced we simply add these equations together:

$$5SO_3^{2-}(aq) + 5H_2O(l) + 2MnO_4^-(aq) + 16H^+(aq) \rightarrow 5SO_4^{2-}(aq) + 10H^+(aq) + 2Mn^{2+}(aq) + 8H_2O(l)$$

The electrons cancel. But we are not quite finished yet because you'll have noticed that H^+ ions and H_2O molecules appear on both sides of the equation. These are cancelled down and we finally arrive at the balanced redox equation.

$$5SO_3^{2-}(aq) + 2MnO_4^-(aq) + 6H^+(aq) \rightarrow 5SO_4^{2-}(aq) + 2Mn^{2+}(aq) + 3H_2O(l)$$

It is important to remember that neither electrons nor spectator ions nor spectator molecules must feature in redox equations.

PROBLEMS FOR 24.1

1 a Insert electrons on the appropriate side of each of the following ion-electron equations in order to balance and complete them.
   i $Mg(s) \rightarrow Mg^{2+}(aq)$
   ii $H_2(g) \rightarrow 2H^+(aq)$
   iii $O_2(g) \rightarrow 2O^{2-}(s)$
   iv $Cu^+(aq) \rightarrow Cu^{2+}(aq)$
   v $Sn^{4+}(aq) \rightarrow Sn^{2+}(aq)$
   vi $F_2(g) \rightarrow 2F^-(aq)$
   vii $O_2(g) + 4H^+(aq) \rightarrow 2H_2O(l)$
   viii $MnO_4^-(aq) + 8H^+(aq) \rightarrow Mn^{2+}(aq) + 4H_2O(l)$

  b For each completed ion-electron equation describe the process as oxidation or reduction.

2 a Write a balanced ion-electron equation for each of the following changes.
   i $ClO_3^-(aq) \rightarrow Cl_2(g)$
   ii $SO_3^{2-}(aq) \rightarrow SO_4^{2-}(aq)$
   iii $ClO^-(aq) \rightarrow Cl^-(aq)$
   iv $FeO_4^{2-}(aq) \rightarrow Fe^{3+}(aq)$
   v $NO_3^-(aq) \rightarrow NH_4^+(aq)$

  b For each completed ion-electron equation describe the process as oxidation or reduction.

3 For each of the following reactions write
  a the oxidation ion-electron equation
  b the reduction ion-electron equation.
   i $Zn(s) + Pb^{2+}(aq) \rightarrow Zn^{2+}(aq) + Pb(s)$
   ii $2Al(s) + 6H^+(aq) \rightarrow 2Al^{3+}(aq) + 3H_2(g)$
   iii $Cl_2(g) + 2I^-(aq) \rightarrow 2Cl^-(aq) + I_2(aq)$
   iv $Zn(s) + S(s) \rightarrow ZnS(s)$

4 In a nickel/cadmium cell the following half-reactions take place:
$Cd(s) + 2OH^-(aq) \rightarrow Cd(OH)_2(s) + 2e^-$
$Ni(OH)_3(s) + e^- \rightarrow Ni(OH)_2(s) + OH^-(aq)$
Combine these to write the equation for the redox reaction.

(SQA)

5 For each of the following redox reactions write
  a the oxidation ion-electron equation
  b the reduction ion-electron equation
  c the redox equation.
   i Copper displaces silver from silver(I) nitrate solution.
   ii Magnesium reacts with sulphuric acid to form hydrogen gas and magnesium sulphate solution.
   iii Tin(II) ions react with permanganate ions (MnO_4^-) to form tin(IV) ions and manganese(II) ions.
   iv Iron(II) ions react with iodine solution to produce iron(III) ions and iodide ions.

6 Iron(II) ions can act as both oxidising and reducing agents.
Write an ion-electron equation for the half-reaction that takes place when iron(II) ions act as
  a an oxidising agent
  b a reducing agent.

7 In each of the following redox reactions identify by formula the
  a oxidising agent
  b reducing agent.
   i $2Ag^+(aq) + Cu(s) \rightarrow 2Ag(s) + Cu^{2+}(aq)$
   ii $2Mg(s) + O_2(g) \rightarrow 2MgO(s)$
   iii $2S_2O_3^{2-}(aq) + I_2(aq) \rightarrow S_4O_6^{2-}(aq) + 2I^-(aq)$
   iv $3Cu(s) + 8H^+(aq) + 2NO_3^-(aq) \rightarrow 3Cu^{2+}(aq) + 2NO(g) + 4H_2O(l)$

24.2 *Redox titrations*

Titration is a procedure which allows chemists to determine the concentration of a solution. It involves adding a solution of known concentration (a standard solution) to one of unknown concentration until the reaction between the two is just complete. An indicator is usually added to the reaction mixture to show the end-point of the titration.

You are already familiar with acid–base titrations but the same procedure can equally well be used with oxidising and reducing agents. Titrating an oxidising agent against a reducing agent or vice versa is called a **redox titration**.

Finding the vitamin C content of orange juice is an interesting example where the redox titration technique can be used. Vitamin C is readily oxidised and so we can determine its concentration by titrating it against a standard solution of an oxidising agent. Iodine solution is a suitable oxidising agent in this case.

As the vitamin C is oxidised the iodine molecules are reduced:

$C_6H_8O_6(aq) \rightarrow C_6H_6O_6(aq) + 2H^+(aq) + 2e^-$ oxidation
vitamin C
$I_2(aq) + 2e^- \rightarrow 2I^-(aq)$ reduction

A fixed volume of orange juice is pipetted into a conical flask along with a few drops of starch solution which act as an indicator. The apparatus is then set up as shown in Figure 3, with the iodine solution in the burette.

The iodine solution, which is brown in colour, is carefully added to the orange juice in the flask. The vitamin C reduces the iodine molecules to colourless iodide ions and so before the end-point of the titration is reached the solution in the flask remains orange. When all the vitamin C molecules have been oxidised the solution in the flask turns blue/black on the addition of just one extra drop of iodine solution. This marks the end-point of the titration. The colour change occurs because no vitamin C is left to reduce the extra iodine molecules and this leaves them to react with the starch indicator.

The titration procedure is normally repeated until two consecutive titre volumes within $0.1\,cm^3$ of each other are obtained.

The results from such a titration can then be used to calculate the concentration of vitamin C in orange juice.

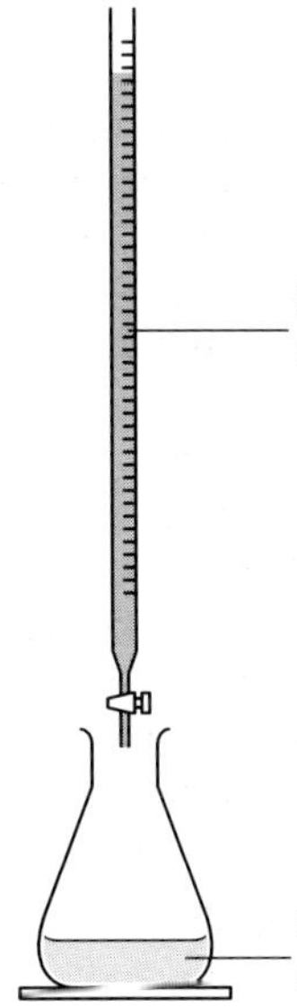

Figure 3 Apparatus for the redox titration

Sample results

Suppose on average $14.1\,cm^3$ of $0.005\,mol\,l^{-1}$ iodine solution are required to oxidise the vitamin C in $25.0\,cm^3$ of orange juice.

Knowing the concentration and volume of the iodine solution we can work out the amount of iodine used in the titration:

$$\text{amount of iodine} = 0.005\,mol\,l^{-1} \times 0.0141\,\text{litre} = 7.05 \times 10^{-5}\,mol$$

The next step is to write the balanced redox equation from the oxidation and reduction ion-electron equations:

$$C_6H_8O_6(aq) + I_2(aq) \rightarrow C_6H_6O_6(aq) + 2H^+(aq) + 2I^-(aq)$$

It shows that 1 mole of iodine oxidises 1 mole of vitamin C and implies that

7.05×10^{-5} mol of iodine will oxidise 7.05×10^{-5} mol of vitamin C

This is the amount of vitamin C in $25.0\,cm^3$ of orange juice. From these data we can calculate the concentration of vitamin C in the orange juice.

$$\text{vitamin C concentration} = \frac{7.05 \times 10^{-5}\,mol}{0.025\,\text{litre}} = 2.82 \times 10^{-3}\,mol\,l^{-1}$$

The concentration can also be expressed in $g\,l^{-1}$ (or $mg\,l^{-1}$). Vitamin C ($C_6H_8O_6$) has a molar mass of $176\,g\,mol^{-1}$ and so the concentration of vitamin C in our orange juice sample is:

$$2.82 \times 10^{-3}\,mol\,l^{-1} \times 176\,g\,mol^{-1} = 0.496\,g\,l^{-1}\ (\text{or } 496\,mg\,l^{-1})$$

PROBLEMS FOR 24.2

1 Hydrogen peroxide is a colourless liquid with molecular formula H_2O_2. Its solution can be used as an antiseptic. A hospital technician was checking the concentration of a hydrogen peroxide solution. She titrated 25.0 cm^3 portions of the solution against an acidified potassium permanganate solution.

The reaction taking place during the titration is:

$2MnO_4^-(aq) + 6H^+(aq) + 5H_2O_2(aq) \rightarrow$ (purple)

$2Mn^{2+}(aq) + 8H_2O(l) + 5O_2(g)$ (colourless)

The technician's results are shown in the table below:

Titration	1	2	3
Titre volume/cm^3	16.5	15.8	15.8

a How would the technician know that the end-point of the titration had been reached?

b Why would the technician ignore the result of the first titration when calculating the average titre volume?

c The concentration of the potassium permanganate solution was 0.101 mol l^{-1}. Use the technician's results to calculate the concentration of the hydrogen peroxide solution.

(SQA)

2 The water in swimming pools can be kept sterile by the addition of chlorine which kills micro-organisms. The chlorine levels in swimming pool water can be determined by titrating samples against iron(II) sulphate solution. The reaction taking place is:

$Cl_2(aq) + 2Fe^{2+}(aq) \rightarrow 2Cl^-(aq) + 2Fe^{3+}(aq)$

a Write the ion-electron equation for the oxidation half-reaction.

b A 100 cm^3 sample of water from a swimming pool required 24.9 cm^3 of 2.82×10^{-4} mol l^{-1} iron(II) sulphate solution to reach the end-point. Calculate the chlorine concentration in

i mol l^{-1} **ii** g l^{-1}.

(SQA)

3 In an experiment to measure the ozone concentration in air, 1×10^5 litres of air were bubbled through a solution of potassium iodide. Ozone reacts with potassium iodide solution releasing iodine:

$2I^-(aq) + O_3(g) + H_2O(l) \rightarrow I_2(aq) + O_2(g) + 2OH^-(aq)$ reaction 1

The iodine formed was titrated against 0.0100 mol l^{-1} sodium thiosulphate solution, $Na_2S_2O_3(aq)$, using starch solution as indicator.

$I_2(aq) + 2S_2O_3^{2-}(aq) \rightarrow 2I^-(aq) + S_4O_6^{2-}(aq)$ reaction 2

22.4 cm^3 of sodium thiosulphate solution were needed to reduce the iodine released in reaction 1.

a Write the formula for the oxidising agent in

i reaction 1 **ii** reaction 2.

b What colour change would show that the titration was complete?

c Use the results to calculate the volume of ozone in one litre of air. (Take the molar volume of ozone at room temperature and pressure to be 24 litres mol^{-1}.)

(SQA)

24.3 *Electrolysis*

Any liquid which contains ions that are free to move and therefore is capable of conducting electricity is known as an **electrolyte**. A molten salt or an aqueous solution of a salt, acid or base could be an electrolyte. When a direct current of electricity is passed through an electrolyte, chemical changes take place resulting in the decomposition of the electrolyte. This process is called **electrolysis**.

Let us consider the electrolysis of molten sodium chloride. The apparatus in which this can be carried out is illustrated in Figure 4. It is known as an **electrolytic cell**.

The cell consists of a container of molten sodium chloride with two carbon electrodes dipping into the electrolyte. When the electrodes are connected to the battery (or some other source of direct current) electrons are supplied to the right-hand electrode, making it negatively charged. The positively charged sodium ions will migrate towards this electrode, where they will gain electrons and be *reduced* to sodium atoms:

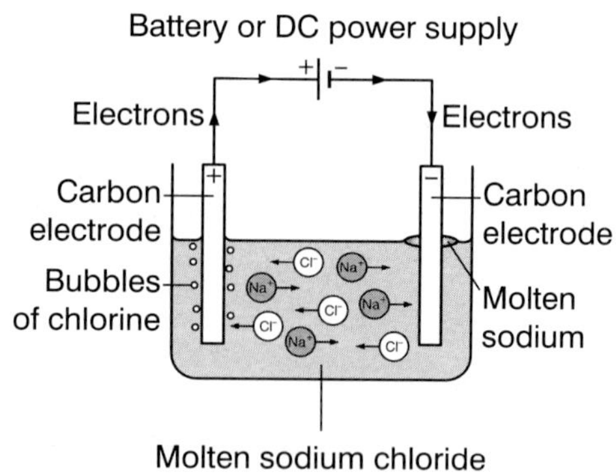

Figure 4 The electrolysis of molten sodium chloride

$Na^+(l) + e^- \rightarrow Na(l)$ reduction

At the positive electrode, the negatively charged chloride ions will lose electrons and be *oxidised* to chlorine molecules:

$2Cl^-(l) \rightarrow Cl_2(g) + 2e^-$ oxidation

The electrons given up by the chloride ions will travel through the external circuit back to the battery. We can build up the overall equation by doubling the reduction ion-electron equation and adding it to the oxidation ion-electron equation. This gives:

$2Na^+Cl^-(l) \rightarrow 2Na(l) + Cl_2(g)$

The overall change involves the transfer of electrons from the chloride ions to the sodium ions. The reaction taking place here, and in all examples of electrolysis, is a redox reaction just like those described in **Section 24.2**. However, there is one important difference. Whereas the reactions described earlier are spontaneous and can be used to generate electricity, the redox reactions associated with electrolysis do not occur of their own free will and require an input of electrical energy to bring them about.

During electrolysis **oxidation** always occurs at the **positive** *electrode* and **reduction** always occurs at the **negative** *electrode*.

Predicting the products of electrolysis

When a *molten salt* is electrolysed between inert electrodes, such as carbon or platinum, it is possible to predict the product at each electrode. Take the example of molten sodium chloride discussed above. Since it contains only sodium ions and chloride ions, the Na^+ ions must be reduced at the negative electrode and the Cl^- ions must be oxidised at the positive electrode – there is no choice.

But when an *aqueous solution* is electrolysed the situation is more complicated. This is because we now have water molecules present in the electrolyte in addition to the ions of the solute. Some of these water molecules dissociate into ions that are capable of being reduced at the negative electrode and oxidised at the positive electrode, just like the ions of the solute. Let's consider sodium chloride again, but this time in solution. The sodium ions and the $H^+(aq)$ ions from the dissociation of water compete for electrons at the negative electrode. In practice bubbles of hydrogen gas are formed at the negative electrode, indicating that it is the hydrogen ions which have gained electrons and been reduced to hydrogen:

$2H^+(aq) + 2e^- \rightarrow H_2(g)$ reduction

At the positive electrode, however, it is the chloride ions, not the $OH^-(aq)$ ions from the dissociation of water, that are oxidised:

$2Cl^-(aq) \rightarrow Cl_2(g) + 2e^-$ oxidation

So changing from molten sodium chloride to an aqueous solution of sodium chloride results in a change of product at the negative electrode.

Changing the material of the electrodes can add yet another complication. For example, when sodium chloride solution is electrolysed with a layer of mercury as the negative electrode, rather than carbon or platinum, the reduction reaction taking place at the negative electrode is:

$Na^+(aq) + e^- \rightarrow Na(l)$

rather than:

$2H^+(aq) + 2e^- \rightarrow H_2(g)$

The sodium dissolves in the mercury, forming a liquid amalgam. Sodium chloride is an important raw material for the chemical industry. By modifying the conditions under which it is electrolysed it can be used to produce a range of chemicals: sodium, chlorine, hydrogen and sodium hydroxide.

At this stage you are not expected to predict the products formed on electrolysing aqueous solutions, but it is important that you are aware of the complications that exist and that the process is not quite as straightforward as one might have first thought.

Quantitative electrolysis

So far we have treated electrolysis in a qualitative way by identifying the products formed during the electrolysis of various electrolytes and the reactions involved in their formation. We have made no attempt to relate the amount of product formed to the amount of electricity passed through the electrolyte. It is to this quantitative aspect of electrolysis that we now turn.

Suppose we set up the electrical circuit shown in Figure 5 in which three electrolytic cells are connected in series.

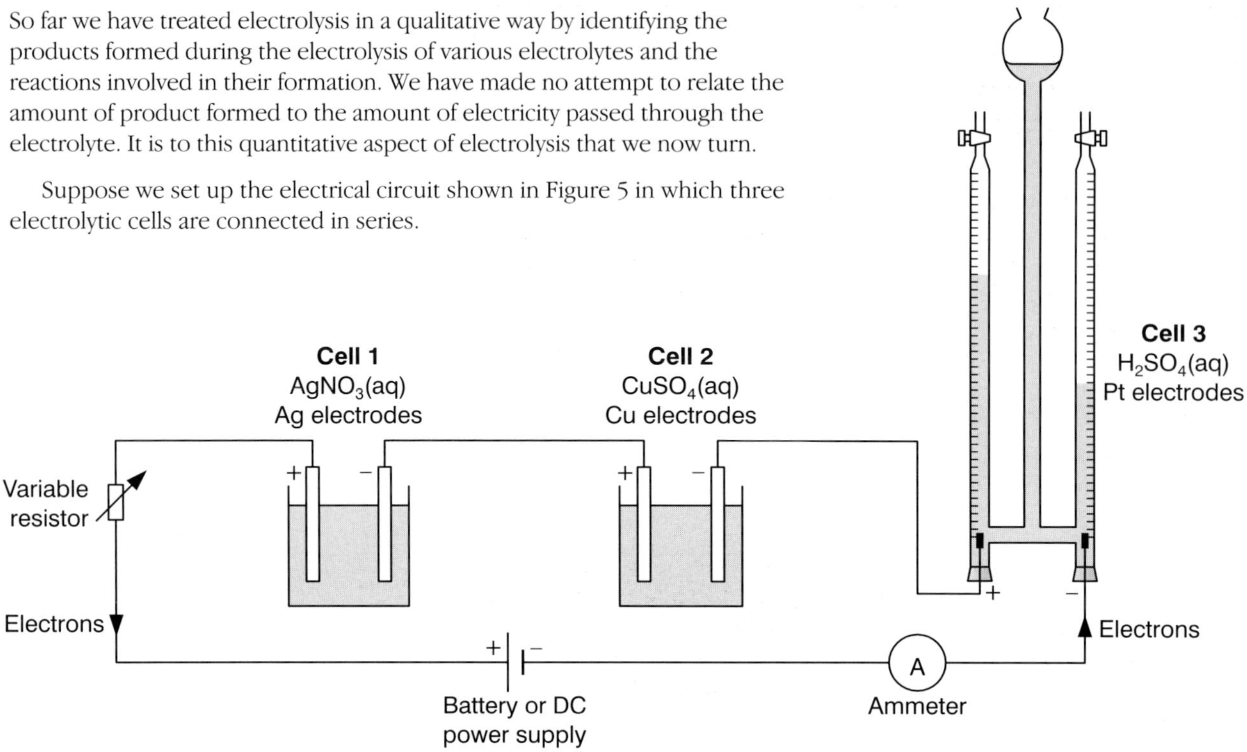

Figure 5 Circuit for quantitative electrolysis of $AgNO_3(aq)$, $CuSO_4(aq)$ and $H_2SO_4(aq)$

The circuit includes an ammeter so that the current can be measured and a variable resistor which can be adjusted to keep the current constant. Because the cells are connected in series, the current will be the same in each cell. Let us suppose that a constant current of 0.50 A is passed through these cells for precisely 10 minutes.

The quantity of electrical charge passing through the cells can be determined using the relationship:

$$Q = It$$

- Q is the quantity of electrical charge and is measured in coulombs (C)
- I is the current in amps (A)
- t is the period of time, in seconds (s), during which the current passes.

In our experiment, $I = 0.50\,\text{A}$ and $t = 10 \times 60 = 600\,\text{s}$. Hence

$$Q = 0.50\,\text{A} \times 600\,\text{s} = 300\,\text{C}$$

In **cell 1**, silver(I) nitrate solution is electrolysed between silver electrodes. During electrolysis, the silver(I) ions are reduced to silver atoms at the negative electrode. The mass of silver deposited can be determined by weighing the negative electrode before and after electrolysis.

In **cell 2**, copper(II) sulphate solution is electrolysed between copper electrodes. The copper(II) ions are reduced to copper atoms at the negative electrode. The mass of copper deposited can again be found by weighing the negative electrode before and after electrolysis.

In **cell 3**, dilute sulphuric acid is electrolysed between platinum electrodes. Hydrogen gas is produced at the negative electrode as the hydrogen ions of the acid are reduced. At the positive electrode $OH^-(aq)$ ions, formed by the dissociation of water molecules, are oxidised, forming oxygen gas. Since the two limbs in the cell are graduated, the volumes of hydrogen and oxygen formed during electrolysis can be measured. The results obtained in this experiment are presented in Table 1.

Cell	Electrode product	Mass or volume of product
1	silver	0.3358 g
2	copper	0.0995 g
3	hydrogen	37.30 cm^3
	oxygen	18.65 cm^3

Table 1 Results from electrolysis experiment

Table 1 shows the masses and volumes of elements formed by passing 300 C of electrical charge through the three electrolytes. There is a pattern here, but it is well hidden! If we use our results to calculate the number of coulombs needed to produce *1 mole* of each element then a relationship becomes more obvious.

Let's take **cell 1**, in which 0.3358 g silver are deposited by 300 C of electrical charge. The molar mass of silver is 108 g mol^{-1}, so 108 g (1 mol) silver are deposited by

$$300\text{ C} \times \frac{108\text{ g}}{0.3358\text{ g}}$$
$$= 96\,500\text{ C}$$

Knowing the molar mass of copper and assuming the molar volume of a gas to be 24.0 litres per mole, we can similarly calculate how much electrical charge is required to produce 1 mole of each of the elements, copper, hydrogen and oxygen. The results are presented in Table 2. The ion-electron equation involved in the formation of 1 mole of each element is also shown in Table 2.

Element	Quantity of electrical charge required to produce 1 mol/C	Ion-electron equation
silver	$1 \times 96\,500$	$Ag^+(aq) + \mathbf{1}e^- \rightarrow Ag(s)$
copper	193 000 or $\mathbf{2} \times 96\,500$	$Cu^{2+}(aq) + \mathbf{2}e^- \rightarrow Cu(s)$
hydrogen	193 000 or $\mathbf{2} \times 96\,500$	$2H^+(aq) + \mathbf{2}e^- \rightarrow H_2(g)$
oxygen	386 000 or $\mathbf{4} \times 96\,500$	$4OH^-(aq) \rightarrow O_2(g) + 2H_2O(l) + \mathbf{4}e^-$

Table 2 Electrical charge per mole and corresponding ion-electron equations

The relationship is now apparent – the quantity of electrical charge required to produce 1 mole of an element is $\boldsymbol{n} \times \mathbf{96\,500\,C}$ where $\boldsymbol{n}$ is the number of moles of electrons in the relevant ion-electron equation.

Our results also reveal a link between electrons and electrical charge. They show that the electrical charge associated with 1 mole of electrons is a constant and takes the value 96 500 C mol^{-1}. This constant is called the **Faraday constant** after Michael Faraday, who did much of the pioneering work on electrolysis.

96 500 C is the charge associated with 1 mole of electrons.

Worked example

Electrolysis has many industrial applications, including the electrolytic refining of copper. This involves electrolysing acidified copper(II) sulphate solution between copper electrodes. The positive electrodes are made of impure copper while the negative electrodes are thin sheets of pure copper. It states in **Section 9.3** that the electrolysis process 'takes about 2 weeks and during this time the negative electrodes increase in mass from as little as 5 kg to 120 kg'.

The electrolytic refining of copper is detailed in **Section 9.3**.

We can use these data to calculate the average current that must have passed through the electrolyte.

mass of copper deposited = 120 kg – 5 kg = 115 kg

Using the relevant ion-electron equation we can now work out the quantity of electrical charge needed to deposit 115 kg (115 000 g) of copper.

$$Cu^{2+}(aq) + 2e^- \rightarrow Cu(s)$$
2 mol 1 mol

$2 \times 96\,500$ C produces 64 g of copper

so $2 \times 96\,500\,C \times \frac{115\,000\,g}{64\,g}$ ($= 3.468 \times 10^8$ C) produces 115 000 g of copper

Using the relationship, $Q = It$, we can finally calculate the current.

$$Q = 3.468 \times 10^8\,C$$

$$t = 2\text{ weeks} = 2 \times 7 \times 24 \times 60 \times 60 = 1.210 \times 10^6\,s$$

$$I = \frac{Q}{t} = \frac{3.468 \times 10^8\,C}{1.210 \times 10^6\,s} = 287\,A$$

PROBLEMS FOR 24.3

1 The following molten salts were electrolysed between inert electrodes:
- a lead(II) iodide
- b magnesium chloride.

For each one
- i name the product at the positive electrode
- ii write the ion-electron equation for the half-reaction taking place at the negative electrode.

2 When copper(II) sulphate solution is electrolysed between carbon electrodes the overall equation for the reaction is:
$2Cu^{2+}(aq) + 2H_2O(l) \rightarrow 2Cu(s) + O_2(g) + 4H^+(aq)$
- a Name the product at the negative electrode.
- b Write the ion-electron equation for the half-reaction taking place at the positive electrode.

3 Calculate
- a the number of coulombs
- b the number of moles of electrons

which pass through an electric heater operating for 2 hours and using a current of 18 A.

4 Aluminium is manufactured by electrolysing molten aluminium oxide.
Calculate the mass of aluminium formed per minute in a cell operating at 300 000 A.

5 In an industrial process, a current of 10 000 A was passed through a gold salt solution for 25 minutes producing 10.21 kg of gold.
Calculate the charge on the gold ions in the solution.

(SQA)

6 The same quantity of electricity was passed through two electrolytic cells. The first contained copper(II) sulphate solution and the second contained silver(I) nitrate solution.
If 1.92 g of copper were deposited in the first cell, calculate the mass of silver that would be deposited in the second cell.

7 60 cm^3 of hydrogen were formed on passing a current of 0.50 A through dilute hydrochloric acid.
For how long did the current flow?
(Take the molar volume of hydrogen at room temperature and pressure to be 24 litres per mole.)

25

NUCLEAR CHEMISTRY

25.1 *Radioactive decay*

Emissions from radioactive substances

Some isotopes of some elements are unstable. Their nuclei break down *spontaneously*, and are described as being **radioactive**. As these nuclei break down, they emit rays and particles called **emissions**. This breakdown (or **radioactive decay**) occurs *of its own accord*: it isn't triggered off by something we do. Some isotopes decay very quickly, for others the process takes thousands of years. Energy is released when nuclei decay. The nuclei formed when these emissions take place are more stable than the original nuclei.

Not all unstable atoms decay in the same way. Three different kinds of emissions have been identified. They are called α, β and γ emissions. All three are capable of knocking electrons off the atoms they collide with, which ionises the atoms. Because of this, these emissions are sometimes referred to as **ionising radiation**. Some of the properties of α, β and γ emissions are summarised in Table 1.

Property	Type of emission		
	α (alpha)	**β (beta)**	**γ (gamma)**
relative charge	+2	−1	0
relative mass	4	0.00055	0
nature	2 protons + 2 neutrons (He nucleus)	electron (produced by nuclear changes)	very high frequency electromagnetic radiation
range in air	few centimetres	few metres	very long
stopped by	paper	aluminium foil	lead sheet
deflection by electrical field	low	high	nil

Table 1 Some properties of α, β and γ emissions

Stable and unstable nuclei

The stability of a nucleus can be related to its neutron-to-proton ratio. The stable nuclei of the lighter elements contain approximately equal numbers of neutrons and protons (1:1). As the atomic number increases and more protons have to be packed into the nucleus, the forces of repulsion between these positively charged protons rises. A larger excess of neutrons is required to reduce these repulsions between protons. The neutron/proton ratio (n:p ratio) for stable heavy nuclei has to be around 1.5:1. Any nuclei which have n:p ratios less than this tend to be radioactive.

Nuclear equations

Nuclear equations summarise the processes which produce α- and β-radiation. Nuclear equations include the **mass number** (number of protons plus neutrons), **nuclear charge** (usually indicated by the **atomic**

number, or number of protons) and **chemical symbol** for each particle involved. Both mass and charge must balance in a nuclear equation.

α-**decay** involves the emission of α-particles, which are helium nuclei, ${}^{4}_{2}He$. α-decay is common among heavier elements with atomic numbers greater than 83. α-decay reduces the mass of these heavy nuclei. The isotope produced from α-decay will have a mass number 4 units lower and a nuclear charge 2 units lower than the original atom. An example is

$${}^{238}_{92}U \rightarrow {}^{234}_{90}Th + {}^{4}_{2}He$$

β-**decay** involves the emission of electrons, written as ${}^{0}_{-1}e$. β-decay is common among lighter elements where the isotopes contain a relatively large numbers of neutrons. For example:

$${}^{14}_{6}C \rightarrow {}^{14}_{7}N + {}^{0}_{-1}e$$

During β-decay, the mass number (protons plus neutrons) remains constant but the nuclear charge (number of protons) increases by one unit. This means that a neutron, ${}^{1}_{0}n$, is converted into a proton, ${}^{1}_{1}p$, plus an electron which is ejected from the nucleus:

$${}^{1}_{0}n \rightarrow {}^{1}_{1}p + {}^{0}_{-1}e$$

Notice that α- and β-decay result in the production of a different element. For example, when uranium-238 undergoes α-decay, it turns into thorium-234; when carbon-14 undergoes β-decay, it turns into nitrogen.

γ-**decay** is different: it is the emission of energy from a nucleus which is changing from a high energy level to a lower one.

Half-lives

Radioactive decay is a random process. Each nucleus in a sample of an isotope decays at random, regardless of what the other nuclei are doing and regardless of outside conditions such as temperature and pressure. Decay is also independent of the chemical state of the isotope. For example, uranium would decay in the same way whether it was in a compound or in the free state as an element. Similarly, it would make no difference if the radioactive substance was solid, liquid or in solution. Gradually, as more and more of the nuclei in the sample decay, the sample becomes less and less radioactive.

The time for half of the nuclei to decay is called the **half-life**. For any given isotope, the half-life is fixed – it doesn't matter how much of the isotope is present, or what is the temperature or pressure. In the course of one half-life, half of the radioactivity of the sample always disappears. But the sample never completely disappears: that's why we can still detect radioactive isotopes even though the Earth was formed over four billion years ago.

Table 2 shows you some examples of isotopes and their half-lives.

For example, strontium-90 has a half-life of 28 years. If you started now with 8 g of strontium-90, in 28 years' time half the nuclei would have decayed, and only 4 g of strontium-90 would be left. In 56 years' time, only 2 g would be left. By that time the isotope would have gone through two half-lives: the original quantity would have been halved and halved again (Figure 1).

Isotope	Half-life
uranium-238	4.5×10^9 years
carbon-14	5.7×10^3 years
strontium-90	28 years
iodine-131	8.1 days
bismuth-214	19.7 minutes
polonium-214	1.5×10^{-4} seconds

Table 2 Some isotopes and their half-lives

Calculating and using half-lives

If a sample of a radioisotope has a count rate of 56 counts min^{-1} and 4 hours later the count rate is 13 counts min^{-1}, the half life of the radioisotope can be calculated.

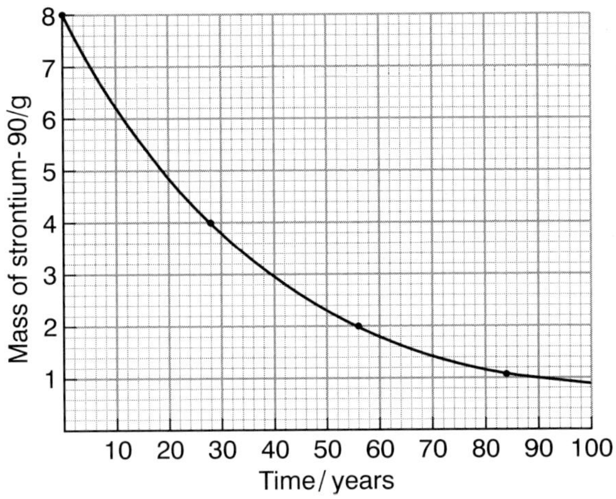

Figure 1 The half-life of strontium-90 is 28 years. This means that the amount of strontium-90 halves every 28 years

At the start the count rate is 56 counts min^{-1}.
After one half-life the count rate would be 26 counts min^{-1}.
After two half-lives the count rate would be 13 counts min^{-1}.
The count rate would have fallen to 13 counts min^{-1} after two half-lives. We are told that this is equivalent to 4 hours.
If two half-lives is equal to 4 hours then one half-life is equal to 2 hours. The half-life of this radioisotope is 2 hours.

If you know the half-life of a radioactive isotope, you can use it to calculate:

- the time needed for the activity or the mass of the isotope to fall to a certain value, or
- the mass of the isotope or its activity after it has decayed for a certain time.

Example 1

The half-life of the radioisotope carbon-14 is 5730 years. How long will it take for the radioactive count rate to drop to 25% of its original value?

At the start it will have 100% activity (the original count rate).
After one half-life it will be 50% of the original count rate.
After two half-lives it will be 25% of the original count rate.

This means that over two half-lives the count drops to 25% of the original. As one half-life is 5730 years, the time taken for the isotope to decay to 25% of its original value is $2 \times 5730 = 11\,460$ years.

Calculations similar to this could be used to date objects containing radioisotopes.

Example 2

Another radioisotope, iodine-131, has a half-life of 8 days. If you start with 8 g of the isotope, how much of it will be left after 24 days?

Number of half-lives $= \frac{24}{8} = 3$

At the start we have 8 g

After one half-life we have $\frac{1}{2}$ of the original amount, ie $\frac{8}{2} = 4$ g

After two half-lives we have $\frac{1}{4}$ of the original amount, ie $\frac{8}{4} = 2$ g

After three half-lives we have $\frac{1}{8}$ of the original amount, ie $\frac{8}{8} = 1$ g

This type of calculation is done when considering the dangers of waste from nuclear processes or the emissions from nuclear accidents.

Radiation from radioisotopes can be extremely dangerous, but under carefully controlled conditions some radioisotopes have important uses in medicine, industry, scientific research and energy production. Details of some uses can be found in **Section 9.4**.

PROBLEMS FOR 25.1

1 Write nuclear equations for the α-decay of
a $^{238}_{94}Pu$ b $^{221}_{87}Fr$ c $^{230}_{90}Th$
(You will need to refer to the Periodic Table (**Unit 1, Section 12.1**) to identify the symbols for the elements produced.)

2 Write nuclear equations for the β-decay of
a $^{90}_{38}Sr$ b $^{131}_{53}I$ c $^{231}_{90}Th$

3 Write a nuclear equation for each of the following.
a A $^{7}_{3}Li$ nucleus absorbs a colliding proton and then disintegrates into two identical fragments.
b The production of carbon-14 by collision of a neutron with an atom of nitrogen-14.
c The collision of an α-particle with an atom of nitrogen-14 releasing a proton and forming another element. (Rutherford observed this reaction in 1919. It was the first nuclear reaction ever observed.)
d The collision of an α-particle with an atom of aluminium-27 to form phosphorus-30.

4 Part of the decay series involving radon-222 is shown below. Copy it out and fill in the missing symbols for atomic number, mass number and type of decay.

$$_{88}Ra \longrightarrow {}^{222}_{86}Rn \xrightarrow[\alpha]{} Po \longrightarrow {}^{214}Pb$$

5 $^{232}_{90}Th$ emits a total of six α- and four β-particles in its natural decay series. What is the atomic number and the mass number of the final product?

6 This question refers to the isotopes in Table 2 and assumes you begin with a 10 g sample of each isotope.
a How much uranium-238 would be left after 4.5×10^9 years?
b How much bismuth-214 would be left after 78.8 minutes?
c How long would it take for you to be left with 1.25 g of polonium-214?
d How much carbon-14 would be left after 4.56×10^4 years?

7 A rock sample from the rugged coast of northern Labrador in Canada has recently been dated at 4.0×10^9 years. As the Earth is believed to have been formed 4.6×10^9 years ago, this sample provides a glimpse of the Earth's crust as it was soon after its formation.
Geochemist Kenneth Collerson of the University of California dated the sample by measuring the ratio of two rare earth element isotopes, samarium-147 and neodymium-143. $^{147}_{62}Sm$ decays to $^{143}_{60}Nd$ by alpha decay with a half-life of 1.06×10^{11} years.
a Write a nuclear equation for the decay of samarium-147.
b Use the half-life to plot a graph of the percentage of samarium-147 remaining against time. Use your graph to estimate the percentage of samarium in the Labrador rock sample. What was the ratio of samarium-147 to neodymium-143 found by Collerson?

25.2 *Nuclear fusion*

In a nuclear fusion reaction, two light atomic nuclei fuse together to form a single heavier nucleus of a new element. The process releases enormous quantities of energy.

Fusion reactions only occur at very high temperatures, such as those found in the Sun and in other stars. For two nuclei to fuse, they must come very close together. At the normal temperatures found on the Earth, the positive nuclei repel one another so strongly that this cannot happen. At very high temperatures, however, the nuclei collide with so much energy that this repulsive energy barrier can be overcome. Once the nuclei are close enough, the strong nuclear forces which hold protons and neutrons together in the nucleus take over and the nuclei fuse.

Nuclear fusion reactions take place in the gas clouds of stars and result in the formation of new elements. For example, when two hydrogen nuclei join together by nuclear fusion, hydrogen turns into helium. The energy released causes the gas cloud to glow.

You have already met this type of nuclear reaction in **Unit 1, Section 1.1**.

Here are two examples of nuclear reactions that take place in the Sun. They involve the three isotopes of hydrogen ($^{2}_{1}H$ is called deuterium, $^{3}_{1}H$ is called tritium).

$$^{1}_{1}H + {}^{2}_{1}H \rightarrow {}^{3}_{2}He \qquad\qquad {}^{2}_{1}H + {}^{3}_{1}H \rightarrow {}^{4}_{2}He + {}^{1}_{0}n$$

Atomic numbers and mass numbers must balance in nuclear equations.

At the high temperatures in the gas clouds, the electrons have enough energy to escape from the nuclei. The gases exist in an ionised form called a **plasma** in which the positive atomic nuclei exist in a 'sea' of delocalised electrons. The gases in the plasma have very different properties from the gases we know at lower temperatures. Although the idea of a plasma may seem strange to us on Earth, the Sun and other stars exist as plasma, so it is the most common form of matter in the universe.

It is thought that our solar system (the Sun and the planets) condensed from a huge gas cloud originating from a supernova explosion. Figure 2 shows the estimated abundances of chemical elements in the solar system. The Sun contains over 99% of the total mass of the solar system and so it is not surprising that hydrogen and helium make up more than 99% of the total mass. But virtually every other element in the Periodic Table is represented in the remaining fraction of a percentage.

All naturally occurring elements, including those found in our bodies, originated in stars as a result of nuclear fusion reactions. You can remind yourself of the way heavier elements are built up in stars by reading **Unit 1, Section 1.1**.

Some of them, such as beryllium (Be) and gold (Au), are present in tiny amounts – less than one part per billion relative to hydrogen. Some elements in the Periodic Table, such as astatine, the fifth halogen, can be regarded as not occurring naturally as all of their isotopes are radioactive and have very short half-lives. All the elements with atomic numbers higher than uranium have been synthesisied, usually in very small quantities, by bombarding heavy nuclei with high energy particles. All the isotopes of these synthetic elements are radioactive.

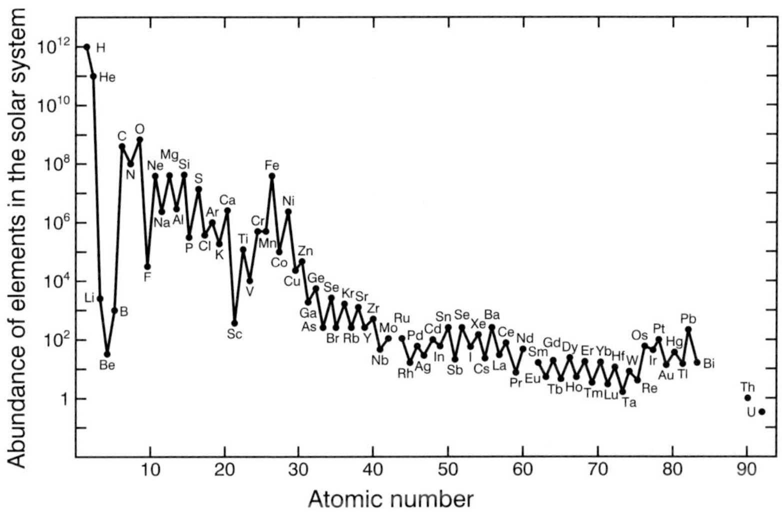

Figure 2 The estimated abundances of elements in the solar system, showing the number of atoms present, for each 10^{12} atoms of hydrogen.
(Note the logarithmic scale on the vertical axis, with each division representing a factor of ten difference from the neighbouring one)

The composition of the Earth is not fixed to just that blend of elements which condensed around the Sun all those billions of years ago. Some of the atoms were unstable and began breaking down into atoms of other elements by radioactive decay. That process is still going on today.

It is hoped that nuclear fusion reactions can be harnessed under carefully controlled conditions to produce electrical energy on Earth. The problem, of course, is to safely generate and contain a plasma at the high temperature needed for fusion to occur (over 100 million °C) and to make the process sustainable and economic. Scientists are still working on this problem.

You can read more about the use of nuclear fusion reactions to generate energy in **Section 9.4**.

PROBLEMS FOR 25.2

1 Identify the isotopes which are missing from the following nuclear equations:

a $^{12}_{6}C + ? \rightarrow ^{16}_{8}O$ **b** $^{14}_{7}N + ^{1}_{1}H \rightarrow ?$

c $^{7}_{3}Li + ? \rightarrow ^{4}_{2}He + ^{4}_{2}He$

2 Explain, as concisely as you can, how the elements in our bodies originated from hydrogen.

EXAM QUESTIONS

Multiple-choice questions

1 For any chemical, the temperature is a measure of
- A the average kinetic energy of the particles which react
- B the average kinetic energy of all the particles
- C the activation energy
- D the minimum kinetic energy required before reaction occurs.

2 An experiment was carried out at four temperatures. The table shows the times taken for the reaction to occur.

Temperature/°C	20	30	40	50
Time/s	60	30	14	5

The results show that
- A the reaction is endothermic
- B the activation energy increases with increasing temperature
- C the rate of the reaction is directly proportional to the temperature
- D a small rise in temperature results in a large increase in reaction rate.

3 Which of the following equations represents the first ionisation energy of fluorine?
- A $F(g) + e \rightarrow F^-(g)$
- B $F(g) \rightarrow F^+(g) + e$
- C $F^-(g) \rightarrow F(g) + e$
- D $F^+(g) + e \rightarrow F(g)$

4 Which compound has a relative formula mass of 84?
- A Hexane
- B Hexene
- C Cyclohexene
- D Benzene

5 Synthesis gas can be made by
- A reacting ethene gas with steam
- B burning coke in excess air
- C burning methane gas in excess air
- D reacting methane gas with steam.

6 Methanol may be converted to methanal by
- A oxidation
- B hydration
- C cracking
- D condensation.

7 In which of the following systems will the equilibrium be **unaffected** by a change in pressure?
- A $2NO(g) + O_2(g) \rightleftharpoons 2NO_2(g)$
- B $2NO_2(g) \rightleftharpoons N_2O_4(g)$
- C $H_2(g) + I_2(g) \rightleftharpoons 2HI(g)$
- D $N_2(g) + 3H_2(g) \rightleftharpoons 2NH_3(g)$

8 The half-life of 1 g bismuth oxide compared to 1 g bismuth sulphate will be
- A greater because the percentage of bismuth is greater
- B less because of the greater stability of the smaller oxide ion
- C the same because the half-life is independent of the percentage of bismuth
- D impossible to predict.

9 The enthalpies of combustion of C(s), $H_2(g)$ and $C_4H_9OH(l)$ (in kJ mol^{-1}) are as follows.

$C(s) + O_2(g) \rightarrow CO_2(g)$ $\quad \Delta H = a$

$H_2(g) + \frac{1}{2}O_2(g) \rightarrow H_2O(l)$ $\quad \Delta H = b$

$C_4H_9OH(l) + 6O_2(g) \rightarrow 4CO_2(g) + 5H_2O(l)$ $\quad \Delta H = c$

What is the enthalpy change for the following reaction?

$4C(s) + 5H_2(g) + \frac{1}{2}O_2(g) \rightarrow C_4H_9OH(l)$
- A $4a + 5b - c$
- B $2a + 10b - c$
- C $c - 4a - 5b$
- D $2a + 5b + c$

Grid questions

10 The value for the Avogadro constant is $6{\cdot}02 \times 10^{23}$ mol^{-1}. Identify the true statement(s).

A There are 6.02×10^{23} atoms in 0.5 mol of neon gas.
B There are 6.02×10^{23} electrons in 0.5 mol of hydrogen gas.
C There are 6.02×10^{23} molecules in 0.5 mol of oxygen gas.
D There are 6.02×10^{23} hydrogen atoms in 0.5 mol of water.
E There are 6.02×10^{23} oxide ions in 0.5 mol of potassium oxide.
F There are 6.02×10^{23} sodium ions in 0.5 mol of sodium chloride.

11 The statements below can be applied to different elements in the solid state.

A discrete molecules
B conducts electricity
C contains van der Waals bonds
D contains covalent bonds
E network solid
F monatomic

a Identify the statement(s) which can be applied to sulphur but **not** to silicon.
b Identify the statement(s) which can be applied to neon.

12 Many organic functional groups contain oxygen.

A
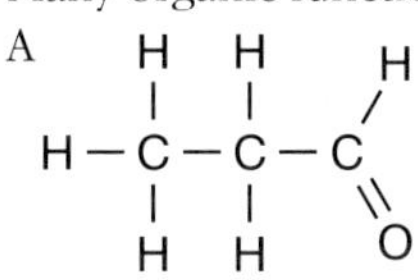

B
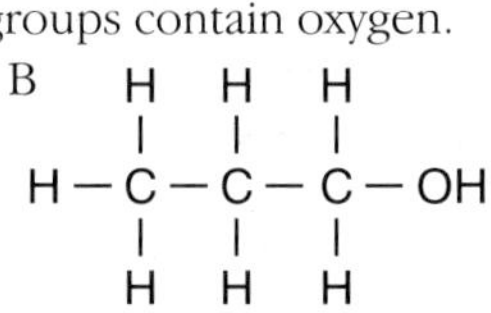

C
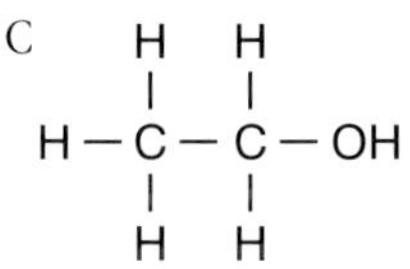

D
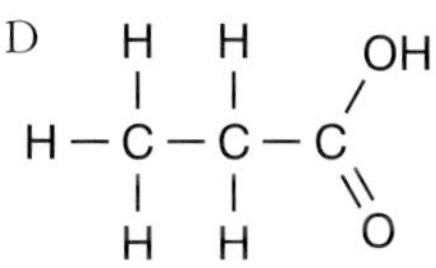

E
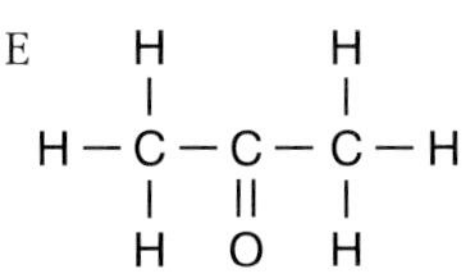

F
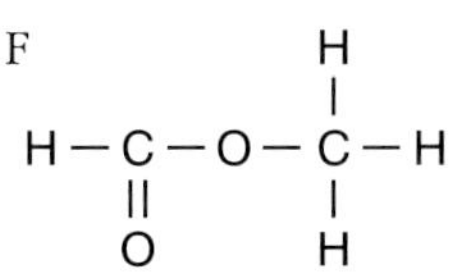

a Identify the aldehyde.
b Identify the compound which could be hydrolysed when warmed with sodium hydroxide solution.
c Identify the **two** compounds which could be oxidised to produce the compound shown in **D**.
d Identify the **two** isomers.

13

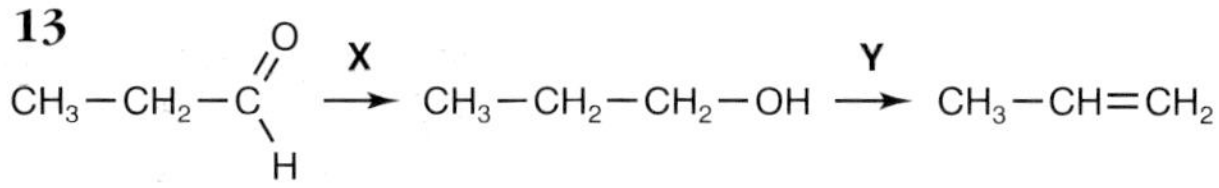

A Oxidation
B Reduction
C Hydrogenation
D Dehydrogenation
E Dehydration
F Condensation

a Identify the name which could be applied to reaction **Y**.
b Identify the name(s) which could be applied to reaction **X**.

14 Changes in concentration can alter the position of an equilibrium.

$Cl_2(aq) + H_2O(l) \rightleftharpoons 2H^+(aq) + ClO^-(aq) + Cl^-(aq)$

A $KCl(s)$
B $KOH(s)$
C $Na_2SO_4(s)$
D $AgNO_3(s)$
E $KF(s)$
F $NaNO_3(s)$

a Identify the compound which if added to the equilibrium mixture would move the equilibrium to the left.
b Identify the compound(s) which if added to the equilibrium mixture would move the equilibrium to the right.

15 Read the following passage from a popular scientific journal. It discusses the nuclear reactions which occur during the explosion of a star.

"Stars produce heavier elements when they explode as supernovae ... The main product (of the supernova) should be radioactive nickel. It forms (from) nuclei of oxygen. Heat from a shock wave 'welds' the oxygen into heavier nuclei. This radioactive nickel decays into cobalt, which in turn decays into iron."

(from *New Scientist*, 11 August 1988)

Identify the reaction(s) which can be found in the passage.

A Nuclear fission
B Nuclear fusion
C Loss of an alpha particle
D Loss of a beta particle
E Loss of a proton
F Loss of a neutron

Structured questions

16 A group of pupils designed the following apparatus to compare the viscosities of different liquids:

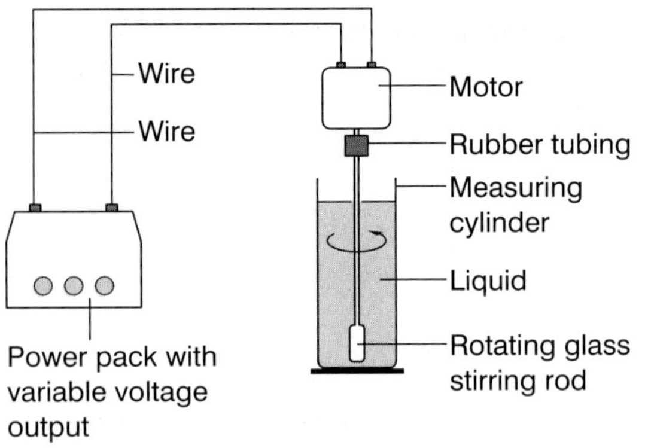

a Suggest how the apparatus could be used to compare the viscosities of different liquids. **(2)**

b The high viscosities of many liquids are due to hydrogen bonding.
What is meant by hydrogen bonding? **(2)**

17 The graph shows the boiling points of the Group 6 hydrides.

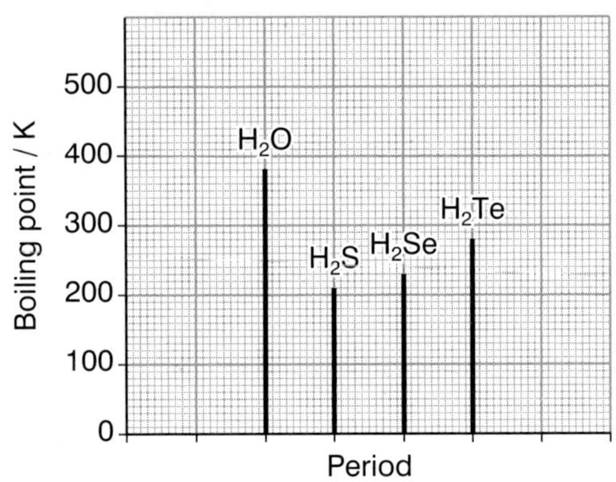

a Explain why the boiling points increase from H_2S to H_2Te. **(2)**

b Why does H_2O have an unusually high boiling point compared to the other Group 6 hydrides? **(1)**

18 At room temperature, a lump of white phosphorus reacts with air forming phosphorus(V) oxide.

$$4P(s) + 5O_2(g) \rightarrow P_4O_{10}(s) \qquad \Delta H = -3000\,\text{kJ mol}^{-1}$$

a Use this information to calculate the enthalpy of combustion of phosphorus. **(1)**

b State the effect of using powdered phosphorus in place of lump phosphorus on

i the rate of the reaction **(1)**

ii the enthalpy change for the reaction. **(1)**

19 Butan-2-ol reacts in different ways.

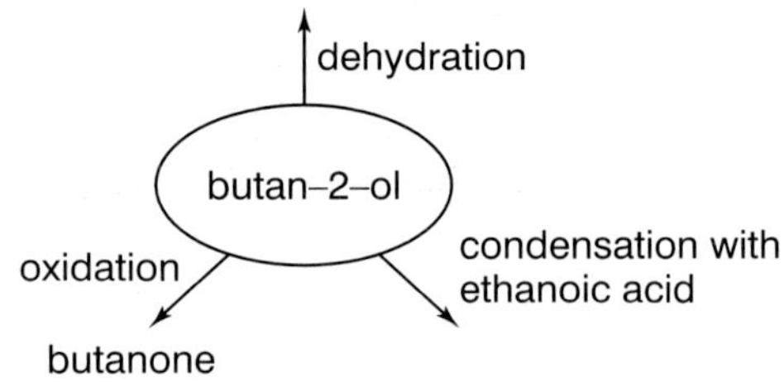

a Name the **two** products formed by the dehydration of butan-2-ol. **(1)**

b Name a reagent which could be used to oxidise butan-2-ol to butanone. **(1)**

c Draw the full structural formula for the ester produced by the condensation of butan-2-ol with ethanoic acid. **(1)**

20 The structure shown is part of a protein molecule.

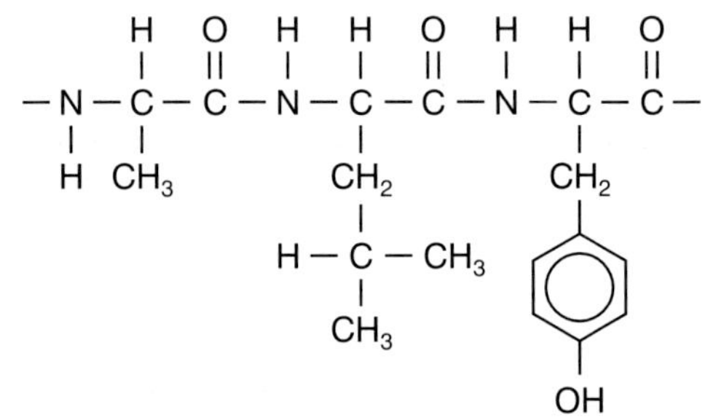

a Circle a peptide (amide) link in the structure. **(1)**

b Draw the structural formula for **one** of the molecules which would be formed if the protein was hydrolysed. **(1)**

c Enzymes are proteins.

i Why does the enzyme maltase catalyse the hydrolysis of maltose, but not the hydrolysis of sucrose? **(1)**

ii Maltase has optimum activity in alkaline conditions.
Why does maltase lose its ability to act as a catalyst in acid conditions? **(1)**

21 Some of the molecules in water are dissociated, giving the equilibrium:

$$H_2O(l) \rightleftharpoons H^+(aq) + OH^-(aq)$$

a State the concentration of hydrogen ions in pure water. **(1)**

b When sodium carbonate dissolves in water, the pH increases.
Explain this change. **(2)**

22 A pupil tried to confirm Hess's law using the reactions shown below.

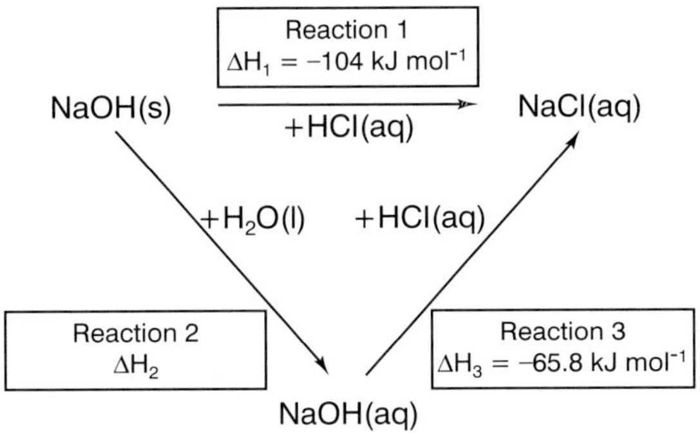

In reaction 1, the pupil measured the mass of NaOH(s) and the temperature change of the reaction mixture.

a Which further measurement would have been taken? **(1)**

b Use the enthalpy changes in the diagram to calculate the enthalpy change for reaction 2. **(2)**

c Write, in words, a statement of Hess's Law. **(1)**

23 The water in swimming pools can be kept sterile by the addition of chlorine which kills micro-organisms. The chlorine levels in swimming pool water can be determined by titrating samples against acidified iron(II) sulphate solution. The reaction taking place is:

$Cl_2(aq) + 2Fe^{2+}(aq) \rightarrow 2Cl^-(aq) + 2Fe^{3+}(aq)$

a Write the ion-electron equation for the oxidation half-reaction. **(1)**

b A 100 cm^3 sample of water from a swimming pool required 24.9 cm^3 of 2.82×10^{-4} mol l^{-1} iron(II) sulphate solution to reach the end-point.
Calculate the chlorine concentration, in g l^{-1}, in the swimming pool water. **(3)**

24 Diamond and graphite are well known forms of the element carbon. New forms of pure carbon have recently been made. They exist as individual molecules of different sizes and are called fullerenes. The main fullerene has the formula C_{60}.

a How does the structure of a fullerene differ from that of diamond? **(1)**

b Fullerenes were first made by passing a high current of electricity through a graphite rod in an atmosphere of helium. This caused the graphite to vaporise.
Suggest why helium gas was used. **(1)**

c Fullerenes can be made into hydrocarbons. One such hydrocarbon has the formula $C_{60}H_{36}$.
Describe a chemical test which could be carried out on a solution of $C_{60}H_{36}$ to show whether the hydrocarbon is saturated or unsaturated. **(1)**

25 Urea is a substance found in human urine. The enzyme urease catalyses the hydrolysis of urea.

$$\underset{\text{urea}}{CO(NH_2)_2} + H_2O \xrightarrow{\text{urease enzyme}} CO_2 + 2NH_3$$

The concentration of urea in a sample can be estimated using an indicator as shown in the diagram.

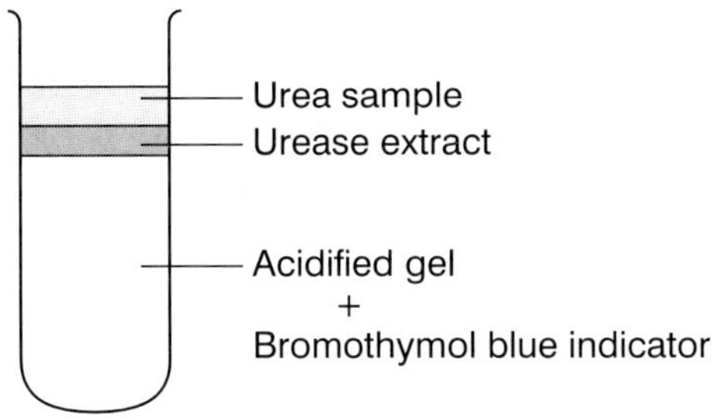

The bromothymol blue indicator is yellow below pH 6 and blue above pH 8.3.

a Draw the full structural formula for urea. **(1)**

b The initial yellow colour of the indicator changed to blue as the experiment proceeded. Explain **fully** the colours observed. **(2)**

c The pH of the gel after one completed experiment was found to be 11.
Calculate the concentration of hydroxide ions. **(1)**

d The graph shows the potential energy diagram for a urease catalysis of urea.

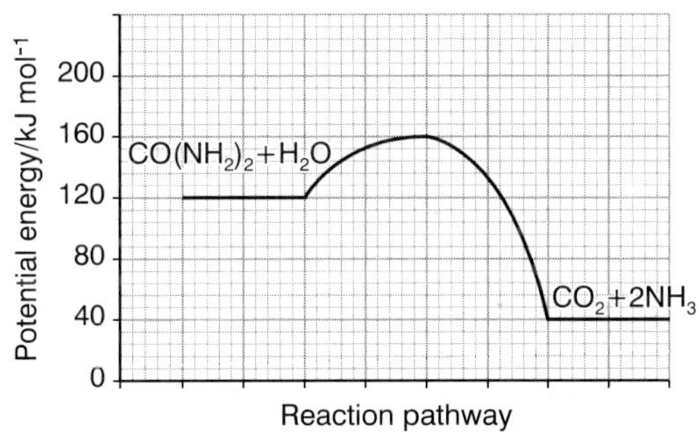

i What is the enthalpy change for the reaction? **(1)**

ii Acid is a **less** effective catalyst than urease for this reaction.
Add a curve to the potential energy diagram to show the hydrolysis when acid is used as the catalyst. **(1)**

INDEX